Springer Proceedings in Mathematics & Statistics

Volume 241

Springer Proceedings in Mathematics & Statistics

This book series features volumes composed of selected contributions from workshops and conferences in all areas of current research in mathematics and statistics, including operation research and optimization. In addition to an overall evaluation of the interest, scientific quality, and timeliness of each proposal at the hands of the publisher, individual contributions are all refereed to the high quality standards of leading journals in the field. Thus, this series provides the research community with well-edited, authoritative reports on developments in the most exciting areas of mathematical and statistical research today.

More information about this series at http://www.springer.com/series/10533

Art B. Owen · Peter W. Glynn
Editors

Monte Carlo and Quasi-Monte Carlo Methods

MCQMC 2016, Stanford, CA, August 14–19

Springer

Editors
Art B. Owen
Department of Statistics
Stanford University
Stanford, CA
USA

Peter W. Glynn
Department of Management Science
and Engineering
Stanford University
Stanford, CA
USA

ISSN 2194-1009 ISSN 2194-1017 (electronic)
Springer Proceedings in Mathematics & Statistics
ISBN 978-3-030-08255-0 ISBN 978-3-319-91436-7 (eBook)
https://doi.org/10.1007/978-3-319-91436-7

Mathematics Subject Classification (2010): 65D32, 65C05, 65C10, 65C20, 65C30, 65C35, 65C40, 65C50, 68U05, 52C07, 65Y20, 65N21, 65N55, 65N75, 65R20, 65R32, 68Q17, 68U20, 91G20, 91G80

Printed on acid-free paper

This Springer imprint is published by the registered company Springer International Publishing AG part of Springer Nature
The registered company address is: Gewerbestrasse 11, 6330 Cham, Switzerland

Preface

This volume represents the refereed proceedings of the Twelfth International Conference on Monte Carlo and Quasi-Monte Carlo Methods in Scientific Computing which was held at the Stanford University from August 14 to 19, 2016. It contains a limited selection of articles based on presentations given at the conference.

The conference was organized by Art Owen (chair), Peter Glynn, Wing Wong, and Kay Giesecke, all of Stanford University. The conference program was arranged by those organizers joined by an international committee consisting of:

- Dmitriy Bilyk (USA, Minnesota)
- Nicolas Chopin (France, ENSAE)
- Ronald Cools (Belgium, KU Leuven)
- Josef Dick (Australia, University of New South Wales)
- Arnaud Doucet (UK, Oxford)
- Henri Faure (France, Aix-Marseille Université)
- Mike Giles (UK, Oxford University)
- Paul Glasserman (USA, Columbia University)
- Michael Gnewuch (Germany, Universität Kaiserslautern)
- Stefan Heinrich (Germany, Universität Kaiserslautern)
- Fred Hickernell (USA, Illinois Institute of Technology)
- Aicke Hinrichs (Germany, Universität Rostock)
- Christophe Hery (USA, Pixar)
- Wenzel Jakob (Switzerland, ETH & Disney)
- Alexander Keller (Germany, NVIDIA)
- Frances Kuo (Australia, University of New South Wales)
- Dirk Kroese (Australia, The University of Queensland)
- Pierre L'Ecuyer (Canada, Université de Montréal)
- Erich Novak (Germany, Friedrich-Schiller-Universität Jena)
- Gareth Peters (UK, University College London)
- Aneta Karaivanova (Bulgaria, Bulgarian Academy of Sciences)
- Dirk Nuyens (Belgium, KU Leuven)

- Faming Liang (USA, U Florida, Gainesville)
- Gerhard Larcher (Austria, Johannes Kepler Universität Linz)
- Christiane Lemieux (Canada, University of Waterloo)
- Makoto Matsumoto (Japan, Hiroshima University)
- Thomas Müeller-Gronbach (Germany, Universität Passau)
- Harald Niederreiter (Austria, Austrian Academy of Sciences)
- Klaus Ritter (Germany, Universität Kaiserslautern)
- Wolfgang Schmid (Austria, Universität Salzburg)
- Steve Scott (USA, Google)
- Xiaoqun Wang (China, Tsinghua University)
- Yazhen Wang (USA, Wisconsin)
- Grzegorz Wasilkowski (USA, University of Kentucky)
- Dawn Woodard (USA, Uber)
- Henryk Woźniakowski (Poland, University of Warsaw)
- Qing Zhou (USA, UCLA)

This conference continued the tradition of biennial MCQMC conferences initiated by Harald Niederreiter, held previously at:

1. Las Vegas, USA (1994)
2. Salzburg, Austria (1996)
3. Claremont, USA (1998)
4. Hong Kong (2000)
5. Singapore (2002)
6. Juan-Les-Pins, France (2004)
7. Ulm, Germany (2006)
8. Montreal, Canada (2008)
9. Warsaw, Poland (2010)
10. Sydney, Australia (2012)
11. Leuven, Belgium (2014)

The next conference will be held in Rennes, France, in July 2018.

The proceedings of these previous conferences were all published by Springer Verlag, under the following titles:

- *Monte Carlo and Quasi-Monte Carlo Methods in Scientific Computing* (H. Niederreiter and P. J.-S. Shiue, eds.)
- *Monte Carlo and Quasi-Monte Carlo Methods 1996* (H. Niederreiter, P. Hellekalek, G. Larcher and P. Zinterhof, eds.)
- *Monte Carlo and Quasi-Monte Carlo Methods 1998* (H. Niederreiter and J. Spanier, eds.)
- *Monte Carlo and Quasi-Monte Carlo Methods 2000* (K.-T. Fang, F. J. Hickernell and H. Niederreiter, eds.)
- *Monte Carlo and Quasi-Monte Carlo Methods 2002* (H. Niederreiter, ed.)
- *Monte Carlo and Quasi-Monte Carlo Methods 2004* (H. Niederreiter and D. Talay, eds.)

- *Monte Carlo and Quasi-Monte Carlo Methods 2006* (A. Keller, S. Heinrich and H. Niederreiter, eds.)
- *Monte Carlo and Quasi-Monte Carlo Methods 2008* (P. L'Ecuyer and A. Owen, eds.)
- *Monte Carlo and Quasi-Monte Carlo Methods 2010* (L. Plaskota and H. Woźniakowski, eds.)
- *Monte Carlo and Quasi-Monte Carlo Methods 2012* (J. Dick, F. Y. Kuo, G. W. Peters and I. H. Sloan, eds.)
- *Monte Carlo and Quasi-Monte Carlo Methods 2014* (R. Cools and D. Nuyens, eds.)

The program included talks on a rich variety of topics centered around Monte Carlo, quasi-Monte Carlo, discrepancy, Markov chain Monte Carlo, and related topics. There were over 220 registrants and over 180 talks including 3 tutorials and 10 plenary talks. The tutorials by Pierre L'Ecuyer, Fred Hickernell, and Frances Kuo introduced basic QMC ideas to participants from other fields. The plenary speakers were Christoph Aistleitner, Jose Blanchet, Nicolas Chopin, Arnaud Doucet, Peter Frazier, Michael Jordan, Frances Kuo, Christiane Lemieux, Dirk Nuyens, and Andrew Stuart. As we write this, slides at `mcqmc2016.stanford.edu` include all plenary and tutorial talks. That site also has a group photograph.

The heavy lifting of running a conference was carried out by many people. Stanford statistics staff members Ellen van Stone, Emily Lauderdale, Heather Murthy, and Joanna Yu looked after many organizational and some artistic details. Students Keli Liu, Amir Sepehri, and Zeyu Zhang rang chimes and lead campus tours. Stanford conference services people Suzette Escobar, Meredith Noe, John Ventrella, and Dixee Kimball, lead by Brigid Neff, kept everything running smoothly. Sunshine Cootauco of Stanford FedEx made sure that our program booklet came out nicely.

Some fond memories include Jojo Styles and Rich Armstrong of Bossa Nuevo performing at the reception, Fred Hickernell receiving the 2016 Joseph F. Traub Prize for Achievement in Information-Based Complexity from Henryk Woźniakowski at the banquet, and Ph.D. student Adrian Ebert of KU Leuven winning the 'know your point sets' prize, a Bossa Nuevo CD provided by Carmen Milagro. Pieterjan Robbe and Christian Robert also submitted notable entries.

The papers in this volume cover theory and applications of Monte Carlo and quasi-Monte Carlo. We thank the reviewers for their careful and extensive reports. We gratefully acknowledge financial support and support in kind from our sponsors: Google, Two Sigma, Uber, SAMSI, Intel, Stanford University's statistics department, and Yi Zhou (Stanford Ph.D. 1998) and Brice Rosenzweig. We had promotional assistance from Xian's Og, SIAM, and the IMS. Finally, we are grateful to Springer Verlag for publishing this volume.

Stanford, CA, USA
December 2017

Art B. Owen
Peter W. Glynn

Contents

Part III Regular Talks

Part I
Tutorials

The Trio Identity for Quasi-Monte Carlo Error

Fred J. Hickernell

Abstract Monte Carlo methods approximate integrals by sample averages of integrand values. The error of Monte Carlo methods may be expressed as a trio identity: the product of the variation of the integrand, the discrepancy of the sampling measure, and the confounding. The trio identity has different versions, depending on whether the integrand is deterministic or Bayesian and whether the sampling measure is deterministic or random. Although the variation and the discrepancy are common in the literature, the confounding is relatively unknown and under-appreciated. Theory and examples are used to show how the cubature error may be reduced by employing the low discrepancy sampling that defines quasi-Monte Carlo methods. The error may also be reduced by rewriting the integral in terms of a different integrand. Finally, the confounding explains why the cubature error might decay at a rate different from that of the discrepancy.

Keywords Bayesian · Confounding · Deterministic · Discrepancy · Error analysis · Randomized · Variation

1 Introduction

Monte Carlo methods are used to approximate multivariate integrals that cannot be evaluated analytically, i.e., integrals of the form

$$\mu = \int_{\mathscr{X}} f(\boldsymbol{x})\, \nu(\mathrm{d}\boldsymbol{x}), \qquad \text{(INT)}$$

F. J. Hickernell (✉)
Department of Applied Mathematics, Illinois Institute of Technology,
10 W. 32nd Street, RE 208, Chicago, IL 60616, USA
e-mail: hickernell@iit.edu

A. B. Owen and P. W. Glynn (eds.), *Monte Carlo and Quasi-Monte Carlo Methods*, Springer Proceedings in Mathematics & Statistics 241,
https://doi.org/10.1007/978-3-319-91436-7_1

where $f : \mathscr{X} \to \mathbb{R}$ is a measurable function, $\mathscr{X}$ is a measurable set, and ν is a *probability* measure. Here, μ is the weighted average of the integrand. Also, $\mu = \mathbb{E}[f(\boldsymbol{X})]$, where the random variable $\boldsymbol{X}$ has probability measure ν. Monte Carlo methods take the form of a weighted average of values of f at a finite number of data sites, $\boldsymbol{x}_1, \dots, \boldsymbol{x}_n$:

$$\widehat{\mu} = \sum_{i=1}^{n} f(\boldsymbol{x}_i) w_i = \int_{\mathscr{X}} f(\boldsymbol{x})\, \widehat{\nu}(\mathrm{d}\boldsymbol{x}). \qquad \text{(MC)}$$

The sampling measure, $\widehat{\nu}$, assigns a weight w_i to the function value at $\boldsymbol{x}_i$ and lies in the vector space

$$\mathscr{M}_{\mathrm{S}} := \left\{ \sum_{i=1}^{n} w_i \delta_{\boldsymbol{x}_i} : w_1, \dots, w_n \in \mathbb{R},\ \boldsymbol{x}_1, \dots, \boldsymbol{x}_n \in \mathscr{X},\ n \in \mathbb{N} \right\}, \qquad (1)$$

where $\delta_{\boldsymbol{t}}$ denotes a Dirac measure concentrated at point $\boldsymbol{t}$. The data sites, the weights, and the sample size may be deterministic or random. Later, we impose some constraints to facilitate the analysis.

We are particularly interested in sampling measures that choose the data sites *more cleverly* than independently and identically distributed (IID) with the aim of obtaining smaller errors for the same computational effort. Such sampling measures are the hallmark of *quasi-Monte Carlo methods*. It is common to choose $w_1 = \cdots = w_n = 1/n$, in which case the sampling quality is determined solely by the choice of the data sites.

This tutorial describes how to characterize and analyze the *cubature error*, $\mu - \widehat{\mu}$, as a trio identity:

$$\mu - \widehat{\mu} = \mathrm{CNF}(f, \nu - \widehat{\nu})\, \mathrm{DSC}(\nu - \widehat{\nu})\, \mathrm{VAR}(f), \qquad \text{(TRIO)}$$

introduced by Xiao-Li Meng [30]. Each term in this identity contributes to the error, and there are ways to decrease each.

$\mathrm{VAR}(f)$ measures the *variation* of the integrand from a typical value. The variation is positively homogeneous, i.e., $\mathrm{VAR}(cf) = |c|\mathrm{VAR}(f)$. The variation is *not* the variance. Expressing μ in terms of a different integrand by means of a variable transformation may decrease the variation.

$\mathrm{DSC}(\nu - \widehat{\nu})$ measures the *discrepancy* of the sampling measure from the probability measure that defines the integral. The convergence rate of the discrepancy to zero as $n \to \infty$ characterizes the quality of the sampling measure.

$\mathrm{CNF}(f, \nu - \widehat{\nu})$ measures the *confounding* between the integrand and the difference between the measure defining the integral and the sampling measure. The magnitude of the confounding is bounded by one in some settings and has an expected square value of one in other settings. When the convergence rate of

Table 1 Different versions of the trio identity

Integrand, f	Sampling Measure, $\widehat{\nu}$	
	Deterministic	Random
Deterministic	Deterministic = ^{D}	Randomized = ^{R}
Gaussian Process	Bayesian = ^{B}	Randomized Bayesian = ^{RB}

$\widehat{\mu} \to \mu$ differs from the convergence rate of $\mathrm{DSC}(\nu - \widehat{\nu}) \to 0$, the confounding is behaving unusually.

There are four versions of the trio identity corresponding to different models for the integrand and for the sampling measure as depicted in Table 1. The integrand may be an arbitrary (deterministic) element of a Banach space or it may be a Gaussian stochastic process. The sampling measure may be an arbitrary (deterministic) element of $\mathcal{M}_{\mathrm{S}}$ or chosen randomly. Here we derive and explain these four different versions of the trio identity and draw a baker's dozen of key lessons, which are repeated at the end of this article.

Lesson 1 *The trio identity* (TRIO) *decomposes the cubature error into a product of three factors: the variation of the integrand, the discrepancy of the sampling measure, and the confounding. This identity shows how the integrand and the sampling measure each contribute to the cubature error.*

2 A Deterministic Trio Identity for Cubature Error

We start by generalizing the error bounds of Koksma [28] and Hlawka [25]. See also the monograph of Niederreiter [32]. Suppose that the integrand lies in some Banach space, $(\mathcal{F}, \|\cdot\|_{\mathcal{F}})$, where function evaluation at any point in the domain, $\mathcal{X}$, is a bounded, linear functional. This means that $\sup_{f \in \mathcal{F}} \left|f(\boldsymbol{t})\right| / \|f\|_{\mathcal{F}} < \infty$ for all $\boldsymbol{t} \in \mathcal{X}$ and that $\int_{\mathcal{X}} f(\boldsymbol{x})\, \delta_{\boldsymbol{t}}(\mathrm{d}\boldsymbol{x}) = f(\boldsymbol{t})$ for all $f \in \mathcal{F}$, $\boldsymbol{t} \in \mathcal{X}$. For example, one might choose $\mathcal{F} = C[0,1]^d$, but $\mathcal{F} = L^2[0,1]^d$ is unacceptable. Let $T : \mathcal{F} \to \mathbb{R}$ be some bounded linear functional providing a typical value of f, e.g., $T(f) = f(\boldsymbol{1})$ or $T(f) = \int_{\mathcal{X}} f(\boldsymbol{x})\, \nu(\mathrm{d}\boldsymbol{x})$. If $\{T(f) : f \in \mathcal{F}\} \neq \{0\}$, then $\mathcal{F}$ is assumed to contain constant functions. The deterministic variation is a semi-norm that is defined as the norm of the function minus its typical value:

$$\mathrm{VAR}^{\mathrm{D}}(f) := \|f - T(f)\|_{\mathcal{F}} \qquad \forall f \in \mathcal{F}. \tag{2}$$

Let $\mathcal{M}$ denote the vector space of signed measures for which integrands in $\mathcal{F}$ have finite integrals: $\mathcal{M} := \left\{\text{signed measures } \eta : \left|\int_{\mathcal{X}} f(\boldsymbol{x})\, \eta(\mathrm{d}\boldsymbol{x})\right| < \infty \forall f \in \mathcal{F}\right\}$.

We assume that our integral of interest is defined, so $\nu \in \mathscr{M}$. Since function evaluation is bounded, $\mathscr{M}$ includes $\mathscr{M}_S$ defined in (1) as well. Define the subspace

$$\mathscr{M}_\perp := \begin{cases} \{\eta \in \mathscr{M} : \eta(\mathscr{X}) = 0\}, & \{T(f) : f \in \mathscr{F}\} \neq \{0\}, \\ \mathscr{M}, & \{T(f) : f \in \mathscr{F}\} = \{0\}. \end{cases} \tag{3}$$

For example, if $\widehat{\nu}(\mathscr{X}) = \nu(\mathscr{X})$, which is common, then $\nu - \widehat{\nu}$ is automatically in $\mathscr{M}_\perp$. However, in some situations $\widehat{\nu}(\mathscr{X}) \neq \nu(\mathscr{X})$, as is noted in the discussion following (8) below. A semi-norm on $\mathscr{M}_\perp$ is induced by the norm on $\mathscr{F}$, which provides the definition of discrepancy:

$$\|\eta\|_{\mathscr{M}_\perp} := \sup_{f \in \mathscr{F} : f \neq 0} \frac{\left| \int_{\mathscr{X}} f(\boldsymbol{x})\, \eta(\mathrm{d}\boldsymbol{x}) \right|}{\|f\|_{\mathscr{F}}}, \qquad \mathrm{DSC}^{\mathrm{D}}(\nu - \widehat{\nu}) := \|\nu - \widehat{\nu}\|_{\mathscr{M}_\perp}. \tag{4}$$

Finally, define the confounding as

$$\mathrm{CNF}^{\mathrm{D}}(f, \nu - \widehat{\nu}) := \begin{cases} \dfrac{\int_{\mathscr{X}} f(\boldsymbol{x})\, (\nu - \widehat{\nu})(\mathrm{d}\boldsymbol{x})}{\mathrm{VAR}^{\mathrm{D}}(f)\mathrm{DSC}^{\mathrm{D}}(\nu - \widehat{\nu})}, & \mathrm{VAR}^{\mathrm{D}}(f)\mathrm{DSC}^{\mathrm{D}}(\nu - \widehat{\nu}) \neq 0, \\ 0, & \text{otherwise}. \end{cases} \tag{5}$$

The above definitions lead to the deterministic trio identity for cubature error.

Theorem 1 (Deterministic Trio Error Identity) *For the spaces of integrands and measures defined above, and for the above definitions of variation, discrepancy, and confounding, the following error identity holds for all $f \in \mathscr{F}$ and $\nu - \widehat{\nu} \in \mathscr{M}_\perp$:*

$$\mu - \widehat{\mu} = \mathrm{CNF}^{\mathrm{D}}(f, \nu - \widehat{\nu})\, \mathrm{DSC}^{\mathrm{D}}(\nu - \widehat{\nu})\, \mathrm{VAR}^{\mathrm{D}}(f). \tag{DTRIO}$$

Moreover, $\left|\mathrm{CNF}^{\mathrm{D}}(f, \nu - \widehat{\nu})\right| \leq 1$.

Proof The proof of this identity follows from the definitions above. It follows from (INT) and (MC) that for all $f \in \mathscr{F}$ and $\nu - \widehat{\nu} \in \mathscr{M}_\perp$, the cubature error can be written as a single integral:

$$\mu - \widehat{\mu} = \int_{\mathscr{X}} f(\boldsymbol{x})\, (\nu - \widehat{\nu})(\mathrm{d}\boldsymbol{x}). \tag{6}$$

If $\mathrm{VAR}^{\mathrm{D}}(f) = 0$, then $f = T(f)$, and the integral above vanishes by the definition of $\mathscr{M}_\perp$. If $\mathrm{DSC}^{\mathrm{D}}(\nu - \widehat{\nu}) = 0$, then the integral above vanishes by (4). Thus, for $\mathrm{VAR}^{\mathrm{D}}(f)\mathrm{DSC}^{\mathrm{D}}(\nu - \widehat{\nu}) = 0$ the trio identity holds. If $\mathrm{VAR}^{\mathrm{D}}(f)\mathrm{DSC}^{\mathrm{D}}(\nu - \widehat{\nu}) \neq 0$, then the trio identity also holds by the definition of the confounding.

Next, we bound the magnitude of the confounding for $\mathrm{VAR}^{\mathrm{D}}(f)\mathrm{DSC}^{\mathrm{D}}(\nu - \widehat{\nu}) \neq 0$:

$$\begin{aligned}\left|\mathrm{CNF}(f, \nu - \widehat{\nu})\right| &= \frac{\left|\int_{\mathscr{X}} f(\boldsymbol{x})\,(\nu - \widehat{\nu})(\mathrm{d}\boldsymbol{x})\right|}{\mathrm{VAR}^{\mathrm{D}}(f)\mathrm{DSC}^{\mathrm{D}}(\nu - \widehat{\nu})} \quad \text{by (5)} \\ &= \frac{\left|\int_{\mathscr{X}} [f(\boldsymbol{x}) - T(f)]\,(\nu - \widehat{\nu})(\mathrm{d}\boldsymbol{x})\right|}{\|f - T(f)\|_{\mathscr{F}}\,\mathrm{DSC}^{\mathrm{D}}(\nu - \widehat{\nu})} \quad \text{by (2) and (3)} \\ &\leq 1 \quad \text{by (4)},\end{aligned}$$

since $\mathrm{VAR}^{\mathrm{D}}(f) \neq 0$ and so $f - T(f) \neq 0$. □

Because $\left|\mathrm{CNF}^{\mathrm{D}}(f, \nu - \widehat{\nu})\right| \leq 1$, the deterministic trio identity implies a deterministic error bound: $|\mu - \widehat{\mu}| \leq \mathrm{DSC}^{\mathrm{D}}(\nu - \widehat{\nu})\,\mathrm{VAR}^{\mathrm{D}}(f)$. However, there is value in keeping the confounding term as noted below in Lesson 5.

The error in approximating the integral of cf is c times that for approximating the integral of f. This is reflected in the fact that $\mathrm{VAR}^{\mathrm{D}}(cf) = |c|\mathrm{VAR}^{\mathrm{D}}(f)$ and $\mathrm{CNF}(cf, \nu - \widehat{\nu}) = \mathrm{sign}(c)\mathrm{CNF}(f, \nu - \widehat{\nu})$, while $\mathrm{DSC}^{\mathrm{D}}(\nu - \widehat{\nu})$ does not depend on the integrand.

When $\mathscr{F}$ is a Hilbert space with reproducing kernel K, the discrepancy has an explicit expression in terms of K. The reproducing kernel is the unique function, $K : \mathscr{X} \times \mathscr{X} \to \mathbb{R}$ satisfying these two properties [3, Sect. 1]:

$$K(\cdot, \boldsymbol{t}) \in \mathscr{F} \quad \text{and} \quad f(\boldsymbol{t}) = \langle K(\cdot, \boldsymbol{t}), f\rangle_{\mathscr{F}} \qquad \forall f \in \mathscr{F},\ \boldsymbol{t} \in \mathscr{X}.$$

The Riesz Representation Theorem implies that the representer of cubature error is

$$\eta_{\mathrm{err}}(\boldsymbol{t}) = \langle K(\cdot, \boldsymbol{t}), \eta_{\mathrm{err}}\rangle_{\mathscr{F}} = \int_{\mathscr{X}} K(\boldsymbol{x}, \boldsymbol{t})\,(\nu - \widehat{\nu})(\mathrm{d}\boldsymbol{x}).$$

Thus, the deterministic trio identity for the reproducing kernel Hilbert space (RKHS) case is

$$\mu - \widehat{\mu} = \langle \eta_{\mathrm{err}}, f\rangle_{\mathscr{F}} = \underbrace{\frac{\langle \eta_{\mathrm{err}}, f\rangle_{\mathscr{F}}}{\|f - T(f)\|_{\mathscr{F}}\,\|\eta_{\mathrm{err}}\|_{\mathscr{F}}}}_{\mathrm{CNF}^{\mathrm{D}}(f, \nu - \widehat{\nu})} \underbrace{\|\eta_{\mathrm{err}}\|_{\mathscr{F}}}_{\mathrm{DSC}^{\mathrm{D}}(\nu - \widehat{\nu})} \underbrace{\|f - T(f)\|_{\mathscr{F}}}_{\mathrm{VAR}^{\mathrm{D}}(f)}$$

provided that

$$T(f)[\nu(\mathscr{X}) - \widehat{\nu}(\mathscr{X})] = 0. \tag{7}$$

The squared discrepancy takes the form [18]

$$
\begin{aligned}
[\mathrm{DSC}^{\mathrm{D}}(\nu - \widehat{\nu})]^2 = \|\eta_{\mathrm{err}}\|_{\mathscr{F}}^2 &= \langle \eta_{\mathrm{err}}, \eta_{\mathrm{err}} \rangle_{\mathscr{F}} \\
&= \int_{\mathscr{X} \times \mathscr{X}} K(\boldsymbol{x}, \boldsymbol{t}) \, (\nu - \widehat{\nu})(\mathrm{d}\boldsymbol{x}) \, (\nu - \widehat{\nu})(\mathrm{d}\boldsymbol{t}) \\
&= \int_{\mathscr{X} \times \mathscr{X}} K(\boldsymbol{x}, \boldsymbol{t}) \, \nu(\mathrm{d}\boldsymbol{x}) \, \nu(\mathrm{d}\boldsymbol{t}) \\
&\quad - 2 \sum_{i=1}^{n} w_i \int_{\mathscr{X}} K(\boldsymbol{x}_i, \boldsymbol{t}) \, \nu(\mathrm{d}\boldsymbol{t}) + \sum_{i,j=1}^{n} w_i w_j K(\boldsymbol{x}_i, \boldsymbol{x}_j).
\end{aligned}
$$

Assuming that the single integral and double integral of the reproducing kernel can be evaluated analytically, the computational cost to evaluate the discrepancy is $\mathscr{O}(n^2)$ unless the kernel has a special form that speeds up the calculation of the double sum.

Lesson 2 *The deterministic discrepancy when $\mathscr{F}$ is an RKHS has a simple, explicit form involving three terms.*

In the RKHS case, the confounding corresponds to the cosine of the angle between $f - T(f)$ and the cubature error representer, η_{err}. This cosine is no greater than one in magnitude, as expected.

The square deterministic discrepancy for an RKHS may be expressed in terms of vectors and matrices:

$$
\boldsymbol{w} = \left(w_i\right)_{i=1}^{n}, \qquad k_0 = \int_{\mathscr{X}} K(\boldsymbol{x}, \boldsymbol{t}) \, \nu(\mathrm{d}\boldsymbol{x}) \, \nu(\mathrm{d}\boldsymbol{t}), \tag{8a}
$$

$$
\boldsymbol{k} = \left(\int_{\mathscr{X}} K(\boldsymbol{x}_i, \boldsymbol{t}) \, \nu(\mathrm{d}\boldsymbol{t}) \right)_{i=1}^{n}, \qquad \mathsf{K} = \left(K(\boldsymbol{x}_i, \boldsymbol{x}_j)\right)_{i,j=1}^{n}, \tag{8b}
$$

$$
[\mathrm{DSC}^{\mathrm{D}}(\nu - \widehat{\nu})]^2 = k_0 - 2\boldsymbol{k}^T \boldsymbol{w} + \boldsymbol{w}^T \mathsf{K} \boldsymbol{w}. \tag{8c}
$$

Given fixed data sites, the optimal cubature weights to minimize the discrepancy are $\boldsymbol{w} = \mathsf{K}^{-1}\boldsymbol{k}$. If $\boldsymbol{1}^T \mathsf{K}^{-1} \boldsymbol{k} = 1$, which is possible but not automatic, then $\widehat{\nu}(\mathscr{X}) = \nu(\mathscr{X}) = 1$ for these optimal weights, and (DTRIO) holds for general T. Otherwise, one must define $T(f) = 0$ for all $f \in \mathscr{F}$ to satisfy condition (7) for these optimal cubature weights.

A particular example of this RKHS setting corresponds to the uniform probability measure ν on the d-dimensional unit cube, $\mathscr{X} = [0, 1]^d$, and the reproducing kernel defined by [17]

$$
K(\boldsymbol{x}, \boldsymbol{t}) = \prod_{k=1}^{d} [2 - \max(x_k, t_k)]. \tag{9}
$$

In this example, $T(f) = f(\boldsymbol{1})$, and the variation is

$$
\mathrm{VAR}^{\mathrm{D}}(f) = \|f - f(\boldsymbol{1})\|_{\mathscr{F}} = \left\| \left(\|\partial^{\mathfrak{u}} f\|_{L^2} \right)_{\emptyset \subsetneq \mathfrak{u} \subseteq 1:d} \right\|_{\ell^2}, \qquad \partial^{\mathfrak{u}} f := \left. \frac{\partial^{|\mathfrak{u}|} f}{\partial \boldsymbol{x}_{\mathfrak{u}}} \right|_{\boldsymbol{x}_{\bar{\mathfrak{u}}} = \boldsymbol{1}}.
$$

Here $1:d$ means $= \{1, \ldots, d\}$, $x_{\mathfrak{u}}$ means $(x_k)_{k\in\mathfrak{u}}$, and $\bar{\mathfrak{u}}$ denotes the complement of $\mathfrak{u}$. The square discrepancy for the equally weighted case with $w_1 = \cdots = w_n = 1/n$ is

$$\begin{aligned}[\mathrm{DSC^D}(\nu - \widehat{\nu})]^2 &= \left(\frac{4}{3}\right)^d - \frac{2}{n}\sum_{i=1}^{n}\prod_{k=1}^{d}\left(\frac{3 - x_{ik}^2}{2}\right) + \frac{1}{n^2}\sum_{i,j=1}^{n}\prod_{k=1}^{d}[2 - \max(x_{ik}, x_{jk})] \\ &= \left\| \left(\| \nu([\mathbf{0}, \cdot_{\mathfrak{u}}]) - \widehat{\nu}([\mathbf{0}, \cdot_{\mathfrak{u}}]) \|_{L^2} \right)_{\emptyset \subsetneq \mathfrak{u} \subseteq 1:d} \right\|_{\ell^2}. \end{aligned} \tag{10}$$

This discrepancy has a geometric interpretation: $\nu([\mathbf{0}, \boldsymbol{x}_{\mathfrak{u}}])$ corresponds to the *volume* of the $|\mathfrak{u}|$-dimensional box $[\mathbf{0}, \boldsymbol{x}_{\mathfrak{u}}]$, and $\widehat{\nu}([\mathbf{0}, \boldsymbol{x}_{\mathfrak{u}}])$ corresponds to the *proportion* of data sites lying in the box $[\mathbf{0}, \boldsymbol{x}_{\mathfrak{u}}]$. The discrepancy in (10), which is called the L^2-discrepancy, depends on the difference between this volume and this proportion for all $\boldsymbol{x} \in [0, 1]^d$ and for all $\emptyset \subsetneq \mathfrak{u} \subseteq 1:d$.

If the data sites $\boldsymbol{x}_1, \ldots, \boldsymbol{x}_n$ are chosen to be IID with probability measure ν, and $w_1 = \cdots = w_n = 1/n$, then the mean square discrepancy for the RKHS case is

$$\mathbb{E}\{[\mathrm{DSC^D}(\nu - \widehat{\nu})]^2\} = \frac{1}{n}\left[\int_{\mathscr{X}} K(\boldsymbol{x}, \boldsymbol{x})\, \nu(\mathrm{d}\boldsymbol{x}) - \int_{\mathscr{X}\times\mathscr{X}} K(\boldsymbol{x}, \boldsymbol{t})\, \nu(\mathrm{d}\boldsymbol{x})\, \nu(\mathrm{d}\boldsymbol{t})\right].$$

For the L^2-discrepancy in (10) this becomes

$$\mathbb{E}\{[\mathrm{DSC^D}(\nu - \widehat{\nu})]^2\} = \frac{1}{n}\left[\left(\frac{3}{2}\right)^d - \left(\frac{4}{3}\right)^d\right]. \tag{11}$$

Quasi-Monte Carlo methods generally employ sampling measures of the form $\widehat{\nu} = n^{-1}\sum_{i=1}^{n}\delta_{\boldsymbol{x}_i}$, but choose the data sites $\{\boldsymbol{x}_i\}_{i=1}^{n}$ to be better than IID in the sense of discrepancy. For integration over $\mathscr{X} = [0, 1]^d$ with respect to the uniform measure, these *low discrepancy* data sites may come from

- a digital sequence [9], such as that proposed by Sobol' [48], Faure [10], Niederreiter [31], or Niederreiter and Xing [33], or
- a sequence of node sets of an integration lattice [46].

The constructions of such sets are described in the references above and L'Ecuyer's tutorial in this volume. The L^2-discrepancy defined in (10) and its relatives are $\mathscr{O}(n^{-1+\varepsilon})$ as $n \to \infty$ for any positive ε for these low discrepancy data sites [32].

Figure 1 displays examples of IID and randomized low discrepancy data sites. Figure 2 shows the rates of decay for the L^2-discrepancy for various dimensions. The scaled discrepancy is the empirically computed root mean square discrepancy divided by its value for $n = 1$. Although the decay for the low discrepancy points is $\mathscr{O}(n^{-1+\varepsilon})$ for large enough n, the decay in Fig. 2 resembles $\mathscr{O}(n^{-1/2})$ for large dimensions and modest n. The scaled discrepancy for IID samples in Fig. 2 does not exhibit a dimension dependence because it is masked by the scaling. The dimension dependence of the convergence of the discrepancy to zero is addressed later in Sect. 8.

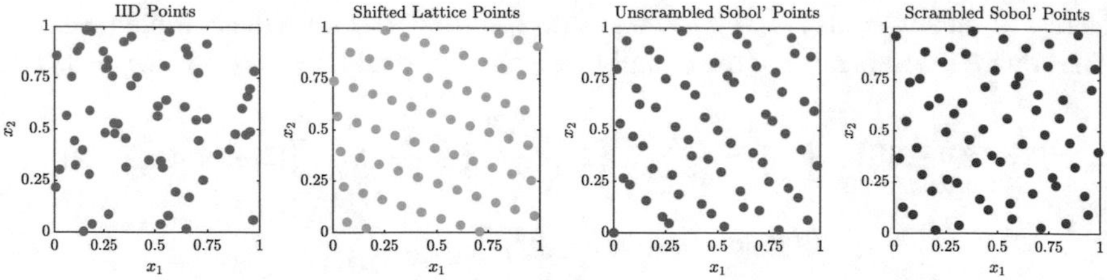

Fig. 1 IID points and three examples of low discrepancy points

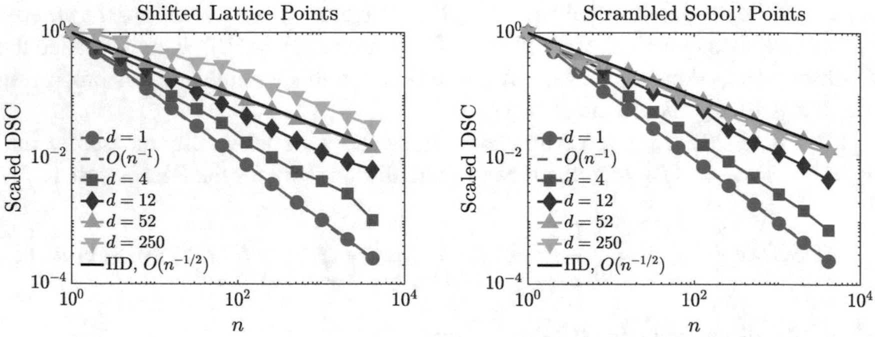

Fig. 2 The root mean square L^2-discrepancies given by (10) for randomly shifted lattice sequence nodesets and randomly scrambled and shifted Sobol' sequences points for a variety of dimensions

Lesson 3 *Quasi-Monte Carlo methods replace IID data sites by low discrepancy data sites, such as Sobol' sequences and integration lattice nodeset sequences. The resulting sampling measures have discrepancies and cubature errors that decay to zero at a faster rate than in the case of IID sampling.*

No sampling scheme can produce a faster convergence rate than $\mathscr{O}(n^{-1})$ for the L^2-discrepancy. This is due to the limited smoothness of the reproducing kernel defined in (9) and the corresponding limited smoothness of the corresponding Hilbert space of integrands.

3 A Randomized Trio Identity for Cubature Error

For the randomized version of the trio identity, we again assume that the integrands lie in a Banach space, $(\mathscr{F}, \|\cdot\|_{\mathscr{F}})$. This space is required to contain constant functions if $\{T(f) : f \in \mathscr{F}\} \neq \{0\}$. We assume that $\int_{\mathscr{X}} f(\boldsymbol{x})\, \nu(\mathrm{d}\boldsymbol{x})$ is defined for all $f \in \mathscr{F}$, however, we do not require function evaluation to be a bounded linear functional on $\mathscr{F}$. The definitions of the bounded linear functional T and the variation in the deterministic case in (2) apply here as well.

Now endow the vector space of all sampling measures, $\mathscr{M}_S$, with a probability distribution. This means that the placement of the data sites, the number of data sites, and the choice of the weights may all be random. We require the following two conditions:

$$\begin{gathered}\mathbb{E}_{\widehat{\nu}}\left|\int_{\mathscr{X}} f(\boldsymbol{x})\,\widehat{\nu}(\mathrm{d}\boldsymbol{x})\right|^2 < \infty \qquad \forall f \in \mathscr{F}, \\ \{T(f) : f \in \mathscr{F}\} = \{0\} \qquad \text{or} \qquad \widehat{\nu}(\mathscr{X}) = \nu(\mathscr{X}) \text{ almost surely.}\end{gathered} \tag{12}$$

The first condition implies that $\int_{\mathscr{X}} f(\boldsymbol{x})\,\widehat{\nu}(\mathrm{d}\boldsymbol{x})$ exists almost surely for every $f \in \mathscr{F}$.

The randomized discrepancy is defined as the worst normalized root mean squared error:

$$\mathrm{DSC}^{\mathrm{R}}(\nu - \widehat{\nu}) := \sup_{f \in \mathscr{F} : f \neq 0} \frac{\sqrt{\mathbb{E}_{\widehat{\nu}}\left|\int_{\mathscr{X}} f(\boldsymbol{x})\,(\nu - \widehat{\nu})(\mathrm{d}\boldsymbol{x})\right|^2}}{\|f\|_{\mathscr{F}}}. \tag{13}$$

The randomized discrepancy does not depend on the particular instance of the sampling measure but on the distribution of the sampling measure.

Finally, define the confounding as

$$\mathrm{CNF}^{\mathrm{R}}(f, \nu - \widehat{\nu}) := \begin{cases} \dfrac{\int_{\mathscr{X}} f(\boldsymbol{x})\,(\nu - \widehat{\nu})(\mathrm{d}\boldsymbol{x})}{\mathrm{VAR}^{\mathrm{D}}(f)\mathrm{DSC}^{\mathrm{R}}(\nu - \widehat{\nu})}, & \mathrm{VAR}^{\mathrm{D}}(f)\mathrm{DSC}^{\mathrm{R}}(\nu - \widehat{\nu}) \neq 0, \\ 0, & \text{otherwise.} \end{cases} \tag{14}$$

Here, the confounding *does* depend on the particular instance of the sampling measure. The above definitions allow us to establish the randomized trio identity for cubature error.

Theorem 2 (Randomized Trio Error Identity) *For the spaces of integrands and measures defined above, and for the above definitions of variation, discrepancy, and confounding, the following error identity holds for all $f \in \mathscr{F}$ and $\widehat{\nu} \in \mathscr{M}_S$:*

$$\mu - \widehat{\mu} = \mathrm{CNF}^{\mathrm{R}}(f, \nu - \widehat{\nu})\,\mathrm{DSC}^{\mathrm{R}}(\nu - \widehat{\nu})\,\mathrm{VAR}^{\mathrm{D}}(f) \quad \textit{almost surely.} \tag{RTRIO}$$

Moreover, $\mathbb{E}_{\widehat{\nu}}\left|\mathrm{CNF}^{\mathrm{R}}(f, \nu - \widehat{\nu})\right|^2 \leq 1$ for all $f \in \mathscr{F}$.

Proof For all $f \in \mathscr{F}$ and $\widehat{\nu} \in \mathscr{M}_S$, the error can be written as the single integral in (6) almost surely. If $\mathrm{VAR}^{\mathrm{D}}(f) = 0$, then $f = T(f)$, and $\mu - \widehat{\mu}$ vanishes almost surely by (12). If $\mathrm{DSC}^{\mathrm{R}}(\nu - \widehat{\nu}) = 0$, then $\mu - \widehat{\mu}$ vanishes almost surely by (13). Thus, for $\mathrm{VAR}^{\mathrm{D}}(f)\mathrm{DSC}^{\mathrm{R}}(\nu - \widehat{\nu}) = 0$ the trio identity holds. If $\mathrm{VAR}^{\mathrm{D}}(f)\mathrm{DSC}^{\mathrm{R}}(\nu - \widehat{\nu}) \neq 0$, then the trio identity also holds by the definition of the confounding.

Next, we analyze the magnitude of the confounding for $\mathrm{VAR}^{\mathrm{D}}(f)\mathrm{DSC}^{\mathrm{D}}(\nu - \widehat{\nu}) \neq 0$:

$$\begin{aligned}\mathbb{E}\left|\mathrm{CNF}^{\mathrm{R}}(f, \nu - \widehat{\nu})\right|^2 &= \frac{\mathbb{E}\left|\int_{\mathscr{X}} f(\boldsymbol{x})\,(\nu - \widehat{\nu})(\mathrm{d}\boldsymbol{x})\right|^2}{[\mathrm{VAR}^{\mathrm{D}}(f)\mathrm{DSC}^{\mathrm{D}}(\nu - \widehat{\nu})]^2} \quad \text{by (14)} \\ &= \frac{\mathbb{E}\left|\int_{\mathscr{X}} [f(\boldsymbol{x}) - T(f)]\,(\nu - \widehat{\nu})(\mathrm{d}\boldsymbol{x})\right|^2}{[\|f - T(f)\|_{\mathscr{F}}\,\mathrm{DSC}^{\mathrm{D}}(\nu - \widehat{\nu})]^2} \quad \text{by (2) and (12)} \\ &\leq 1 \quad \text{by (13)},\end{aligned}$$

since $\mathrm{VAR}^{\mathrm{D}}(f) \neq 0$ and so $f - T(f) \neq 0$. □

Consider simple Monte Carlo, where the approximation to the integral is an equally weighted average using IID sampling $\boldsymbol{x}_1, \boldsymbol{x}_2, \ldots \sim \nu$. Let the sample size be fixed at n. Let $\mathscr{F} = L^{2,\nu}(\mathscr{X})$, the space of functions that are square integrable with respect to the measure ν, and let $T(f)$ be the mean of f. Then the variation of f is just its standard deviation, $\mathrm{std}(f)$. The randomized discrepancy is $1/\sqrt{n}$. The randomized confounding is

$$\mathrm{CNF}^{\mathrm{R}}(f, \nu - \widehat{\nu}) = \frac{-1}{\sqrt{n}\,\mathrm{std}(f)} \sum_{i=1}^{n} [f(\boldsymbol{x}_i) - \mu].$$

Unlike the deterministic setting, there is no simple expression for the randomized discrepancy under general sampling measures and RKHSs. The randomized discrepancy can sometimes be conveniently calculated or bounded for spaces of integrands that are represented by series expansions, and where the randomized sampling measures for the bases of these expansions have special properties [16, 19].

It is instructive to contrast the variation, discrepancy, and confounding in the deterministic and randomized settings. For some integrand, f, and some sampling measure, $\widehat{\nu}$, satisfying the conditions defining both (DTRIO) and (RTRIO):

- the variation in both settings is the same,
- the randomized discrepancy must be *no greater* than the deterministic discrepancy by definition, and thus
- the randomized confounding must be *no less* than the deterministic confounding.

The deterministic confounding is never greater than one in magnitude. By contrast, the randomized confounding may be arbitrarily large. However, Markov's inequality implies that it may be larger than $1/\sqrt{\alpha}$ with probability no greater than α. The next section illustrates the differences in the deterministic and randomized trio identities.

4 Multivariate Gaussian Probabilities

Consider the d-variate integral corresponding to the probability of a $\mathcal{N}(\mathbf{0}, \Sigma)$ random variable lying inside the box $[\boldsymbol{a}, \boldsymbol{b}]$:

$$\mu = \int_{[\boldsymbol{a},\boldsymbol{b}]} \frac{\exp(-\frac{1}{2}z^T\Sigma^{-1}z)}{\sqrt{(2\pi)^d \det(\Sigma)}}\,\mathrm{d}z = \int_{[0,1]^{d-1}} f_{\mathrm{Genz}}(\boldsymbol{x})\,\mathrm{d}\boldsymbol{x}, \tag{15}$$

where $\Sigma = \mathsf{L}\mathsf{L}^T$ is the Cholesky decomposition of the covariance matrix, $\mathsf{L} = \left(l_{jk}\right)_{j,k=1}^{d}$, is a lower triangular matrix, and

$$\alpha_1 = \Phi(a_1), \qquad \beta_1 = \Phi(b_1),$$

$$\alpha_j(x_1,\ldots,x_{j-1}) = \Phi\left(\frac{1}{l_{jj}}\left(a_j - \sum_{k=1}^{j-1} l_{jk}\Phi^{-1}(\alpha_k + x_k(\beta_k - \alpha_k))\right)\right), \quad j = 2,\ldots,d,$$

$$\beta_j(x_1,\ldots,x_{j-1}) = \Phi\left(\frac{1}{l_{jj}}\left(b_j - \sum_{k=1}^{j-1} l_{jk}\Phi^{-1}(\alpha_k + x_k(\beta_k - \alpha_k))\right)\right), \quad j = 2,\ldots,d,$$

$$f_{\mathrm{Genz}}(\boldsymbol{x}) = \prod_{j=1}^{d}[\beta_j(\boldsymbol{x}) - \alpha_j(\boldsymbol{x})].$$

Here, Φ represents the cumulative distribution function for a standard normal random variable. Genz [11] developed this clever transformation of variables above. Not only is the dimension decreased by one, but the integrand is typically made less peaky and more favorable to cubature methods.

The left plot of Fig. 3 shows the absolute errors in computing the multivariate Gaussian probability via the Genz transformation for

$$\boldsymbol{a} = \begin{pmatrix} -6 \\ -2 \\ -2 \end{pmatrix}, \quad \boldsymbol{b} = \begin{pmatrix} 5 \\ 2 \\ 1 \end{pmatrix}, \quad \Sigma = \begin{pmatrix} 16 & 4 & 4 \\ 4 & 2 & 1.5 \\ 4 & 1.5 & 1.3125 \end{pmatrix}, \quad \mathsf{L} = \begin{pmatrix} 4 & 0 & 0 \\ 1 & 1 & 0 \\ 1 & 0.5 & 0.25 \end{pmatrix},$$

by IID sampling, unscrambled Sobol' sampling, and scrambled Sobol' sampling [39–41]. Multiple random scramblings of a very large scrambled Sobol' set were used to infer that $\mu \approx 0.6763373243578$. For the two randomized sampling measures 100 replications were taken. The marker denotes the median error and the top of the stem extending above the marker denotes the 90% quantile of the error.

Empirically, the error for scrambled Sobol' sampling appears to be tending towards a convergence rate of $\mathcal{O}(n^{-2})$. This is a puzzle. It is unknown why this should be or whether this effect is only transient. In the discussion below we assume the expected rate of $\mathcal{O}(n^{-3/2+\varepsilon})$.

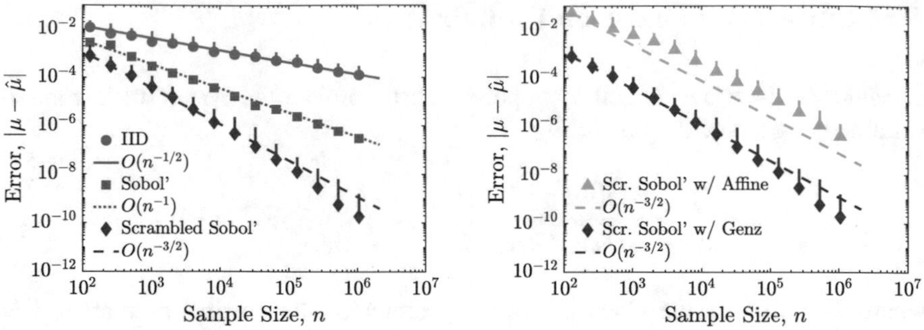

Fig. 3 The error of an example of the multivariate Gaussian probability in (15). The left side shows the result of Genz's transformation and different sampling measures. The right side shows scrambled Sobol' sampling using different transformations

The orders of the discrepancy and confounding in Table 2 explain the rates of decay of the error and the benefits of randomization. Note that in all cases

$$\mu - \widehat{\mu} \text{ decay rate} = \text{CNF decay/growth rate} \times \text{DSC decay rate}.$$

We consider equally weighted cubature rules for two kinds of random sampling measures, IID and scrambled Sobol', and for both the deterministic and randomized settings. Here, $\mathscr{F}$ is assumed to be the RKHS used to define the L^2-discrepancy.

For IID sampling both the root mean square L^2-discrepancy and the randomized discrepancy are $\mathscr{O}(n^{-1/2})$. The confounding for typical IID sampling is $\mathscr{O}(1)$. In the randomized setting one may have an atypically poor instance of data sites that leads to an atypically high confounding of $\mathscr{O}(n^{1/2})$. On the other hand, unscrambled Sobol' sampling and scrambled Sobol' sampling are atypically superior instances of data sites under an IID sampling measure that yield atypically small confoundings of $\mathscr{O}(n^{-1/2+\varepsilon})$ and $\mathscr{O}(n^{-1+\varepsilon})$, respectively.

Table 2 Confounding orders for deterministic randomized settings and two different sets of equi-weighted random sampling measures. Sufficient smoothness of the integrand is assumed. The order of the error equals the order of the discrepancy times the order of the confounding

Deterministic setting	RMS $L^2 - \text{DSC}^\text{D}$	$\widehat{\nu}$	Worst	Typical IID	Unscr. Sobol'	Typical Scr. Sobol'
IID Sampling	$\mathscr{O}(n^{-1/2})$	CNF^D		$\mathscr{O}(1)$	$\mathscr{O}(n^{-1/2+\varepsilon})$	$\mathscr{O}(n^{-1+\varepsilon})$
Scr. Sobol' Sampling	$\mathscr{O}(n^{-1+\varepsilon})$	CNF^D			$\mathscr{O}(1)$	$\mathscr{O}(n^{-1/2+\varepsilon})$
		$\mu - \widehat{\mu}$	$\mathscr{O}(1)$	$\mathscr{O}(n^{-1/2})$	$\mathscr{O}(n^{-1+\varepsilon})$	$\mathscr{O}(n^{-3/2+\varepsilon})$
Randomized setting	DSC^R					
IID Sampling	$\mathscr{O}(n^{-1/2})$	CNF^R	$\mathscr{O}(n^{1/2})$	$\mathscr{O}(1)$	$\mathscr{O}(n^{-1/2+\varepsilon})$	$\mathscr{O}(n^{-1+\varepsilon})$
Scr. Sobol' Sampling	$\mathscr{O}(n^{-3/2+\varepsilon})$	CNF^R			$\mathscr{O}(n^{1/2+\varepsilon})$	$\mathscr{O}(1)$

For scrambled Sobol' sampling, the root mean square L^2-discrepancy is now only $\mathscr{O}(n^{-1+\varepsilon})$, an improvement over IID sampling. However, the randomized discrepancy is an even smaller $\mathscr{O}(n^{-3/2+\varepsilon})$ [16, 41]. In the deterministic setting, unscrambled Sobol' sampling has a typical $\mathscr{O}(1)$ confounding, whereas typical scrambled Sobol' sampling has an atypically low $\mathscr{O}(n^{-1/2})$ confounding. This is because scrambled Sobol' sampling can take advantage of the additional smoothness of the given integrand, which is not reflected in the definition of $\mathscr{F}$. In the randomized setting, unscrambled Sobol' sampling has an atypically high $\mathscr{O}(n^{1/2})$ confounding. Thus, unscrambled Sobol' sampling is among the awful minority of sampling measures under scrambled Sobol' sampling.

Lesson 4 *Randomizing the sampling measure may not only eliminate bias, but it may help improve accuracy by avoiding the awful minority of possible sampling measures.*

Lesson 5 *Although it has traditionally been ignored, the confounding helps explain why the cubature error may decay to zero much faster or more slowly than the discrepancy.*

An alternative to the Genz transformation above is an affine transformation to compute the multivariate Gaussian probability:

$$z = \boldsymbol{a} + (\boldsymbol{b} - \boldsymbol{a}) \circ \boldsymbol{x}, \quad f_{\mathrm{aff}}(\boldsymbol{x}) = \frac{\exp\left(-\frac{1}{2}z^T \Sigma^{-1} z\right)}{\sqrt{(2\pi)^d \det(\Sigma)}} \prod_{j=1}^{d} (b_j - a_j),$$
$$\mu = \int_{[0,1]^d} f_{\mathrm{aff}}(\boldsymbol{x}) \, \mathrm{d}\boldsymbol{x},$$

where $\circ$ denotes the Hadamard (term-by-term) product. The right plot in Fig. 3 shows that the error using the affine transformation is much worse than that using the Genz transformation even though the two convergence rates are the same. The difference in the magnitudes of the errors is primarily because $\mathrm{VAR}^{\mathrm{D}}(f_{\mathrm{aff}})$ is greater than $\mathrm{VAR}^{\mathrm{D}}(f_{\mathrm{Genz}})$.

Lesson 6 *Well-chosen variable transformations may reduce cubature error by producing an integrand with a smaller variation than otherwise.*

5 Option Pricing

The prices of financial derivatives can often be modeled by high dimensional integrals. If the underlying asset is described in terms of a Brownian motion, B, at times $t_1, \ldots, t_d$, then $\boldsymbol{Z} = (B(t_1), \ldots, B(t_d)) \sim \mathscr{N}(\boldsymbol{0}, \Sigma)$, where $\Sigma = \left(\min(t_j, t_k)\right)_{j,k=1}^{d}$, and the fair price of the option is

$$\mu = \int_{\mathbb{R}^d} \text{payoff}(z) \frac{\exp(-\frac{1}{2}z^T \Sigma^{-1} z)}{\sqrt{(2\pi)^d \det(\Sigma)}} \, dz = \int_{[0,1]^d} f(\boldsymbol{x}) \, d\boldsymbol{x},$$

where the function payoff(·) describes the discounted payoff of the option,

$$f(\boldsymbol{x}) = \text{payoff}(z), \qquad z = \mathsf{L} \begin{pmatrix} \Phi^{-1}(x_1) \\ \vdots \\ \Phi^{-1}(x_d) \end{pmatrix}.$$

In this example, L may be any square matrix satisfying $\Sigma = \mathsf{L}\mathsf{L}^T$.

Figure 4 shows the cubature error using IID sampling, unscrambled Sobol' sampling, and scrambled Sobol' sampling for the Asian arithmetic mean call option with the following parameters:

$$\text{payoff}(z) = \max\left(\frac{1}{d}\sum_{j=1}^{d} S_j - K, 0\right) e^{-r\tau}, \quad S_j = S_0 \exp\bigl((r - \sigma^2/2)t_j + \sigma z_j\bigr),$$

$$\tau = 1, \quad d = 12, \quad S_0 = K = 100, \quad r = 0.05, \quad \sigma = 0.5,$$

$$t_j = j\tau/d, \quad j = 1 : d.$$

The convergence rates for IID and unscrambled Sobol' sampling are the same as in Fig. 3 for the previous example of multivariate probabilities. However, for this example scrambling the Sobol' set improves the accuracy but not the convergence rate. The convergence rate for scrambled Sobol' sampling, $\widehat{\nu}$, is poorer than hoped for because f is not smooth enough for $\text{VAR}^{\text{D}}(f)$ to be finite in the case where $\text{DSC}^{\text{R}}(\nu - \widehat{\nu}) = \mathcal{O}(n^{-3/2+\varepsilon})$.

Lesson 7 *The benefits of sampling measures with asymptotically smaller discrepancies are limited to those integrands with finite variation.*

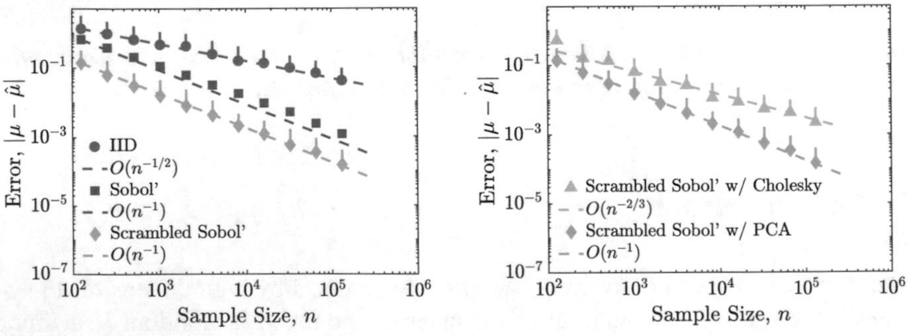

Fig. 4 Cubature error for the price of an Asian arithmetic mean option using different sampling measures. The left side uses the PCA decomposition and the right side contrasts the PCA with the Cholesky decomposition

The left plot in Fig. 4 chooses $\mathsf{L} = \mathsf{V}\Lambda^{1/2}$, where the columns of V are the normalized eigenvectors of Σ, and the diagonal elements of the diagonal matrix Λ are the eigenvalues of Σ. This is also called a principal component analysis (PCA) construction. The advantage is that the main part of the Brownian motion affecting the option payoff is concentrated in the smaller dimensions. The right plot of Fig. 4 contrasts the cubature error for two choices of L: one chosen by the PCA construction and the other coming from the Cholesky decomposition of Σ. This latter choice corresponds to constructing the Brownian motion by time differences. The Cholesky decomposition of Σ gives a poorer rate of convergence, illustrating again Lesson 6. The superiority of the PCA construction was observed in [1].

6 A Bayesian Trio Identity for Cubature Error

An alternative to the deterministic integrand considered thus far is to assume that the integrand is a stochastic process. Random input functions have been hypothesized by Diaconis [8], O'Hagan [38], Ritter [45], Rasmussen and Ghahramani [43], and others. Specifically, suppose that $f \sim \mathcal{GP}(0, s^2 C_{\boldsymbol{\theta}})$, a zero mean Gaussian process. The covariance of this Gaussian process is $s^2 C_{\boldsymbol{\theta}}$, where s is a scale parameter, and $C_{\boldsymbol{\theta}} : \mathcal{X} \times \mathcal{X} \to \mathbb{R}$ is defined by a shape parameter $\boldsymbol{\theta}$. The sample space for this Gaussian process, $\mathcal{F}$, does not enter significantly into the analysis. Define the vector space of measures

$$\mathcal{M} = \left\{\eta : \left|\int_{\mathcal{X}^2} C_{\boldsymbol{\theta}}(\boldsymbol{x},\boldsymbol{t})\,\eta(\mathrm{d}\boldsymbol{x})\eta(\mathrm{d}\boldsymbol{t})\right| < \infty, \ \left|\int_{\mathcal{X}} C_{\boldsymbol{\theta}}(\boldsymbol{x},\boldsymbol{t})\,\eta(\mathrm{d}\boldsymbol{t})\right| < \infty \ \forall \boldsymbol{x} \in \mathcal{X}\right\},$$

and let $C_{\boldsymbol{\theta}}$ be such that $\mathcal{M}$ contains both ν and the Dirac measures $\delta_{\boldsymbol{t}}$ for all $\boldsymbol{t} \in \mathcal{X}$.

For a Gaussian process, all vectors of linear functionals of f have a multivariate Gaussian distribution. It then follows that for a deterministic sampling measure, $\widehat{\nu} = \sum_{i=1}^n w_i \delta_{\boldsymbol{x}_i}$, the cubature error, $\mu - \widehat{\mu}$, is distributed as $\mathcal{N}\big(0, s^2(c_0 - 2\boldsymbol{c}^T\boldsymbol{w} + \boldsymbol{w}^T\mathsf{C}\boldsymbol{w})\big)$, where

$$c_0 = \int_{\mathcal{X}^2} C_{\boldsymbol{\theta}}(\boldsymbol{x},\boldsymbol{t})\,\nu(\mathrm{d}\boldsymbol{x})\nu(\mathrm{d}\boldsymbol{t}), \qquad \boldsymbol{c} = \left(\int_{\mathcal{X}} C_{\boldsymbol{\theta}}(\boldsymbol{x}_i,\boldsymbol{t})\,\nu(\mathrm{d}\boldsymbol{t})\right)_{i=1}^n, \tag{16a}$$

$$\mathsf{C} = \big(C_{\boldsymbol{\theta}}(\boldsymbol{x}_i,\boldsymbol{x}_j)\big)_{i,j=1}^n, \qquad \boldsymbol{w} = \big(w_i\big)_{i=1}^n. \tag{16b}$$

The dependence of c_0, $\boldsymbol{c}$, and C on the shape parameter $\boldsymbol{\theta}$ is suppressed in the notation for simplicity. We define the Bayesian variation, discrepancy and confounding as

$$\mathrm{VAR}^{\mathrm{B}}(f) = s, \qquad \mathrm{DSC}^{\mathrm{B}}(\nu - \widehat{\nu}) = \sqrt{c_0 - 2\boldsymbol{c}^T\boldsymbol{w} + \boldsymbol{w}^T\mathsf{C}\boldsymbol{w}}, \tag{17a}$$

$$\mathrm{CNF}^{\mathrm{B}}(f, \nu - \widehat{\nu}) := \frac{\displaystyle\int_{\mathcal{X}} f(\boldsymbol{x})\,(\nu - \widehat{\nu})(\mathrm{d}\boldsymbol{x})}{s\sqrt{c_0 - 2\boldsymbol{c}^T\boldsymbol{w} + \boldsymbol{w}^T\mathsf{C}\boldsymbol{w}}}. \tag{17b}$$

Theorem 3 (Bayesian Trio Error Identity) *Let the integrand be an instance of a zero mean Gaussian process with covariance $s^2 C_{\boldsymbol{\theta}}$ and that is drawn from a sample space $\mathscr{F}$. For the variation, discrepancy, and confounding defined in (17), the following error identity holds:*

$$\mu - \widehat{\mu} = \mathrm{CNF}^{\mathrm{B}}(f, \nu - \widehat{\nu})\, \mathrm{DSC}^{\mathrm{B}}(\nu - \widehat{\nu})\, \mathrm{VAR}^{\mathrm{B}}(f) \quad \textit{almost surely.} \qquad \text{(BTRIO)}$$

Moreover, $\mathrm{CNF}^{\mathrm{B}}(f, \nu - \widehat{\nu}) \sim \mathscr{N}(0, 1)$.

Proof Although $\int_{\mathscr{X}} f(\boldsymbol{x})\, \nu(\mathrm{d}\boldsymbol{x})$ and $f(\boldsymbol{t}) = \int_{\mathscr{X}} f(\boldsymbol{x})\, \delta_{\boldsymbol{t}}(\mathrm{d}\boldsymbol{x})$ may not exist for all $f \in \mathscr{F}$, these two quantities exist almost surely because $\mathbb{E}_f[\int_{\mathscr{X}} f(\boldsymbol{x})\, \nu(\mathrm{d}\boldsymbol{x})]^2 = s^2 c_0$, and $\mathbb{E}_f[f(\boldsymbol{x})]^2 = s^2 C_{\boldsymbol{\theta}}(\boldsymbol{x}, \boldsymbol{x})$ are both well-defined and finite. The proof of the Bayesian trio identity follows directly from the definitions above. The distribution of the confounding follows from the distribution of the cubature error. □

The choice of cubature weights that minimizes the Bayesian discrepancy in (17a) is $\boldsymbol{w} = \mathsf{C}^{-1}\boldsymbol{c}$, which results in $\mathrm{DSC}^{\mathrm{B}}(\nu - \widehat{\nu}) = \sqrt{c_0 - \boldsymbol{c}^T \mathsf{C}^{-1} \boldsymbol{c}}$ and $\mu - \widehat{\mu} \sim \mathscr{N}\big(0, s^2(c_0 - \boldsymbol{c}^T \mathsf{C}^{-1}\boldsymbol{c})\big)$. However, computing the weights requires $\mathscr{O}(n^3)$ operations unless C has some special structure. This computational cost is significant and may be a deterrent to the use of optimal weights unless the weights are precomputed. For smoother covariance functions, $C_{\boldsymbol{\theta}}$, there is often a challenge of C being ill-conditioned.

The *conditional* distribution of the cubature error, $\mu - \widehat{\mu}$, given the observed data $\{f(\boldsymbol{x}_i) = y_i\}_{i=1}^n$ is $\mathscr{N}\big(\boldsymbol{y}^T(\mathsf{C}^{-1}\boldsymbol{c} - \boldsymbol{w}), s^2(c_0 - \boldsymbol{c}^T \mathsf{C}^{-1}\boldsymbol{c})\big)$. To remove the bias one should again choose $\boldsymbol{w} = \mathsf{C}^{-1}\boldsymbol{c}$. This also makes the conditional distribution of the cubature error the same as the unconditional distribution of the cubature error.

Because the cubature error is a normal random variable, we may use function values to perform useful inference, namely,

$$\mathbb{P}_f\big[|\mu - \widehat{\mu}| \le 2.58\, \mathrm{DSC}^{\mathrm{B}}(\nu - \widehat{\nu}) \mathrm{VAR}^{\mathrm{B}}(f)\big] = 99\%. \tag{18}$$

However, unlike our use of random sampling measures that are constructed via carefully crafted random number generators, there is no assurance that our integrand is actually drawn from a Gaussian process whose covariance we have assumed.

The covariance function, $s^2 C_{\boldsymbol{\theta}}$, should be estimated, and one way to do so is through maximum likelihood estimation (MLE), using the function values drawn for the purpose of estimating the integral. The log-likelihood function for the data $\{f(\boldsymbol{x}_i) = y_i\}_{i=1}^n$ is

$$\begin{aligned}
\ell(s, \boldsymbol{\theta} | \boldsymbol{y}) &= \log\left(\frac{\exp\left(-\frac{1}{2}s^{-2}\boldsymbol{y}^T \mathsf{C}_{\boldsymbol{\theta}}^{-1}\boldsymbol{y}\right)}{\sqrt{(2\pi)^n \det(s^2 \mathsf{C}_{\boldsymbol{\theta}})}}\right) \\
&= -\frac{1}{2}s^{-2}\boldsymbol{y}^T \mathsf{C}_{\boldsymbol{\theta}}^{-1}\boldsymbol{y} - \frac{1}{2}\log\big(\det(\mathsf{C}_{\boldsymbol{\theta}})\big) - \frac{n}{2}\log(s^2) + \text{constants}.
\end{aligned}$$

Maximizing with respect to s^2, yields the MLE scale parameter:

$$s_{\mathrm{MLE}} = \sqrt{\frac{1}{n} \boldsymbol{y}^T \mathsf{C}_{\boldsymbol{\theta}_{\mathrm{MLE}}}^{-1} \boldsymbol{y}}.$$

Plugging this into the log likelihood leads to the MLE shape parameter:

$$\boldsymbol{\theta}_{\mathrm{MLE}} = \underset{\boldsymbol{\theta}}{\operatorname{argmin}} \left[\frac{1}{n} \log\bigl(\det(\mathsf{C}_{\boldsymbol{\theta}})\bigr) + \log\bigl(\boldsymbol{y}^T \mathsf{C}_{\boldsymbol{\theta}}^{-1} \boldsymbol{y}\bigr) \right],$$

which requires numerical optimization to evaluate. Using MLE estimates, the probabilistic error bound in (18) becomes

$$\mathbb{P}_f \left[|\mu - \widehat{\mu}| \le 2.58 \sqrt{\frac{1}{n} \left(c_{0,\boldsymbol{\theta}_{\mathrm{MLE}}} - \boldsymbol{c}_{\boldsymbol{\theta}_{\mathrm{MLE}}}^T \mathsf{C}_{\boldsymbol{\theta}_{\mathrm{MLE}}}^{-1} \boldsymbol{c}_{\boldsymbol{\theta}_{\mathrm{MLE}}}\right) \left(\boldsymbol{y}^T \mathsf{C}_{\boldsymbol{\theta}_{\mathrm{MLE}}}^{-1} \boldsymbol{y}\right)} \right] = 99\%. \qquad (19)$$

Note that the value of $\boldsymbol{\theta}_{\mathrm{MLE}}$ and the above Bayesian cubature error bound is unchanged by replacing $C_{\boldsymbol{\theta}}$ by a positive multiple of itself.

Let's revisit the multivariate normal probability example of Sect. 4, and perform Bayesian cubature with a covariance kernel with modest smoothness from the Matérn family:

$$C_\theta(\boldsymbol{x}, \boldsymbol{t}) = \prod_{j=1}^{d} \left(1 + \theta |x_j - t_j|\right) \exp\left(-\theta |x_j - t_j|\right) \qquad (20)$$

Using 100 randomly scrambled Sobol' samples, the Bayesian cubature method outlined above was used to compute the multivariate normal probability μ. We used MLE scale and shape parameters and optimal cubature weights $\boldsymbol{w} = \mathsf{C}_{\theta_{\mathrm{MLE}}}^{-1} \boldsymbol{c}_{\theta_{\mathrm{MLE}}}$. The actual errors are plotted in Fig. 5, which also provides a contrast of the actual error and the probabilistic error bound. This bound was correct about 83% of the time. Based on the smoothness of the integrand and the kernel, one might expect $\mathscr{O}(n^{-2})$ convergence of the answer, but this is not clear from the numerical computations.

Lesson 8 *Bayesian cubature provides data-based probabilistic error bounds under the assumption that the integrand is a Gaussian process.*

Bayesian cubature offers hope with a dose of caution. The theory is solid, but as this example shows, one cannot know if the actual integrand under consideration is a typical instance of the Gaussian process being assumed, even when using MLE to determine the parameters of the distribution. The success rate of the probabilistic error bound for this example is high, but not as high as the theory would suggest. One may ask whether a larger candidate family of Gaussian processes needs to be considered, but then this might increase the time required for estimation of the parameters. This example was carried out to only a rather modest sample size because of the $\mathscr{O}(n^3)$ operations required to compute each $\widehat{\mu}$. Efforts to reduce this operation count

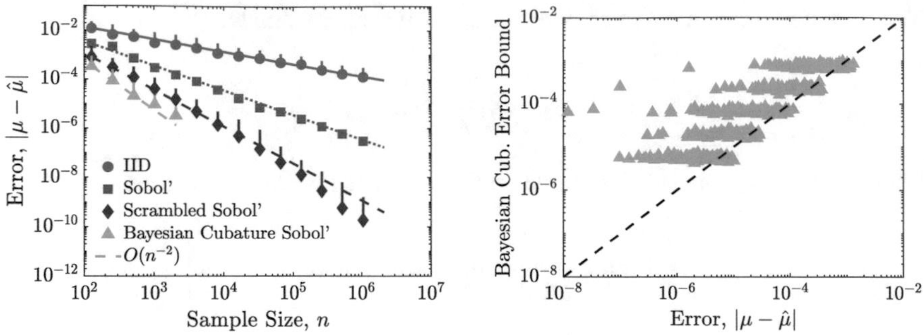

Fig. 5 The cubature errors for the multivariate normal probability example using Bayesian cubature (left), and the Bayesian cubature error versus the probabilistic error bound in (19) (right)

have been made by Anitescu, Chen, and Stein [2], Parker, Reich and Gotwalt [42], and others. Probabilistic numerics, http://www.probabilistic-numerics.org, of which Bayesian cubature is an example, holds promise that deserves further exploration.

The formulas for the Bayesian trio identity are analogous to those for the deterministic trio identity for reproducing kernel Hilbert spaces when $T(f) = 0$ for all $f \in \mathscr{F}$. Suppose that the reproducing kernel $K_{\boldsymbol{\theta}}$ in the deterministic case is numerically equivalent to the covariance function $C_{\boldsymbol{\theta}}$ used in Bayesian cubature. The optimal cubature weights in the Bayesian case then mirror those in the deterministic case. Likewise, for these optimal weights $\mathrm{DSC}^{\mathrm{D}}(\nu - \widehat{\nu})$ is numerically the same as $\mathrm{DSC}^{\mathrm{B}}(\nu - \widehat{\nu})$.

Lesson 9 *The formula for the Bayesian discrepancy mimics that for the deterministic discrepancy with $\mathscr{F}$ an RKHS.*

7 A Randomized Bayesian Trio Identity for Cubature Error

So far, we have presented three versions of the trio identity: a deterministic version in Theorem 1, a randomized version in Theorem 2, and a Bayesian version in Theorem 3. The fourth and final version is a randomized Bayesian trio identity. The variation remains unchanged from the Bayesian definition in (17a). The randomized Bayesian discrepancy and confounding are defined as follows:

$$\mathrm{DSC}^{\mathrm{RB}}(\nu - \widehat{\nu}) = \sqrt{\mathbb{E}_{\widehat{\nu}}\left(c_0 - 2\boldsymbol{c}^T\boldsymbol{w} + \boldsymbol{w}^T\mathsf{C}\boldsymbol{w}\right)}, \tag{21a}$$

$$\mathrm{CNF}^{\mathrm{RB}}(f, \nu - \widehat{\nu}) := \frac{\displaystyle\int_{\mathscr{X}} f(\boldsymbol{x})\,(\nu - \widehat{\nu})(\mathrm{d}\boldsymbol{x})}{s\sqrt{\mathbb{E}_{\widehat{\nu}}\left(c_0 - 2\boldsymbol{c}^T\boldsymbol{w} + \boldsymbol{w}^T\mathsf{C}\boldsymbol{w}\right)}}. \tag{21b}$$

The proof of the randomized Bayesian trio error identity is similar to the proofs of the other trio identities and is omitted.

Theorem 4 (Randomized Bayesian Trio Error Identity) *Let the integrand be an instance of a zero mean Gaussian process with covariance $s^2 C_{\boldsymbol{\theta}}$ and that is drawn from a sample space $\mathcal{F}$. Let the sampling measure be drawn randomly from $\mathcal{M}_S$ according to some probability distribution. For the variation defined in (17a), and the discrepancy and confounding defined in (21), the following error identity holds:*

$$\mu - \widehat{\mu} = \mathrm{CNF}^{\mathrm{RB}}(f, \nu - \widehat{\nu})\, \mathrm{DSC}^{\mathrm{RB}}(\nu - \widehat{\nu})\, \mathrm{VAR}^{\mathrm{B}}(f) \quad \textit{almost surely}. \qquad \text{(RBTRIO)}$$

Moreover, $\mathrm{CNF}^{\mathrm{RB}}(f, \nu - \widehat{\nu}) \sim \mathcal{N}(0, 1)$.

Lesson 10 *The trio identity has four versions,* (DTRIO), (RTRIO), (BTRIO), *and* (RBTRIO), *depending on whether the integrand is deterministic or Bayesian and whether the sampling measure is deterministic or random.*

8 Dimension Dependence of the Discrepancy, Cubature Error and Computational Cost

The statements about the rates of decay of discrepancy and cubature error as the sample size increases have so far hidden the dependence on the dimension of the integration domain. Figure 4 on the left shows a clear error decay rate of $\mathcal{O}(n^{-1+\varepsilon})$ for low discrepancy sampling for the option pricing problem with dimension 12. However, Fig. 2 shows that the discrepancy for these scrambled Sobol' points does not decay as quickly as $\mathcal{O}(n^{-1+\varepsilon})$ for moderate n.

There has been a tremendous effort to understand the effect of the dimension of the integration problem on the convergence rate. Sloan and Woźniakowski [47] pointed out how the sample size required to achieve a desired error tolerance could grow exponentially with dimension. Such problems are called *intractable*. This led to a search for settings where the sample size required to achieve a desired error tolerance only grows polynomially with dimension (*tractable* problems) or is independent of the dimension (*strongly tractable* problems). The three volume masterpiece by Novak and Woźniakowski [35–37] and the references cited therein contain necessary and sufficient conditions for tractability. The parallel idea of *effective dimension* was introduced by Caflisch, Morokoff, and Owen [5] and developed further in [29].

Here we provide a glimpse into those situations where the dimension of the problem does not have an adverse effect on the convergence rate of the cubature error and the discrepancy. Let's generalize the reproducing kernel used to define the L^2-discrepancy in (9), as well as the corresponding variation and the discrepancy for equi-weighted sampling measures by introducing *coordinate weights* $\gamma_1, \gamma_2, \ldots$:

$$K(\boldsymbol{x},\boldsymbol{t}) = \prod_{k=1}^{d}[1+\gamma_k^2\{1-\max(x_k,t_k)\}],$$

$$\mathrm{VAR}^{\mathrm{D}}(f) = \left\| \left(\gamma_{\mathfrak{u}}^{-1}\|\partial^{\mathfrak{u}} f\|_{L^2}\right)_{\mathfrak{u}\neq\emptyset} \right\|_2 \qquad \gamma_{\mathfrak{u}} = \prod_{k\in\mathfrak{u}} \gamma_k,$$

$$\left[\mathrm{DSC}^{\mathrm{D}}(\nu-\widehat{\nu})\right]^2 = \prod_{k=1}^{d}\left(1+\frac{\gamma_k^2}{3}\right) - \frac{2}{n}\sum_{i=1}^{n}\prod_{k=1}^{d}\left(1+\frac{\gamma_k^2(1-x_{ik}^2)}{2}\right) + \frac{1}{n^2}\sum_{i,j=1}^{n}\prod_{k=1}^{d}[1+\gamma_k^2(1-\max(x_{ik},x_{jk}))]. \tag{22}$$

For $\gamma_1 = \cdots = \gamma_d = 1$, we recover the situation in Sect. 2, where the decay rate of the discrepancy is dimension dependent for moderate sample sizes. However if $\gamma_k^2 = k^{-3}$, then the discrepancies for randomly shifted lattice nodesets and scrambled Sobol' sequences show only a slight dimension dependence, as shown in Fig. 6.

When the weights γ_k decay with k, the discrepancy depends *less* on how evenly the data sites appear in projections involving the higher numbered coordinates. On the other hand, the variation in this case gives *heavier* weight to the $\partial^{\mathfrak{u}} f$ with $\mathfrak{u}$ containing large k. For the cubature error decay to mirror the decay of the discrepancy shown in Fig. 6, the integrand must depend only slightly on the coordinates with higher indices, so that the variation will be modest.

Lesson 11 *The cubature error for high dimensional problems can often be reduced by arranging for the integrand to depend primarily on those coordinates with lower indices.*

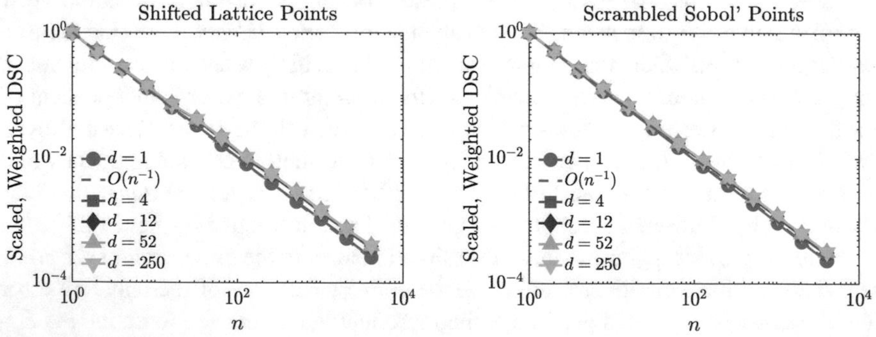

Fig. 6 The root mean square weighted L^2-discrepancies given by (22) with $\gamma_k^2 = k^{-3}$ for randomly shifted lattice sequence nodesets and randomly scrambled and shifted Sobol' sequences points. A variety of dimensions is shown

For some integration problems the dimension is infinite and so our problem (INT) becomes

$$\mu = \lim_{d\to\infty} \mu^{(d)}, \qquad \mu^{(d)} = \int_{\mathcal{X}^{(d)}} f^{(d)}(\boldsymbol{x})\, \nu^{(d)}(\mathrm{d}\boldsymbol{x}), \qquad (\infty\text{INT})$$

where $\mathcal{X}^{(d)}=\mathcal{X}_1 \times \cdots \times \mathcal{X}_d$, $\nu^{(d)}$ is a measure on $\mathcal{X}^{(d)}$ with independent marginals ν_k on $\mathcal{X}_k$, and $f^{(1)}, f^{(2)}, \ldots$ are approximations to an infinite-dimensional integrand. The discrepancy and cubature error analysis for $d \to \infty$ is similar to the large d situation, but now the *computational cost* of the approximate integrand is a concern [7, 20, 22, 34].

One could approximate μ by $\widehat{\mu}^{(d)}$, the approximation to $\mu^{(d)}$, for some large d. However, the computational cost of evaluating $f^{(d)}(\boldsymbol{x})$ for a single $\boldsymbol{x}$ typically requires $\mathcal{O}(d)$ operations. So this approach would require a high computational cost of $\mathcal{O}(nd)$ operations to compute $\widehat{\mu}^{(d)}$.

The often better alternative is to decompose the $f^{(d)}$ into pieces $f_{\mathfrak{u}}$, for $\mathfrak{u} \subseteq 1{:}d$, such that $f^{(d)} = \sum_{\mathfrak{u}\subseteq 1:d} f_{\mathfrak{u}}$ and the $f_{\mathfrak{u}}$ depend on $\mathfrak{u}$ but *not on d*. Multi-level Monte Carlo approximates (∞INT) by

$$\widehat{\mu} := \widehat{\mu}\big(f^{(d_1)}\big) + \widehat{\mu}\big(f^{(d_2)} - f^{(d_1)}\big) + \cdots + \widehat{\mu}\big(f^{(d_L)} - f^{(d_{L-1})}\big),$$

for some choice of d_l with $d_1 < \cdots < d_L$. This works well when $\mathrm{VAR}\big(f^{(d_l)} - f^{(d_{l-1})}\big)$ decreases as l increases and when $\mu - \mu^{(d_L)}$ is small [12–15, 22, 34]. The computational cost of $\widehat{\mu}\big(f^{(d_l)} - f^{(d_{l-1})}\big)$ is $\mathcal{O}(n_l d_l)$, and as d_l increases, n_l decreases, thus moderating the cost. There is bias, since $\mu - \mu^{(d_L)}$ is not approximated at all, but this can be removed by a clever randomized sampling method [44].

The Multivariate Decomposition Method approximates (∞INT) by

$$\widehat{\mu} = \widehat{\mu}(f_{\mathfrak{u}_1}) + \widehat{\mu}(f_{\mathfrak{u}_2}) + \cdots + \widehat{\mu}(f_{\mathfrak{u}_L}),$$

where the $\mathfrak{u}_l$ are the important sets of coordinate indices as judged by $\mathrm{VAR}^{\mathrm{D}}(f_{\mathfrak{u}})$ to ensure that $\mu - \sum_{\mathfrak{u}\notin\{\mathfrak{u}_1,\ldots,\mathfrak{u}_L\}} \mu(f_{\mathfrak{u}})$ is small [49]. The computational cost of each $\widehat{\mu}(f_{\mathfrak{u}_l})$ is $\mathcal{O}(n_l|\mathfrak{u}_l|)$. If the important sets have small cardinality, $|\mathfrak{u}_l|$, the computational cost is moderate.

Lesson 12 *Infinite dimensional problems may be efficiently solved by multi-level methods or multivariate decomposition methods, which approximate the integral by a sum of finite dimensional integrals.*

9 Automatic Stopping Criteria for Cubature

The trio identity decomposes the cubature error into three factors. By improving the sampling scheme, the discrepancy may be made smaller. By re-writing the integral, the variation of the integrand might be made smaller. For certain situations, we may find that the confounding is small. While the trio identity helps us understand what contributes to the cubature error, it does not directly answer the question of how many samples are required to achieve the desired accuracy, i.e., how to ensure that

$$|\mu - \widehat{\mu}| \leq \varepsilon \qquad \text{(ErrCrit)}$$

for some predetermined ε.

Bayesian cubature, as described in Sect. 6, provides data-based cubature error bounds. These can be used to determine how large n must be to satisfy (ErrCrit) with high probability.

For IID Monte Carlo the Central Limit Theorem may be used to construct an approximate confidence interval for μ, however, this approach relies on believing that n is large enough to have (i) reached the asymptotic limit, and (ii) obtained a reliable upper bound on the standard deviation in terms of a sample standard deviation. There have been recent efforts to develop a more robust approach to fixed width confidence intervals [4, 23, 26]. An upper bound on the standard deviation may be computed by assuming an upper bound on the kurtosis or estimating the kurtosis from data. The standard deviation of an integrand can be confidently bounded in terms of the sample standard deviation if it lies in the cone of functions with a known bound on their kurtosis. A bound on the kurtosis also allows one to use a Berry–Esseen inequality, which is a finite sample version of the Central Limit Theorem, to determine a sufficient sample size for computing the integral with the desired accuracy.

For low discrepancy sampling, independent random replications may be used to estimate the error, but this approach lacks a rigorous justification. An alternative proposed by the author and his collaborators is to decompose the integrand into a Fourier series and estimate the decay rate of the Fourier coefficients that contribute to the error [21, 24, 27]. This approach may also be used to satisfy relative error criteria or error criteria involving a function of several integrals [24]. Our automatic stopping criteria have been implemented in the Guaranteed Automatic Integration Library (GAIL) [6].

Lesson 13 *Automatic stopping criteria for (quasi-)Monte Carlo simulations have been developed for integrands that lie in a cone of functions that are not too wild.*

10 Summary

To conclude, we repeat the lessons highlighted above. The order may be somewhat different.

The trio identity (TRIO) decomposes the cubature error into a product of three factors: the variation of the integrand, the discrepancy of the sampling measure, and the confounding. This identity shows how the integrand and the sampling measure each contribute to the cubature error. The trio identity has four versions, (DTRIO),

(RTRIO), (BTRIO), and (RBTRIO), depending on whether the integrand is deterministic or Bayesian and whether the sampling measure is deterministic or random. The deterministic discrepancy when $\mathscr{F}$ is an RKHS has a simple, explicit form involving three terms. The formula for the Bayesian discrepancy mimics that for the deterministic discrepancy with $\mathscr{F}$ an RKHS. Although it has traditionally been ignored, the confounding helps explain why the cubature error may decay to zero much faster or more slowly than the discrepancy.

How do good sampling measures, $\widehat{\nu}$, make the error smaller? Quasi-Monte Carlo methods replace IID data sites by low discrepancy data sites, such as Sobol' sequences and integration lattice nodeset sequences. The resulting sampling measures have discrepancies and cubature errors that decay to zero at a faster rate than in the case of IID sampling. Randomizing the sampling measure may not only eliminate bias, but it may help improve accuracy by avoiding the awful minority of possible sampling measures. The benefits of sampling measures with asymptotically smaller discrepancies are limited to those integrands with finite variation.

How can the error be decreased by re-casting the problem with a different integrand, f? Well-chosen variable transformations may reduce cubature error by producing an integrand with a smaller variation than otherwise. The cubature error for high dimensional problems can often be reduced by arranging for the integrand to depend primarily on those coordinates with lower indices. Infinite dimensional problems may be efficiently solved by multi-level methods or multivariate decomposition methods, which approximate the integral by a sum of finite dimensional integrals.

How many samples, n, are required to meet a specified error tolerance? Bayesian cubature provides data-based probabilistic error bounds under the assumption that the integrand is a Gaussian process. Automatic stopping criteria for (quasi-)Monte Carlo simulations have been developed for integrands that lie in a cone of functions that are not too wild.

Acknowledgements The author would like to thank the organizers of MCQMC 2016 for an exceptional conference. The author is indebted to his colleagues in the MCQMC community for all that he has learned from them. In particular, the author thanks Xiao-Li Meng for introducing the trio identity and for discussions related to its development. The author also thanks Lluís Antoni Jiménez Rugama for helpful comments in preparing this tutorial. This work is partially supported by the National Science Foundation grant DMS-1522687.

References

1. Acworth, P., Broadie, M., Glasserman, P.: A comparison of some Monte Carlo techniques for option pricing. In: Niederreiter, H., Hellekalek, P., Larcher, G., Zinterhof, P. (eds.) Monte Carlo and Quasi-Monte Carlo Methods 1996. Lecture Notes in Statistics, vol. 127, pp. 1–18. Springer, New York (1998)
2. Anitescu, M., Chen, J., Stein, M.: An inversion-free estimating equation approach for Gaussian process models. J. Comput. Graph. Stat. (2016)

3. Aronszajn, N.: Theory of reproducing kernels. Trans. Am. Math. Soc. **68**, 337–404 (1950)
4. Bayer, C., Hoel, H., von Schwerin, E., Tempone, R.: On nonasymptotic optimal stopping criteria in Monte Carlo simulations. SIAM J. Sci. Comput. **36**, A869–A885 (2014)
5. Caflisch, R.E., Morokoff, W., Owen, A.: Valuation of mortgage backed securities using Brownian bridges to reduce effective dimension. J. Comput. Financ. **1**, 27–46 (1997)
6. Choi, S.C.T., Ding, Y., Hickernell, F.J., Jiang, L., Jiménez Rugama, Ll.A., Li, D., Rathinavel, J., Tong, X., Zhang, K., Zhang, Y., Zhou, X.: GAIL: Guaranteed Automatic Integration Library (versions 1.0–2.2). MATLAB software (2013–2017). http://gailgithub.github.io/GAIL_Dev/
7. Creutzig, J., Dereich, S., Müller-Gronbach, T., Ritter, K.: Infinite-dimensional quadrature and approximation of distributions. Found. Comput. Math. **9**, 391–429 (2009)
8. Diaconis, P.: Bayesian numerical analysis. In: Gupta, S.S., Berger, J.O. (eds.) Statistical Decision Theory and Related Topics IV, Papers from the 4th Purdue Symposium, West Lafayette, Indiana 1986, vol. 1, pp. 163–175. Springer, New York (1988)
9. Dick, J., Pillichshammer, F.: Digital Nets and Sequences: Discrepancy Theory and Quasi-Monte Carlo Integration. Cambridge University Press, Cambridge (2010)
10. Faure, H.: Discrépance de suites associées à un système de numération (en dimension s). Acta Arith. **41**, 337–351 (1982)
11. Genz, A.: Comparison of methods for the computation of multivariate normal probabilities. Comput. Sci. Stat. **25**, 400–405 (1993)
12. Giles, M.: Multilevel Monte Carlo methods. In: Dick, J., Kuo, F.Y., Peters, G.W., Sloan, I.H. (eds.) Monte Carlo and Quasi-Monte Carlo Methods 2012. Springer Proceedings in Mathematics and Statistics, vol. 65. Springer, Berlin (2013)
13. Giles, M.: Multilevel Monte Carlo methods. Acta Numer. **24**, 259–328 (2015)
14. Giles, M.B.: Multilevel Monte Carlo path simulation. Oper. Res. **56**, 607–617 (2008)
15. Heinrich, S.: Multilevel Monte Carlo methods. In: Margenov, S., Wasniewski, J., Yalamov, P.Y. (eds.) Large-Scale Scientific Computing, Third International Conference, LSSC 2001. Lecture Notes in Computer Science, vol. 2179, pp. 58–67. Springer, Berlin (2001)
16. Heinrich, S., Hickernell, F.J., Yue, R.X.: Optimal quadrature for Haar wavelet spaces. Math. Comput. **73**, 259–277 (2004)
17. Hickernell, F.J.: A generalized discrepancy and quadrature error bound. Math. Comput. **67**, 299–322 (1998)
18. Hickernell, F.J.: Goodness-of-fit statistics, discrepancies and robust designs. Stat. Probab. Lett. **44**, 73–78 (1999)
19. Hickernell, F.J., Woźniakowski, H.: The price of pessimism for multidimensional quadrature. J. Complex. **17**, 625–659 (2001)
20. Hickernell, F.J., Wang, X.: The error bounds and tractability of quasi-Monte Carlo algorithms in infinite dimension. Math. Comput. **71**, 1641–1661 (2002)
21. Hickernell, F.J., Jiménez Rugama, Ll.A.: Reliable adaptive cubature using digital sequences. In: Cools, R., Nuyens, D. (eds.) Monte Carlo and Quasi-Monte Carlo Methods: MCQMC, Leuven, Belgium, April 2014. Springer Proceedings in Mathematics and Statistics, vol. 163, pp. 367–383. Springer, Berlin (2016)
22. Hickernell, F.J., Müller-Gronbach, T., Niu, B., Ritter, K.: Multi-level Monte Carlo algorithms for infinite-dimensional integration on $\mathbb{R}^{\mathbb{N}}$. J. Complex. **26**, 229–254 (2010)
23. Hickernell, F.J., Jiang, L., Liu, Y., Owen, A.B.: Guaranteed conservative fixed width confidence intervals via Monte Carlo sampling. In: Dick, J., Kuo, F.Y., Peters, G.W., Sloan, I.H. (eds.) Monte Carlo and Quasi-Monte Carlo Methods 2012. Springer Proceedings in Mathematics and Statistics, vol. 65, pp. 105–128. Springer, Berlin (2013)
24. Hickernell, F.J., Jiménez Rugama, Ll.A., Li, D.: Adaptive quasi-Monte Carlo methods for cubature (2017+). Submitted for publication, arXiv:1702.01491 [math.NA]
25. Hlawka, E.: Funktionen von beschränkter Variation in der Theorie der Gleichverteilung. Ann. Mat. Pura Appl. **54**, 325–333 (1961)
26. Jiang, L.: Guaranteed adaptive Monte Carlo methods for estimating means of random variables. Ph.D. thesis, Illinois Institute of Technology (2016)

27. Jiménez Rugama, Ll.A., Hickernell, F.J.: Adaptive multidimensional integration based on rank-1 lattices. In: Cools, R., Nuyens, D. (eds.) Monte Carlo and Quasi-Monte Carlo Methods: MCQMC, Leuven, Belgium, April 2014. Springer Proceedings in Mathematics and Statistics, vol. 163, pp. 407–422. Springer, Berlin (2016)
28. Koksma, J.F.: Een algemeene stelling uit de theorie der gelijkmatige verdeeling modulo 1. Mathematica B (Zutphen) **11**, 7–11 (1942/1943)
29. Liu, R., Owen, A.B.: Estimating mean dimensionality. J. Am. Stat. Assoc. **101**, 712–721 (2006)
30. Meng, X.: Statistical paradises and paradoxes in big data (2017+). In preparation
31. Niederreiter, H.: Low-discrepancy and low-dispersion sequences. J. Number Theory **30**, 51–70 (1988)
32. Niederreiter, H.: Random Number Generation and Quasi-Monte Carlo Methods. CBMS-NSF Regional Conference Series in Applied Mathematics. SIAM, Philadelphia (1992)
33. Niederreiter, H., Xing, C.: Quasirandom points and global function fields. In: Cohen, S., Niederreiter, H. (eds.) Finite Fields and Applications. London Mathematical Society Lecture Note Series, vol. 233, pp. 269–296. Cambridge University Press, Cambridge (1996)
34. Niu, B., Hickernell, F.J., Müller-Gronbach, T., Ritter, K.: Deterministic multi-level algorithms for infinite-dimensional integration on $\mathbb{R}^{\mathbb{N}}$. J. Complex. **27**, 331–351 (2011)
35. Novak, E., Woźniakowski, H.: Tractability of Multivariate Problems Volume I: Linear Information. EMS Tracts in Mathematics, vol. 6. European Mathematical Society, Zürich (2008)
36. Novak, E., Woźniakowski, H.: Tractability of Multivariate Problems Volume II: Standard Information for Functionals. EMS Tracts in Mathematics, vol. 12. European Mathematical Society, Zürich (2010)
37. Novak, E., Woźniakowski, H.: Tractability of Multivariate Problems: Volume III: Standard Information for Operators. EMS Tracts in Mathematics, vol. 18. European Mathematical Society, Zürich (2012)
38. O'Hagan, A.: Bayes–Hermite quadrature. J. Stat. Plan. Inference **29**, 245–260 (1991)
39. Owen, A.B.: Randomly permuted (t, m, s)-nets and (t, s)-sequences. In: Niederreiter, H., Shiue, P.J.S. (eds.) Monte Carlo and Quasi-Monte Carlo Methods in Scientific Computing. Lecture Notes in Statistics, vol. 106, pp. 299–317. Springer, New York (1995)
40. Owen, A.B.: Monte Carlo variance of scrambled net quadrature. SIAM J. Numer. Anal. **34**, 1884–1910 (1997)
41. Owen, A.B.: Scrambled net variance for integrals of smooth functions. Ann. Stat. **25**, 1541–1562 (1997)
42. Parker, R.J., Reigh, B.J., Gotwalt, C.M.: Approximate likelihood methods for estimation and prediction in Gaussian process regression models for computer experiments (2016+). Submitted for publication
43. Rasmussen, C.E., Ghahramani, Z.: Bayesian Monte Carlo. In: Thrun, S., Saul, L.K., Obermayer, K. (eds.) Advances in Neural Information Processing Systems, vol. 15, pp. 489–496. MIT Press, Cambridge (2003)
44. Rhee, C., Glynn, P.: A new approach to unbiased estimation for SDE's. In: Laroque, C., Himmelspach, J., Pasupathy, R., Rose, O., Uhrmacher, A.M. (eds.) Proceedings of the 2012 Winter Simulation Conference (2012)
45. Ritter, K.: Average-Case Analysis of Numerical Problems. Lecture Notes in Mathematics, vol. 1733. Springer, Berlin (2000)
46. Sloan, I.H., Joe, S.: Lattice Methods for Multiple Integration. Oxford University Press, Oxford (1994)
47. Sloan, I.H., Woźniakowski, H.: An intractability result for multiple integration. Math. Comput. **66**, 1119–1124 (1997)
48. Sobol', I.M.: The distribution of points in a cube and the approximate evaluation of integrals. U.S.S.R. Comput. Math. Math. Phys. **7**, 86–112 (1967)
49. Wasilkowski, G.W.: On tractability of linear tensor product problems for ∞-variate classes of functions. J. Complex. **29**, 351–369 (2013)

Randomized Quasi-Monte Carlo: An Introduction for Practitioners

Pierre L'Ecuyer

Abstract We survey basic ideas and results on randomized quasi-Monte Carlo (RQMC) methods, discuss their practical aspects, and give numerical illustrations. RQMC can improve accuracy compared with standard Monte Carlo (MC) when estimating an integral interpreted as a mathematical expectation. RQMC estimators are unbiased and their variance converges at a faster rate (under certain conditions) than MC estimators, as a function of the sample size. Variants of RQMC also work for the simulation of Markov chains, for function approximation and optimization, for solving partial differential equations, etc. In this introductory survey, we look at how RQMC point sets and sequences are constructed, how we measure their uniformity, why they can work for high-dimensional integrals, and how can they work when simulating Markov chains over a large number of steps.

Keywords QMC · Variance reduction · Multivariate integration · Stochastic simulation · Simulation in finance · Array-RQMC

1 Introduction

We consider a setting in which Monte Carlo (MC) or quasi-Monte Carlo (QMC) is used to estimate the expectation $\mu = \mathbb{E}[X]$ of a random variable X defined over a probability space $(\Omega, \mathcal{F}, \mathbb{P})$. We assume that $\boldsymbol{\omega} \in \Omega$ can be identified with a sequence of s independent $\mathcal{U}(0, 1)$ random variables (uniform over $(0, 1)$) for some integer $s > 0$, so we can write $X = f(\mathbf{U})$ and

$$\mu = \int_0^1 \cdots \int_0^1 f(u_1, \ldots, u_s)\, du_1 \cdots du_s = \int_{(0,1)^s} f(\mathbf{u})\, d\mathbf{u} = \mathbb{E}[f(\mathbf{U})], \quad (1)$$

P. L'Ecuyer (✉)
Département d'Informatique et de Recherche Opérationnelle, Université de Montréal, Canada and Inria, Rennes, France
e-mail: lecuyer@iro.umontreal.ca

A. B. Owen and P. W. Glynn (eds.), *Monte Carlo and Quasi-Monte Carlo Methods*, Springer Proceedings in Mathematics & Statistics 241,
https://doi.org/10.1007/978-3-319-91436-7_2

for some function $f : \Omega = (0,1)^s \to \mathbb{R}$, where $\mathbf{u} = (u_1, \ldots, u_s) \in (0,1)^s$, and $\mathbf{U} \sim \mathcal{U}(0,1)^s$ (uniform over the unit hypercube). We can allow s to be random and unbounded; then this model is very general [28].

The standard *Monte Carlo* (MC) method estimates μ by

$$\bar{X}_n = \frac{1}{n}\sum_{i=0}^{n-1} X_i \tag{2}$$

where $X_i = f(\mathbf{U}_i)$ and $\mathbf{U}_0, \ldots, \mathbf{U}_{n-1}$ are n independent $\mathcal{U}(0,1)^s$ random vectors. In implementations, these $\mathbf{U}_i$ are replaced by vectors of "random numbers" that drive the simulation, but the MC theory developed under the above probabilistic assumptions still works well. We have $\mathbb{E}[\bar{X}_n] = \mu$ and $\mathrm{Var}[\bar{X}_n] = \sigma^2/n$ where $\sigma^2 := \int_{(0,1)^s} f^2(\mathbf{u})\,d\mathbf{u} - \mu^2$. If $\sigma^2 < \infty$ then when $n \to \infty$ we have $\bar{X}_n \to \mu$ with probability 1 by the strong law of large numbers and $\sqrt{n}(\bar{X}_n - \mu)/S_n \Rightarrow \mathcal{N}(0,1)$ (the standard normal distribution) by the usual central limit theorem (CLT), where $S_n^2 = \frac{1}{n-1}\sum_{i=0}^{n-1}(X_i - \bar{X}_n)^2$. This CLT is invoked routinely to compute a confidence interval on μ based on a normal approximation. The width of this confidence interval is asymptotically proportional to $\sigma/\sqrt{n}$, which means that for each additional decimal digit of accuracy on μ, one must multiply the sample size n by 100.

Quasi-Monte Carlo (QMC) replaces the independent random points $\mathbf{U}_i$ by a set of *deterministic* points $P_n = \{\mathbf{u}_0, \ldots, \mathbf{u}_{n-1}\}$ that cover $[0,1)^s$ more evenly. (Here we include 0 in the interval because some of these deterministic points often have coordinates at 0.) It estimates μ by

$$\bar{\mu}_n = \frac{1}{n}\sum_{i=0}^{n-1} f(\mathbf{u}_i).$$

Roughly speaking, P_n is called a *highly-uniform point set* or *low-discrepancy point set* if some measure of *discrepancy* between the empirical distribution of P_n and the uniform distribution converges to 0 faster than $\mathcal{O}(n^{-1/2})$, which is the typical rate for independent random points, when $n \to \infty$. QMC theory defines several types of discrepancies, usually by defining function spaces (often Hilbert spaces) $\mathcal{H}$ in which by applying the Cauchy–Schwarz inequality, one obtains the worst-case error bound

$$|\bar{\mu}_n - \mu| \le D(P_n)V(f) \tag{3}$$

for all $f \in \mathcal{H}$, where $V(f) = \|f - \mu\|_{\mathcal{H}}$ is the norm of $f - \mu$ in $\mathcal{H}$ (it measures the *variation* of f), and $D(P_n)$ is the *discrepancy* measure of P_n associated with this space [9, 18, 50]. For any fixed $f \in \mathcal{H}$ with $V(f) \neq 0$, this error bound converges at the same rate as $D(P_n)$. The error itself sometimes converges faster than the bound. To capture this, Hickernell [21] considers a setting in which (3) is transformed into an equality by introducing a third multiplicative factor on the right side. This new factor is the *confounding* between f and the empirical distribution of the points $\mathbf{u}_i$;

it is always in [0, 1] and is equal to the left term divided by the right term in (3). The resulting equality is named a *trio identity*.

In a well-known special case, $D(P_n)$ is the *star discrepancy* $D^*(P_n)$, defined as follows: for each rectangular box $[\mathbf{0}, \mathbf{u})$ with opposite corners at $\mathbf{0}$ and $\mathbf{u}$, let $\Delta(\mathbf{u})$ be the absolute difference between the volume of the box and the fraction of P_n that fall in that box. Then define $D^*(P_n)$ as the supremum of $\Delta(\mathbf{u})$ over all $\mathbf{u} \in (0, 1)^s$. There are known explicit constructions that can provide a P_n for each n, for which $D^*(P_n) = \mathcal{O}(n^{-1}(\ln n)^{s-1})$. Variants of these constructions provide an infinite sequence of points for which $D^*(P_n) = \mathcal{O}(n^{-1}(\ln n)^s)$ if P_n comprises the first n points of the sequence. One important limitation of (3) for this case is that the corresponding $V(f)$, known as the Hardy–Krause variation of f, is infinite as soon as f has a discontinuity that is not aligned with the axes. Computing $D^*(P_n)$ explicitly is also difficult: the best known algorithms are polynomial in n but exponential in s; see [12] for a coverage of various types of discrepancies and their computation.

There are other interesting (Hilbert) spaces of periodic smooth functions for which the corresponding $D(P_n)$ in (3) converges as $\mathcal{O}(n^{-\alpha+\epsilon})$ for any $\epsilon > 0$, for some $\alpha > 0$ that depends on the space and can be arbitrarily large. The main construction methods for P_n are *lattice rules* and *digital nets*. We discuss them later and give examples.

With deterministic QMC, it is hard to estimate the integration error in practice. *Randomized quasi-Monte Carlo* (RQMC) randomizes the QMC points in a way that for the RQMC point set $P_n = \{\mathbf{U}_0, \ldots, \mathbf{U}_{n-1}\} \subset (0, 1)^s$ (which is now random),

(i) each point $\mathbf{U}_i$ has the uniform distribution over $(0, 1)^s$;
(ii) P_n as a whole is a low-discrepancy point set.

This turns QMC into a variance-reduction method. The RQMC estimator

$$\hat{\mu}_{n,\text{rqmc}} = \frac{1}{n}\sum_{i=0}^{n-1} f(\mathbf{U}_i) \tag{4}$$

is an unbiased estimator of μ, with variance

$$\text{Var}[\hat{\mu}_{n,\text{rqmc}}] = \frac{\text{Var}[f(\mathbf{U}_i)]}{n} + \frac{2}{n^2}\sum_{i<j}\text{Cov}[f(\mathbf{U}_i), f(\mathbf{U}_j)]. \tag{5}$$

We want to make the last sum as negative as possible, by inducing pairwise negative covariance. Well-known ways of creating such negative correlation include antithetic variates (with $n = 2$), Latin hypercube sampling (LHS), and stratification. The first two can reduce the variance under some conditions, but do not improve the $\mathcal{O}(n^{-1/2})$ MC rate. Stratification and other RQMC methods based for example on lattice rules and digital nets can improve the rate; see Sect. 2. Some RQMC methods provide a better convergence rate than the squared worst-case error, because the average over the randomizations is sometimes better than the worst case [17, 48, 56].

Note that because of the nonzero covariances, we cannot use the sample variance of the $f(\mathbf{U}_i)$ to estimate $\mathrm{Var}[\hat{\mu}_{n,\mathrm{rqmc}}]$ as for MC. To estimate the variance, we can simulate m independent realizations $X_1, \ldots, X_m$ of $\hat{\mu}_{n,\mathrm{rqmc}}$, then estimate μ and $\mathrm{Var}[\hat{\mu}_{n,\mathrm{rqmc}}]$ by their sample mean $\bar{X}_m$ and sample variance S_m^2. This can be used to estimate the integration error. It may seem natural to compute a confidence interval by assuming that $\bar{X}_m$ is approximately normally distributed, but one should be careful: The CLT holds in general for $m \to \infty$, but for fixed m and $n \to \infty$ it holds only for a few RQMC methods [38, 45]. When applying RQMC to estimate μ, for a given total computing budget mn, we prefer n as large as possible to benefit from the faster convergence rate in n, and then m is small (e.g., 10 or 20) and $\bar{X}_m$ may be far from normally distributed. We give an example in Sect. 4.1.

In the remainder of this tutorial, we focus on RQMC to estimate an integral on the unit cube, and we assume that the goal is to reduce the variance. For simplicity, we assume that s and n are fixed. There are constructions (not discussed here) in which the point sets can be expanded dynamically in dimension by adding new coordinates and in size by adding new points without changing the existing points. There are settings in which the criterion to minimize is not the variance. It can be the worst-case error or the average error for some class of functions, for example. This may give rise to various kinds of Hilbert (or Banach) spaces and discrepancies that we do not examine here. We look at just a few RQMC settings to give illustrations. Another interesting fact for practitioners is that the faster convergence rates of RQMC are proved under conditions on f that are often not satisfied in applications, but despite this, RQMC often reduces the variance by significant factors in those applications, so it is worth giving it a try and comparing RQMC vs MC empirical variances.

RQMC methods are best-known for estimating an integral, but they can effectively apply much more widely, for example to estimate the derivative of an expectation, or a function of several expectations, or a quantile, or a density, or the optimizer of a parameterized expectation or function, etc. This tutorial does not (and cannot) cover every aspect of RQMC. It gives more emphasis to what the author knows best. For more on QMC and RQMC, see for example [9, 28, 33, 34, 41, 50, 55, 58]. Tutorials on more advanced QMC topics can be found elsewhere in this book.

2 RQMC Point Set Constructions

2.1 Stratification

A first approach to obtain a negative sum of covariances in (5) is to stratify the unit hypercube [16]. We partition axis j in $k_j \geq 1$ equal parts, for $j = 1, \ldots, s$. This determines $n = k_1 \cdots k_s$ rectangular boxes of volume $1/n$. Then we draw n random points, one per box, independently and uniformly in each box. Fig. 1 (left) gives an illustration with $s = 2$, $k_1 = 12$, $k_2 = 8$, and $n = 12 \times 8 = 96$.

The stratified estimator in the general case is

$$X_{s,n} = \frac{1}{n}\sum_{j=0}^{n-1} f(\mathbf{U}_j).$$

The crude MC variance with n points can be decomposed as

$$\mathrm{Var}[\bar{X}_n] = \mathrm{Var}[X_{s,n}] + \frac{1}{n}\sum_{j=0}^{n-1}(\mu_j - \mu)^2$$

where μ_j is the mean over box j. The more the μ_j's differ, the more the variance is reduced. Stratification provides an unbiased estimator and never increases the variance. One can estimate the variance by replicating the scheme $m \geq 2$ times, computing the empirical variance in each box, and averaging. If f' is continuous and bounded, and all k_j are equal to k (so $n = k^s$), then by using a Taylor expansion in each box one can show [16] that $\mathrm{Var}[X_{s,n}] = \mathcal{O}(n^{-1-2/s})$. This may provide a significant improvement when s is small, but for large s, the rate is not much better than for MC, and the method quickly becomes impractical because n increases exponentially with k. Nevertheless, it is sometimes effective to apply stratification to just a few important (selected) coordinates.

It is interesting to note that for f' continuous and bounded, a (deterministic) multivariate midpoint rule which takes one point at the center of each box, as shown in the right panel of Fig. 1 for $k_1 = k_2 = 8$, gives the same rate as stratification for the worst-case square integration error. For the midpoint rule, each one-dimensional projection of P_n has only d distinct points, each two-dimensional projection has only d^2 distinct points, etc. This means that for integrating functions that depend (mostly) on just a few of the s coordinates, many of the n points are identical with respect to those important coordinates, so the scheme becomes inefficient.

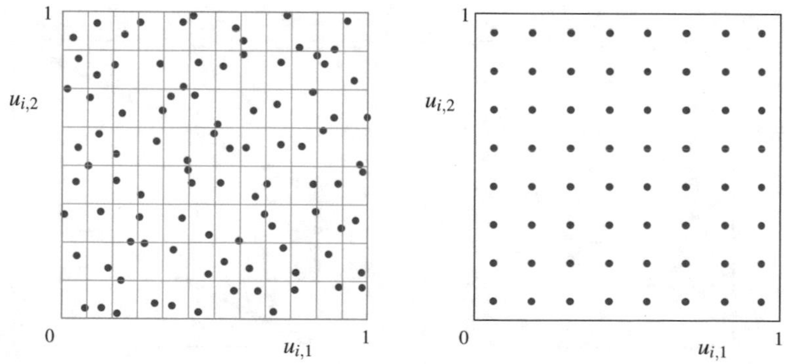

Fig. 1 An illustration of stratified sampling (left) and a midpoint rule (right) over $(0, 1)^2$

2.2 Lattice Rules

An *integration lattice* is a vector space of the form

$$L_s = \left\{ \mathbf{v} = \sum_{j=1}^{s} z_j \mathbf{v}_j \text{ such that each } z_j \in \mathbb{Z} \right\},$$

where $\mathbf{v}_1, \ldots, \mathbf{v}_s \in \mathbb{R}^s$ are linearly independent over $\mathbb{R}$ and where L_s contains $\mathbb{Z}^s$, the vectors of integers. A *lattice rule* is a QMC method that takes $P_n = \{\mathbf{u}_0, \ldots, \mathbf{u}_{n-1}\} = L_s \cap [0, 1)^s$. It has *rank 1* if we can write $\mathbf{u}_i = i\mathbf{v}_1 \mod 1$ for $i = 0, \ldots, n-1$, where $n\mathbf{v}_1 = \mathbf{a} = (a_1, \ldots, a_s) \in \{0, 1, \ldots, n-1\}^s$. These are the most widely used rules in practice. We have a *Korobov rule* if $\mathbf{a} = (1, a, a^2 \mod n, \ldots, a^{s-1} \mod n)$ for some integer a such that $1 \le a < n$.

Figure 2 shows the points of a two-dimensional Korobov lattice rule with $n = 101$ and $\mathbf{a} = (1, 12)$ on the left and $\mathbf{a} = (1, 51)$ on the right. In both cases the points have a lattice structure. They are very evenly spread over the unit square for $\mathbf{a} = (1, 12)$, but for $\mathbf{a} = (1, 51)$ all the points of P_n lie on two parallel straight lines! Thus, the joint choice of n and $\mathbf{a}$ must be done carefully.

A lattice rule can be turned into an RQMC method simply by shifting the lattice randomly, modulo 1, with respect to each coordinate [7, 33]. One generates a single point $\mathbf{U}$ uniformly in $(0, 1)^s$, and adds it to each point of P_n modulo 1, coordinate-wise. This satisfies the two RQMC conditions. Figure 3 gives an example in which $\mathbf{U} = (0.40, 0.08)$, for the lattice rule of Fig. 2.

A good randomly-shifted lattice rule provides an unbiased estimator with points that seem to cover the unit cube more evenly than independent points, but does it make the variance converge faster with n than MC? The answer is yes, under some conditions on the integrand f. Suppose f has Fourier expansion

$$f(\mathbf{u}) = \sum_{\mathbf{h} \in \mathbb{Z}^s} \hat{f}(\mathbf{h}) e^{2\pi\sqrt{-1}\mathbf{h}^t\mathbf{u}}.$$

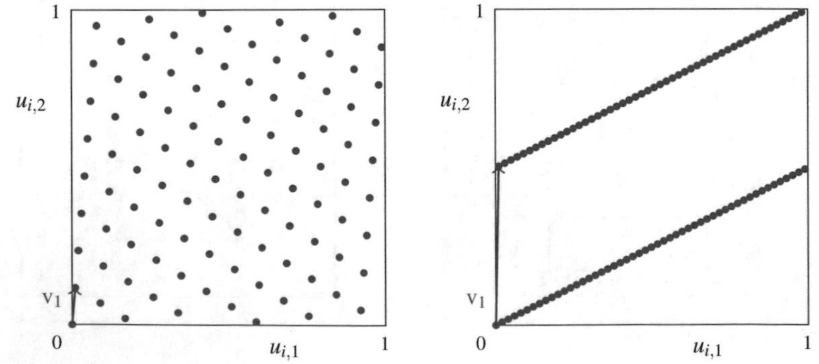

Fig. 2 An integration lattice with $s = 2$, $n = 101$, with $\mathbf{v}_1 = (1, 12)/n$ on the left and $\mathbf{v}_1 = (1, 51)/n$ on the right. In both cases, we can take $\mathbf{v}_2 = (0, 1)$

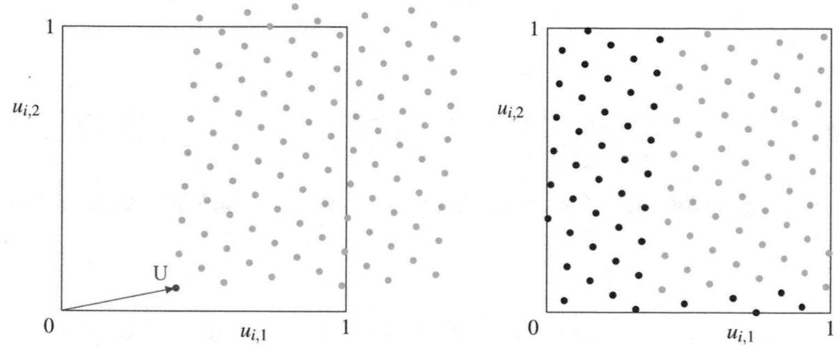

Fig. 3 Random shift modulo 1 for the lattice of Fig. 2 (left), with $\mathbf{U} = (0.40, 0.08)$

For a *randomly shifted lattice*, the exact variance is always (see [33]):

$$\mathrm{Var}[\hat{\mu}_{n,\mathrm{rqmc}}] = \sum_{\mathbf{0} \neq \mathbf{h} \in L_s^*} |\hat{f}(\mathbf{h})|^2, \tag{6}$$

where $L_s^* = \{\mathbf{h} \in \mathbb{R}^s : \mathbf{h}^{\mathsf{t}}\mathbf{v} \in \mathbb{Z} \text{ for all } \mathbf{v} \in L_s\} \subseteq \mathbb{Z}^s$ is the *dual lattice*. Thus, from the viewpoint of variance reduction, an optimal lattice for any given f is one that minimizes (6). But finding it for a given f is generally too hard and unrealistic.

Let $\alpha > 0$ be an even integer. If f has *square-integrable mixed partial derivatives* up to order $\alpha/2 > 0$, and the periodic continuation of its derivatives up to order $\alpha/2 - 1$ is *continuous across the boundaries of the unit cube modulo 1*, then it is known that $|\hat{f}(\mathbf{h})|^2 = \mathcal{O}((\max(1, h_1) \cdots \max(1, h_s))^{-\alpha})$. It is also known that for any $\epsilon > 0$, there is always a vector $\mathbf{v}_1 = \mathbf{v}_1(n)$ such that

$$\mathcal{P}_\alpha := \sum_{\mathbf{0} \neq \mathbf{h} \in L_s^*} (\max(1, h_1) \cdots \max(1, h_s))^{-\alpha} = \mathcal{O}(n^{-\alpha+\epsilon}). \tag{7}$$

This $\mathcal{P}_\alpha$ has been proposed long ago as a figure of merit, often with $\alpha = 2$ [58]. It is the variance for a *worst-case* f having $|\hat{f}(\mathbf{h})|^2 = (\max(1, |h_1|) \cdots \max(1, |h_s|))^{-\alpha}$. A larger α means a smoother f and a faster convergence rate. This $\mathcal{P}_\alpha$ is defined by an infinite sum, which cannot be computed exactly in general. However, when α is an even integer, the worst-case f is

$$f^*(\mathbf{u}) = \sum_{\mathfrak{u} \subseteq \{1,\ldots,s\}} \prod_{j \in \mathfrak{u}} \frac{(2\pi)^{\alpha/2}}{(\alpha/2)!} B_{\alpha/2}(u_j)$$

where $B_{\alpha/2}$ is the Bernoulli polynomial of degree $\alpha/2$ (e.g., $B_1(u) = u - 1/2$), and $\mathcal{P}_\alpha$ can be written as a finite sum that is easy to compute and can be used as a criterion to search for good lattices; see (9) in Sect. 3, where we give a more general version.

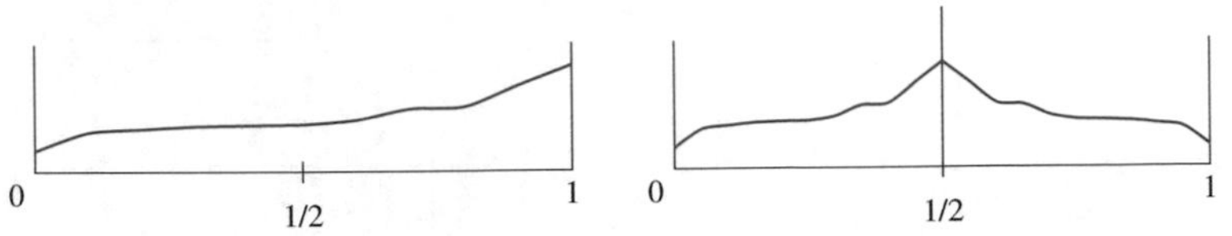

Fig. 4 Applying the baker transformation to the points $\mathbf{U}_i$ is equivalent to transforming f as shown

Thus, under the above conditions on the periodic continuation of f, for all $\epsilon > 0$ there is a sequence of integration lattices indexed by n for which the variance converges as $\mathcal{O}(n^{-\alpha+\epsilon})$. What if f does not satisfy these conditions? One can often change f to a continuous and/or smoother function that integrates to the same μ over $[0, 1)^s$, usually via a change of variable. For example, suppose f is continuous in $[0, 1)$, but discontinuous at the boundary, i.e., $f(\ldots, u_j = 0, \ldots) \neq f(\cdots, u_j = 1, \ldots)$. To simplify the exposition, suppose f is a function of a single real variable u (the other ones are fixed). Consider the change of variable $v = \varphi(u) = 2u$ if $u \leq 1/2$ and $1 - 2u$ if $u > 1/2$. It is known as the *baker* transformation [20]: it stretches the points by a factor of 2, from $[0, 1)$ to $[0, 2)$, and folds back the segment $[1, 2)$ to $[1, 0)$. Its impact on the RQMC estimator (4) is exactly equivalent to compressing the graph of f horizontally by a factor of 1/2, and then making a mirror copy on the interval $[1/2, 1)$. The transformed f is a continuous function. Figure 4 gives an illustration. In practice, it is more convenient to apply the baker transformation to the randomized points $\mathbf{U}_i$ instead of changing f. Higher-order transformations can also make the derivatives (up any given order) continuous and improve the asymptotic rate even further, but they often increases the variance for "practical" values of n by increasing $V(f)$, so are not necessarily better in the end.

Note that the worst-case function for the bounds we discussed is not necessarily representative of what happens in applications. Also, the *hidden factor in the* $\mathcal{O}$ *may increase quickly with* s, so the rate result in (7) is not very useful for large s. To get a bound that is uniform in s, the Fourier coefficients must decrease faster with the dimension and "size" of vectors $\mathbf{h}$; that is, f must be "smoother" in high-dimensional projections [10, 59, 60]. This is typically what happens in applications for which RQMC is effective. The criterion (9) will take that into account.

2.3 Digital Nets

Niederreiter [50] defines a *digital net in base b* as follows. Choose the base b, usually a prime number or a power of a prime, and an integer $k > 0$. For $i = 0, \ldots, b^k - 1$ and $j = 1, \ldots, s$, put

$$i = a_{i,0} + a_{i,1}b + \cdots + a_{i,k-1}b^{k-1} = a_{i,k-1}\cdots a_{i,1}a_{i,0},$$

$$\begin{pmatrix} u_{i,j,1} \\ \vdots \\ u_{i,j,w} \end{pmatrix} = \mathbf{C}_j \begin{pmatrix} a_{i,0} \\ \vdots \\ a_{i,k-1} \end{pmatrix} \bmod b,$$

$$u_{i,j} = \sum_{\ell=1}^{w} u_{i,j,\ell} b^{-\ell}, \qquad \mathbf{u}_i = (u_{i,1}, \ldots, u_{i,s}),$$

where the *generating matrices* $\mathbf{C}_j$ are $w \times k$ with elements in $\mathbb{Z}_b$. This gives $n = b^k$ points. In practice, w and k are finite, but there is no limit. The definition in [50] is actually more general: One can define bijections between $\mathbb{Z}_b$ and some ring R, and perform the multiplication in R. Assuming that each $\mathbf{C}_j$ has full rank, each one-dimensional projection truncated to its first k digits is $\mathbb{Z}_n/n = \{0, 1/n, \ldots, (n-1)/n\}$. That is, each $\mathbf{C}_j$ defines a permutation of $\mathbb{Z}_n/n$.

If each $\mathbf{C}_j$ is defined with an infinite number of columns, then we have an infinite sequence of points, called a *digital sequence in base b*. One can always take the first $n = b^k$ points of a digital sequence to define a digital net, for any k.

Measuring uniformity. A standard way of measuring the uniformity of a digital net in base b with $n = b^k$ points is as follows [34, 50]. Suppose we divide axis j in b^{q_j} equal parts for some integer $q_j \geq 0$, for each j. This determines a partition of $[0, 1)^s$ into $2^{q_1+\cdots+q_s}$ rectangles of equal sizes. If each rectangle contains exactly the same number of points from P_n, we say that P_n is $(q_1, \ldots, q_s)$*-equidistributed in base b*. This occurs if and only if the matrix formed by the first q_1 rows of $\mathbf{C}_1$, the first q_2 rows of $\mathbf{C}_2$, ..., the first q_s rows of $\mathbf{C}_s$, is of full rank (mod b). The $(q_1, \ldots, q_s)$-equidistribution can be verified by constructing this matrix and checking its rank. We say that P_n is a (t, k, s)*-net in base b* if and only if it is $(q_1, \ldots, q_s)$-equidistributed whenever $q_1 + \cdots + q_s = k - t$. This is possible for $t = 0$ only if $b \geq s - 1$. The *t-value* of a digital net is the smallest t for which it is a (t, k, s)-net.

Figure 5 gives an example of a $(0, 6, 2)$-net in base 2. The equidistribution can be observed on the left with $q_1 = q_2 = 3$ and on the right with $q_1 = 4$ and $q_2 = 2$. This point set is obtained by taking $\mathbf{C}_2$ as the identity matrix and $\mathbf{C}_1$ as the reverse identity (with 1's on the descending diagonal). The points are enumerated by their first coordinate and the second coordinate follows the van der Corput sequence in base 2. Many of the points sit exactly on the left or bottom boundary of their rectangle in Fig. 5, because only the first six bits of each coordinate can be nonzero. For any integer $k > 0$, this construction (with these $\mathbf{C}_1$ and $\mathbf{C}_2$) is a $(0, k, 2)$-net in base 2; it is the two-dimensional *Hammersley point set*.

An infinite sequence $\{\mathbf{u}_0, \mathbf{u}_1, \ldots\}$ in $[0, 1)^s$ is a (t, s)*-sequence in base b* if for all $k > 0$ and $\nu \geq 0$, $Q(k, \nu) = \{\mathbf{u}_i : i = \nu b^k, \ldots, (\nu + 1)b^k - 1\}$, is a (t, k, s)-net in base b. This is possible for $t = 0$ only if $b \geq s$.

A key property that connects digital nets with the star discrepancy and QMC error bounds is that for fixed s, if P_n is a (t, k, s)-net in base b for $k = 1, 2, 3, \ldots$ and t is bounded, then $D^*(P_n) = \mathcal{O}(n^{-1}(\log n)^{s-1})$, and for any f having finite Hardy-Krause variation $V(f)$, the error bound (3) with these point sets converges at this same rate.

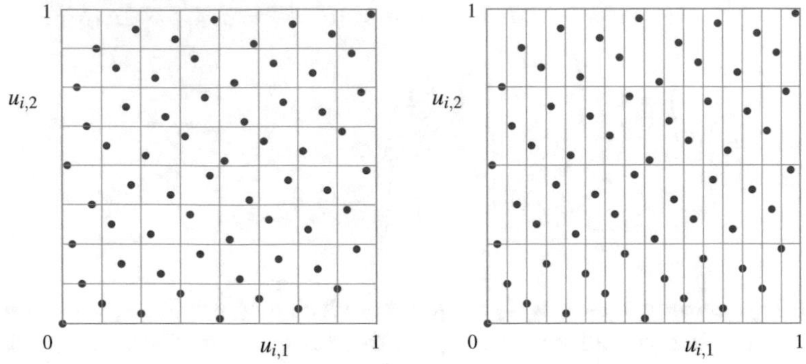

Fig. 5 The Hammersley point set (or Sobol net with appended first coordinate) for $s = 2$ and $n = 64$

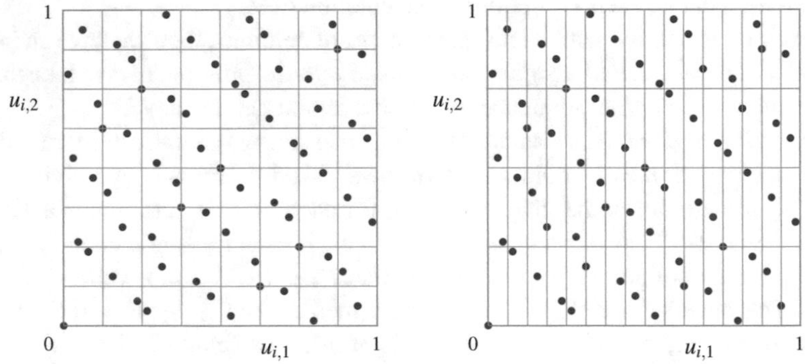

Fig. 6 The first $n = 64$ Sobol points in $s = 2$ dimensions

Specific constructions. The most popular (and oldest) specific instance of digital sequence was proposed by Sobol' [61], in base $b = 2$. Each binary matrix $\mathbf{C}_j$ is upper triangular with ones on the main diagonal. The bits above the diagonal in any given column form the binary expansion of an integer called a *direction number*. The first few direction numbers are selected and the following columns are determined by a bitwise linear recurrence across the columns. The choice of the initial direction numbers is important for the quality of the points. For original values proposed by Sobol', in particular, the uniformity of several low-dimensional projections are very poor. Better direction numbers are proposed in [26, 42], for example. Figure 6 shows the first 64 points of the Sobol sequence in base 2, for which $\mathbf{C}_1$ is the identity matrix. These points form a (0, 6, 2)-net in base 2, just like the Hammersley points of Fig. 5.

Faure [14] proposed digital sequences in base b for any prime b by taking $\mathbf{C}_j$ as the $(j-1)$th power of the Pascal matrix, modulo b. He proved that the resulting sequence is a $(0, s)$-sequence in base b for any $s \leq b$. The latter condition is a practical limitation, because it imposes the choice of a large base when s is large. Also, the arithmetic modulo a prime $b > 2$ is less convenient and generally slower on computers than arithmetic modulo 2. Other sequences and nets for arbitrary prime power bases and nets in base 2 with better t-values than those of Sobol' are also available (see, e.g., [9, 51]), but they rarely outperform Sobol' points in applications.

For all these sequences, if we fix $n = b^k$, we can take the first coordinate of point i as i/n, which corresponds to taking $\mathbf{C}_1$ as the reflected identity matrix, and then take $\mathbf{C}_{j+1}$ as the old $\mathbf{C}_j$, for $j \geq 1$. It turns out that by doing this with a (t, s)-sequence, we can obtain a $(t, k, s + 1)$-net for any k. That is, we gain one coordinate in the net. By doing this with the Sobol' sequence with $s = 1$, we obtain the Hammersley net illustrated in Fig. 5. Other types of digital net constructions can be found in [9, 43, 52] and the references given there.

Randomization. If we apply a random shift modulo 1 to a digital net, the equidistribution properties are not preserved in general. However, a *random digital shift* in base b preserves them and also satisfies the two criteria that define RQMC. It works as follows. As for the ordinary random shift, we generate a single $\mathbf{U} = (U_1, \ldots, U_s) \sim \mathcal{U}[0, 1)^s$ where $U_j = \sum_{\ell=1}^{w} U_{j,\ell}\, b^{-\ell}$. For coordinate j of point i, before the shift we have $u_{i,j} = \sum_{\ell=1}^{w} u_{i,j,\ell} b^{-\ell}$. The digital shift replaces each $u_{i,j}$ by $\tilde{U}_{i,j} = \sum_{\ell=1}^{w}[(u_{i,j,\ell} + U_{j,\ell}) \bmod b]b^{-\ell}$. It is not difficult to show that if P_n is $(q_1, \ldots, q_s)$-equidistributed in base b before the shift, it retains this property after the shift. Moreover, if $w = \infty$, each randomized point $\tilde{\mathbf{U}}_i$ has the uniform distribution over $(0, 1)^s$. As a result, if f has finite Hardy-Krause variation and we use (t, k, s)-nets with fixed s and bounded t for RQMC with a random digital shift, the estimator $\hat{\mu}_{n,\mathrm{rqmc}}$ is unbiased and by squaring the worst-case error we immediately find that its variance converges as $\mathcal{O}(n^{-2}(\log n)^{2(s-1)})$. (Better rates are obtained for certain classes of smooth functions and certain types of randomizations; see below.)

In base $b = 2$, the digital shift consists in applying a bitwise XOR between $\mathbf{U}$ and each $\mathbf{u}_i$. To illustrate how the equidistribution is preserved, take for example $k_1 = 3$ and $k_2 = 5$. For the given $\mathbf{U}$, the bits marked as "C" in the result have been flipped and those still marked with $*$ are unchanged:

$$\begin{aligned} \mathbf{u}_i &= (\texttt{0.***,}\quad \texttt{0.*****})_2 \\ \mathbf{U} &= (\texttt{0.101,}\quad \texttt{0.01011})_2 \\ \mathbf{u}_i \oplus \mathbf{U} &= (\texttt{0.C*C,}\quad \texttt{0.*C*CC})_2 \end{aligned}$$

If the eight considered bits for $\mathbf{u}_i$ take each of the 2^8 possible configurations exactly the same number of times when $i = 0, \ldots, n-1$, then this also holds for $\mathbf{u}_i \oplus \mathbf{U}$. More generally, for a digital net in base 2 with $n = 2^k$ points in s dimensions, this preservation holds for any $\mathbf{U}$ and any non-negative integers $k_1, \ldots, k_s$ such that $k_1 + \cdots + k_s \leq k$. Figure 7 shows a digital shift in base 2 with

$$\mathbf{U} = (0.10100101100\ldots,\quad 0.01011001100\ldots)_2$$

applied to the Hammersley points (Sobol' net with one extra coordinate) of Fig. 5. For this given $\mathbf{U}$, for each point $\mathbf{u}_i$ we flip the first, third, sixth, ..., bits of the first coordinate, and we flip the second, fourth, fifth, ..., bits of the second coordinate.

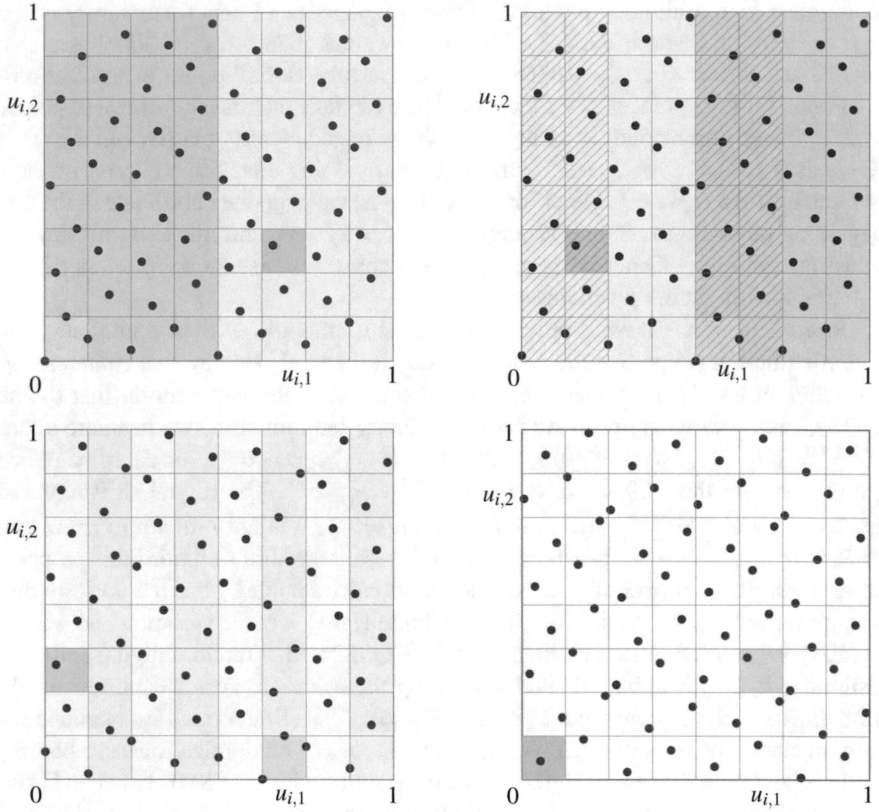

Fig. 7 A random digital shift with $\mathbf{U} = (0.10100101100\ldots,\ 0.01011001100\ldots)_2$ applied to a Sobol net with 64 points. The original points are in upper left. Flipping the first bit of the first coordinate permutes the left half with the right one, giving the upper right plot. Then flipping the third bit of the first coordinate gives the lower left plot. The lower right shows the points after the full digital shift of the two coordinates. One can follow the movement of the green and blue boxes during the permutations. In the end, those two square boxes are permuted

The figure shows what happens when we flip the first bit of the first coordinate (top right); it permutes the left half with the right half. If the second bit in $\mathbf{U}$ was a 1 (here it is not), we would also permute the first (shaded) quarter with the second and the third (shaded) with the fourth. Since the third bit in $\mathbf{U}$ is 1, we flip the third bit of the first coordinate of each $\mathbf{u}_i$, The lower left plot shows the points after this flip, which has permuted each lightly colored vertical stripe with the yellow one on its right. After doing all the permutations specified by $\mathbf{U}$ for the first coordinate, we do the same with the second coordinate. The points after the full digital shift are shown in the lower right. Equidistribution is preserved because for each relevant partition in dyadic rectangular boxes, the digital shift only permutes those boxes (by pairs).

A more powerful but more costly randomization for digital nets is the *nested uniform scramble* (NUS) of Owen [53, 54]. The difference with the digital shift (in base 2) is as follows. For any given coordinate, with probability 1/2 we flip the first bit of all points, just as before, then for the left half, we flip the second bit of all points with probability 1/2, and we do the same for the right half, but *independently*. That is, in the top right of Fig. 7, we would permute the first shaded area (first quarter) with the light area on its right (second quarter) with probability 1/2, and independently we would permute the second shaded area (third quarter) with the pink area on its right (fourth quarter). Then we do this recursively, and we repeat for each coordinate. For instance, in the lower left Fig. 7, we would use four independent random bits, one for each pair of successive (light, yellow) columns. Doing permutations like this for 31 bits or more would be very time-consuming, but in fact one can do it for the first k bits only, and then generate the other bits randomly and independently across points and coordinates. From a statistical viewpoint this is equivalent to NUS and less time-consuming [48]. NUS also works in general base b: for each digit of each coordinate, we generate a random permutation of b elements to permute the points according to this digit. Owen proved that under sufficient smoothness condition on f, for digital nets in base b with fixed s and bounded t, with this scrambling the variance converges as $\mathcal{O}(n^{-3}(\log n)^{s-1})$, which is better than for the random digital shift.

Simpler and less costly permutations than NUS have been proposed. One popular example is the (left) *linear matrix scramble* [48, 56, 64]: left-multiply each matrix $\mathbf{C}_j$ by a random non-singular and lower triangular $w \times w$ matrix $\mathbf{L}_j$, mod b. With this scrambling just by itself, each point does not have the uniform distribution (e.g., the point $\mathbf{0}$ is unchanged), but one can apply a random digital shift in base b after the matrix scramble to obtain an RQMC scheme. There are other types of linear matrix scrambles, such as the stripe scramble, ibinomial scramble, etc.; see [23, 48, 49, 56]. These non-nested scrambles use less randomization than NUS and none has a proved convergence rate of $\mathcal{O}(n^{-3}(\log n)^{s-1})$ for the variance for $s > 1$, to my knowledge.

3 Anova Decomposition

Filling the unit hypercube very evenly requires an excessive number of points in large dimension. For example, in $s = 100$ dimensions, it takes $n = 2^{100}$ points to have one in each quadrant; this is unrealistic. The reason why RQMC might still work for large s is because f can often be well approximated by a sum of low-dimensional functions, and RQMC with a well-chosen point set can integrate these low-dimensional functions with small error. A standard way of formalizing this is as follows [44, 62].

An *ANOVA decomposition* of $f(\mathbf{u}) = f(u_1, \ldots, u_s)$ can be written as

$$f(\mathbf{u}) = \sum_{\mathfrak{u} \subseteq \{1,\ldots,s\}} f_{\mathfrak{u}}(\mathbf{u}) = \mu + \sum_{i=1}^{s} f_{\{i\}}(u_i) + \sum_{i,j=1}^{s} f_{\{i,j\}}(u_i, u_j) + \cdots \quad (8)$$

where

$$f_{\mathfrak{u}}(\mathbf{u}) = \int_{[0,1)^{|\bar{\mathfrak{u}}|}} f(\mathbf{u})\, d\mathbf{u}_{\bar{\mathfrak{u}}} - \sum_{\mathfrak{v} \subset \mathfrak{u}} f_{\mathfrak{v}}(\mathbf{u}_{\mathfrak{v}}),$$

and the *Monte Carlo variance* decomposes accordingly as

$$\sigma^2 = \sum_{\mathfrak{u} \subseteq \{1,\ldots,s\}} \sigma_{\mathfrak{u}}^2, \quad \text{where } \sigma_{\mathfrak{u}}^2 = \text{Var}[f_{\mathfrak{u}}(\mathbf{U})].$$

Getting a rough estimate of the variance $\sigma_{\mathfrak{u}}^2$ captured by each subset $\mathfrak{u}$ of coordinates suffices to define relevant uniformity criteria that give more weight to the more important projections. The $\sigma_{\mathfrak{u}}^2$ can be estimated by MC or RQMC; see [47, 57].

One example of this is the following weighted version of $\mathcal{P}_\alpha$, with projection-dependent weights $\gamma_{\mathfrak{u}}$, in which $\mathfrak{u}(\mathbf{h}) = \mathfrak{u}(h_1, \ldots, h_s) = \{j : h_j \neq 0\}$:

$$\mathcal{P}_{\gamma,\alpha} = \sum_{\mathbf{0} \neq \mathbf{h} \in L_s^*} \gamma_{\mathfrak{u}(\mathbf{h})} (\max(1, |h_1|) \cdots \max(1, |h_s|))^{-\alpha}.$$

If $\alpha/2$ is a positive integer, for a lattice rule with $\mathbf{u}_i = (u_{i,1}, \ldots, u_{i,s})$, we have

$$\mathcal{P}_{\gamma,\alpha} = \sum_{\emptyset \neq \mathfrak{u} \subseteq \{1,\ldots,s\}} \frac{1}{n} \sum_{i=0}^{n-1} \gamma_{\mathfrak{u}} \left[\frac{-(-4\pi^2)^{\alpha/2}}{(\alpha)!} \right]^{|\mathfrak{u}|} \prod_{j \in \mathfrak{u}} B_\alpha(u_{i,j}), \tag{9}$$

and the corresponding *variation* (squared) is

$$V_\gamma^2(f) = \sum_{\emptyset \neq \mathfrak{u} \subseteq \{1,\ldots,s\}} \frac{1}{\gamma_{\mathfrak{u}} (4\pi^2)^{\alpha|\mathfrak{u}|/2}} \int_{[0,1]^{|\mathfrak{u}|}} \left| \frac{\partial^{\alpha|\mathfrak{u}|/2}}{\partial \mathbf{u}^{\alpha/2}} f_{\mathfrak{u}}(\mathbf{u}) \right|^2 d\mathbf{u},$$

for $f : [0, 1)^s \to \mathbb{R}$ smooth enough. Then,

$$\text{Var}[\hat{\mu}_{n,\text{rqmc}}] = \sum_{\mathfrak{u} \subseteq \{1,\ldots,s\}} \text{Var}[\hat{\mu}_{n,\text{rqmc}}(f_{\mathfrak{u}})] \leq V_\gamma^2(f) \mathcal{P}_{\gamma,\alpha}. \tag{10}$$

This $\mathcal{P}_{\gamma,\alpha}$ with properly chosen α and weights $\gamma_{\mathfrak{u}}$ is a good practical choice of figure of merit for lattice rules [10, 19]. The weights $\gamma_{\mathfrak{u}}$ are usually chosen to have a specific form with just a few parameters, such as order-dependent or product weights [35, 60]. The *Lattice Builder* software [36] permits one to search for good lattices for arbitrary n, s, and weights, using various figures of merit, under various constraints.

4 Examples

The numerical results reported in this paper were obtained using the Java library SSJ [29], which offers tools for RQMC and stochastic simulation in general. In the examples, all random variates are generated by inversion.

4.1 A Stochastic Activity Network

This example is from [2, 33]. We consider the stochastic activity network in Fig. 8, in which Y_j is the length of arc j, for $j = 0, \ldots, 12$. The Y_j are assumed independent with cdf's F_j given in [2, Sect. 4.1] and [35], and we generate them by inversion: $Y_j = F_j^{-1}(U_j)$ where $U_j \sim U(0, 1)$. Let T be the (random) length of the longest path from node 0 to node 8. We compare RQMC and MC for two estimation problems: (a) estimating $\mathbb{P}[T > x]$ for some constant x and (b) estimating $\mathbb{E}[T]$.

To estimate $\mathbb{E}[T]$ we simply use T, so $s = 13$. To estimate $\mathbb{P}[T > x]$, we consider two base MC estimators. The first one is $X = \mathbb{I}[T > x]$ (where $\mathbb{I}[\cdot]$ is the indicator function) and the second one uses conditional Monte Carlo (CMC) as follows. We generate the Y_j's only for the 8 arcs that *do not* belong to the cut $\mathscr{L} = \{4, 5, 6, 8, 9\}$, and replace $\mathbb{I}[T > x]$ by its *conditional expectation* given those Y_j's,

$$X_{\text{CMC}} = \mathbb{P}\left[T > x \mid \{Y_j : j \notin \mathscr{L}\}\right] = 1 - \prod_{j \in \mathscr{L}} \mathbb{P}[Y_j \leq x - P_j] \tag{11}$$

where P_j is the length of the longest path that goes through edge j when we put $Y_j = 0$. This X_{CMC} is easy to compute, as explained in [33], and is guaranteed to have smaller variance than the indicator (the first one) under MC. A more important advantage under RQMC is that CMC makes the estimator continuous as a function of the U_j's, whereas the first one is discontinuous. It also reduces the dimension s from 13 to 8. Figure 9 shows the impact of CMC on the ANOVA decomposition. For each estimator, the length of the white box is proportional to the variance captured by one-dimensional projections, the second lightest box is for the two-dimensional projections, etc. CMC pushes much of the variance to the projections over one and

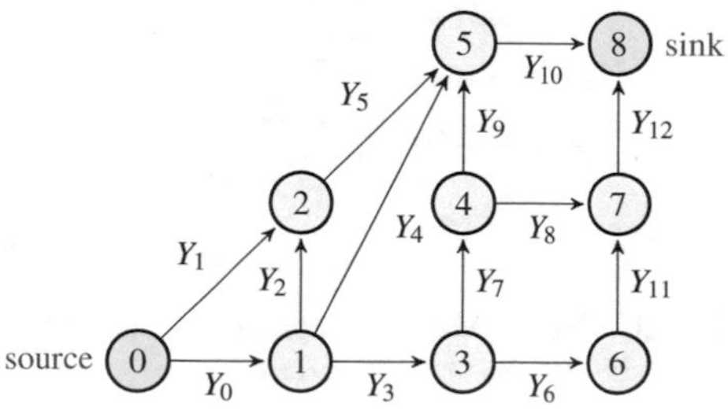

Fig. 8 A stochastic activity network

Table 1 Empirical convergence rate ν and $\log_{10}$ of empirical variance with $n \approx 2^{20}$ (in parentheses) for the stochastic activity network example

	$\mathbb{P}[T > 60]$, MC		$\mathbb{P}[T > 60]$, CMC		$\mathbb{E}[T]$, MC	
	ν	$\log_{10}$ Var	ν	$\log_{10}$ Var	ν	$\log_{10}$ Var
Independent	1.00	(−6.6)	1.00	(−7.3)	1.04	(−3.5)
Lattice	1.20	(−8.0)	1.51	(−10.9)	1.47	(−6.3)
Sobol+LMS+DS	1.20	(−7.5)	1.68	(−11.2)	1.36	(−6.2)
Sobol+NUS	1.18	(−8.0)	1.65	(−11.2)	1.37	(−6.3)

two dimensions. The variance components $\sigma_{\mathfrak{u}}^2$ were estimated using the algorithm of [63] with RQMC (100 independent shifts of a lattice rule with $n = 2^{20} - 3$ points); see [35, Sect. 8] for the details. The bounds (3) and (10) are useless here because the mixed derivatives are not defined everywhere, so the variation is infinite. It will be interesting to see that RQMC nevertheless improves the variance. This is frequent in applications.

In an experiment reported in [35], an integration lattice of rank 1 was constructed for 50 different prime values of n ranging from $2^5 - 1$ to $2^{22} - 3$, roughly uniformly spread in log scale. For each n a generating vector $\mathbf{a}$ was found based on the criterion (9), with weights selected based on estimates of the ANOVA components. The variance was estimated for each n and a linear regression was fitted for $\log \text{Var}[\hat{\mu}_{n,\text{rqmc}}]$ vs $\log n$ to estimate the rate ν for which $\text{Var}[\hat{\mu}_{n,\text{rqmc}}] \propto n^{-\nu}$ in this range of values of n. For the estimation of $\mathbb{P}[T > x]$ with $x = 60$, for example, we found $\nu \approx 1.2$ for the standard estimator (indicator) and $\nu \approx 1.5$ with CMC. The log-log plot follows pretty much a straight line. Based on an interpolation from the regression model, for $n \approx 2^{20}$, RQMC reduces the variance approximately by a factor of 27 with the standard estimator and 4400 with CMC. This shows that the combined smoothing of f and dimension reduction provided by CMC has a large impact on the effectiveness of RQMC. For the estimation of $\mathbb{E}[T]$ (without CMC), we had $\nu \approx 1.47$. Table 1 reports the results for other RQMC schemes, namely Sobol' points with a linear matrix scramble and a random digital shift (LMS+DS), and Sobol' points with NUS. We see that the choice of scrambling makes almost no difference in this example.

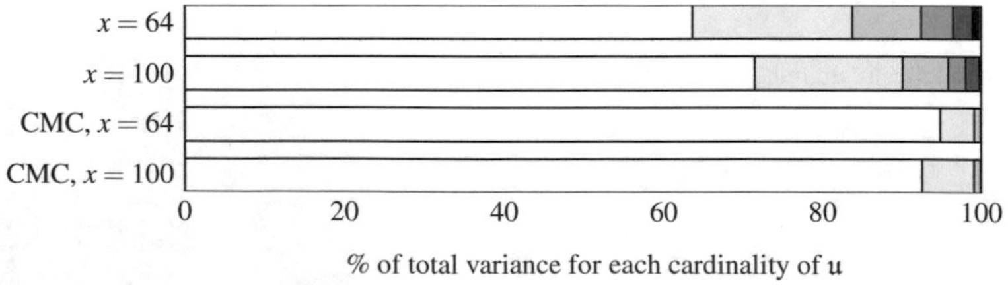

Fig. 9 ANOVA Variance captured by each projection order for estimators of $\mathbb{P}[T > x]$ for the stochastic activity network example

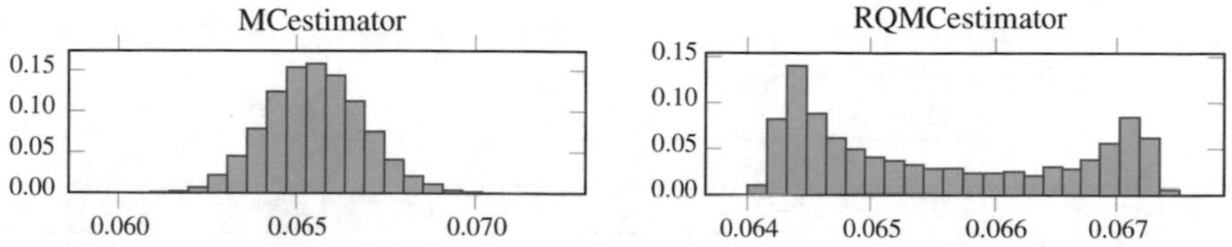

Fig. 10 Histogram of MC and RQMC estimators with $n = 8191$ for the stochastic activity network with CMC, for $x = 100$, based on $m = 10000$ replications of each estimator. The RQMC estimator is a randomly shifted lattice rule with $n = 8191$

Figure 10 shows histograms for $m = 10000$ replications of the MC estimator with $n = 8191$ (left) and for a particular randomly shifted lattice rule with $n = 8191$ (right, taken from [38]). The MC histogram resembles a normal distribution, as expected, but the second one is far from normal. A confidence interval based on a normality assumption is certainly inappropriate in this case. The limiting distribution of an RQMC estimator based on a randomly-shifted lattice rule when $n \to \infty$ is analyzed in [38]; the properly scaled limiting distribution is usually a spline, not a normal distribution. For a digital net with a digital random shift, the CLT does not apply either (in one dimension it is equivalent to a randomly-shifted lattice), but the CLT does apply for a digital net with NUS [45].

4.2 A Financial Option Under a Variance-Gamma Process

Alternative sampling schemes. Consider an asset price that evolves according to a *geometric variance-gamma* (GVG) process S defined by [3, 4, 46]:

$$S(t) = S(0) \exp\left[rt + X(G(t; 1, v), \mu, \sigma) + \omega t\right],$$

where X is a *Brownian process* with drift and variance parameters μ and σ, G is a *gamma process* with mean and variance parameters 1 and v, X and G are independent, and $\omega = \ln(1 - \mu v - \sigma^2 v/2)/v$. The process $Y(\cdot) = X(G(\cdot))$ is a *variance-gamma* process. We want to estimate by simulation the value of an *Asian call option*, given by $\mathbb{E}[e^{-rT} \max(\bar{S} - K, 0)]$, where $\bar{S} = (1/d) \sum_{j=1}^{d} S(t_j)$ and $t_j = jT/d$ for $0 \le j \le d$.

The realization of $Y(t)$ (and of $S(t)$) at the d observation points t_j can be generated in the following three ways (among others), as explained in [3]: Brownian and gamma sequential sampling (BGSS), Brownian and gamma bridge sampling (BGBS), and difference of gammas bridge sampling (DGBS). BGSS generates $\tau_1 = G(t_1)$, then $X(\tau_1)$, then $\tau_2 - \tau_1 = G(t_2) - G(t_1)$, then $X(\tau_2) - X(\tau_1)$, etc., in that order. This requires d gamma variates and d normal variates, so the dimension is $s = 2d$. BGBS generates $\tau_d = G(t_d)$, $X(\tau_d)$, $\tau_{d/2} = G(t_{d/2})$, $X(\tau_{d/2})$, $\tau_{d/4} = G(t_{d/4})$, $X(\tau_{d/4})$, $\tau_{3d/4} = G(t_{3d/4})$, $X(\tau_{3d/4})$, ..., in that order, exploiting the fact

Table 2 Empirical convergence rate ν and $\log_{10}$ of empirical variance for $n = 2^{20}$ (in parentheses) for the option pricing example under a GVG process

	BGSS		BGBS		DGBS	
	ν	$\log_{10}$ Var	ν	$\log_{10}$ Var	ν	$\log_{10}$ Var
Independent	1.00	(−4.5)	1.00	(−4.5)	1.00	(−4.5)
Sobol+LMS+DS	1.26	(−6.5)	1.42	(−7.6)	1.34	(−7.6)

that for any given values $t_a < t < t_b$ and $\tau_a < \tau < \tau_b$, the distribution of $G(t)$ conditional on $(G(t_a), G(t_b))$ is beta with known parameters, and the distribution of $X(\tau)$ conditional on $(X(\tau_a), X(\tau_b))$ is normal with known parameters. BGBS requires one gamma variate, $d - 1$ beta variates, and d normal variates, so $s = 2d$. DGBS uses the fact that $\{S(t),\ t \geq 0\}$ can be written as a difference of two gamma processes, which can be simulated via a bridge (conditional) approach as for BGBS. This requires two gamma variates and $2d - 2$ beta variates. When d is a power of 2, all the beta variates are symmetric, and for that case there is a fast inversion algorithm [40]. The idea of the bridge constructions is to reformulate the integrand f in a way that more of the variance is captured by the low-dimensional projections on the first few coordinates of the points (the first few coordinates already determine a sketch of the trajectory), to make RQMC more effective.

For a numerical illustration, we take the following parameters from [4]: $\theta = -0.1436$, $\boldsymbol{\sigma} = 0.12136$, $\nu = 0.3$, $r = 0.1$, $T = 1$, $K = 101$, and $S(0) = 100$. The exact value and the MC variance are $\mu \approx 5.725$ and $\boldsymbol{\sigma}^2 \approx 29.89$. Table 2 compares the variance rates ν for RQMC (estimated from experiments with $n = 2^9, \ldots, 2^{21}$) and MC (for which $\nu = 1$ and the variance is the same for all sampling schemes). RQMC improves ν and reduces the variance in all three cases, BGBS gives the best rate empirically, and for $n = 2^{20}$ it reduces the variance by a factor of about 1000.

For comparison, we ran a similar experiment with the GVG process replaced by a geometric Brownian motion (GBM) process, for which

$$S(t) = S(0) \exp\left[(r - \boldsymbol{\sigma}^2/2)t + X(t, 0, \boldsymbol{\sigma})\right],$$

with the same parameter values (ν and θ are not used). We tried three sampling methods: sequential sampling (SS), Brownian bridge sampling (BBS), and Brownian sampling with principal component analysis (PCA). SS and BBS work as in the GVG case. For PCA, to generate the multivariate normal vector $(X(t_1), \ldots, X(t_d))$, we do a principal component decomposition of its covariance matrix, say $\boldsymbol{\sigma} = \mathbf{A}\mathbf{A}^{\mathsf{t}}$, and return $\mathbf{X} = \mathbf{A}\mathbf{Z}$ where $\mathbf{Z}$ is a vector of d independent standard normals. With this decomposition, the first few coordinates of $\mathbf{Z}$ (i.e., the first few coordinates of the RQMC points) capture a large fraction of the variance, even larger than with BBS. The results are in Table 3. SS does not improve the variance rate very much, BSS does better, and PCA much better. For PCA, $\nu \approx 1.9$ and the variance is reduced by a factor of over two millions for $n = 2^{20}$. Similar improvements were obtained in [35] with lattice rules constructed by Lattice Builder [36].

Table 3 Empirical convergence rate ν and $\log_{10}$ of empirical variance (in parentheses) for the option pricing example under a GBM process

		BSS		BBS		BPCA	
		ν	$\log_{10}$ Var	ν	$\log_{10}$ Var	ν	$\log_{10}$ Var
No CV	Independent	1.00	(−4.5)	1.00	(−4.5)	1.00	(−4.5)
	Sobol+LMS+DS	1.19	(−7.2)	1.42	(−8.8)	1.90	(−11.0)
With CV	Independent	1.00	(−7.8)	1.00	(−7.8)	1.00	(−7.8)
	Sobol+LMS+DS	1.21	(−9.5)	1.17	(−10.1)	1.37	(−11.1)

Option pricing examples with multiple correlated assets, in up to 250 dimensions, and in which PCA is very effective, can be found in [24, 27, 28]. For more on changing the sampling method of a Brownian motion to reduce the effective dimension and make RQMC more effective, see, e.g., [5, 25].

Control variates. There are various ways of reducing the RQMC variance by making the integrand smoother. Using control variates (CVs) is one of them. For the Asian option under the GBM process, for instance, if we replace the arithmetic average $\bar{S}$ by a geometric average $\tilde{S} = (\prod_{j=1}^{d} S(t_j))^{1/d}$, there is a closed-form formula for the expected payoff, so this payoff can be used as a CV with either MC or RQMC [22, 33]. This is very effective in this example, especially when T and the volatility σ are not too large. Table 3 show some results for when we add this CV.

Importance sampling. Financial options for which the payoff is zero most of the time and takes a large or significant value only once in a while are not rare. Many options are indeed like insurance contracts, in which a nonzero claim is a *rare event* (it has a small probability). The contract value is then an integral whose integrand (the payoff function) has a peak in a small region and is zero elsewhere. MC performs poorly in this situation, because an excessively large sample size n might be required to obtain a sufficiently large number of occurrences of the rare event, and RQMC does not solve this problem. One appropriate tool then is *importance sampling* (IS), which can be seen as a change of variable that, when done properly, flattens out the integrand to reduce its variation. It can help both MC and RQMC.

We illustrate this with an ordinary European call option under the GVG model. We have $d = 1$ and $\bar{S} = S(t_1)$ in the payoff. We also take $t_1 = 1$ and $K = 180$, so a positive payoff is a rare event. We simulate the VG process via DGBS. To apply IS, we will increase the mean of G^+ and reduce the mean of G^- so $G^+(1) - G^-(1)$ takes larger values. A standard IS strategy for this is exponential twisting (see [1]): we multiply the density $g^+(x)$ of $G^+(1)$ by $e^{\theta x}$ and the density $g^-(y)$ of $G^-(1)$ by $e^{-\theta y}$, for some constant $\theta \geq 0$, and the re-normalize those densities. They are still gamma, but with different means. Then the (unbiased) IS estimator is the payoff multiplied by the likelihood ratio L of the old density product $g^+(y)g^-(y)$ divided by the new one. Since the payoff is nonzero only when $S(1) \geq K$, i.e., when $r + \omega + G^+(1) - G^-(1) = \ln[S(1)/S(0)] > \ln(K/S(0))$, we (heuristically) choose θ so that $\mathbb{E}[G^+(1) - G^-(1)] = \ln(K/S(0)) - r - \omega$ under IS. We write the expectation as a (monotone) function of θ and find the root θ^* of the equation numerically. For $K = 180$, we find $\theta^* = 25.56$ and $\rho \approx 0.26$. We can do even better, as follows.

Table 4 Empirical variances per run for the European call option under a GVG process, with and without IS, with MC and RQMC

	MC	Sobol+LMS+DS	Sobol+NUS	Lattice+S+B
no-IS	1.8×10^{-3}	6.3×10^{-4}	6.0×10^{-4}	7.8×10^{-4}
IS-twist	1.0×10^{-7}	1.9×10^{-11}	1.3×10^{-11}	1.1×10^{-11}
IS-twist-cond	2.8×10^{-8}	7.1×10^{-13}	7.4×10^{-13}	7.4×10^{-13}

First generate $G^-(1)$ under IS with θ^*, and then generate $G^+(1)$ from its original gamma distribution but conditional on $G^+(1) > G^-(1) + \ln(K/S(0)) - r - \omega$ (a truncated gamma). This way the payoff is positive with probability 1 and the resulting IS estimator has an even smaller variation. We call the previous one IS-twist, this one IS-twist-cond, and the original estimator no-IS. We can use RQMC with any of these three estimators. We try a Sobol' net with one extra coordinate (i.e., the Hammersley point set) with $n = 2^{16}$ randomized by LMS+DS, the same point set with NUS, and also a randomly shifted lattice rule with a baker's transformation, in two dimensions, with the same n. With this we find that the option price is 1.601×10^{-4} (the given digits here are significant). Table 4 reports the empirical variances for all the estimators discussed above, with both MC and RQMC. For RQMC, the variance of $\hat{\mu}_{\mathrm{rqmc},n}$ is multiplied by n to obtain the *variance per run* (for a fair comparison with MC). We see that without IS, RQMC does not reduce the variance by much, and the standard deviation is more than 300 times the mean in all cases. The combination of IS and RQMC has a synergistic effect: IS first makes the function flatter, then RQMC can shine. The difference between the three RQMC methods here is not significant.

5 RQMC for Markov Chains

When simulating a Markov chain over a large number of steps, if we use RQMC in a standard way, the dimension s will be large and RQMC is then likely to become ineffective. If each step requires d uniform random numbers and we simulate τ steps, then we have $s = \tau d$. The *Array-RQMC* method, which we now summarize, has been designed for this situation [31, 32, 37].

Consider a Markov chain with state space $\mathscr{X} \subseteq \mathbb{R}^\ell$, which evolves as

$$X_0 = x_0, \qquad X_j = \varphi_j(X_{j-1}, \mathbf{U}_j), \ j \geq 1,$$

where the $\mathbf{U}_j$ are i.i.d. uniform over $(0, 1)^d$. Want to estimate

$$\mu = \mathbb{E}[Y] \ \text{ where } \ Y = \sum_{j=1}^{\tau} g_j(X_j).$$

Ordinary MC or RQMC would produce n realizations of Y and take the average. Each realization requires $s = \tau d$ uniforms.

The idea of Array-RQMC is to simulate an *array* (or population) of n chains in parallel, in a way that at any given step j, there is small discrepancy between the empirical distribution of the n states $S_{n,j} = \{X_{0,j}, \ldots, X_{n-1,j}\}$ and the theoretical distribution of X_j. At each step, an RQMC point set is used to advance all the chains by one step.

To provide insight about the method, it is useful to assume for simplification that $X_j \sim U(0,1)^\ell$ for all j. This can be achieved conceptually by a change of variable, and is not necessary for implementing the method. We estimate

$$\mu_j = \mathbb{E}[g_j(X_j)] = \mathbb{E}[g_j(\varphi_j(X_{j-1}, \mathbf{U}))] = \int_{[0,1)^{\ell+d}} g_j(\varphi_j(\mathbf{x}, \mathbf{u}))d\mathbf{x}d\mathbf{u}$$

by

$$\hat{\mu}_{\mathrm{arqmc},j,n} = \frac{1}{n}\sum_{i=0}^{n-1} g_j(X_{i,j}) = \frac{1}{n}\sum_{i=0}^{n-1} g_j(\varphi_j(X_{i,j-1}, \mathbf{U}_{i,j})).$$

This is (roughly) RQMC with the point set $Q_n = \{(X_{i,j-1}, \mathbf{U}_{i,j}),\ 0 \le i < n\}$. We want Q_n to have low discrepancy (LD) over $[0,1)^{\ell+d}$.

However, we do not choose the $X_{i,j-1}$'s in Q_n: they come from the simulation. We can select a low discrepancy point set $\tilde{Q}_n = \{(\mathbf{w}_0, \mathbf{U}_{0,j}), \ldots, (\mathbf{w}_{n-1}, \mathbf{U}_{n-1,j})\}$, in which the $\mathbf{w}_i \in [0,1)^\ell$ are fixed and each $\mathbf{U}_{i,j} \sim U(0,1)^d$. Then a key operation is to permute the states $X_{i,j-1}$ so that $X_{\pi_j(i),j-1}$ is "close" to $\mathbf{w}_i$ for each i (low discrepancy between the two sets), and compute $X_{i,j} = \varphi_j(X_{\pi_j(i),j-1}, \mathbf{U}_{i,j})$ for each i. In particular, if $\ell = 1$, we can take $\mathbf{w}_i = (i+0.5)/n$ and just sort the states in increasing order. For $\ell > 1$, there are various ways to define the matching (multivariate sorts). At the end, we return the average $\bar{Y}_n = \hat{\mu}_{\mathrm{arqmc},n} = \sum_{j=1}^{\tau} \hat{\mu}_{\mathrm{arqmc},j,n}$ as an estimator of μ.

The array-RQMC estimator satisfies [32]: (i) $\bar{Y}_n$ is an *unbiased* estimator of μ, and (ii) the *empirical variance* of m independent realizations of $\bar{Y}_n$ is an unbiased estimator of $\mathrm{Var}[\bar{Y}_n]$. Known convergence rate results for special cases are summarized in [36]. For example, it is proved in [32] that for $\ell = 1$ and if the RQMC points form a stratified sample, the variance converges as $\mathcal{O}(n^{-3/2})$. In higher dimensions, it is show in [13] under some conditions that the worst-case error converges as $\mathcal{O}(n^{-1/(\ell+1)})$. In a sequential QMC setting (with particle filters) in which a digital net with NUS is used for RQMC and a Hilbert curve sort for mapping the states to the points, Gerber and Chopin [15] show that the variance is $o(n^{-1})$. Applications in finance and computer graphics can be found in [31, 66]. There are combinations with splitting techniques (multilevel and without levels), with importance sampling, and with weight windows (related to particle filters) [8, 30], combination with "coupling from the past" for exact sampling [39], and combination with approximate dynamic programming and for optimal stopping problems [11].

Examples in which the observed convergence rate for the empirical variance is $\mathcal{O}(n^{-2})$, or even $\mathcal{O}(n^{-3})$ in some cases, and does not depend on τ, can be found in [31, 37]. For example, when pricing an Asian call option under a geometric

Brownian motion, we observe $\mathcal{O}(n^{-2})$ convergence for the variance regardless of the number τ of observation times that determine the payoff [31], both with lattice rules and Sobol' points, while with stratification the observed rate is more like $\mathcal{O}(n^{-3/2})$. For this example, the state is in $d = 2$ dimensions and the RQMC points are in 3 dimensions.

A different way of using QMC to simulate Markov chains is studied in [6, 65]. The main idea is to use an approximation of a completely uniformly distributed (CUD) sequence, implemented by taking successive overlapping vectors of output values produced by a small linear random number generator as a source of "randomness" to simulate the chain (one vector per step).

Acknowledgements This work has been supported by a Canada Research Chair, an Inria International Chair, a NSERC Discovery Grant, to the author. It was presented at the MCQMC conference with the help of SAMSI funding. David Munger made Figs. 9 and 10. Several comments from the Guest Editor Art Owen helped improving the paper.

References

1. Asmussen, S., Glynn, P.W.: Stochastic Simulation. Springer, New York (2007)
2. Avramidis, A.N., Wilson, J.R.: Correlation-induction techniques for estimating quantiles in simulation experiments. Oper. Res. **46**(4), 574–591 (1998)
3. Avramidis, A.N., L'Ecuyer, P.: Efficient Monte Carlo and quasi-Monte Carlo option pricing under the variance-gamma model. Manag. Sci. **52**(12), 1930–1944 (2006)
4. Avramidis, A.N., L'Ecuyer, P., Tremblay, P.A.: Efficient simulation of gamma and variance-gamma processes. In: Proceedings of the 2003 Winter Simulation Conference, pp. 319–326. IEEE Press, Piscataway, New Jersey (2003)
5. Caflisch, R.E., Morokoff, W., Owen, A.: Valuation of mortgage-backed securities using Brownian bridges to reduce effective dimension. J. Comput. Financ. **1**(1), 27–46 (1997)
6. Chen, S., Dick, J., Owen, A.B.: Consistency of Markov chain quasi-Monte Carlo on continuous state spaces. Ann. Stat. **39**(2), 673–701 (2011)
7. Cranley, R., Patterson, T.N.L.: Randomization of number theoretic methods for multiple integration. SIAM J. Numer. Anal. **13**(6), 904–914 (1976)
8. Demers, V., L'Ecuyer, P., Tuffin, B.: A combination of randomized quasi-Monte Carlo with splitting for rare-event simulation. In: Proceedings of the 2005 European Simulation and Modeling Conference, pp. 25–32. EUROSIS, Ghent, Belgium (2005)
9. Dick, J., Pillichshammer, F.: Digital Nets and Sequences: Discrepancy Theory and Quasi-Monte Carlo Integration. Cambridge University Press, Cambridge, U.K. (2010)
10. Dick, J., Sloan, I.H., Wang, X., Woźniakowski, H.: Good lattice rules in weighted Korobov spaces with general weights. Numer. Math. **103**, 63–97 (2006)
11. Dion, M., L'Ecuyer, P.: American option pricing with randomized quasi-Monte Carlo simulations. In: Proceedings of the 2010 Winter Simulation Conference, pp. 2705–2720 (2010)
12. Doerr, C., Gnewuch, M., Wahlström, M.: Calculation of discrepancy measures and applications. In: Chen W., Srivastav A., Travaglini G. (eds.) A Panorama of Discrepancy Theory, pp. 621–678. Springer, Berlin (2014)
13. El Haddad, R., Lécot, C., L'Ecuyer, P., Nassif, N.: Quasi-Monte Carlo methods for Markov chains with continuous multidimensional state space. Math. Comput. Simul. **81**, 560–567 (2010)
14. Faure, H.: Discrépance des suites associées à un système de numération en dimension s. Acta Arith. **61**, 337–351 (1982)

15. Gerber, M., Chopin, N.: Sequential quasi-Monte Carlo. J. R. Stat. Soc. Ser. B **77**(Part 3), 509–579 (2015)
16. Haber, S.: A modified Monte Carlo quadrature. Math. Comput. **19**, 361–368 (1966)
17. Hickernell, F.J.: The mean square discrepancy of randomized nets. ACM Trans. Model. Comput. Simul. **6**(4), 274–296 (1996)
18. Hickernell, F.J.: A generalized discrepancy and quadrature error bound. Math. Comput. **67**(221), 299–322 (1998)
19. Hickernell, F.J.: What affects the accuracy of quasi-Monte Carlo quadrature? In: Niederreiter, H., Spanier, J. (eds.) Monte Carlo and Quasi-Monte Carlo Methods 1998, pp. 16–55. Springer, Berlin (2000)
20. Hickernell, F.J.: Obtaining $O(N^{-2+\epsilon})$ convergence for lattice quadrature rules. In: Fang, K.T., Hickernell, F.J., Niederreiter, H. (eds.) Monte Carlo and Quasi-Monte Carlo Methods 2000, pp. 274–289. Springer, Berlin (2002)
21. Hickernell, F.J.: Error analysis for quasi-Monte Carlo methods. In: Glynn P.W., Owen A.B. (eds.) Monte Carlo and Quasi-Monte Carlo Methods 2016 (2017)
22. Hickernell, F.J., Lemieux, C., Owen, A.B.: Control variates for quasi-Monte Carlo. Stat. Sci. **20**(1), 1–31 (2005)
23. Hong, H.S., Hickernell, F.H.: Algorithm 823: implementing scrambled digital sequences. ACM Trans. Math. Softw. **29**, 95–109 (2003)
24. Imai, J., Tan, K.S.: Enhanced quasi-Monte Carlo methods with dimension reduction. In: Yücesan E., Chen C.H., Snowdon J.L., Charnes J.M. (eds.) Proceedings of the 2002 Winter Simulation Conference, pp. 1502–1510. IEEE Press, Piscataway, New Jersey (2002)
25. Imai, J., Tan, K.S.: A general dimension reduction technique for derivative pricing. J. Comput. Financ. **10**(2), 129–155 (2006)
26. Joe, S., Kuo, F.Y.: Constructing Sobol sequences with better two-dimensional projections. SIAM J. Sci. Comput. **30**(5), 2635–2654 (2008)
27. L'Ecuyer, P.: Quasi-Monte Carlo methods in finance. In: Proceedings of the 2004 Winter Simulation Conference, pp. 1645–1655. IEEE Press, Piscataway, New Jersey (2004)
28. L'Ecuyer, P.: Quasi-Monte Carlo methods with applications in finance. Financ. Stoch. **13**(3), 307–349 (2009)
29. L'Ecuyer, P.: SSJ: Stochastic simulation in Java (2016). http://simul.iro.umontreal.ca/ssj/
30. L'Ecuyer, P., Demers, V., Tuffin, B.: Rare-events, splitting, and quasi-Monte Carlo. ACM Trans. Model. Comput. Simul. **17**(2), Article 9 (2007)
31. L'Ecuyer, P., Lécot, C., L'Archevêque-Gaudet, A.: On array-RQMC for Markov chains: mapping alternatives and convergence rates. In: L'Ecuyer, P., Owen, A.B. (eds.) Monte Carlo and Quasi-Monte Carlo Methods 2008, pp. 485–500. Springer, Berlin (2009)
32. L'Ecuyer, P., Lécot, C., Tuffin, B.: A randomized quasi-Monte Carlo simulation method for Markov chains. Oper. Res. **56**(4), 958–975 (2008)
33. L'Ecuyer, P., Lemieux, C.: Variance reduction via lattice rules. Manag. Sci. **46**(9), 1214–1235 (2000)
34. L'Ecuyer, P., Lemieux, C.: Recent advances in randomized quasi-Monte Carlo methods. In: Dror, M., L'Ecuyer, P., Szidarovszky, F. (eds.) Modeling Uncertainty: An Examination of Stochastic Theory, Methods, and Applications, pp. 419–474. Kluwer Academic, Boston (2002)
35. L'Ecuyer, P., Munger, D.: On figures of merit for randomly-shifted lattice rules. In: Woźniakowski, H., Plaskota, L. (eds.) Monte Carlo and Quasi-Monte Carlo Methods 2010, pp. 133–159. Springer, Berlin (2012)
36. L'Ecuyer, P., Munger, D.: Algorithm 958: lattice builder: A general software tool for constructing rank-1 lattice rules. ACM Trans. Math. Softw. **42**(2), Article 15 (2016)
37. L'Ecuyer, P., Munger, D., Lécot, C., Tuffin, B.: Sorting methods and convergence rates for array-rqmc: some empirical comparisons. Math. Comput. Simul. **143**, 191–201 (2018). https://doi.org/10.1016/j.matcom.2016.07.010
38. L'Ecuyer, P., Munger, D., Tuffin, B.: On the distribution of integration error by randomly-shifted lattice rules. Electron. J. Stat. **4**, 950–993 (2010)

39. L'Ecuyer, P., Sanvido, C.: Coupling from the past with randomized quasi-Monte Carlo. Math. Comput. Simul. **81**(3), 476–489 (2010)
40. L'Ecuyer, P., Simard, R.: Inverting the symmetrical beta distribution. ACM Trans. Math. Softw. **32**(4), 509–520 (2006)
41. Lemieux, C.: Monte Carlo and Quasi-Monte Carlo Sampling. Springer, New York (2009)
42. Lemieux, C., Cieslak, M., Luttmer, K.: RandQMC user's guide: a package for randomized Quasi-Monte Carlo methods in C (2004). Software user's guide. http://www.math.uwaterloo.ca/~clemieux/randqmc.html
43. Lemieux, C., L'Ecuyer, P.: Randomized polynomial lattice rules for multivariate integration and simulation. SIAM J. Sci. Comput. **24**(5), 1768–1789 (2003)
44. Liu, R., Owen, A.B.: Estimating mean dimensionality of analysis of variance decompositions. J. Am. Stat. Assoc. **101**(474), 712–721 (2006)
45. Loh, W.L.: On the asymptotic distribution of scramble nets quadratures. Ann. Stat. **31**, 1282–1324 (2003)
46. Madan, D.B., Carr, P.P., Chang, E.C.: The variance gamma process and option pricing. Eur. Financ. Rev. **2**, 79–105 (1998)
47. Mara, T.A., Rakoto, J.O.: Comparison of some efficient methods to evaluate the main effect of computer model factors. J. Stat. Comput. Simul. **78**(2), 167–178 (2008)
48. Matousěk, J.: On the L_2-discrepancy for anchored boxes. J. Complex. **14**, 527–556 (1998)
49. Matoušek, J.: Geometric Discrepancy: An Illustrated Guide. Springer, Berlin (1999)
50. Niederreiter, H.: Random Number Generation and Quasi-Monte Carlo Methods. SIAM CBMS-NSF Regional Conference Series in Applied Mathematics, vol. 63. SIAM, Philadelphia (1992)
51. Niederreiter, H., Xing, C.: Nets, (t, s)-sequences, and algebraic geometry. In: Hellekalek, P., Larcher, G. (eds.) Random and Quasi-Random Point Sets. Lecture Notes in Statistics, vol. 138, pp. 267–302. Springer, New York (1998)
52. Nuyens, D.: The construction of good lattice rules and polynomial lattice rules. In: Kritzer, P., Niederreiter, H., Pillichshammer, F., Winterhof, A. (eds.) Uniform Distribution and Quasi-Monte Carlo Methods: Discrepancy, Integration and Applications, pp. 223–255. De Gruyter, Berlin (2014)
53. Owen, A.B.: Monte Carlo variance of scrambled equidistribution quadrature. SIAM J. Numer. Anal. **34**(5), 1884–1910 (1997)
54. Owen, A.B.: Scrambled net variance for integrals of smooth functions. Ann. Stat. **25**(4), 1541–1562 (1997)
55. Owen, A.B.: Latin supercube sampling for very high-dimensional simulations. ACM Trans. Model. Comput. Simul. **8**(1), 71–102 (1998)
56. Owen, A.B.: Variance with alternative scramblings of digital nets. ACM Trans. Model. Comput. Simul. **13**(4), 363–378 (2003)
57. Owen, A.B.: Better estimation of small sobol sensitivity indices. ACM Trans. Model. Comput. Simul. **23**(2), Article 11 (2013)
58. Sloan, I.H., Joe, S.: Lattice Methods for Multiple Integration. Clarendon Press, Oxford (1994)
59. Sloan, I.H., Woźniakowski, H.: When are quasi-Monte Carlo algorithms efficient for high-dimensional integrals. J. Complex. **14**, 1–33 (1998)
60. Sloan, I.H., Woźniakowski, H.: Tractability of multivariate integration for weighted Korobov classes. J. Complex. **17**(4), 697–721 (2001)
61. Sobol', I.M.: The distribution of points in a cube and the approximate evaluation of integrals. U.S.S.R. Comput. Math. Math. Phys. **7**(4), 86–112 (1967)
62. Sobol', I.M.: Sensitivity indices for nonlinear mathematical models. Math. Model. Comput. Exp. **1**, 407–414 (1993)
63. Sobol', I.M., Myshetskaya, E.E.: Monte Carlo estimators for small sensitivity indices. Monte Carlo Methods Appl. **13**(5–6), 455–465 (2007)
64. Tezuka, S.: Uniform Random Numbers: Theory and Practice. Kluwer Academic, Boston (1995)
65. Tribble, S.D., Owen, A.B.: Constructions of weakly CUD sequences for MCMC. Electron. J. Stat. **2**, 634–660 (2008)
66. Wächter, C., Keller, A.: Efficient simultaneous simulation of Markov chains. In: Keller, A., Heinrich, S., Niederreiter, H. (eds.) Monte Carlo and Quasi-Monte Carlo Methods 2006, pp. 669–684. Springer, Berlin (2008)

Application of Quasi-Monte Carlo Methods to PDEs with Random Coefficients – An Overview and Tutorial

Frances Y. Kuo and Dirk Nuyens

Abstract This article provides a high-level overview of some recent works on the application of quasi-Monte Carlo (QMC) methods to PDEs with random coefficients. It is based on an in-depth survey of a similar title by the same authors, with an accompanying software package which is also briefly discussed here. Embedded in this article is a step-by-step tutorial of the required analysis for the setting known as the uniform case with first order QMC rules. The aim of this article is to provide an easy entry point for QMC experts wanting to start research in this direction and for PDE analysts and practitioners wanting to tap into contemporary QMC theory and methods.

Keywords Quasi-Monte Carlo methods · PDEs with random coefficients · Uniform · Lognormal · Multi-level · Randomly shifted lattice rules

1 Introduction

Uncertainty quantification is the science of quantitative characterization and reduction of uncertainties in both computational and real world applications, and it is the source of many challenging high dimensional integration and approximation problems. Often the high dimensionality comes from uncertainty or randomness in the data, e.g., in groundwater flow from permeability that is rapidly varying and uncertain, or in financial mathematics from the rapid and often unpredictable changes

F. Y. Kuo (✉)
School of Mathematics and Statistics, University of New South Wales,
Sydney, NSW 2052, Australia
e-mail: f.kuo@unsw.edu.au

D. Nuyens
Department of Computer Science, KU Leuven, Celestijnenlaan 200A,
3001 Leuven, Belgium
e-mail: dirk.nuyens@cs.kuleuven.be

A. B. Owen and P. W. Glynn (eds.), *Monte Carlo and Quasi-Monte Carlo Methods*, Springer Proceedings in Mathematics & Statistics 241,
https://doi.org/10.1007/978-3-319-91436-7_3

within markets. The input data may be a random variable or a random field, in which case the derived quantity of interest will in general also be a random variable or a random field. The computational goal is usually to find the expected value or other statistics of these derived quantities.

A popular example is the flow of water through a disordered porous medium, modeled by Darcy's law coupled with the mass conservation law, i.e.,

$$\begin{aligned} q(\boldsymbol{x}, \omega) + a(\boldsymbol{x}, \omega)\, \nabla p(\boldsymbol{x}, \omega) &= 0\,, \\ \nabla \cdot q(\boldsymbol{x}, \omega) &= 0\,, \end{aligned}$$

for $\boldsymbol{x}$ in a bounded domain $D \subset \mathbb{R}^d$, $d \in \{1, 2, 3\}$, and for almost all events ω in the probability space $(\Omega, \mathscr{A}, \mathbb{P})$. Here $q(\boldsymbol{x}, \omega)$ is the velocity (also called the specific discharge) and $p(\boldsymbol{x}, \omega)$ is the residual pressure, while $a(\boldsymbol{x}, \omega)$ is the permeability (or more precisely, the ratio of permeability to dynamic viscosity) which is modelled as a random field. Uncertainty in $a(\boldsymbol{x}, \omega)$ leads to uncertainty in $q(\boldsymbol{x}, \omega)$ and $p(\boldsymbol{x}, \omega)$. Quantities of interest include for example the breakthrough time of a plume of pollution moving through the medium.

QMC for PDEs with Random Coefficients

There is a huge literature on treating these PDEs with random coefficients using various methods, see e.g., the surveys [1, 23, 43] and the references therein. Here we are interested in the application of *quasi-Monte Carlo* (*QMC*) *methods*, which are equal-weight quadrature rules for high dimensional integrals, see e.g., [3, 5, 36–39, 44].

QMC methods are still relatively new to these PDE problems. It began with the 2011 paper [21] which included comprehensive numerical experiments showing promising QMC results, but without any theoretical justification. The first fully justified theory was provided in the 2012 paper [32], and this has lead to a flood of research activities. We will follow the recent survey [31] to provide a high-level overview of how QMC theory can be applied to PDEs with random coefficients. The survey [31] covered the detailed analysis from six papers [6, 8, 22, 32–34] in a unified view. Different algorithms have been analyzed: *single-level* vs *multi-level*, *deterministic* vs *randomized*, and *first order* vs *higher order*, and they were considered under different models for the randomness as we explain below.

It is popular to assume that $a(\boldsymbol{x}, \omega)$ is a *lognormal* random field, that is, $\log(a(\boldsymbol{x}, \omega))$ is a Gaussian random field on the spatial domain D with a specified mean and covariance function. Then one can use the *Karhunen–Loève* (*KL*) *expansion* to write $\log(a(\boldsymbol{x}, \omega))$ as an infinite series parametrised by a sequence $y_j = y_j(\omega)$, $j \geq 1$, of i.i.d. standard normal random numbers from $\mathbb{R}$. Aside from the lognormal case, often the simpler *uniform* case is considered, where $a(\boldsymbol{x}, \omega)$ is written as an infinite series that depends linearly on a sequence $y_j = y_j(\omega)$, $j \geq 1$, of i.i.d. uniform random numbers from a bounded interval of $[-1, 1]$ or $[-\frac{1}{2}, \frac{1}{2}]$. In both the lognormal and uniform cases the infinite series is truncated in practice to, say, s terms. The expected value of any quantity of interest is then approximated by an s-dimensional integral with respect to the parameters y_j, which can in turn be approximated by QMC methods, combined with finite element methods for solving the PDE.

The six papers surveyed in [31] all followed this KL-based general direction. With respect to the QMC method they can be either *first order* or *higher order*, which refers to the rate of convergence being close to $\mathcal{O}(n^{-1})$ or $\mathcal{O}(n^{-\alpha})$, $\alpha > 1$, with n being the number of integrand evaluations. With respect to the approximation of the integrand function they can be either *single-level* or *multi-level*, which refers to how spatial discretization and dimension truncation are performed. A summary of the results is given in the table below:

	Uniform case	Lognormal case
First order single-level analysis	[32]	[22]
First order multi-level analysis	[33]	[34]
Higher order single-level analysis	[6]	
Higher order multi-level analysis	[8]	

The first order results [22, 32–34] are based on *randomly shifted lattice rules* and are accompanied by probabilistic error bounds. The higher order results [6, 8] are based on *interlaced polynomial lattice rules* and are accompanied by deterministic error bounds. The lognormal results [22, 34] require a non-standard function space setting for integrands with domain $\mathbb{R}^s$. A key feature in all these analysis is that the QMC error bounds are independent of the number of integration variables s. There is as yet no satisfactory QMC theory that can give higher order convergence for the lognormal case with error bound independent of s.

Plan of this Article

In Sect. 2 we provide an overview of the different settings and algorithms covered in the survey [31], with the goal to convey the overall picture while keeping the exposition as simple and accessible as possible. In Sect. 3 we take a change of pace and style to give a step-by-step tutorial of the required analysis for the uniform case with first order QMC rules. That is, we zoom in and focus on the essence of the paper [32] in such a way that the tutorial can be used to extend the analysis to other cases by interested readers. Then in Sect. 4 we zoom out again and continue to provide insights to the key analysis required for the six papers surveyed in [31]. In Sect. 5 we briefly discuss the software accompanying [31]. Finally in Sect. 6 we give a short conclusion.

Beyond the Survey

There have been many developments beyond the scope of the survey [31].

Instead of using the KL expansion, in the lognormal case one can sample the random field only at a discrete set of points with respect to the covariance matrix inherited from the given covariance function of the continuous field. The random field is then represented exactly at these points, thus eliminating completely the truncation error associated with the KL-based approach. (Note that interpolation may be required at the finite element quadrature nodes.) The resulting large matrix factorization problem could potentially be handled by *circulant embedding* and FFT, if the covariance function is *stationary* and the grid is regular, see [10]. In fact, this was the approach taken in the first QMC paper for PDEs with random coefficients [21], and the corresponding analysis is being considered in [19, 20].

Another way to tackle the large matrix factorization is to make use of *H-matrix techniques*, see [24], and this has been considered in [11].

The uniform framework can be extended from the elliptic PDE to the general framework of *affine* parametric operator equations, see [42] as well as [6, 8]. A different QMC theory for the lognormal case is offered in [26]. Further PDE computations with higher order QMC are reported in [14], and with multi-level and multi-index QMC in [40]. QMC has also been applied to PDEs on the sphere [35], holomorphic equations [9], Bayesian inversion [4, 41], stochastic wave propagation [12, 13], and eigenproblems [16].

Moreover, there has been some significant development in the use of functions with local support in the expansions of $a(\boldsymbol{x}, \omega)$ which leads to a simplified norm estimate for the integrand and a reduced construction cost (pre-computation) for QMC, see [15, 27, 29].

2 Overview

Throughout this article we refer to the number of integration variables s as the *stochastic dimension*, which can be in the hundreds or thousands or more (and controls the truncation error), in contrast to the *spatial dimension* d which is just 1, 2 or 3.

2.1 Uniform Versus Lognormal

For a given parameter $\boldsymbol{y}$ we consider the parametric elliptic Dirichlet problem

$$-\nabla\cdot(a(\boldsymbol{x},\boldsymbol{y})\,\nabla u(\boldsymbol{x},\boldsymbol{y})) = \kappa(\boldsymbol{x}) \quad \text{for } \boldsymbol{x} \text{ in } D\,, \quad u(\boldsymbol{x},\boldsymbol{y}) = 0 \quad \text{for } \boldsymbol{x} \text{ on } \partial D\,, \tag{1}$$

for domain $D \subset \mathbb{R}^d$ a bounded, convex, Lipschitz polyhedron with boundary ∂D, where the spatial dimension $d = 1, 2,$ or 3 is assumed given and fixed. The differential operators in (1) are understood to be with respect to the physical variable $\boldsymbol{x}$ which belongs to D. The parametric variable $\boldsymbol{y} = (y_j)_{j\geq 1}$ belongs to either a bounded or unbounded domain, depending on which of the two popular formulations of the parametric coefficient $a(\boldsymbol{x}, \boldsymbol{y})$ is being considered.

Uniform Case

In the uniform case, we assume that the y_j are independent and uniformly distributed on $[-\frac{1}{2}, \frac{1}{2}]$, and

$$a(\boldsymbol{x}, \boldsymbol{y}) = a_0(\boldsymbol{x}) + \sum_{j\geq 1} y_j \psi_j(\boldsymbol{x}), \tag{2}$$

with $0 < a_{\min} \leq a(\boldsymbol{x}, \boldsymbol{y}) \leq a_{\max} < \infty$ for all $\boldsymbol{x}$ and $\boldsymbol{y}$. We need further assumptions on a_0 and ψ_j, see [31] for details. Here we mention only one important assumption that there exists $p_0 \in (0, 1)$ such that

$$\sum_{j\geq 1} \|\psi_j\|_{L_\infty}^{p_0} < \infty. \tag{3}$$

The value of p_0 reflects the rate of decay of the fluctuations in (2); later we will see that it directly affects the QMC convergence rate.

Our goal is to compute the integral, i.e., the expected value, with respect to $\boldsymbol{y}$, of a bounded linear functional G applied to the solution $u(\cdot, \boldsymbol{y})$ of the PDE (1)

$$\int_{\left[-\frac{1}{2},\frac{1}{2}\right]^{\mathbb{N}}} G(u(\cdot, \boldsymbol{y}))\,\mathrm{d}\boldsymbol{y} := \lim_{s\to\infty} \int_{\left[-\frac{1}{2},\frac{1}{2}\right]^s} G(u(\cdot, (y_1, \ldots, y_s, 0, 0, \ldots)))\,\mathrm{d}y_1 \cdots \mathrm{d}y_s\,. \tag{4}$$

Lognormal Case

In the lognormal case, we assume that the y_j are independent standard normal random numbers on $\mathbb{R}$, and

$$a(\boldsymbol{x}, \boldsymbol{y}) = a_0(\boldsymbol{x}) \exp\left(\sum_{j\geq 1} y_j \sqrt{\mu_j}\, \xi_j(\boldsymbol{x})\right), \tag{5}$$

where $a_0(\boldsymbol{x}) > 0$, the $\mu_j > 0$ are non-increasing, and the ξ_j are orthonormal in $L_2(D)$. This can arise from the KL expansion in the case where $\log(a)$ is a stationary Gaussian random field with a specified mean and covariance function; a popular choice is the *Matérn* covariance.

Our goal now is the integral of $G(u(\cdot, \boldsymbol{y}))$ over $\boldsymbol{y} \in \mathbb{R}^{\mathbb{N}}$ with a countable product Gaussian measure $\mu_G(\mathrm{d}\boldsymbol{y})$ (formally, we restrict the domain to some $Y \subset \mathbb{R}^{\mathbb{N}}$ with full measure $\mu_G(Y) = 1$, but we omit this in the notation)

$$\int_{\mathbb{R}^{\mathbb{N}}} G(u(\cdot, \boldsymbol{y})) \prod_{j\geq 1} \phi_{\text{nor}}(y_j)\,\mathrm{d}\boldsymbol{y} = \int_{[0,1]^{\mathbb{N}}} G(u(\cdot, \Phi_{\text{nor}}^{-1}(\boldsymbol{w})))\,\mathrm{d}\boldsymbol{w}\,, \tag{6}$$

where $\phi_{\rm nor}(y) := \exp(-y^2/2)/\sqrt{2\pi}$ is the univariate standard normal probability density function, while $\Phi_{\rm nor}^{-1}$ denotes the inverse of the corresponding cumulative distribution function, and is applied component-wise to a vector. The transformed integral over the unit cube on the right-hand side of (6) is obtained by the change of variables $\boldsymbol{y} = \Phi_{\rm nor}^{-1}(\boldsymbol{w})$.

2.2 *Single-Level Versus Multi-level*

Single-Level Algorithms

We approximate the integral (4) or (6) in three steps:

i. Dimension truncation: the infinite sum in (2) or (5) is truncated to s terms.
ii. Finite element discretization: the PDE (1) in weak formulation (see (13) below) is solved using a finite element method with meshwidth h.
iii. QMC quadrature: the integral of the finite element solution for the truncated problem is estimated using a deterministic or randomized QMC method.

The deterministic version of this algorithm is

$$\frac{1}{n}\sum_{i=1}^{n} G(u_h^s(\cdot,\boldsymbol{y}_i)), \qquad \boldsymbol{y}_i = \begin{cases} \boldsymbol{t}_i - \frac{1}{2} & \text{for uniform,} \\ \Phi_{\rm nor}^{-1}(\boldsymbol{t}_i) & \text{for lognormal,} \end{cases} \tag{7}$$

where $\boldsymbol{t}_1,\ldots,\boldsymbol{t}_n \in [0,1]^s$ are n QMC points from the s-dimensional standard unit cube. In the uniform case, these points are translated to the unit cube $[-\frac{1}{2},\frac{1}{2}]^s$. In the lognormal case, these points are mapped to the Euclidean space $\mathbb{R}^s$ by applying the inverse of the cumulative normal distribution function component-wise.

A randomized version of this algorithm with *random shifting* is given by

$$\frac{1}{r}\sum_{k=1}^{r}\frac{1}{n}\sum_{i=1}^{n} G(u_h^s(\cdot,\boldsymbol{y}_{i,k})), \qquad \boldsymbol{y}_{i,k} = \begin{cases} \{\boldsymbol{t}_i + \boldsymbol{\Delta}_k\} - \frac{1}{2} & \text{for uniform,} \\ \Phi_{\rm nor}^{-1}(\{\boldsymbol{t}_i + \boldsymbol{\Delta}_k\}) & \text{for lognormal,} \end{cases} \tag{8}$$

where $\boldsymbol{t}_1,\ldots,\boldsymbol{t}_n \in [0,1]^s$ are n QMC points as above, and $\boldsymbol{\Delta}_1,\ldots,\boldsymbol{\Delta}_r \in [0,1]^s$ are r independent *random shifts* generated from the uniform distribution on $[0,1]^s$. The braces in $\{\boldsymbol{t}_i + \boldsymbol{\Delta}_k\}$ mean that we take the fractional part of each component in the vector $\boldsymbol{t}_i + \boldsymbol{\Delta}_k$. Other randomization strategies can be used analogously but need to be chosen appropriately to preserve the special properties of the QMC points. Randomized algorithms have the advantages of being unbiased as well as providing a practical error estimate.

Multi-level Algorithms

The general concept of *multi-level* can be explained as follows (see e.g., [17]): if we denote the integral (4) or (6) by I_∞ and define a sequence $I_0, I_1, \ldots$ of approximations converging to I_∞, then we can write I_∞ as a telescoping sum $I_\infty = (I_\infty - I_L) + \sum_{\ell=0}^{L}(I_\ell - I_{\ell-1})$, $I_{-1} := 0$, and then apply different quadrature rules to the differences $I_\ell - I_{\ell-1}$, which we anticipate to get smaller as ℓ increases. Here we define I_ℓ to be the integral of $G(u_{h_\ell}^{s_\ell})$ corresponding to the finite element solution with meshwidth h_ℓ, for the truncated problem with s_ℓ terms, where $1 \le s_0 \le s_1 \le s_2 \le \cdots \le s_L \le \cdots$ and $h_0 \ge h_1 \ge h_2 \ge \cdots \ge h_L \ge \cdots > 0$, so that I_ℓ becomes a better approximation to I_∞ as ℓ increases.

The deterministic version of our multi-level algorithm takes the form (remembering the linearity of G)

$$\sum_{\ell=0}^{L}\left(\frac{1}{n_\ell}\sum_{i=1}^{n_\ell} G((u_{h_\ell}^{s_\ell} - u_{h_{\ell-1}}^{s_{\ell-1}})(\cdot, \boldsymbol{y}_i^\ell))\right), \qquad \boldsymbol{y}_i^\ell = \begin{cases} \boldsymbol{t}_i^\ell - \frac{\boldsymbol{1}}{\boldsymbol{2}} & \text{for uniform,} \\ \Phi_{\rm nor}^{-1}(\boldsymbol{t}_i^\ell) & \text{for lognormal,} \end{cases} \tag{9}$$

where we apply an s_ℓ-dimensional QMC rule with n_ℓ points $\boldsymbol{t}_1^\ell, \ldots, \boldsymbol{t}_{n_\ell}^\ell \in [0,1]^{s_\ell}$ to the integrand $G(u_{h_\ell}^{s_\ell} - u_{h_{\ell-1}}^{s_{\ell-1}})$, and we define $u_{h_{-1}}^{s_{-1}} := 0$.

The corresponding randomized version can be obtained analogously to (8) by taking r_ℓ random shifts at each level, noting that all shifts from all levels should be independent.

2.3 *First-Order Versus Higher-Order*

Up to this point we have said very little about QMC methods, other than noting that they are equal-weight quadrature rules as seen in (7). Actually, we will not say much about QMC methods in this article at all. In this subsection we will mention three different QMC theoretical settings which have been used for PDEs applications, giving just enough details in the first setting needed for the tutorial in Sect. 3. These three settings are discussed in slightly more detail in [30] in this volume, and more comprehensively in [31]; see also the references in these papers.

First Order QMC Over the Unit Cube – Randomly Shifted Lattice Rules for Weighted Sobolev Spaces

Suppose we wish to approximate the s-dimensional integral over the unit cube $[0,1]^s$

$$\int_{[0,1]^s} f(\boldsymbol{y})\,\mathrm{d}\boldsymbol{y}\,, \tag{10}$$

where the integrand f belongs to a *weighted Sobolev space of smoothness one*, with the *unanchored* norm defined by (see e.g., [45])

$$\|f\|_{\boldsymbol{\gamma}} = \left[\sum_{\mathrm{u}\subseteq\{1:s\}} \frac{1}{\gamma_\mathrm{u}} \int_{[0,1]^{|\mathrm{u}|}} \left(\int_{[0,1]^{s-|\mathrm{u}|}} \frac{\partial^{|\mathrm{u}|} f}{\partial \boldsymbol{y}_\mathrm{u}}(\boldsymbol{y})\,\mathrm{d}\boldsymbol{y}_{\{1:s\}\setminus\mathrm{u}}\right)^2 \mathrm{d}\boldsymbol{y}_\mathrm{u}\right]^{1/2}. \tag{11}$$

Here $\{1:s\}$ is a shorthand notation for the set of indices $\{1, 2, \ldots, s\}$, $(\partial^{|\mathrm{u}|} f)/(\partial \boldsymbol{y}_{\mathrm{u}})$ denotes the mixed first derivative of f with respect to the "active" variables $\boldsymbol{y}_{\mathrm{u}} = (y_j)_{j\in\mathrm{u}}$, while $\boldsymbol{y}_{\{1:s\}\setminus\mathrm{u}} = (y_j)_{j\in\{1:s\}\setminus\mathrm{u}}$ denotes the "inactive" variables. There is a weight parameter $\gamma_{\mathrm{u}} \geq 0$ associated with each subset of variables $\boldsymbol{y}_{\mathrm{u}}$ to moderate the relative importance between the different sets of variables. We denote the weights collectively by $\boldsymbol{\gamma}$.

In this setting we pair the weighted Sobolev space with *randomly shifted lattice rules*; the complete theory can be found in [5]. They approximate the integral (10) by

$$\frac{1}{n}\sum_{i=1}^{n} f(\boldsymbol{t}_i), \qquad \boldsymbol{t}_i = \left\{\frac{i\,\boldsymbol{z}}{n} + \boldsymbol{\Delta}\right\},$$

where $\boldsymbol{z} \in \mathbb{Z}^s$ is known as the *generating vector*, $\boldsymbol{\Delta}$ is a *random shift* drawn from the uniform distribution over $[0, 1]^s$, and as in (8) the braces indicate that we take the fractional parts of each component in a vector. It is known that good generating vectors can be obtained using a *CBC construction* (*component-by-component construction*), determining the components of $\boldsymbol{z}$ one at a time sequentially, to achieve first order convergence in this setting, where the implied constant can be independent of s under appropriate conditions on the weights $\boldsymbol{\gamma}$.

Specifically, if n is a power of 2 then we know that the CBC construction yields the root-mean-square error bound (with respect to the uniform random shift), for all $\lambda \in (1/2, 1]$,

$$\text{r.m.s. error} \leq \left(\frac{2}{n}\sum_{\emptyset\neq\mathrm{u}\subseteq\{1:s\}} \gamma_{\mathrm{u}}^{\lambda}\,[\vartheta(\lambda)]^{|\mathrm{u}|}\right)^{1/(2\lambda)} \|f\|_{\boldsymbol{\gamma}}\,, \tag{12}$$

where $\vartheta(\lambda) := 2\zeta(2\lambda)/(2\pi^2)^{\lambda}$, with $\zeta(a) := \sum_{k=1}^{\infty} k^{-a}$ denoting the Riemann zeta function. A similar result holds for general n. The best rate of convergence clearly comes from choosing λ close to $1/2$.

We need some structure in the weights $\boldsymbol{\gamma}$ for the CBC construction cost to be feasible in practice. *Fast* CBC algorithms (using FFT) can find a generating vector of a good n-point lattice rule in s dimensions in $\mathcal{O}(s\,n\log n)$ operations in the case of *product weights*, and in $\mathcal{O}(s\,n\log n + s^2\,n)$ operations in the case of *POD weights* (see (25) ahead).

First Order QMC Over $\mathbb{R}^s$

We can pair randomly shifted lattice rules with a special function space setting over $\mathbb{R}^s$ to achieve first order convergence. The norm in this function space setting includes some additional weight functions to control the behavior of the derivatives of the functions as the components go to $\pm\infty$. The root-mean-square error bound takes the same form as (12), but with a different definition of the norm and $\vartheta(\lambda)$.

Higher Order QMC Over the Unit Cube

We can pair a family of QMC methods called *interlaced polynomial lattice rules* with another special function space setting over the unit cube to achieve higher order convergence. The norm in this function space setting involves higher order mixed derivatives of the functions. The key advantage of this family of QMC methods over other higher order QMC methods is that, in the cost of finding a generating vector which achieves the best theoretical convergence rate, the *order* or the *interlacing factor* appears as a multiplying factor rather than sitting in the exponent of the number of points n.

3 Tutorial

We conclude from the error bound (12) that the first step in applying QMC theory is to estimate the norm of the practical integrand. We see from (7), (8), and (9) that this means we need to estimate the norms

$$\|G(u_h^s)\|_\gamma \qquad \text{and} \qquad \|G(u_{h_\ell}^{s_\ell} - u_{h_{\ell-1}}^{s_{\ell-1}})\|_\gamma \,,$$

for the single-level and the multi-level algorithms, respectively.

In this section we provide a step-by-step tutorial on the analysis for the single-level algorithm in the uniform case with first order QMC rules.

Differentiate the PDE

1. We start with the variational formulation of the PDE (1): find $u(\cdot, \boldsymbol{y}) \in H_0^1(D)$ such that

$$\int_D a(\boldsymbol{x}, \boldsymbol{y})\, \nabla u(\boldsymbol{x}, \boldsymbol{y}) \cdot \nabla w(\boldsymbol{x})\, \mathrm{d}\boldsymbol{x} \;=\; \int_D \kappa(\boldsymbol{x})\, w(\boldsymbol{x})\, \mathrm{d}\boldsymbol{x} \quad \forall w \in H_0^1(D)\,. \tag{13}$$

   Here we consider the Sobolev space $H_0^1(D)$ of functions which vanish on the boundary of D, with norm $\|w\|_{H_0^1} := \|\nabla w\|_{L_2}$, and together with the dual space $H^{-1}(D)$ and pivot space $L_2(D)$.
2. We take the mixed partial derivatives $\partial^{\boldsymbol{\nu}}$ with respect to $\boldsymbol{y}$ with multi-index $\boldsymbol{\nu} \neq \boldsymbol{0}$ (i.e., we differentiate ν_j times with respect to y_j for each j) on both sides of (13) to obtain

$$\int_D \partial^{\boldsymbol{\nu}} \Big(a(\boldsymbol{x}, \boldsymbol{y})\, \nabla u(\boldsymbol{x}, \boldsymbol{y}) \cdot \nabla w(\boldsymbol{x}) \Big)\, \mathrm{d}\boldsymbol{x} \;=\; 0 \quad \forall w \in H_0^1(D)\,. \tag{14}$$

   We can move the derivatives inside the integrals because they operate on different variables $\boldsymbol{y}$ and $\boldsymbol{x}$, respectively. The right-hand side vanishes because it does not depend on $\boldsymbol{y}$.

3. Next we apply the Leibniz product rule on the left-hand side of (14) to obtain

$$\int_D \Big(\sum_{\boldsymbol{m}\le\boldsymbol{\nu}} \binom{\boldsymbol{\nu}}{\boldsymbol{m}} (\partial^{\boldsymbol{m}} a)(\boldsymbol{x},\boldsymbol{y})\, \nabla(\partial^{\boldsymbol{\nu}-\boldsymbol{m}} u)(\boldsymbol{x},\boldsymbol{y}) \cdot \nabla w(\boldsymbol{x})\Big)\, \mathrm{d}\boldsymbol{x} = 0 \quad \forall w \in H_0^1(D)\,, \tag{15}$$

where the sum is over all multi-indices $\boldsymbol{m}$ satisfying $\boldsymbol{m} \le \boldsymbol{\nu}$ (i.e., $m_j \le \nu_j$ for all j), and $\binom{\boldsymbol{\nu}}{\boldsymbol{m}} := \prod_{j\ge1}\binom{\nu_j}{m_j}$. So far we have made no use of any assumption on $a(\boldsymbol{x},\boldsymbol{y})$.

4. For the uniform case, it is easy to see from the formula (2) of $a(\boldsymbol{x},\boldsymbol{y})$ that

$$(\partial^{\boldsymbol{m}} a)(\boldsymbol{x},\boldsymbol{y}) = \begin{cases} a(\boldsymbol{x},\boldsymbol{y}) & \text{if } \boldsymbol{m} = \boldsymbol{0}\,,\\ \psi_j(\boldsymbol{x}) & \text{if } \boldsymbol{m} = \boldsymbol{e}_j\,,\\ 0 & \text{otherwise}\,, \end{cases} \tag{16}$$

where $\boldsymbol{e}_j$ denotes the multi-index whose jth component is 1 and all other components are 0. Essentially, due to the linearity of a with respect to each y_j, if we differentiate once then we obtain ψ_j, and if we differentiate a second time with respect to any variable we get 0.

5. Substituting (16) into (15) and separating out the $\boldsymbol{m} = \boldsymbol{0}$ term, we obtain

$$\begin{aligned} &\int_D a(\boldsymbol{x},\boldsymbol{y})\, \nabla(\partial^{\boldsymbol{\nu}} u)(\boldsymbol{x},\boldsymbol{y}) \cdot \nabla w(\boldsymbol{x})\, \mathrm{d}\boldsymbol{x} \\ &= -\sum_{j\ge1} \nu_j \int_D \psi_j(\boldsymbol{x})\, \nabla(\partial^{\boldsymbol{\nu}-\boldsymbol{e}_j} u)(\boldsymbol{x},\boldsymbol{y}) \cdot \nabla w(\boldsymbol{x})\, \mathrm{d}\boldsymbol{x} \quad \forall w \in H_0^1(D)\,. \end{aligned} \tag{17}$$

6. Note that (17) holds for all test functions in $H_0^1(D)$. We now take the particular choice of $w = (\partial^{\boldsymbol{\nu}} u)(\cdot,\boldsymbol{y})$ (yes, it is allowed to depend on $\boldsymbol{y}$) in (17). Applying $a(\boldsymbol{x},\boldsymbol{y}) \ge a_{\min}$ to the left-hand side, and $|\psi_j(\boldsymbol{x})| \le \|\psi_j\|_{L_\infty}$ and the Cauchy–Schwarz inequality to the right-hand side, we obtain

$$\begin{aligned} &a_{\min} \int_D |\nabla(\partial^{\boldsymbol{\nu}} u)(\boldsymbol{x},\boldsymbol{y})|^2\, \mathrm{d}\boldsymbol{x} \\ &\le \sum_{j\ge1} \nu_j\, \|\psi_j\|_{L_\infty} \Big(\int_D |\nabla(\partial^{\boldsymbol{\nu}-\boldsymbol{e}_j} u)(\boldsymbol{x},\boldsymbol{y})|^2\, \mathrm{d}\boldsymbol{x}\Big)^{1/2} \Big(\int_D |\nabla(\partial^{\boldsymbol{\nu}} u)(\boldsymbol{x},\boldsymbol{y})|^2\, \mathrm{d}\boldsymbol{x}\Big)^{1/2}. \end{aligned} \tag{18}$$

7. Canceling one common factor from both sides of (18) and then dividing through by $a_{\min}$, we obtain the recurrence

$$\|\nabla(\partial^{\boldsymbol{\nu}} u)(\cdot,\boldsymbol{y})\|_{L_2} \le \sum_{j\ge1} \nu_j\, b_j\, \|\nabla(\partial^{\boldsymbol{\nu}-\boldsymbol{e}_j} u)(\cdot,\boldsymbol{y})\|_{L_2}\,, \qquad b_j := \frac{\|\psi_j\|_{L_\infty}}{a_{\min}}\,. \tag{19}$$

8. Finally we prove by induction that

$$\|\nabla(\partial^{\boldsymbol{\nu}} u)(\cdot,\boldsymbol{y})\|_{L_2} \le |\boldsymbol{\nu}|!\,\boldsymbol{b}^{\boldsymbol{\nu}}\,\frac{\|\kappa\|_{H^{-1}}}{a_{\min}}, \tag{20}$$

where $|\boldsymbol{\nu}| := \sum_{j\ge1}\nu_j$ and $\boldsymbol{b}^{\boldsymbol{\nu}} := \prod_{j\ge1} b_j^{\nu_j}$.

   a. BASE STEP. We return to the variational form (13) and take $w = u(\cdot,\boldsymbol{y})$. Applying $a(\boldsymbol{x},\boldsymbol{y}) \ge a_{\min}$ to the left-hand side and estimating the right-hand side using duality pairing $|\langle\kappa, u(\cdot,\boldsymbol{y})\rangle| \le \|\kappa\|_{H^{-1}}\,\|u(\cdot,\boldsymbol{y})\|_{H_0^1}$, we obtain

   $$a_{\min}\,\|\nabla u(\cdot,\boldsymbol{y})\|_{L_2}^2 \le \|\kappa\|_{H^{-1}}\,\|\nabla u(\cdot,\boldsymbol{y})\|_{L_2},$$

   which can be rearranged to yield the case $\boldsymbol{\nu} = \boldsymbol{0}$ in (20).

   b. INDUCTION STEP. As the induction hypothesis, we assume that (20) holds for all multi-indices of order $< |\boldsymbol{\nu}|$. Then we have

   $$\|\nabla(\partial^{\boldsymbol{\nu}-\boldsymbol{e}_j} u)(\cdot,\boldsymbol{y})\|_{L_2} \le |\boldsymbol{\nu}-\boldsymbol{e}_j|!\,\boldsymbol{b}^{\boldsymbol{\nu}-\boldsymbol{e}_j}\,\frac{\|\kappa\|_{H^{-1}}}{a_{\min}}.$$

   Substituting this into (19) and noting that $\nu_j\,|\boldsymbol{\nu}-\boldsymbol{e}_j|! = |\boldsymbol{\nu}|!$ and $b_j\,\boldsymbol{b}^{\boldsymbol{\nu}-\boldsymbol{e}_j} = \boldsymbol{b}^{\boldsymbol{\nu}}$, we obtain (20) and conclude the induction.

Estimate the Norm

9. We want to estimate the norm $\|G(u_h^s)\|_{\boldsymbol{\gamma}}$. We see from the definition of the norm in (11) that we need to obtain estimates on the mixed first derivatives of $G(u_h^s(\cdot,\boldsymbol{y}))$ with respect to $\boldsymbol{y}$. Using linearity and boundedness of G, we have

$$\left|\frac{\partial^{|\mathfrak{u}|}}{\partial \boldsymbol{y}_{\mathfrak{u}}} G(u_h^s(\cdot,\boldsymbol{y}))\right| = \left|G\left(\frac{\partial^{|\mathfrak{u}|}}{\partial \boldsymbol{y}_{\mathfrak{u}}} u_h^s(\cdot,\boldsymbol{y})\right)\right| \le \|G\|_{H^{-1}}\left\|\frac{\partial^{|\mathfrak{u}|}}{\partial \boldsymbol{y}_{\mathfrak{u}}} u_h^s(\cdot,\boldsymbol{y})\right\|_{H_0^1}. \tag{21}$$

10. We can repeat the above proof of (20) for the truncated finite element solution u_h^s instead of the true solution u. Then we restrict the result to mixed first derivatives (i.e., $\nu_j \le 1$ for all j) and deduce that

$$\left\|\frac{\partial^{|\mathfrak{u}|}}{\partial \boldsymbol{y}_{\mathfrak{u}}} u_h^s(\cdot,\boldsymbol{y})\right\|_{H_0^1} \le |\mathfrak{u}|!\left(\prod_{j\in\mathfrak{u}} b_j\right)\frac{\|\kappa\|_{H^{-1}}}{a_{\min}}, \quad \mathfrak{u} \subseteq \{1:s\}. \tag{22}$$

11. Combining (21) with (22) and substituting the upper bound into the definition of the norm (11), we conclude that

$$\|G(u_h^s)\|_{\boldsymbol{\gamma}} \le \frac{\|\kappa\|_{H^{-1}}\|G\|_{H^{-1}}}{a_{\min}}\left(\sum_{\mathfrak{u}\subseteq\{1:s\}} \frac{(|\mathfrak{u}|!)^2\prod_{j\in\mathfrak{u}} b_j^2}{\gamma_{\mathfrak{u}}}\right)^{1/2}. \tag{23}$$

Choose the Weights

12. Now we apply the upper bound on the norm (23) in the error bound for randomly shifted lattice rules (12), to yield (leaving out some constants as indicated by $\lesssim$) for all $\lambda \in (1/2, 1]$,

$$\text{r.m.s. error} \lesssim \left(\frac{2}{n} \sum_{\mathfrak{u}\subseteq\{1:s\}} \gamma_{\mathfrak{u}}^{\lambda} \, [\vartheta(\lambda)]^{|\mathfrak{u}|}\right)^{1/(2\lambda)} \left(\sum_{\mathfrak{u}\subseteq\{1:s\}} \frac{(|\mathfrak{u}|!)^2 \prod_{j\in\mathfrak{u}} b_j^2}{\gamma_{\mathfrak{u}}}\right)^{1/2}. \tag{24}$$

13. With elementary calculus, for any λ, we can minimize the the upper bound in (24) with respect to the weights $\gamma_{\mathfrak{u}}$ to yield the formula

$$\gamma_{\mathfrak{u}} = \left(|\mathfrak{u}|! \prod_{j\in\mathfrak{u}} \frac{b_j}{\sqrt{\vartheta(\lambda)}}\right)^{2/(1+\lambda)}. \tag{25}$$

This form of weights is called *product and order dependent weights*, or *POD weights* in short, because of the presence of some product factors as well as the cardinality of $\mathfrak{u}$.
14. We substitute (25) into (24) and simplify the expression to

$$\text{r.m.s. error} \tag{26}$$
$$\lesssim \left(\frac{2}{n}\right)^{1/(2\lambda)} \left[\sum_{\mathfrak{u}\subseteq\{1:s\}} \left(|\mathfrak{u}|! \prod_{j\in\mathfrak{u}} \left(b_j \, [\vartheta(\lambda)]^{1/(2\lambda)}\right)\right)^{2\lambda/(1+\lambda)}\right]^{(1+\lambda)/(2\lambda)}.$$

15. We now derive a condition on λ for which the sum in (26) is bounded independently of s. In an abstract form, we have

$$\sum_{\mathfrak{u}\subseteq\{1:s\}} \left(|\mathfrak{u}|! \prod_{j\in\mathfrak{u}} \alpha_j\right)^k = \sum_{\ell=0}^{s} (\ell!)^k \sum_{\mathfrak{u}\subseteq\{1:s\},\, |\mathfrak{u}|=\ell} \prod_{j\in\mathfrak{u}} \alpha_j^k \le \sum_{\ell=0}^{s} (\ell!)^{k-1} \left(\sum_{j=1}^{s} \alpha_j^k\right)^{\ell},$$

where the inequality holds because each term $\prod_{j\in\mathfrak{u}} \alpha_j^k$ from the left-hand side of the inequality appears in the expansion $(\sum_{j=1}^{s} \alpha_j^k)^\ell$ exactly $\ell!$ times and yet the expansion contains other terms. The right-hand side is bounded independently of s if $\sum_{j=1}^{\infty} \alpha_j^k < \infty$ and $k < 1$, which can be verified by the ratio test. In our case, we have $k = 2\lambda/(1+\lambda)$ and $\sum_{j=1}^{\infty} \alpha_j^k = [\vartheta(\lambda)]^{1/(1+\lambda)} \sum_{j=1}^{\infty} b_j^k < \infty$ if $k \ge p_0$, where we recall that b_j is defined in (19) and p_0 is defined in (3). Hence we require

$$p_0 \le \frac{2\lambda}{1+\lambda} < 1 \iff \frac{p_0}{2-p_0} \le \lambda < 1. \tag{27}$$

16. Clearly the best rate of convergence is obtained by taking λ as small as possible. Combining the original constraint of $\lambda \in (1/2, 1]$ with (27), we now take

$$\lambda = \begin{cases} \dfrac{1}{2-2\delta} \text{ for } \delta \in \left(0, \frac{1}{2}\right) & \text{when } p_0 \in \left(0, \frac{2}{3}\right], \\ \dfrac{p_0}{2-p_0} & \text{when } p_0 \in \left(\frac{2}{3}, 1\right). \end{cases} \tag{28}$$

Fast CBC Construction

17. The chosen weights (25) with λ given by (28) are then fed into the CBC construction to produce tailored randomly shifted lattice rules that achieve a root-mean-square error of order

$$n^{-\min(1/p_0 - 1/2, 1-\delta)}, \quad \delta \in \left(0, \tfrac{1}{2}\right),$$

with the implied constant independent of s, where p_0 is given by (3). The fast CBC construction with POD weights can then find a good generating vector in $\mathcal{O}(s\, n \log n + s^2\, n)$ operations.

4 Key Analysis

Having completed our embedded tutorial in the previous section, we now continue to provide our overview of the analysis required in applying QMC to PDEs with random coefficients.

Some Hints at the Technical Difficulties for the Multi-level Analysis

We have seen in the uniform case with the single-level algorithm that the key is to estimate $\|G(u_h^s)\|_{\gamma}$, and this is achieved by estimating (see (20) and [31, Lemma 6.1])

$$\|\nabla \partial^{\boldsymbol{\nu}} u(\cdot, \boldsymbol{y})\|_{L_2}.$$

For the multi-level algorithm, the key estimate is $\|G(u_{h_\ell}^{s_\ell} - u_{h_{\ell-1}}^{s_{\ell-1}})\|_{\gamma}$, and we need to estimate in turn (see [31, Lemmas 6.2–6.4])

$$\|\Delta \partial^{\boldsymbol{\nu}} u(\cdot, \boldsymbol{y})\|_{L_2}, \quad \|\nabla \partial^{\boldsymbol{\nu}} (u - u_h)(\cdot, \boldsymbol{y})\|_{L_2}, \quad \text{and} \quad |\partial^{\boldsymbol{\nu}} G((u - u_h)(\cdot, \boldsymbol{y}))|.$$

All three bounds involve factors of the form $(|\boldsymbol{\nu}|+a_1)!\,\overline{\boldsymbol{b}}^{\boldsymbol{\nu}}$ for $a_1 \geq 0$ and a sequence $\overline{b}_j$ similar to the previously defined b_j. Assuming that both the forcing term κ and the linear functional G are in $L_2(D)$, we obtain that the second bound is of order h and the third bound is of order h^2. The difficulty is that we need to establish these regularity estimates simultaneously in $\boldsymbol{x}$ and $\boldsymbol{y}$. We also use duality tricks to gain on the convergence rate due to the linear functional G.

Some Hints at the Technical Difficulties for the Lognormal Case

For the lognormal case the argument is quite technical due to the more complicated form of $a(\boldsymbol{x}, \boldsymbol{y})$. In the single-level algorithm we need to estimate (see [31, Lemma 6.5])

$$\|\nabla \partial^{\boldsymbol{\nu}} u(\cdot, \boldsymbol{y})\|_{L_2} \quad \text{by first estimating} \quad \|a^{1/2}(\cdot, \boldsymbol{y})\, \nabla \partial^{\boldsymbol{\nu}} u(\cdot, \boldsymbol{y})\|_{L_2}\,.$$

In the multi-level algorithm we need to estimate (see [31, Lemma 6.6])

$$\|\Delta \partial^{\boldsymbol{\nu}} u(\cdot, \boldsymbol{y})\|_{L_2} \quad \text{by first estimating} \quad \|a^{-1/2}(\cdot, \boldsymbol{y}) \nabla \cdot (a(\cdot, \boldsymbol{y})\, \nabla \partial^{\boldsymbol{\nu}} u(\cdot, \boldsymbol{y}))\|_{L_2}\,,$$

and then estimate in turn (see [31, Lemmas 6.7–6.8])

$$\|a^{1/2}(\cdot, \boldsymbol{y}) \nabla \partial^{\boldsymbol{\nu}} (u - u_h)(\cdot, \boldsymbol{y})\|_{L_2} \quad \text{and} \quad |\partial^{\boldsymbol{\nu}} G((u - u_h)(\cdot, \boldsymbol{y}))|\,.$$

All bounds involve factors of the form $J(\boldsymbol{y})\,(|\boldsymbol{\nu}| + a_1)!\,\boldsymbol{\beta}^{\boldsymbol{\nu}}$ for $a_1 \geq 0$ and some sequence β_j, where $J(\boldsymbol{y})$ indicates some factor depending on $\boldsymbol{y}$ which is not present in the uniform case. The proofs are by induction, and the tricky part is knowing what multiplying factor of $a(\cdot, \boldsymbol{y})$ should be included in the recursion. The growth of $J(\boldsymbol{y})$ needs to be taken into account when estimating the norm.

Summary of Results

Now we summarize and compare the results from [6, 8, 32, 33] for the uniform case:

First-order single-level [33]

$$s^{-2(1/p_0 - 1)} + h^{t+t'} + n^{-\min(1/p_0 - 1/2, 1 - \delta)} \quad \text{(r.m.s.)}$$

First-order multi-level [34]

$$s_L^{-2(1/p_0 - 1)} + h_L^{t+t'} + \sum_{\ell=0}^{L} n_\ell^{-\min(1/p_1 - 1/2, 1 - \delta)} \big(\theta_{\ell-1}\, s_{\ell-1}^{-(1/p_0 - 1/p_1)} + h_{\ell-1}^{t+t'}\big) \quad \text{(r.m.s.)}$$

Higher-order single-level [4]

$$s^{-2(1/p_0 - 1)} + h^{t+t'} + n^{-1/p_0}$$

Higher-order multi-level [6]

$$s_L^{-2(1/p_0 - 1)} + h_L^{t+t'} + \sum_{\ell=0}^{L} n_\ell^{-1/p_t} \big(\theta_{\ell-1}\, s_{\ell-1}^{-(1/p_0 - 1/p_t)} + h_{\ell-1}^{t+t'}\big)$$

For the first-order results, the "r.m.s." in brackets indicates that the error is in the root-mean-square sense since we use a randomized QMC method. The higher-order results are deterministic. Without giving the full details, we simply say that the results include general parameters t and t' for the regularity of κ and G, respectively. Recall that p_0 corresponds to the summability of $\|\psi_j\|_{L_\infty}$, see (3). Here p_1 corresponds essentially to the summability of $\|\nabla \psi_j\|_{L_\infty}$, while p_t corresponds analogously to higher derivatives of ψ_j. For the multi-level results we include the analysis for potentially taking different s_ℓ at each level: $\theta_{\ell-1}$ is 0 if $s_\ell = s_{\ell-1}$ and is 1 otherwise.

In the single-level algorithms, the error is the sum of three terms. In the multi-level algorithms, we see the multiplicative effect between the finite element error and the QMC error. However, comparing p_1 and p_t with p_0, we see that multi-level algorithms need stronger regularity in $\boldsymbol{x}$ than single-level algorithms.

Going from first-order to higher-order results, we see that the cap of $n^{-(1-\delta)}$ is removed. We also see a gain of an extra factor of $n^{-1/2}$; this benefit appears to arise from the switch of function space setting to a non-Hilbert space.

The error versus cost analysis depends crucially on the cost assumptions. For the single-level algorithms, we simply choose n, s and h to balance three errors. In the multi-level algorithms, we choose n_ℓ, s_ℓ, h_ℓ to minimize the total cost for a fixed total error using Lagrange multiplier arguments.

For the lognormal case we have similar first order results, see [22, 34]. There is no higher order results for the lognormal case because presently there is no QMC theory in this setting.

5 Software

The software package QMC4PDE accompanies the survey [31], see https://people.cs.kuleuven.be/~dirk.nuyens/qmc4pde/. Here we very briefly outline its usage.

Construction of the Generating Vector in Python

In the analysis for the PDE problems we obtain generic bounds on mixed derivatives of the form

$$|\partial^{\boldsymbol{\nu}} F(\boldsymbol{y})| \lesssim \big((|\boldsymbol{\nu}| + a_1)!\big)^{d_1} \prod_{j=1}^{s} (a_2 \mathscr{B}_j)^{\nu_j} \exp(a_3 \mathscr{B}_j |y_j|),$$

for some constants a_1, a_2, a_3, d_1 and some sequence $\mathscr{B}_j$, where

$$F(\boldsymbol{y}) = \begin{cases} G(u_h^s) & \text{for single-level algorithms,} \\ G(u_{h_\ell}^s - u_{h_{\ell-1}}^s) & \text{for multi-level algorithms,} \end{cases}$$

and in particular

$$\begin{cases} a_3 = 0 & \text{for the uniform case,} \\ a_3 > 0 & \text{for the lognormal case.} \end{cases}$$

The Python construction script takes the number of points (as a power of 2), the dimension, and all these parameters as input from the user, works out the appropriate weights $\gamma_{\mathfrak{u}}$, and then constructs a good generating vector for the QMC rule. This is either a lattice sequence (constructed following a minimax strategy as described in [2]) or an interlaced polynomial lattice rule. In the latter case the script also assembles the interlaced generating matrices, because this is the most convenient way to generate the points.

- To construct a generating vector for a lattice sequence (output written to file `z.txt`)

```
## uniform case, 100-dim, 2^10 points, with specified bounds b:
./lat-cbc.py --s=100 --m=10 --d2=3 --b="0.1 * j**-3 / log(j+1)"
```

```
## lognormal case, 100-dim, 2^10 points, with algebraic decay:
./lat-cbc.py --s=100 --m=10 --a2="1/log(2)" --a3=1 --d2=3 --c=0.1
```

- To construct generating matrices for an interlaced polynomial lattice rule (output written to file `Bs53.col`)

```
## 100-dim, 2^10 points, interlacing 3, with bounds from file:
./polylat-cbc.py --s=100 --m=10 --alpha=3 --a1=5 --b_file=in.txt
```

Point Generators in Matlab/Octave (also available in C++ and Python)

Here are some Matlab/Octave usage examples for generating the actual QMC point sets from the output files of the Python construction script.

- To generate a lattice sequence (specified by the file `z.txt`)

```
load z.txt                        % load generating vector
latticeseq_b2('init0', z)         % initialize the generator
Pa = latticeseq_b2(20, 512);      % first 512 20-dim points
Pb = latticeseq_b2(20, 512);      % next 512 20-dim points
```

- To generate an interlaced polynomial lattice rule (specified by the file `Bs53.col`)

```
load Bs53.col                      % load generating matrices
digitalseq_b2g('init0', Bs53)      % initialize the generator
Pa = digitalseq_b2g(100, 512);     % first 512 100-dim points
Pb = digitalseq_b2g(100, 512);     % next 512 100-dim points
```

The same function `digitalseq_b2g` can also be used to generate interlaced Sobol′ points by specifying the corresponding interlaced generating matrices. The parameters for generating Sobol′ points are taken from [28].

- To generate an interlaced Sobol′ sequence (interlaced matrices specified by the file `sobol_alpha3_Bs53.col`)

```
load sobol_alpha3_Bs53.col              % load generating matrices
digitalseq_b2g('init0', sobol_alpha3_Bs53)    % initialize
Pa = digitalseq_b2g(50, 512);}                % first 512 50-dim
Pb = digitalseq_b2g(50, 512);}                % next 512 50-dim
```

The last example produces interlaced Sobol′ points with interlacing factor $\alpha = 3$. They can provide third order convergence if the integrand has sufficient smoothness.

6 Concluding Remarks

QMC (deterministic or randomized) convergence rate and implied constant can be independent of the dimension. This is achieved by working in a weighted function space setting. To apply QMC theory, we need an estimate of the norm of the integrand, and in turn this can help us to choose appropriate weights for the function space. The chosen weights then enter the fast CBC construction of the generating vector for the QMC points. The pairing between the function space setting and the QMC method is very important, in the sense that we want to achieve the best possible convergence rate under the weakest assumption on the problem. In practice, it may be that an off-the-shelf QMC rule works just as well, barring no theory.

In this article we considered multi-level algorithms. There are other cost saving strategies for the lognormal case and for other general situations, see e.g., [7, 25] as well as [18, 30] in this volume. Moreover, there have been many others developments on the application of QMC to PDEs with random coefficients, for some examples see the last part of the introduction.

Acknowledgements The authors acknowledge the financial supports from the Australian Research Council (FT130100655 and DP150101770) and the KU Leuven research fund (OT:3E130287 and C3:3E150478).

References

1. Cohen, A., DeVore, R.: Approximation of high-dimensional parametric PDEs. Acta Numer. **24**, 1–159 (2015)
2. Cools, R., Kuo, F.Y., Nuyens, D.: Constructing embedded lattice rules for multivariate integration. SIAM J. Sci. Comput. **28**, 2162–2188 (2006)
3. Dick, J., Pillichshammer, F.: Digital Nets and Sequences. Cambridge University Press, Cambridge (2010)
4. Dick, J., Gantner, R.N., Le Gia, Q.T., Schwab, Ch.: Higher order Quasi-Monte Carlo integration for Bayesian estimation (in review)

5. Dick, J., Kuo, F.Y., Sloan, I.H.: High-dimensional integration: the Quasi-Monte Carlo way. Acta Numer. **22**, 133–288 (2013)
6. Dick, J., Kuo, F.Y., Le Gia, Q.T., Nuyens, D., Schwab, Ch.: Higher order QMC Galerkin discretization for parametric operator equations. SIAM J. Numer. Anal. **52**, 2676–2702 (2014)
7. Dick, J., Kuo, F.Y., Le Gia, Q.T., Schwab, Ch.: Fast QMC matrix-vector multiplication. SIAM J. Sci. Comput. **37**, A1436–A1450 (2015)
8. Dick, J., Kuo, F.Y., Le Gia, Q.T., Schwab, Ch.: Multi-level higher order QMC Galerkin discretization for affine parametric operator equations. SIAM J. Numer. Anal. **54**, 2541–2568 (2016)
9. Dick, J., Le Gia, Q.T., Schwab, Ch.: Higher order Quasi-Monte Carlo integration for holomorphic, parametric operator equations. SIAM/ASA J. Uncertain. Quantif. **4**, 48–79 (2016)
10. Dietrich, C.R., Newsam, G.H.: Fast and exact simulation of stationary Gaussian processes through circulant embedding of the covariance matrix. SIAM J. Sci. Comput. **18**, 1088–1107 (1997)
11. Feischl, M., Kuo, F.Y., Sloan, I.H.: Fast random field generation with H-matrices. Numer. Math. (to appear)
12. Ganesh, M., Hawkins, S.C.: A high performance computing and sensitivity analysis algorithm for stochastic many-particle wave scattering. SIAM J. Sci. Comput. **37**, A1475–A1503 (2015)
13. Ganesh, M., Kuo, F.Y., Sloan, I.H.: Quasi-Monte Carlo finite element wave propagation in heterogeneous random media (in preparation)
14. Gantner, R.N., Schwab, Ch.: Computational higher order quasi-Monte Carlo integration. In: Nuyens, D., Cools, R. (eds.) Monte Carlo and Quasi-Monte Carlo Methods 2014, pp. 271–288. Springer, Berlin (2016)
15. Gantner, R.N., Herrmann, L., Schwab, Ch.: Quasi-Monte Carlo integration for affine-parametric, elliptic PDEs: local supports and product weights SIAM. J. Numer. Anal. **56**(1), 111–135 (2018)
16. Gilbert, A.D., Graham, I.G., Kuo, F.Y., Scheichl, R., Sloan, I.H.: Analysis of quasi-Monte Carlo methods for elliptic eigenvalue problems with stochastic coefficients (in preparation)
17. Giles, M.B.: Multilevel Monte Carlo methods. Acta Numer. **24**, 259–328 (2015)
18. Giles, M.B., Kuo, F.Y., Sloan, I.H.: Combining sparse grids, multilevel MC and QMC for elliptic PDEs with random coefficients (in this volume)
19. Graham, I.G., Kuo, F.Y., Nuyens, D., Scheichl, R., Sloan, I.H.: Analysis of circulant embedding methods for sampling stationary random fields. SIAM J. Numer. Anal. (to appear)
20. Graham, I.G., Kuo, F.Y., Nuyens, D., Scheichl, R., Sloan, I.H.: Circulant embedding with QMC – analysis for elliptic PDE with lognormal coefficients. Numer. Math. (to appear)
21. Graham, I.G., Kuo, F.Y., Nuyens, D., Scheichl, R., Sloan, I.H.: Quasi-Monte Carlo methods for elliptic PDEs with random coefficients and applications. J. Comput. Phys. **230**, 3668–3694 (2011)
22. Graham, I.G., Kuo, F.Y., Nichols, J.A., Scheichl, R., Schwab, Ch., Sloan, I.H.: QMC FE methods for PDEs with log-normal random coefficients. Numer. Math. **131**, 329–368 (2015)
23. Gunzburger, M., Webster, C., Zhang, G.: Stochastic finite element methods for partial differential equations with random input data. Acta Numer. **23**, 521–650 (2014)
24. Hackbusch, W.: Hierarchical Matrices: Algorithms and Analysis. Springer, Heidelberg (2015)
25. Haji-Ali, A.L., Nobile, F., Tempone, R.: Multi-index Monte Carlo: when sparsity meets sampling. Numer. Math. **132**, 767–806 (2016)
26. Harbrecht, H., Peters, M., Siebenmorgen, M.: On the quasi-Monte Carlo method with Halton points for elliptic PDEs with log-normal diffusion. Math. Comput. **86**, 771–797 (2017)
27. Herrmann, L., Schwab, Ch.: QMC integration for lognormal-parametric, elliptic PDEs: local supports imply product weights (in review)
28. Joe, S., Kuo, F.Y.: Constructing Sobol' sequences with better two-dimensional projections. SIAM J. Sci. Comput. **30**, 2635–2654 (2008)
29. Kazashi, Y.: Quasi-Monte Carlo integration with product weights for elliptic PDEs with log-normal coefficients. IMA J. Numer. Anal. (to appear)

30. Kuo, F.Y., Nuyens, D.: Hot new directions for quasi-Monte Carlo research in step with applications (in this volume)
31. Kuo, F.Y., Nuyens, D.: Application of quasi-Monte Carlo methods to elliptic PDEs with random diffusion coefficients - a survey of analysis and implementation. Found. Comput. Math. **16**, 1631–1696 (2016)
32. Kuo, F.Y., Schwab, Ch., Sloan, I.H.: Quasi-Monte Carlo finite element methods for a class of elliptic partial differential equations with random coefficient. SIAM J. Numer. Anal. **50**, 3351–3374 (2012)
33. Kuo, F.Y., Schwab, Ch., Sloan, I.H.: Multi-level quasi-Monte Carlo finite element methods for a class of elliptic partial differential equations with random coefficient. Found. Comput. Math. **15**, 411–449 (2015)
34. Kuo, F.Y., Scheichl, R., Schwab, Ch., Sloan, I.H., Ullmann, E.: Multilevel Quasi-Monte Carlo methods for lognormal diffusion problems. Math. Comput. **86**, 2827–2860 (2017)
35. Le Gia, Q.T.: A QMC-spectral method for elliptic PDEs with random coefficients on the unit sphere. In: Dick, J., Kuo, F.Y., Peters, G.W., Sloan, I.H. (eds.) Monte Carlo and Quasi-Monte Carlo Methods 2012, pp. 491–508. Springer, Heidelberg (2013)
36. Lemieux, C.: Monte Carlo and Quasi-Monte Carlo Sampling. Springer, New York (2009)
37. Leobacher, G., Pillichshammer, F.: Introduction to Quasi-Monte Carlo Integration and Applications. Springer, Berlin (2014)
38. Niederreiter, H.: Random Number Generation and Quasi-Monte Carlo Methods. SIAM, Philadelphia (1992)
39. Nuyens, D.: The construction of good lattice rules and polynomial lattice rules. In: Kritzer, P., Niederreiter, H., Pillichshammer, F., Winterhof, A. (eds.) Uniform Distribution and Quasi-Monte Carlo Methods. Radon Series on Computational and Applied Mathematics, vol. 15, pp. 223–256. De Gruyter, Berlin (2014)
40. Robbe, P., Nuyens, D., Vandewalle, S.: A multi-index quasi-Monte Carlo algorithm for lognormal diffusion problems. SIAM J. Sci. Comput. Â **39**, S851–S872 (2017)
41. Scheichl, R., Stuart, A., Teckentrup, A.L.: Quasi-Monte Carlo and multilevel Monte Carlo methods for computing posterior expectations in elliptic inverse problems. SIAM/ASA J. Uncertain. Quantif. **5**, 493–518 (2017)
42. Schwab, Ch.: QMC Galerkin discretizations of parametric operator equations. In: Dick, J., Kuo, F.Y., Peters, G.W., Sloan, I.H. (eds.) Monte Carlo and Quasi-Monte Carlo Methods 2012, pp. 613–629. Springer, Heidelberg (2013)
43. Schwab, Ch., Gittelson, C.J.: Sparse tensor discretizations of high-dimensional parametric and stoch astic PDEs. Acta Numer. **20**, 291–467 (2011)
44. Sloan, I.H., Joe, S.: Lattice Methods for Multiple Integration. Oxford University Press, Oxford (1994)
45. Sloan, I.H., Wang, X., Woźniakowski, H.: Finite-order weights imply tractability of multivariate integration. J. Complex. **20**, 46–74 (2004)

Part II
Invited Talks

Malliavin-Based Multilevel Monte Carlo Estimators for Densities of Max-Stable Processes

Jose Blanchet and Zhipeng Liu

Abstract We introduce a class of unbiased Monte Carlo estimators for multivariate densities of max-stable fields generated by Gaussian processes. Our estimators take advantage of recent results on the exact simulation of max-stable fields combined with identities studied in the Malliavin calculus literature and ideas developed in the multilevel Monte Carlo literature. Our approach allows estimating multivariate densities of max-stable fields with precision ε at a computational cost of order $O\left(\varepsilon^{-2}\log\log\log(1/\varepsilon)\right)$.

Keywords Max-stable process · Density estimation · Malliavin calculus

1 Introduction

Max-stable random fields arise as the asymptotic limit of suitably normalized maxima of many i.i.d. random fields. Intuitively, max-stable fields are utilized to study the extreme behavior of spatial statistics. For instance, if the logarithm of a precipitation field during a relatively short time span follows a Gaussian random field, then extreme precipitations over a long time horizon (which are obtained by taking the maximum at each location of many precipitation fields) can be argued as long as enough temporal

J. Blanchet—Support from NSF grant DMS-132055, NSF grant CMMI-1538217 and NSF grant DMS-1720451 is gratefully acknowledged by J. Blanchet. We also are grateful to the referees for their comments which helped us improve our manuscript.

J. Blanchet (✉) · Z. Liu
Department of Industrial Engineering and Operations Research,
Columbia University, 500 West 120th Street, New York, NY, USA
e-mail: jose.blanchet@columbia.edu

Z. Liu
e-mail: zl2337@columbia.edu

A. B. Owen and P. W. Glynn (eds.), *Monte Carlo and Quasi-Monte Carlo Methods*, Springer Proceedings in Mathematics & Statistics 241,
https://doi.org/10.1007/978-3-319-91436-7_4

independence can be assumed to follow a suitable max-stable process. It is these types of applications in environmental science that motivate the study of max-stable processes (see, for example, [4] for a recent study of this type).

In order to estimate the parameters of a max-stable random field, for instance using maximum-likelihood estimation, it is desirable to evaluate the density over a finite set of locations (i.e. multivariate density of finite-dimensional coordinates of the max-stable field). The recent work in [3] discusses the challenges involved in applying maximum likelihood estimation of max-stable fields. We believe that the algorithms and techniques that we develop in this paper can be used to study maximum likelihood estimators for max-stable fields. For example, our representations can be used to obtain convenient expressions for the derivative of the density. In turn, as explained in [3], differentiability of the density is useful in the asymptotic analysis of maximum likelihood estimators. In addition, our algorithms can be implemented using common random numbers under a wide range of parameters of the max-stable fields. Thus, our algorithms can be used to perform approximate maximum-likelihood estimation by optimizing over a wide range of parameters.

In order to precisely explain our estimators, we now introduce some basic facts about max-stable processes.

We will focus on a class of max-stable random fields which are driven by Gaussian processes. These max-stable fields are popular in practice because their spatial dependence structure is inherited from the underlying Gaussian covariance structure.

To introduce the max-stable field of interest, let us first fix its domain $T \subseteq \mathbb{R}^m$, for $m \geq 1$. We introduce a sequence, $(X_n(\cdot))$, of independent and identically distributed copies of a centered Gaussian random field, $X(\cdot) = (X(t) : t \in T)$. We let (A_n) be the sequence of arrivals in a Poisson process, with unit rate and independent of $(X_n(\cdot))$.

Finally, given a deterministic and bounded function, $\mu : T \longrightarrow \mathbb{R}$, we will focus on developing Monte Carlo methods for the finite dimensional densities of the max-stable field

$$M(t) = \sup_{n\geq 1} \big\{ -\log A_n + X_n(t) + \mu(t) \big\}, \qquad t \in T. \tag{1}$$

(The name max-stable is justified because $M(\cdot)$ turns out to satisfy a distributional equation involving the maximum of i.i.d. centered and normalized copies of $M(\cdot)$, see [9, Theorem 2].)

An elegant argument involving Poisson point processes (see [15] and [2, Lemma 5.1]) allows us to conclude that

$$\begin{aligned} &P(M(t_1) \leq x_1, \ldots, M(t_d) \leq x_d) \qquad (2)\\ &= \exp\left(-E\left[\max_{i=1}^{d} \{\exp(X(t_i) + \mu(t_i) - x_i)\} \right] \right). \end{aligned}$$

By redefining x_i as $x_i - \mu(t_i)$, we might assume without loss of generality, for the purpose of computing the density of $M = (M(t_1), \ldots, M(t_d))^T$, that $\mu(t_i) = 0$. We will keep imposing this assumption throughout the rest of the paper.

Throughout the paper we will keep the number of locations, d, over which $M(\cdot)$ is observed, fixed. So, M will remain a d-dimensional vector throughout our discussion. To avoid confusion between M and $M(\cdot)$, note that we use $M(\cdot)$ when discussing the whole max-stable field. We will maintain this convention throughout the rest of the paper for the field $M(\cdot)$ as well as the fields $X_n(\cdot), n \geq 1$.

The joint density of M can be obtained by subsequent differentiation of (2) with respect to $x_1, \ldots, x_d$. However, the final expression obtained for the density contains exponentially many terms. So, computing the density of M using this direct approach becomes quickly intractable, even for moderate values of d. For example, [15] argues that even for $d = 10$, one obtains a sum of more than 10^5 terms.

We will construct an unbiased estimator for the density, $f(x)$, of M evaluated at $x = (x_1, \ldots, x_d)$ for $d \geq 3$. The construction of our estimator, denoted as $V(x)$, is explained in Sect. 2.5. Implementing our estimator avoids the exponential growth issues which arise if one attempts to directly evaluate the density. We concentrate on $d \geq 3$ because the case $d = 2$ leads to only four terms which can be easily computed as explained in [8]. More precisely, our contributions are as follows:

1. The properties of $V(x)$ are summarized in Sect. 3. In particular as shown in (16), $f(x) = E(V(x))$, $Var(V(x)) < \infty$, and given a computational budget of size b, we provide a limit theorem which can be used to estimate $f(x)$ with complexity $O\left((b \cdot \log\log\log(b))^2\right)$ for an error of order $O(1/b)$ – see Theorem 1 and its discussion.
2. As far as we know, this is the first estimator which uses Malliavin calculus in the context of max-stable density estimation. We believe that the techniques that we introduce are of independent interest in other areas in which Malliavin calculus has been used to construct Monte Carlo estimators. For example, we highlight the following techniques in this regard,
   a. We introduce a technique which can be used to estimate the density of the (coordinate-wise) maximum of multivariate variables. We apply this technique to the case of independent Gaussian vectors, but the technique can be used more generally, see the development in Sect. 2.2.
   b. We explain how to extend the technique in item 2(a) to the case of the maxima of infinitely many variables. This extension, which is explained in Sect. 2.3, highlights the role of a recently introduced record-breaking technique for the exact sampling of variables such as M.
   c. We introduce a perturbation technique which controls the variance of so-called Malliavin–Thalmaier estimators (which are explained in Sect. 2.1). These types of estimators have been used to compute densities of multivariate diffusions (see [10]). Our perturbation technique, introduced in Sect. 2.4, can be directly used to improve upon the density estimators in [10], enabling a close-to-optimal Monte Carlo rate of convergence for density estimation of multivariate diffusions.
3. The perturbation technique in Sect. 2.4 is combined with randomized multilevel Monte Carlo techniques (see [13, 14]) in order to achieve the following: Starting

from an infinite variance estimator, we introduce a perturbation which makes the estimator biased, but with finite variance. The randomized multilevel Monte Carlo technique is then used to remove the bias while keeping the variance finite. The price to pay is a small degradation in the rate of convergence in the associated Central Limit Theorem for confidence interval estimation. Instead of an error rate of order $O(1/b^{1/2})$ as a function of the computational budget b, which is the typical rate, we obtain a rate of order $O\left((\log\log\log(b))^{1/2}/b^{1/2}\right)$. The Central Limit Theorem is obtained using recently developed results in [18].

The rest of the paper is organized as follows: In Sect. 2 we explain step-by-step, at a high level, the construction of our estimator. The final form of our estimator is given in Sect. 2.5. The properties of our estimator are summarized in Sect. 3. A numerical experiment is given in Sect. 4. Finally, the details of the implementation of our estimator, in the form of pseudo-codes, are given in the Sect. 5 Appendix.

2 General Strategy and Background

The general strategy is explained in several steps. We first review the Malliavin–Thalmaier identity by providing a brief explanation of its origins and connections to classical potential theory. We finish the first step by noting that there are several disadvantages of the identity having to do with variance properties of the estimator and the implicit assumption that a great degree of information is assumed about the density which we want to estimate. The subsequent steps in our construction are designed to address these disadvantages.

In the second step of our construction, we introduce a series of manipulations which enable the application of the Malliavin–Thalmaier indirectly, by working only with the X_n's. These manipulations are performed assuming that only finitely many Gaussian elements are considered in the description of M.

The third step deals with the fact that the description of M contains infinitely many Gaussian elements. So, first, we need to explain how to sample M exactly. We utilize a recently developed algorithm (see [11]). Based on this algorithm, we explain how to extend the construction from the second step in order to obtain a direct Malliavin–Thalmaier estimator for the density of M.

The fourth step of our construction deals with the fact that a direct Malliavin–Thalmaier estimator will generally have infinite variance. We introduce a small random perturbation to remove the singularity appearing in the Malliavin–Thalmaier estimator, which is the source of the poor variance performance. Unfortunately, such a perturbation also introduces bias in the estimator.

In order to remove the bias, we then apply randomized multilevel Monte Carlo methods (see [13, 14]). Our resulting estimator then is unbiased and has finite variance as we explain in Sect. 3. The price to pay is a small degradation in the rate of convergence of the associated Central Limit Theorem to obtain confidence intervals.

2.1 Step 1: The Malliavin–Thalmaier Identity

The initial idea behind the construction of our estimator comes from the Malliavin–Thalmaier identity, which we shall briefly explain. First, recall the Newtonian potential, given by

$$G(x) = \kappa_d \frac{1}{\|x\|_2^{d-2}},$$

with $\kappa_d = (d(2-d)\omega_d)^{-1}$, where ω_d is the volume of a unit ball in d dimensions, for $d \geq 3$. It is well known, see [5], that $G(\cdot)$ satisfies the equation

$$\Delta G(x-y) = \delta(x-y)$$

in the sense of distributions (where $\delta(x)$ is the delta function). Therefore, if $M \in R^d$ has density $f(\cdot)$ we can write

$$f(x) = \int f(y)\, \Delta G(x-y)\, dy = E(\Delta G(x-M)). \tag{3}$$

But the previous identity cannot be implemented directly because $G(\cdot)$ is harmonic, that is, one can easily verify that $\Delta G(x) = 0$ for $x \neq 0$ (which is not surprising given that one expects ΔG to act as a delta function). The key insight of Malliavin and Thalmaier is to apply integration by parts in the expression (3). So, let us define

$$G_i(x) = \frac{\partial G(x)}{\partial x_i} = (2-d)\, \kappa_d \frac{x_i}{\|x\|_2^d},$$

and therefore write

$$\Delta G(x-y) = \sum_{i=1}^{d} \frac{\partial^2 G(x-y)}{\partial x_i^2} = \sum_{i=1}^{d} \frac{\partial G_i(x-y)}{\partial x_i}.$$

Consequently, because

$$\frac{\partial G_i(x-y)}{\partial x_i} = -\frac{\partial G_i(x-y)}{\partial y_i},$$

we have that

$$E\left(\frac{\partial G_i(x-M)}{\partial x_i}\right) = \int \cdots \int \frac{\partial G_i(x-y)}{\partial x_i} f(y_1, \ldots, y_d)\, dy_1 dy_2 \ldots dy_d$$
$$= -\int \cdots \int \frac{\partial G_i(x-y)}{\partial y_i} f(y_1, \ldots, y_d)\, dy_1 dy_2 \ldots dy_d$$

$$= \int \cdots \int G_i (x - y) \frac{\partial f (y_1, \ldots, y_d)}{\partial y_i} dy_1 dy_2 \ldots dy_d$$
$$= E \left(G_i (x - M) \frac{\partial}{\partial y_i} \log f (M) \right).$$

Therefore, using (3), we arrive at the following Malliavin–Thalmaier identity,

$$f (x) = E \left(\sum_{i=1}^{d} \frac{\partial G_i (x - M)}{\partial x_i} \right) = \sum_{i=1}^{d} E \left(G_i (x - M) \frac{\partial}{\partial y_i} \log f (M) \right). \quad (4)$$

Refer to [10, 12] for rigorous proof of this identity.

There are two immediate concerns when applying the Malliavin–Thalmaier identity. First, a direct use of the identity requires some basic knowledge of the density of interest, which is precisely the quantity that we wish to estimate. The second issue, which is not evident from (4), is that the singularity which arises when $x = M$ in the definition of $G_i (x - M)$, causes the estimator (4) to typically have infinite variance.

2.2 Step 2: Applying the Malliavin–Thalmaier Identity to Finite Maxima

We now shall explain how to address the first issue discussed at the end of the previous subsection. Define

$$M_n (t) = \max_{1 \le k \le n} \{- \log (A_k) + X_k (t)\},$$

where X_k's are i.i.d. centered Gaussian vectors with covariance matrix Σ, and put $M_n = (M_n (t_1), \ldots, M_n (t_d))^T$. Note that

$$\frac{\partial G_i (x - M_n)}{\partial x_i} = - \frac{\partial G_i (x - M_n)}{\partial M_n (t_i)}. \quad (5)$$

In turn, by the chain rule,

$$\frac{\partial G_i (x - M_n)}{\partial X_k (t_i)} = \frac{\partial G_i (x - M_n)}{\partial M_n (t_i)} \frac{\partial M_n (t_i)}{\partial X_k (t_i)}. \quad (6)$$

Further, with probability one (due to the fact that $(A_1, A_2, \ldots, A_k)$ has a density),

$$\sum_{k=1}^{n} \frac{\partial M_n (t_i)}{\partial X_k (t_i)} = \sum_{k=1}^{n} I (M_n (t_i) = X_k (t_i) - \log (A_k)) = 1.$$

Consequently, from Eq. (6) we conclude that

$$\sum_{k=1}^{n} \frac{\partial G_i\left(x - M_n\right)}{\partial X_k\left(t_i\right)} = \frac{\partial G_i\left(x - M_n\right)}{\partial M_n\left(t_i\right)},$$

and therefore, from (5), we obtain

$$\frac{\partial G_i\left(x - M_n\right)}{\partial x_i} = -\sum_{k=1}^{n} \frac{\partial G_i\left(x - M_n\right)}{\partial X_k\left(t_i\right)}.$$

We now can apply integration by parts as we did in our derivation of (4). The difference is that the density of $X_k = \left(X_k\left(t_1\right), \ldots, X_k\left(t_d\right)\right)^T$ is known and therefore we obtain that

$$E\left(\frac{\partial G_i\left(x - M_n\right)}{\partial X_k\left(t_i\right)}\right) = E\left(G_i\left(x - M_n\right) \cdot e_i^T \Sigma^{-1} X_k\right),$$

where e_i is the ith vector in the canonical basis in Euclidean space.

Consequently, we conclude that

$$\begin{aligned} E\left(\frac{\partial G_i\left(x - M_n\right)}{\partial x_i}\right) &= -\sum_{k=1}^{n} E\left(\frac{\partial G_i\left(x - M_n\right)}{\partial X_k\left(t_i\right)}\right) \\ &= -E\left(G_i\left(x - M_n\right) \cdot e_i^T \Sigma^{-1} \sum_{k=1}^{n} X_k\right). \end{aligned}$$

In summary, if f_n is the density of M_n we have that

$$\begin{aligned} f_n\left(x_1, \ldots, x_d\right) &= E\left(\sum_{i=1}^{d} \frac{\partial G_i\left(x - M_n\right)}{\partial x_i}\right) \qquad (7) \\ &= -E\left(\sum_{i=1}^{d} \sum_{k=1}^{n} G_i\left(x - M_n\right) \cdot e_i^T \Sigma^{-1} X_k\right). \end{aligned}$$

The verification of this identity follows a very similar argument as that provided for the proof of (4) in [12].

2.3 Step 3: Extending the Malliavin–Thalmaier Identity to Infinite Maxima

In order to extend the definition of the estimator (7), we wish to send $n \to \infty$ and obtain a simulatable expression of an estimator. Because we will be using a recently developed estimator for M in [11], we need to impose the following assumptions on $X_n(\cdot)$.

(B1) In addition to assuming $E[X_n(t)] = 0$, we write $\sigma^2(t) = Var\,(X\,(t))$.
(B2) Assume that $\bar{\sigma} = \sup_{t\in T} \sigma(t) < \infty$.
(B3) Suppose that $E\left(\exp\left(\sup_{t\in T} X\,(t)\right)\right) < \infty$.

A key element of the algorithm in [11] is the idea of record breakers. The general idea is to utilize the properties of those record breakers to construct a Malliavin–Thalmaier estimator for the infinite maxima with finite but random number of Gaussian vectors. In order to describe this idea, let us write $\|X_n\|_\infty = \max_{i=1,\ldots,d} |X_n(t_i)|$.

Following the development in [11], we can identify three random times as follows.

The first is $N_X = N_X(a) < \infty$, defined for any $a \in (0, 1)$, and satisfying that for all $n > N_X$,

$$\|X_n\|_\infty \leq a \log n.$$

The time N_X is finite with probability one because $\|X_n\|_\infty$ is well known to grow at rate $O_p\left((\log n)^{1/2}\right)$ as $n \to \infty$.

The second is $N_A = N_A(\gamma) < \infty$ chosen for any given $\gamma < E\,(A_1)$, satisfying that for $n > N_A$

$$A_n \geq \gamma n. \tag{8}$$

The time N_A is finite with probability one because of the Strong Law of Large Numbers.

The third is N_a such that, for all $n > N_a$, we have

$$n\gamma \geq A_1\, n^a \exp(\|X_1\|_\infty). \tag{9}$$

It is immediate that N_a is finite almost surely because $a \in (0, 1)$.

By successively applying the preceding three equations, we find that for $n > N := \max(N_A, N_X, N_a)$ and any $t = t_1, \ldots, t_d$, we have

$$\begin{aligned} -\log A_n + X_n(t) &\leq -\log A_n + \|X_n\|_\infty \\ &\leq -\log A_n + a \log n \\ &\leq -\log(n\gamma) + a \log n \\ &\leq -\log A_1 - \|X_1\|_\infty \leq -\log A_1 + X_1(t). \end{aligned}$$

Therefore, we conclude that, for $t = t_1, \ldots, t_d$,

$$\sup_{n\geq 1} \{-\log A_n + X_n(t)\} = \max_{1\leq n\leq N} \{-\log A_n + X_n(t)\}.$$

The work in [11] explains how to simulate the random variables N_X, N_A, and N_a, jointly with the sequence $(A_n)_{n\leq N}$ as well as $(X_n)_{n\leq N}$. Moreover, it is also shown in

[11] that the number of random variables required to simulate N_X, N_A and N_a (jointly with $X_1, \ldots, X_N$ and $A_1, \ldots, A_N$) has finite moments of any order. Therefore, N has finite moments of any order. Moreover, $E(N) = O(d^{\varepsilon})$ for any $\varepsilon > 0$. In the appendix, we reproduce the simulation procedure developed in [11].

Now, observe that conditional on $X_1, \ldots, X_{N_X}, N_X$, for $n > N_X$ the random vectors $(X_k)_{k \geq n}$ are independent, but they no longer follow a Gaussian distribution. Nevertheless, the X_n 's still have zero conditional means, given that $n > N$. This is because

$$\begin{aligned} &E\left(X_n \mid \|X_n\|_\infty \leq a \log n\right) \\ &= E\left(-X_n \mid \|-X_n\|_\infty \leq a \log n\right) = E\left(-X_n \mid \|X_n\|_\infty \leq a \log n\right). \end{aligned}$$

Consequently, we have that

$$E\left(e_i^T \Sigma^{-1} X_n | n > N\right) = 0.$$

Therefore, because M is independent of X_n conditional on $n > N$, we obtain that

$$\begin{aligned} &E\left(G_i\left(x - M\right) \cdot e_i^T \Sigma^{-1} X_n | n > N\right) \\ &= E\left(G_i\left(x - M\right) | n > N\right) \cdot E\left(e_i^T \Sigma^{-1} X_n | n > N\right) = 0. \end{aligned} \tag{10}$$

One can let $n \to \infty$ in (7) and formally apply (10) leading to the following result, which is rigorously established in [1].

Proposition 1 *For any* $(x_1, \ldots, x_d) \in R^d$,

$$f\left(x_1, \ldots, x_d\right) = -E\left(\sum_{i=1}^{d} \sum_{k=1}^{N} G_i\left(x - M\right) \cdot e_i^T \Sigma^{-1} X_k\right). \tag{11}$$

2.4 Step 4: Variance Control in Malliavin–Thalmaier Estimators

We now explain how to address the second issue discussed in Sect. 2.1, namely, controlling the variance when using the Malliavin–Thalmaier estimator (11).

Let us write

$$W\left(x\right) = -\sum_{i=1}^{d} \sum_{k=1}^{N} G_i\left(x - M\right) \cdot e_i^T \Sigma^{-1} X_k,$$

and observe that

$$W\left(x\right) = \frac{\left\langle M - x, \sum_{i=1}^{N} \Sigma^{-1} X_k \right\rangle}{d w_d ||M - x||^d}.$$

It turns out that the variance of $W(x)$ blows up because of the singularity in the denominator when $M = x$. This is verified in [1], but a similar calculation is also given in the setting of diffusions in [10]. So instead, we consider an approximating sequence defined via $\bar{W}_0(x) = 0$, and

$$\bar{W}_n(x) = \frac{\left\langle M - x, \sum_{i=1}^{N} \Sigma^{-1} X_k \right\rangle}{dw_d ||M - x||^d + dw_d \delta_n ||M - x||}, \quad n \geq 1,$$

where

$$\delta_n = 1/\log\log\log\left(n + e^e\right). \tag{12}$$

It is apparent that $\lim_{n\to\infty} \bar{W}_n(x) = W(x)$ almost surely. The use of a perturbation in the denominator of the Malliavin–Thalmaier estimator is not new. In [10] also, a small positive perturbation in the denominator is added, but such perturbation is, in their case, deterministic. The difference here is that our perturbation contains the factor $\delta_n \|M - x\|$. We have chosen our perturbation in order to ultimately control both the variance and the bias of our estimator.

In order to quickly motivate the variance implications of our choice, note that

$$\left| \frac{\left\langle M - x, \sum_{i=1}^{N} \Sigma^{-1} X_k \right\rangle}{dw_d ||M - x||^d + dw_d \delta_n ||M - x||} \right| \leq \left| \frac{\left\langle M - x, \sum_{i=1}^{N} \Sigma^{-1} X_k \right\rangle}{dw_d \delta_n ||M - x||} \right| \leq \frac{1}{dw_d \delta_n} \left\| \sum_{i=1}^{N} \Sigma^{-1} X_k \right\|_2,$$

leading to a bound that does not explicitly contain M. Moreover, we mentioned before that N has finite moments of any order and X_k is normally distributed, therefore, one can easily verify that $\left\| \sum_{i=1}^{N} \Sigma^{-1} X_k \right\|_2$ has finite moments of any order, in particular finite second moment and therefore $\bar{W}_n(x)$ has finite variance.

The reader might wonder why choosing δ_n in the definition of $\bar{W}_n(x)$, because any function of n decreasing to zero will ensure the convergence almost surely of $\bar{W}_n(x)$ toward $W(x)$. The previous upper bound, although not sharp when n is large, might also hint to the fact that it is desirable to choose a slowly varying function of n in the denominator (at least the reader notices a bound which deteriorates slowly as n grows).

The precise reason for the selection of our perturbation in the denominator obeys to a detailed variance calculation which can be seen in [1]. A more in-depth discussion is given in Sect. 2.5 below. For the moment, let us continue with our development in order to give the final form of our estimator.

Even though $\bar{W}_n(x)$ has finite variance and is close to $W(x)$, unfortunately, we have that $\bar{W}_n(x)$ is no longer an unbiased estimator of $f(x)$. In order to remove the bias, we take advantage of a randomization idea from [13, 14], which is related to the multilevel Monte Carlo method in [7], as we shall explain next.

2.5 Final Form of Our Estimator

Let us define $\bar{W}_0(x) = 0$ and for $n \geq 1$ let us write

$$\Delta_n(x) = \bar{W}_n(x) - \bar{W}_{n-1}(x).$$

In order to facilitate the variance analysis of our randomized multilevel Monte Carlo estimator, we further consider a sequence $\left(\bar{\Delta}_n(x)\right)_{n\geq 1}$ of independent random variables so that $\Delta_n(x)$ and $\bar{\Delta}_n(x)$ are equal in distribution.

We let L be a random variable taking values on $n \geq 1$, independent of everything else. Moreover, we let $g(n) = P(L \geq n)$ and assume that

$$g(n) = n^{-1}(\log(n + e - 1))^{-1}\left(\log\left(\log\left(n + e^e - 1\right)\right)\right)^{-1}. \tag{13}$$

Then, the final form of our estimator is

$$V(x) = \sum_{k=1}^{L} \frac{\bar{\Delta}_k(x)}{g(k)}. \tag{14}$$

The randomized multilevel Monte Carlo idea applied formally yields that

$$E(V(x)) = E\left(\sum_{k=1}^{\infty} \frac{\bar{\Delta}_k(x) I(L \geq k)}{g(k)}\right) = \sum_{k=1}^{\infty} E\left(\frac{\bar{\Delta}_k(x) I(L \geq k)}{g(k)}\right) \tag{15}$$

$$= \sum_{k=1}^{\infty} E\left(\bar{\Delta}_k(x)\right) = E(W(x)) - E\left(\bar{W}_0(x)\right) = E(W(x)) = f(x).$$

In order to make the previous manipulations rigorous, we must justify exchanging the summation in (15). In addition, we also need to guarantee that $V(x)$ has finite variance. These and other properties will be used to obtain confidence intervals for our estimates, given a computational budget via CLT for renewal processes. In turn, it suffices to make sure the following two conditions hold:

(C1) $\sum_{k\geq 1} E\left(\left|\bar{\Delta}_k(x)\right|\right) < \infty$,

(C2) $\sum_{n\geq 1} \frac{E\left(\left|\bar{\Delta}_n(0)\right|^2\right)}{g(n)} < \infty$.

We shall summarize the properties of $V(x)$ in our main result given in the next section. In particular, we will show that (C1) and (C2) hold with our choice of δ_n and $g(n)$. We also provide a discussion of the running time analysis, which is affected by the choice of $g(n)$.

3 Main Result

Our main contribution is summarized in the following result, which is fully proved in [1]. Our objective now is to give the gist of the technical development in order to have at least an intuitive understanding of the choices behind the design of our estimator (14). We measure computational cost in terms of the elementary random variables simulated.

Theorem 1 *Let ρ be the cost required to generate M so that $V(x)$, defined in (14), has a computational cost equal to $C = \sum_{i=1}^{L} \rho_i + 1$ (where L is independent of $\rho_1, \rho_2, \ldots$, which are i.i.d. copies of ρ). Let $(V_1(x), C_1), (V_2(x), C_2), \ldots$ be i.i.d. copies of $(V(x), C)$ and set $T_n = C_1 + \cdots + C_n$ with $T_0 = 0$. For each $b > 0$ define, $B(b) = \max\{n \geq 0 : T_n \leq b\}$, then we have that*

$$f(x) = E(V(x)) \text{ and } Var(V(x)) < \infty. \tag{16}$$

Moreover,

$$\sqrt{\frac{b}{E(\rho_1) \cdot \log\log\log(b)}} \left(\frac{1}{B(b)} \sum_{i=1}^{B(b)} V_i(x) - f(x) \right) \Rightarrow N(0, Var(V(x))).$$

Before we discuss the analysis of the proof of Theorem 1, it is instructive to note that the previous result can be used to obtain confidence intervals for the value of the density $f(x)$ with precision ε at a computational cost of order $O\left(\varepsilon^{-2} \log\log\log(1/\varepsilon)\right)$, given a fixed confidence level (see Sect. 4 for an example of how to produce such a confidence interval).

The quantity $B(b)$ denotes the number of i.i.d. copies of $V(x)$ which can be simulated with a computational budget b, so the pointwise estimator given in Theorem 1 simply is the empirical average of $B(b)$ i.i.d. copies of $V(x)$.

The rate of convergence implied by Theorem 1 is, for all practical purposes, the same as the highly desirable canonical rate $O\left(\varepsilon^{-2}\right)$, which is rarely achieved in complex density estimation problems, such as the one that we consider in this paper.

3.1 Sketching the Proof of Theorem 1

At the heart of the proof of Theorem 1 lies a bound on the size of $|\Delta_n(x)|$. For notational simplicity, let us concentrate on $|\Delta_n(0)|$ and note that for any $\beta \geq 1$

$$|\Delta_n(0)|^\beta \leq \frac{\left\|\Sigma^{-1}\right\|^\beta}{(dw_d)^\beta} \left(\sum_{i=1}^{N} \|X_k\| \right)^\beta \times \left| \frac{\|M\|}{\|M\|^d + \|M\|\delta_{n+1}} - \frac{\|M\|}{\|M\|^d + \|M\|\delta_n} \right|^\beta . \tag{17}$$

We have argued that, because N has finite moments of any order, the random variable $\sum_{i=1}^{N}\|X_k\|_2$ is easily seen to have finite moments of any order. So, after applying Hölder's inequality to the right-hand-side of (17), it suffices to concentrate on estimating, for any $q > 1$,

$$E\left(\left|\frac{1}{||M||^{d-1}+\delta_{n+1}}-\frac{1}{||M||^{d-1}+\delta_n}\right|^{\beta q}\right)^{1/q}$$
$$=E\left(\left|\frac{\delta_n-\delta_{n+1}}{\left(||M||^{d-1}+\delta_{n+1}\right)\left(||M||^{d-1}+\delta_n\right)}\right|^{\beta q}\right)^{1/q}.$$

Let us define

$$a(n) := \delta_n-\delta_{n+1}\sim\delta_n^2\frac{1}{\log\log(n)\cdot\log(n)\cdot n}, \tag{18}$$

and focus on

$$D_{n,\beta}(0) := E\left(\left|\frac{1}{\left(||M||^{d-1}+\delta_{n+1}\right)\left(||M||^{d-1}+\delta_n\right)}\right|^{\beta q}\right)^{1/q}. \tag{19}$$

Assuming that M has a continuous density in a neighborhood of the origin (a fact which can be shown, for example, from (2), using the Gaussian property of the X_ns), we can directly analyze (19) using a polar coordinates transformation, obtaining that for some $\kappa > 0$

$$D_{n,\beta}^q(0)\lesssim\kappa\int_0^\infty\int_{\theta\in\mathcal{S}^{d-1}}\frac{f(r\cdot\theta)\,r^{d-1}}{\left(r^{d-1}+\delta_{n+1}\right)^{\beta q}\left(r^{d-1}+\delta_n\right)^{\beta q}}dr d\theta, \tag{20}$$

where $\mathcal{S}^{d-1}$ represents the surface of the unit ball in d dimensions. Further study of the decay properties of $f(r\cdot\theta)$ as r grows large, uniformly over $\theta\in\mathcal{S}^{d-1}$, allows us to conclude that

$$D_{n,\beta}^q(0)\leq\kappa'\int_0^\infty\frac{r^{d-1}}{\left(r^{d-1}+\delta_{n+1}\right)^{\beta q}\left(r^{d-1}+\delta_n\right)^{\beta q}}dr, \tag{21}$$

for some $\kappa' > 0$. Applying the change of variables $r = u\delta_n^{1/(d-1)}$ to the right-hand side of (21), allows us to conclude, after elementary algebraic manipulations, that

$$D_{n,\beta}^q(0)=O\left(\delta_n^{d/(d-1)-2\beta q}\right),$$

therefore concluding that

$$E\left(|\Delta_n(0)|^\beta\right) = O\left(\left(\frac{\delta_n - \delta_{n+1}}{\delta_n^2}\right)^\beta \delta_n^{d/(q(d-1))-2\beta}\right). \tag{22}$$

Setting $\beta = 1$ we have (from (18) and the definition of δ_n) that

$$\sum_{n\geq 1} E\left(|\Delta_n(0)|\right) = O\left(\sum_{n\geq 1} \frac{1}{\log\log(n)\cdot\log(n)\cdot n}\delta_n^{d/(q(d-1))}\right) < \infty, \tag{23}$$

because $d/(d-1) > 1$ and $q > 1$ can be chosen arbitrarily close to one. This estimate justifies the formal development in (15) and the fact that $EV(x) = f(x)$.

Now, the analysis in [14] states that $Var(V(x)) < \infty$ if

$$\sum_{n\geq 1} \frac{E\left|\bar{\Delta}_n(0)\right|^2}{g(n)} < \infty. \tag{24}$$

Once again, using (22) and our choice of $g(n)$, we find that (24) holds because of the estimate

$$\sum_{n\geq 1} \frac{n\cdot\log(n)\cdot\log\log(n)}{(\log\log(n)\cdot\log(n)\cdot n)^2}\delta_n^{d/(q(d-1))} < \infty, \tag{25}$$

which is, after immediate cancellations, completely analogous to (23).

Finally, because the cost of sampling M (in terms of the number of elementary random variables, such as multivariate Gaussian random variables) has been shown to have finite moments of any order [11, Theorem 2.2], one can use standard results from the theory of regular variation (see [16, Theorem 3.2]) to conclude that

$$P\left(\sum_{i=1}^{L} \rho_i + 1 > t\right) \sim P\left(L > t/E(\rho_1)\right) \sim E(\rho_1)\, t^{-1}\log(t)^{-1}\log\log(t)^{-1},$$

as $t \to \infty$. Now, the form of the Central Limit Theorem is an immediate application of Theorem 1 in [18].

4 Numerical Examples

In this section, we implement our estimator and compare it against a conventional kernel density estimator. We measure the computational cost in terms of the number of independent samples drawn from **Algorithm M**. This convention translates into assuming that $E(\rho_1) = 1$ in Theorem 1. Given a computational budget b, the estimated density is given by

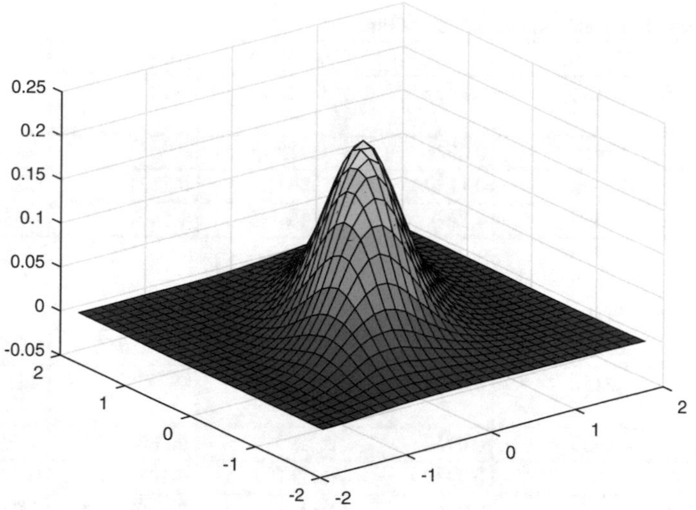

Fig. 1 The estimated three-dimensional joint density of a max-stable process using our algorithm

$$\hat{f}_b(x) = \frac{\sum_{i=1}^{B(b)} V_i(x)}{B(b)}.$$

According to Theorem 1, we can construct the confidence interval for underlying density $f(x)$ with significance level α as

$$\left(\hat{f}_b(x) - z_{\alpha/2}\hat{s}\sqrt{a(b)},\, \hat{f}_b(y) + z_{\alpha/2}\hat{s}\sqrt{a(b)}\right),$$

where $z_{\alpha/2}$ is the quantile corresponding to the $1 - \alpha/2$ percentile,

$$\hat{s}^2 = \frac{\sum_{i=1}^{B(b)} \left(V_i(x) - \hat{f}_b(x)\right)^2}{B(b)},$$

and

$$a(b) = \frac{\log\log\log(b)}{b}.$$

We perform our algorithm to estimate the density of the max-stable process. We assume that $T = [0, 1]$ and $X_n(\cdot)$ is a standard Brownian motion. We are interested in estimating the density of $M = (M(1/3), M(2/3), M(1))^T$. That is, the spatial grid is $(1/3, 2/3, 1)$. The graph in Fig. 1 shows a plot of the density on the set $\{x \in \mathbb{R}^3 : x_1 \in (-2, 2), x_2 \in (-2, 2), x_3 = 0\}$. Our estimation of this three-dimensional density has a computation budget of $B = 10^6$ samples from **Algorithm M**.

Table 1 Density estimation with our algorithm

Values (x)	(0, 0, 0)	(0, 0.5, 0)	(0.5, 0, 0)	(0, −0.5, 0)	(−0.5, 0, 0)
est. density $\hat{f}_b(x)$	0.2126	0.106	0.1292	0.1039	0.1439
Lower CI	0.1916	0.0971	0.1180	0.0947	0.1311
Upper CI	0.2336	0.1149	0.14036	0.1131	0.1567
Relative error	5.05%	4.29%	4.41%	4.54%	4.53%

Table 2 Density estimation with KDE

Values (x)	(0,0,0)	(0,0.5,0)	(0.5,0,0)	(0,−0.5,0)	(-0.5,0,0)
est. density $\hat{f}_b^{KDE}(x)$	0.2163	0.0846	0.1143	0.0938	0.1084
Lower CI	0.1953	0.0712	0.0999	0.0800	0.0934
Upper CI	0.2373	0.0980	0.1287	0.1076	0.1234
Relative error	4.94%	8.07%	6.43%	7.51%	7.05%

We calculate the 95% confidence interval of the density on several selected values of the process $M(\cdot)$.

As a comparison, we also calculate the 95% confidence interval of the density using the plug-in kernel density estimation (KDE) method with the same amount ($b = 10^6$) of i.i.d. samples of M. We use the normal density function as the kernel function and select the bandwidth according to [17]. The estimator is obtained as follows. Sample $M^{(1)}, M^{(2)}, \ldots, M^{(d)}$ i.i.d. copies of M, let $h_b = b^{-1/(2d+1)}$ and compute the sample covariance matrix, $\hat{\Sigma}$, based on $(M^{(1)}, M^{(2)}, \ldots, M^{(d)})$. Then, let

$$\hat{f}_b^{KDE}(x) = \frac{1}{bh_b^d} \sum_{i=1}^{b} \phi\left(A^{-\frac{1}{2}} \frac{x - M^{(i)}}{h_b}\right),$$

where $A = \hat{\Sigma}/\det|\hat{\Sigma}|$. We apply the method from [6] to evaluate the corresponding confidence interval, thereby obtaining the estimates shown in Tables 1 and 2.

From the above tables, we can see that our algorithm provides similar estimates to those obtained using the KDE. However, our estimator also has a smaller relative error when the estimated value is relatively small. Also, as discussed in [6], KDE is biased, and one must carefully choose the bandwidth to obtain the optimal convergence rate of mean squared error. In contrast, the construction of confidence intervals with our estimator is straightforward, with a significantly better convergence rate.

5 Appendix: A Detailed Algorithmic Implementation

In order to make this paper as self-contained as possible, we reproduce here the algorithms from [11] which allow us to simulate the random variables N_X, N_A, and N_a, jointly with $(A_n)_{n\leq N}$ and $(X_n)_{n\leq N}$.

5.1 *Simulating Last Passage Times of Random Walks*

Define the random walk $S_n = \gamma n - A_n$ for $n \geq 0$. Note that $ES_n < 0$, by our choice of $\gamma < E(A_1)$. The authors in [11], argue that the choice of γ is not too consequential so we shall assume that $\gamma = 1/2$.

Here we review an algorithm from [11] for finding a random time N_S such that $S_n < 0$ for all $n > N_S$. Observe that $N_S = N_A$.

The algorithm is based on alternately sampling upcrossings and downcrossings of the level 0. We write $\xi_0^+ = 0$ and, for $i \geq 1$, we recursively define

$$\xi_i^- = \begin{cases} \inf\{n \geq \xi_{i-1}^+ : S_n < 0\} & \text{if } \xi_{i-1}^+ < \infty \\ \infty & \text{otherwise} \end{cases}$$

together with

$$\xi_i^+ = \begin{cases} \inf\{n \geq \xi_i^- : S_n \geq 0\} & \text{if } \xi_i^- < \infty \\ \infty & \text{otherwise.} \end{cases}$$

As usual, in these definitions, the infimum of an empty set should be interpreted as ∞. Writing

$$N_S = \sup\{\xi_n^- : \xi_n^- < \infty\},$$

we have by construction $S_n < 0$ for $n > N_S$. The random variable $N_S - 1$ is an upward last passage time:

$$N_S - 1 = \sup\{n \geq 0 : S_n \geq 0\}.$$

Note that $0 \leq N_S < \infty$ almost surely under P since $(S_n)_{n\geq 0}$ starts at the origin and has negative drift. We will provide pseudo-codes for simulating $(S_1, \ldots, S_{N_S+\ell})$ for any fixed $\ell \geq 0$, but first we need a few definitions.

First, we assume that the *Cramér's root*, $\theta > 0$, satisfying $E(\exp(\theta S_1)) = 1$ has been computed. We shall use $\mathbb{P}_x$ to denote the measure under which $(A_n)_{n\geq 1}$ are arrivals of a Poisson process with unit rate and $S_0 = x$. Then, we define P_x^θ through an exponential change of measure. In particular, on the σ-field generated by $S_1, \ldots, S_n$ we have

$$\frac{dP_x}{dP_x^\theta} = \exp(-\theta(S_n - x)).$$

It turns out that under P_x^θ, $(A_n)_{n\geq 1}$ corresponds to the arrivals of a Poisson process with rate $1-\theta$ and the random walk $(S_n)_{n\geq 1}$ has a positive drift.

To introduce the algorithm to sample $(S_1, \ldots, S_{N_S+\ell})$ we first need the following definitions:

$$\tau^- = \inf\{n \geq 0 : S_n < 0\}, \qquad \tau^+ = \inf\{n \geq 0 : S_n \geq 0\}.$$

For $x \geq 0$, it is immediate that we can sample a downcrossing segment $S_1, \ldots, S_{\tau^-}$ under P_x due to the negative drift, and we record this for later use in a pseudocode function. *Throughout our discussion, 'sample' in pseudocode stands for 'sample independently of anything that has been sampled already'.*

Function SAMPLEDOWNCROSSING(x)**: Samples** $(S_1, \ldots, S_{\tau^-})$ **under** P_x **for** $x \geq 0$
Step 1: Return sample $S_1, \ldots, S_{\tau^-}$ under P_x.
Step 2: EndFunction

Sampling an upcrossing segment is more interesting because it is possible that $\tau^+ = \infty$. So, an algorithm needs to be able to detect this event within a finite amount of computing resources. For this reason, we understand sampling an upcrossing segment under P_x for $x < 0$ to mean that an algorithm outputs $S_1, \ldots, S_{\tau^+}$ if $\tau^+ < \infty$, and otherwise it outputs 'degenerate'. The following pseudo-code samples an upcrossing under P_x for $x < 0$.

Function SAMPLEUPCROSSING(x)**: Samples** $(S_1, \ldots, S_{\tau^+})$ **under** P_x **for** $x < 0$
Step 1: $S \leftarrow$ sample $S_1, \ldots, S_{\tau^+}$ under P_x^θ
Step 2: $U \leftarrow$ sample a standard uniform random variable
Step 3: If $U \leq \exp(-\theta(S_{\tau^+} - x))$
Step 4: Return S
Step 5: Else
Step 6: Return 'degenerate'
Step 7: EndIf
Step 8: EndFunction

We next describe how to sample $(S_k)_{k=1,\ldots,n}$ from P_x conditionally on $\tau^+ = \infty$ for $x < 0$. Since $\tau^+ = \infty$ is equivalent to $\sup_{k\leq\ell} S_k < 0$ and $\sup_{k>\ell} S_k < 0$ for any $\ell \geq 1$, after sampling $S_1, \ldots, S_\ell$, by the Markov property we can use SAMPLEUPCROSSING(S_ℓ) to verify whether or not $\sup_{k>\ell} S_k < 0$.

Function SAMPLEWITHOUTRECORDS(x, ℓ)**: Samples** $(S_k)_{k=1,\ldots,\ell}$ **from** P_x **given** $\tau^+ = \infty$ **for** $\ell \geq 1, x < 0$
Step 1: Repeat
Step 2: $S \leftarrow$ sample $(S_k)_{k=1,\ldots,\ell}$ under P_x
Step 3: Until $\sup_{1\leq k\leq\ell} S_k < 0$ and SAMPLEUPCROSSING (S_ℓ) is 'degenerate'
Step 4: Return S
Step 5: EndFunction

We summarize our discussion with the full algorithm for sampling $(S_0, \ldots, S_{N_S+\ell})$ under P given some $\ell \geq 0$.

Algorithm S: Samples $S = (S_0, \ldots, S_{N_S+\ell})$ ***under*** P ***for*** $\ell \geq 0$
\# We use S_{end} to denote the last element of S.
Step 1: $S \leftarrow [0]$
Step 2: Repeat
Step 3: DowncrossingSegment $\leftarrow$ SAMPLEDOWNCROSSING(S_{end})
Step 4: $S \leftarrow [S,$DowncrossingSegment$]$
Step 5: UpcrossingSegment $\leftarrow$SAMPLEUPCROSSING(S_{end})
Step 6: If UpcrossingSegment is not 'degenerate'
Step 7: $S \leftarrow [S,$ upcrossingSegment$]$
Step 8: EndIf
Step 9: Until UpcrossingSegment is 'degenerate'
Step 10: If $\ell > 0$
Step 11: $S \leftarrow [S,$SAMPLEWITHOUTRECORDS$(S_{\text{end}}, \ell)]$
Step 12: EndIf

5.2 *Simulating Last Passage Times for Maxima of Gaussian Vectors*

The technique is similar to the random walk case using a sequence of record-breaking times. The parameter $a \in (0, 1)$ can be chosen arbitrarily, but [11] suggests selecting a such that

$$\exp\left(\frac{\overline{\sigma}}{a}\overline{\Phi}^{-1}\left(\delta\sqrt{2\pi}\frac{\phi(\overline{\sigma}/a)}{d\overline{\sigma}/a}\right)+\frac{\overline{\sigma}^2}{a^2}\right) = E\left[\left(\frac{A_1\exp(\|X\|_\infty)}{\gamma}\right)^{\frac{1}{1-a}}\right],$$

where $\Phi(\cdot)$ is the cumulative distribution function of a standard Gaussian random variable and $\overline{\Phi} = 1 - \Phi$.

Now, assume that $\eta_0 \geq 0$ is given (we will choose it specifically in the sequel). Let $(X_n)_{n\geq 1}$ be i.i.d. copies of X and define, for $i \geq 1$, a sequence of *record breaking times* (η_i) through

$$\eta_i = \begin{cases} \inf\{n > \eta_{i-1} : \|X_n\|_\infty > a\log n\} & \text{if } \eta_{i-1} < \infty \\ \infty & \text{otherwise.} \end{cases}.$$

We provide pseudo-codes which ultimately will allow us to sample $(X_1, \ldots, X_{N_X+\ell})$ for any fixed $\ell \geq 0$, where

$$N_X = \max\{\eta_i : \eta_i < \infty\}.$$

First, we shall discuss how to sample (X_n) up to a η_1. In order to sample η_1, $\eta_0 = n_0$ needs to be chosen so that $P(\|X\|_\infty > a\log n)$ is controlled for every $n > n_0$. Given the choice of $a \in (0, 1)$, select n_0 such that

$$d\overline{\Phi}\left(\frac{a\log n_0}{\overline{\sigma}} - \frac{\overline{\sigma}}{a}\right) \le \frac{1}{2}\sqrt{\frac{\pi}{2}}\frac{\phi(\overline{\sigma}/a)}{\overline{\sigma}/a}.$$

Define

$$T_{n_0} = \inf\{k \ge 1 : \|X_k\|_\infty > a\log(n_0 + k)\}. \tag{26}$$

We describe an algorithm that outputs 'degenerate' if $T_{n_0} = \infty$ and $(X_1, \ldots, X_{T_{n_0}})$ if $T_{n_0} < \infty$.

First, we describe a simple algorithm to simulate from X conditioned on $\|X\|_\infty > a\log n$. Our algorithm makes use of a probability measure $P^{(n)}$ defined through

$$\frac{dP^{(n)}}{dP}(x) = \frac{\sum_{i=1}^d \mathbf{1}(|x(t_i)| > a\ \log n)}{\sum_{i=1}^d P(|X(t_i)| > a\ \log n)}.$$

It turns out that the measure $P^{(n)}$ approximates the conditional distribution of X given that $\|X\|_\infty > a\log n$ for n large.

Now, define $w^j(t) = \mathrm{Cov}(X(t), X(t_j))/\mathrm{Var}\left(X\left(t_j\right)\right)$ and note that $X(\cdot) - w^\nu(\cdot)X(t_\nu)$ is independent of $X(t_\nu)$ given ν. This property is used in [11] to show that the following algorithm outputs from $P^{(n)}$. We will let U be a uniform random variable in (0, 1) and J is independent of U and such that $P(J = 1) = 1/2 = P(J = -1)$.)

Function ConditionedSampleX (a, n)**: Samples** X **from** $P^{(n)}$
Step 1: $\nu \leftarrow$ sample with probability mass function

$$P(\nu = j) = \frac{P(\left|X(t_j)\right| > a\log n)}{\sum_{i=1}^d P(|X(t_i)| > a\log n)}$$

Step 2: $U \leftarrow$ sample a standard uniform random variable
Step 3: $X(t_\nu) \leftarrow \sigma(t_\nu) \cdot J \cdot \Phi^{-1}(U + (1-U)\Phi(a(\log n)/\sigma(t_\nu)))$ #Conditions on $|X(t_\nu)| > a\log n$
Step 4: $Y \leftarrow$ sample of X under P
Step 5: Return $Y(t) - w^\nu(t)Y(t_\nu) + X(t_\nu)$
Step 6: EndFunction

We now explain how ConditionedSampleX is used to sample T_{n_0}. Define, for $k \ge 1$,

$$g_{n_0}(k) = \frac{\int_{k-1}^k \phi((a\log(n_0+s))/\overline{\sigma})ds}{\int_0^\infty \phi((a\log(n_0+s))/\overline{\sigma})ds},$$

where $\phi(x) = d\Phi(x)/dx$. Note that $g_{n_0}(\cdot) \geq 0$ defines the probability mass function of some random variable K. It turns out that if $U \sim U(0, 1)$ then we can sample

$$K = \left\lceil \exp\left\{\frac{\overline{\sigma}^2}{a^2} + \frac{\overline{\sigma}}{a}\overline{\Phi}^{-1}\left(U\,\overline{\Phi}\left(\frac{a\log n_0}{\overline{\sigma}} - \frac{\overline{\sigma}}{a}\right)\right)\right\} - n_0\right\rceil.$$

The next function samples $(X_1, \ldots, X_{T_{n_1}})$ for $n_1 \geq n_0$.

Function SAMPLESINGLERECORD (a, n_0, n_1)**: Samples** $(X_1, \ldots, X_{T_{n_1}})$ **for** $a \in (0, 1), n_1 \geq n_0 \geq 0$

Step 1: Sample K

Step 2: $[X_1, \ldots, X_{K-1}] \leftarrow$ i.i.d. sample from P

Step 3: $X_K \leftarrow$ CONDITIONEDSAMPLEX $(a, n_1 + K)$

Step 4: $U \leftarrow$ sample a standard uniform random variable

Step 5: If $\|X_k\|_\infty \leq a\log(n_1 + k)$ for $k = 1, \ldots, K-1$ and $U\,g_{n_0}(K) \leq dP/dP^{(n_1+K)}(X_K)$

Step 6: Return $(X_1, \ldots, X_K)$

Step 7: Else

Step 8: Return 'degenerate'

Step 9: EndIf

Step 10: EndFunction

We next describe how to sample $(X_k)_{k=1,\ldots,n}$ conditionally on $T_{n_0} = \infty$. This is a simple task because the X_ns are independent.

Function SAMPLEWITHOUTRECORDX (n_1, ℓ)**: Samples** $(X_k)_{k=1,\ldots,\ell}$ **conditionally on** $T_{n_1} = \infty$ for $\ell \geq 1$

Step 1: Repeat

Step 2: $X \leftarrow$ sample $(X_k)_{k=1,\ldots,\ell}$ under P

Step 3: Until $\sup_{1\leq k\leq \ell}[X_k - a\log(n_1 + k)] < 0$

Step 4: Return X

Step 5: EndFunction

We now can explain how to sample $(X_1, \ldots, X_{N_X+\ell})$ under P given some $\ell \geq 0$. The idea is to successively apply SAMPLESINGLERECORD to generate the sequence $(\eta_i : i \geq 1)$ defined at the beginning of this section. Starting from $\eta_0 = n_0$, then n_1 is replaced by each of the subsequent η_is.

Algorithm X: Samples $(X_1, \ldots, X_{N_X+\ell})$ ***given*** $a \in (0, 1), \overline{\sigma} > 0, \ell \geq 0$

Step 1: $X \leftarrow [\,], \eta \leftarrow n_0$

Step 2: $X \leftarrow$ sample $(X_k)_{k=1,\ldots,\eta}$ under P

Step 3: Repeat

Step 4: segment $\leftarrow$ SAMPLESINGLERECORD (a, n_0, η)

Step 5: If segment is not 'degenerate'

Step 6: $X \leftarrow [X, \text{segment}]$

Step 7: $\eta \leftarrow \text{length}(X)$

Step 8: EndIf

Step 9: Until segment is 'degenerate'
Step 10: If $\ell > 0$
Step 11: $X \leftarrow [X, \textsc{SampleWithoutRecordX}(\eta, \ell)]$
Step 12: EndIf

5.3 Algorithm to Sample $X_1, \ldots, X_N, N$

The final algorithm for sampling $M, X_1, \ldots, X_N, N$ is given next.

Algorithm M: Samples $M, X_1, \ldots, X_N, N$ ***given*** $a \in (0, 1)$, $\gamma < E(A_1)$, and $\overline{\sigma}$
Step 1: Sample $A_1, \ldots, A_{N_A}$ using Steps 1–9 from **Algorithm S** with $S_n = \gamma n - A_n$.
Step 2: Sample $X_1, \ldots, X_{N_X}$ using Steps 1–9 from **Algorithm X**.
Step 3: Calculate N_a with (9) and set $N = \max(N_A, N_X, N_a)$.
Step 4: If $N > N_A$
Step 5: Sample $A_{N_A+1}, \ldots, A_N$ as in Step 10–12 from **Algorithm S** with $S_n = \gamma n - A_n$.
Step 6: EndIf
Step 7: If $N > N_X$
Step 8: Sample $X_{N_X+1}, \ldots, X_N$ as in Step 10–12 from **Algorithm X**.
Step 9: EndIf
Step 10: Return $M(t_i) = \max_{1 \le n \le N} \{-\log A_n + X_n(t_i) + \mu(t_i)\}$ for $i = 1, \ldots, d$, and $X_1, \ldots, X_N, N$.

References

1. Blanchet, J.H., Liu, Z.: Efficient conditional density estimation for max-stable fields. Submitted
2. Dieker, A.B., Mikosch, T.: Exact simulation of Brown–Resnick random fields at a finite number of locations. Extremes **18**(2), 301–314 (2015)
3. Dombry, C., Engelke, S., Oesting, M.: Asymptotic properties of the maximum likelihood estimator for multivariate extreme value distributions. arXiv:1612.05178
4. Embrechts, P., Klüppelberg, C., Mikosch, T.: Modelling Extremal Events for Insurance and Finance. Springer, New York (1997)
5. Evans, L.C.: Partial Differential Equations: Second Edition. American Mathematical Society, Providence (2010)
6. Fiorio, C.V.: Confidence intervals for kernel density estimation. Stata J. **4**, 168–179 (2004)
7. Giles, M.B.: Multilevel Monte Carlo path simulation. Oper. Res. **56**(3), 607–617 (2008)
8. Huser, R., Davison, A.C.: Composite likelihood estimation for the Brown–Resnick process. Biometrika **100**(2), 511–518 (2013)
9. Kabluchko, Z., Schlather, M., de Haan, L.: Stationary max-stable fields associated to negative definite functions. Ann. Probab. **37**(5), 2042–2065
10. Kohatsu-Higa, A., Yasuda, K.: Estimating multidimensional density functions using the Malliavin–Thalmaier formula. SIAM J. Numer. Anal. **47**, 1546–1575 (2009)
11. Liu, Z., Blanchet, J.H., Dieker, A.B., Mikosch, T.: On optimal exact simulation of max-stable and related random fields. Submitted, arXiv:1609.06001

12. Malliavin, P., Thalmaier, A.: Stochastic Calculus of Variations in Mathematical Finance. Springer, New York (2006)
13. McLeish, D.: A general method for debiasing a Monte Carlo estimator. Monte Carlo Methods Appl. **17**(4) (2011)
14. Rhee, C.H., Glynn, P.W.: Unbiased estimation with square root convergence for SDE models. Oper. Res. **63**(5), 1026–1043 (2015)
15. Ribatet, M.: Spatial extremes: max-stable processes at work. Journal de La Société Française de Statistique **154**(2), 156–177 (2013)
16. Robert, C.Y., Segers, J.: Tails of random sums of a heavy-tailed number of light-tailed terms. Insur. Math. Econ. **43**(1), 85–92 (2008)
17. Scott, D.W., Sain, S.R.: Multidimensional density estimation. Handb. Stat. **24**, 229–261 (2005)
18. Zheng, Z., Blanchet, J., Glynn, P.W.: Rates of convergence and CLTs for subcanonical debiased MLMC. Submitted

Sequential Quasi-Monte Carlo: Introduction for Non-experts, Dimension Reduction, Application to Partly Observed Diffusion Processes

Nicolas Chopin and Mathieu Gerber

Abstract SMC (Sequential Monte Carlo) is a class of Monte Carlo algorithms for filtering and related sequential problems. Gerber and Chopin (J R Stat Soc Ser B Stat Methodol 77(3):509–579, 2015, [16]) introduced SQMC (Sequential quasi-Monte Carlo), a QMC version of SMC. This paper has two objectives: (a) to introduce Sequential Monte Carlo to the QMC community, whose members are usually less familiar with state-space models and particle filtering; (b) to extend SQMC to the filtering of continuous-time state-space models, where the latent process is a diffusion. A recurring point in the paper will be the notion of dimension reduction, that is how to implement SQMC in such a way that it provides good performance despite the high dimension of the problem.

Keywords Diffusion models · Particle filtering · Randomised quasi-Monte Carlo · Sequential Monte Carlo · State-space models

1 Introduction

SMC (Sequential Monte Carlo) is a class of algorithms that provide Monte Carlo approximations of a sequence of distributions. The main application of SMC is the filtering problem: a phenomenon of interest is modelled as a Markov chain $\{X_t\}$, which is not observed directly; instead one collects sequentially data such as e.g. $Y_t = f(X_t) + V_t$, where V_t is a noise term. Filtering amounts to computing the distribution of X_t given $Y_{0:t} = (Y_0, \ldots, Y_t)$, the data collected up to time t. Filtering and related problems play an important role in target tracking (where X_t is the position of the target, say a ship), robotic mapping (where X_t is the position of the robot), Epidemiology (where X_t is e.g. the number of infected cases), Finance (X_t is the volatility of a given asset) and many other fields. See e.g. the book of [12].

N. Chopin (✉)
CREST-ENSAE, 3 Av. Pierre Larousse, 92245 Malakoff, France
e-mail: nicolas.chopin@ensae.fr

M. Gerber
School of Mathematics, University of Bristol, University Walk, Clifton, Bristol BS8 1TW, UK
e-mail: mathieu.gerber@bristol.ac.uk

A. B. Owen and P. W. Glynn (eds.), *Monte Carlo and Quasi-Monte Carlo Methods*, Springer Proceedings in Mathematics & Statistics 241,
https://doi.org/10.1007/978-3-319-91436-7_5

In [16], we introduced SQMC (Sequential quasi-Monte Carlo), a QMC version of Sequential Monte Carlo. As other types of QMC algorithms, the main advantage of SQMC is the better rate of convergence one may expect, relative to SMC methods.

It is difficult to write a paper that bridges the gap between two scientific communities; in this case, QMC experts on one side, and Statisticians working on Monte Carlo methods (MCMC and SMC) on the other side. We realise now that [16] may be more approachable by the latter than by the former. In particular, that paper spends time explaining basic QMC notions to non-experts, but it does not do the same for SMC.

To address this short-coming, and hopefully generate some interest about SQMC in the QMC community, we decided to devote the first part of this paper to introducing the motivation and basic principles of SMC. We do so using the so-called Feynman–Kac formalism, which is deemed to be abstract, but may be actually more approachable to non-Statisticians.

The second part of this paper discusses how to extend SQMC to the filtering of continuous-time state-space models; i.e. models where the underlying signal is e.g. a diffusion process. These models are popular in Finance and in Biology. What makes this extension interesting is that the inherent dimension of such models is infinity, whereas the performance of SQMC seems to deteriorate with the dimension (according to the numerical studies in [16]). However, by using the Markov property of the latent process, we are able to make some parts of SQMC operate in a low dimension, and, as result, to make it perform well (and significantly better than SMC) despite the infinite dimension of the problem.

2 SMC

2.1 Basic Notions and Definitions

The state space $\mathfrak{X}$ of interest in the paper is always an open subset of $\mathbb{R}^d$, which we equip with the Lebesgue measure.

We use the standard colon short-hand for collections of random variables and related quantities: e.g. $Y_{0:t}$ denote $Y_0, \ldots, Y_t$, $X_t^{1:N}$ denote $X_t^1, \ldots, X_t^N$, and so on. When such variables are vectors, we denote by $X_t(k)$ their kth component.

2.2 Feynman–Kac Formalism

The phrase 'Feynman–Kac model' comes from Probability theory, where 'model' means distributions for variables of interest, and not specifically observed variables (i.e. data, as in Statistics). A Feynman–Kac model consists of:

1. The law of a (discrete-time) Markov process $\{X_t\}$, specified through an initial distribution $\mathbb{M}_0(\mathrm{d}x_0)$, and a sequence of Markov kernels $M_t(x_{t-1}, \mathrm{d}x_t)$; i.e. $M_t(x_{t-1}, \mathrm{d}x_t)$ is the distribution of X_t, conditional on $X_{t-1} = x_{t-1}$;
2. A sequence of so-called potential (measurable) functions, $G_0 : \mathcal{X} \to \mathbb{R}^+$, $G_t : \mathcal{X} \times \mathcal{X} \to \mathbb{R}^+$. ($\mathbb{R}^+ = [0, +\infty)$.)

From these objects, one defines the following sequence of probability distributions:

$$\mathbb{Q}_t(\mathrm{d}x_{0:t}) = \frac{1}{L_t} \left\{ G_0(x_0) \prod_{s=1}^{t} G_s(x_{s-1}, x_s) \right\} \mathbb{M}_0(\mathrm{d}x_0) \prod_{s=1}^{t} M_s(x_{s-1}, \mathrm{d}x_s)$$

where L_t is simply the normalising constant:

$$L_t = \int_{\mathcal{X}^{T+1}} \left\{ G_0(x_0) \prod_{s=1}^{t} G_s(x_{s-1}, x_s) \right\} \mathbb{M}_0(\mathrm{d}x_0) \prod_{s=1}^{t} M_s(x_{s-1}, \mathrm{d}x_s).$$

(We assume that $0 < L_t < +\infty$.) A good way to think of Feynman–Kac models is that of a sequential change of measure, from the law of the Markov process $\{X_t\}$, to some modified law $\mathbb{Q}_t$, where the modification applied at time t is given by function G_t. In computational terms, one can also think of (sequential) importance sampling: we would like to approximate $\mathbb{Q}_t$ by simulating process $\{X_t\}$, and re-weight realisations at time t by function G_t. Unfortunately the performance of this basic approach would quickly deteriorate with time.

Example 1 Consider a Gaussian auto-regressive process, $X_0 \sim N(0, 1)$, $X_t = \phi X_{t-1} + V_t$, $V_t \sim N(0, 1)$, for $t \geq 1$, and take $G_t(x_{t-1}, x_t) = \mathbb{1}_{\mathbb{R}^+}(x_t)$. Then, if we use sequential importance sampling, the number of simulated trajectories that would get a non-zero weight would decrease quickly with time. In particular, the probability of 'survival' at time t would be $2^{-(t+1)}$ for $\phi = 0$.

The successive distributions $\mathbb{Q}_t$ are related as follows:

$$\mathbb{Q}_t(\mathrm{d}x_{0:t}) = \frac{1}{\ell_t} \mathbb{Q}_{t-1}(\mathrm{d}x_{0:t-1}) M_t(x_{t-1}, \mathrm{d}x_t) G_t(x_{t-1}, x_t) \tag{1}$$

where $\ell_t = L_t / L_{t-1}$. There are many practical settings (as discussed in the next section) where one is interested only in approximating the *marginal* distribution $\mathbb{Q}_t(\mathrm{d}x_t)$, i.e. the marginal distribution of variable X_t relative to the joint distribution $\mathbb{Q}_t(\mathrm{d}x_{0:t})$. One can deduce from (1) the following recursion for these marginals: $\mathbb{Q}_t(\mathrm{d}x_t)$ is the marginal distribution of variable X_t with respect to the bi-variate distribution

$$\mathbb{Q}_t(\mathrm{d}x_{t-1:t}) = \frac{1}{\ell_t} \mathbb{Q}_{t-1}(\mathrm{d}x_{t-1}) M_t(x_{t-1}, \mathrm{d}x_t) G_t(x_{t-1}, x_t). \tag{2}$$

Note the dramatic dimension reduction: the initial definition of $\mathbb{Q}_t$ involved integrals with respect to $\mathcal{X}^{t+1}$, but with the above recursion one may obtain expectations with respect to $\mathbb{Q}_t(\mathrm{d}x_t)$ by computing $t + 1$ integrals with respect to $\mathcal{X}^2$.

2.3 Feynman–Kac in Practice

The main application of the Feynman–Kac formalism is the filtering of a state-space model (also known as a hidden Markov model). This time, 'model' has its standard (statistical) meaning, i.e. a probability distribution for observed data.

A state-space model involves two discrete-time processes $\{X_t\}$ and $\{Y_t\}$; $\{X_t\}$ is Markov, and unobserved, $\{Y_t\}$ is observed, and is such that variable Y_t conditional on X_t and all (X_s, Y_s), $s \neq t$ depends only on X_t. The standard way to specify this model is through:

1. The initial distribution $\mathbb{P}_0(\mathrm{d}x_0)$ and the Markov kernels $P_t(x_{t-1}, \mathrm{d}x_t)$ that define the law of the process $\{X_t\}$;
2. The probability density $f_t(y_t|x_t)$ of $Y_t|X_t = x_t$.

Example 2 The stochastic volatility model is a state-space model popular in Finance (e.g., [19]). One observes the log-return Y_t of a given asset, which is distributed according to $Y_t|X_t = x_t \sim N(0, e^{x_t})$. The quantity X_t represents the (unobserved) market volatility, and evolves according to an auto-regressive process:

$$X_t - \mu = \phi(X_{t-1} - \mu) + \sigma V_t, \qquad V_t \sim N(0, 1).$$

For X_0, one may take $X_0 \sim N\left(\mu, \sigma^2/(1-\phi^2)\right)$ to make the process $\{X_t\}$ stationary.

Example 3 The bearings-only model is a basic model in target tracking, where X_t represents the current position (in $\mathbb{R}^2$) of a target, and Y_t is a noisy angular measurement obtained by some device (such as a radar):

$$Y_t = \arctan\left(\frac{X_t(2)}{X_t(1)}\right) + V_t, \quad V_t \sim N(0, \sigma^2),$$

where $X_t(1)$, $X_t(2)$ denote the two components of vector X_t. There are several standard ways to model the motion of the target; the most basic one is that of a random walk. See e.g. [2] for more background on target tracking.

Filtering is the task of computing the distribution of variable X_t, conditional on the data acquired until time t, $Y_{0:t}$. It is easy to check that, by taking a Feynman–Kac model such that

- the process $\{X_t\}$ has the same distribution as in the considered model; i.e. $\mathbb{M}_0(\mathrm{d}x_0) = \mathbb{P}_0(\mathrm{d}x_0)$, $M_t(x_{t-1}, \mathrm{d}x_t) = P_t(x_{t-1}, \mathrm{d}x_t)$ for any $x_{t-1} \in \mathfrak{X}$;
- the potential functions are set to $G_t(x_{t-1}, x_t) = f_t(y_t|x_t)$;

then one recovers as $\mathbb{Q}_t(\mathrm{d}x_{0:t})$ the distribution of variables $X_{0:t}$, conditional on $Y_{0:t} = y_{0:t}$; in particular $\mathbb{Q}_t(\mathrm{d}x_t)$ is the filtering distribution of the model.

We call this particular Feynman–Kac representation of the filtering problem the bootstrap model. Consider now a Feynman–Kac model with an *arbitrary* distribution for the Markov process $\{X_t\}$, and with potential

$$G_t(x_{t-1}, x_t) = \frac{P_t(x_{t-1}, \mathrm{d}x_t) f_t(y_t|x_t)}{M_t(x_{t-1}, \mathrm{d}x_t)},$$

the Radon–Nikodym derivative of $P_t(x_{t-1}, \mathrm{d}x_t) f_t(y_t|x_t)$ with respect to $M_t(x_{t-1}, \mathrm{d}x_t)$ (assuming the latter dominates the former). Whenever kernels P_t and M_t admit conditional probability densities (with respect to a common dominating measure), this expression simplifies to:

$$G_t(x_{t-1}, x_t) = \frac{p_t(x_t|x_{t-1}) f_t(y_t|x_t)}{m_t(x_t|x_{t-1})}. \tag{3}$$

Then again it is a simple exercise to check that one recovers as $\mathbb{Q}_t(\mathrm{d}x_t)$ the filtering distribution of the considered model. We call any Feynman–Kac model of this form a *guided* model. The bootstrap model corresponds to the special case where $P_t = M_t$.

We shall see in the following section that each Feynman–Kac model generates a different SMC algorithm. Thus, for a given state-space model, we have potentially an infinite number of SMC algorithms that may be used to approximate its sequence of filtering distributions. Which one to choose? We return to this point in Sect. 2.5.

2.4 Sequential Monte Carlo

Consider a given Feynman–Kac model. Sequential Monte Carlo amounts to compute recursive Monte Carlo approximations to the marginal distributions $\mathbb{Q}_t(\mathrm{d}x_t)$ of that model. At time 0, we simulate $X_0^n \sim \mathbb{M}_0(\mathrm{d}x_0)$ for $n = 1, \ldots, N$, and weight these 'particles' according to function G_0. Then

$$\mathbb{Q}_0^N(\mathrm{d}x_0) = \sum_{n=1}^N W_0^n \delta_{X_0^n}(\mathrm{d}x_0), \qquad W_0^n = \frac{G_0(X_0^n)}{\sum_{m=1}^N G_0(X_0^m)}$$

is an importance sampling approximation of $\mathbb{Q}_0(\mathrm{d}x_0)$, in the sense that

$$\mathbb{Q}_0^N(\varphi) = \sum_{n=1}^N W_0^n \varphi(X_0^n) \approx \mathbb{Q}_0(\varphi)$$

for any suitable test function φ.

To progress to time 1, recall from (2) that

$$\mathbb{Q}_1(\mathrm{d}x_{0:1}) = \frac{1}{\ell_1} \mathbb{Q}_0(\mathrm{d}x_0) M_1(x_0, \mathrm{d}x_1) G_1(x_0, x_1)$$

which suggests to perform importance sampling, with proposal $\mathbb{Q}_0(\mathrm{d}x_0) M_1(x_0, \mathrm{d}x_1)$, and weight function G_1. But since $\mathbb{Q}_0(\mathrm{d}x_0)$ is not available, we use instead $\mathbb{Q}_0^N$: that is, we sample N times from

$$\sum_{n=1}^{N} W_0^n \delta_{X_0^n}(\mathrm{d}x_0) M_1(X_0^n, \mathrm{d}x_1).$$

To do so, for each n, we draw $A_1^n \sim \mathscr{M}(W_0^{1:N})$, the multinomial distribution which generates value m with probability W_0^m; then we sample $X_1^n \sim M_1(X_0^{A_1^n}, \mathrm{d}x_1)$. We obtain in this way N pairs $(X_0^{A_1^n}, X_1^n)$, and we re-weight them according to function G_1. In particular

$$\mathbb{Q}_1^N(\mathrm{d}x_1) = \sum_{n=1}^{N} W_1^n \delta_{X_1^n}(\mathrm{d}x_1), \qquad W_1^n = \frac{G_1(X_0^{A_1^n}, X_1^n)}{\sum_{m=1}^{N} G_1(X_0^{A_1^m}, X_1^m)}$$

is our approximation of $\mathbb{Q}_1(\mathrm{d}x_1)$.

We proceed similarly at times $2, 3, \ldots$; see Algorithm 1. At every time t, we sample N points from

$$\sum_{n=1}^{N} W_{t-1}^n \delta_{X_{t-1}^n}(\mathrm{d}x_{t-1}) M_t(X_{t-1}^n, \mathrm{d}x_t)$$

and assign weights $W_t^n \propto G_t(X_{t-1}^{A_t^n}, X_t^n)$ to the so-obtained pairs $(X_{t-1}^{A_t^n}, X_t^n)$. Then we may use

$$\sum_{n=1}^{N} W_t^n \varphi(X_t^n)$$

as an approximation of $\mathbb{Q}_t(\varphi)$, for any test function $\varphi : \mathfrak{X} \to \mathbb{R}$. The approximation error of $\mathbb{Q}_t(\varphi)$ converges to zero at rate $\mathscr{O}_P(N^{-1/2})$, under appropriate conditions [7, 9].

Algorithm 1 Generic SMC sampler, for a given Feynman–Kac model

Step 0:

(a) Sample $X_0^n \sim \mathbb{M}_0(\mathrm{d}x_0)$ for $n = 1, \ldots, N$.
(b) Compute weight $W_0^n = G_0(X_0^n) / \sum_{m=1}^{N} G_0(X_0^m)$ for $n = 1, \ldots, N$.

Recursively, for $t = 1, \ldots, T$:

(a) Sample $A_t^{1:N} \sim \mathscr{M}(W_{t-1}^{1:N})$; see Appendix A.
(b) Sample $X_t^n \sim M_t(X_{t-1}^{A_t^n}, \mathrm{d}x_t)$ for $n = 1, \ldots, N$.
(c) Compute weight $W_t^n = G_t(X_{t-1}^{A_t^n}, X_t^n) / \sum_{m=1}^{N} G_t(X_{t-1}^{A_t^m}, X_t^m)$ for $n = 1, \ldots, N$.

2.5 Back to State-Space Models

We have explained in Sect. 2.3 that, for a given state-space model, there is an infinite number of Feynman–Kac models such that $\mathbb{Q}_t(\mathrm{d}x_t)$ is the filtering distribution. Thus, there is also an infinite number of SMC algorithms that may be used to approximate this filtering distribution.

Example 4 The Feynman–Kac model defined in Example 1 is such that $\mathbb{Q}_t(\mathrm{d}x_t)$ is the distribution of X_t conditional on $X_s \geq 0$ for all $0 \leq s \leq t$, where $\{X_t\}$ is a Gaussian auto-regressive process: $X_t = \phi X_{t-1} + V_t$, $V_t \sim N(0, 1)$. We may interpret $\mathbb{Q}_t(\mathrm{d}x_t)$ as the filtering distribution of a state-space model, where $\{X_t\}$ is the same auto-regressive process, $Y_t = \mathbb{1}_{\mathbb{R}^+}(X_t)$, and $y_t = 1$ for all t. Consider now the following alternative Feynman–Kac model: $M_t(x_{t-1}, \mathrm{d}x_t)$ is the Normal distribution $N(\phi x_{t-1}, 1)$ truncated to $\mathbb{R}^+$, i.e. the distribution with probability density

$$m_t(x_t|x_{t-1}) = \frac{\varphi(x_t - \phi x_{t-1})}{\Phi(\phi x_{t-1})}\mathbb{1}_{\mathbb{R}^+}(x_t)$$

where φ and Φ are respectively the PDF and CDF of a $N(0, 1)$ distribution; and $G_t(x_{t-1}, x_t) = \Phi(\phi x_{t-1})$, as per (3). Again, quick calculations show that we recover exactly the same distributions $\mathbb{Q}_t(\mathrm{d}x_t)$. Hence we have two SMC algorithms that approximate the same sequence of distributions (one for each Feynman–Kac model). Observe however that the latter SMC algorithm simulates all particles directly inside the region of interest ($\mathbb{R}^+$), while the former (bootstrap) algorithm simulates particles 'blindly', and assigns zero weight to those particles that fall outside $\mathbb{R}^+$. As a result, the latter algorithm tends to perform better. Note also that, under both Feynman–Kac formulations, L_t is the probability that $X_s \geq 0$ for all $0 \leq s \leq t$, hence both algorithms may be used to approximate this rare-event probability (see Sect. 2.8 below), but again the latter algorithm should typically give lower variance estimates for L_t.

Of course, the previous example is a bit simplistic, as far as state-space models are concerned. Recall from Sect. 2.3 that, for a given state-space model, any Feynman–Kac model such that G_t is set to (3) recovers the filtering distribution of that model for $\mathbb{Q}_t$. The usual recommendation is to choose one such Feynman–Kac model in a way that the variance of the weights of the corresponding SMC algorithm is low. To minimise the variance of the weights at iteration t, one should take [11] the guided Feynman–Kac model such that

$$M_t^{\mathrm{opt}}(x_{t-1}, \mathrm{d}x_t) \propto P_t(x_{t-1}, \mathrm{d}x_t) f_t(y_t|x_t),$$

the distribution of $X_t|\big(X_{t-1} = x_{t-1}, Y_t = y_t\big)$. In words, one should *guide* particles to a part of space $\mathfrak{X}$ where likelihood $x_t \to f_t(y_t|x_t)$ is high.

In fact, in the previous example, the second Feynman–Kac model corresponds precisely to this optimal kernel. Unfortunately, for most models sampling from the optimal kernel is not easy. One may instead derive an easy-to-sample kernel M_t that approximates M_t^{opt} in some way. Again, provided G_t is set to (3), one will recover the exact filtering distribution as $\mathbb{Q}_t$.

Example 5 In Example 2, [25] observed that the bootstrap filter performs poorly at iterations t where the data-point y_t is an outlier (i.e. takes a large absolute value). A potential remedy is to take into account y_t in some way when simulating X_t. To simplify the discussion, take $\mu = 0$, and consider the probability density of $X_t|X_{t-1}, Y_t$:

$$p_t(x_t|x_{t-1}, y_t) \propto \varphi((x_t - \phi x_{t-1})/\sigma)\varphi(y_t; 0, e^{x_t})$$
$$\propto \exp\left\{-\frac{1}{2\sigma^2}(x_t - \phi x_{t-1})^2 - \frac{x_t}{2} - \frac{y_t^2}{2e^{x_t}}\right\}.$$

It is not easy to simulate from this density, but [25] suggested to approximate it by linearizing $\exp(-x_t)$ around $x_t = \phi x_{t-1}$: $\exp(-x_t) \approx \exp(-\phi x_{t-1})(1 + \phi x_{t-1} - x_t)$. This leads to proposal density

$$m_t(x_t|x_{t-1}) \propto \exp\left\{-\frac{1}{2\sigma^2}(x_t - \phi x_{t-1})^2 - \frac{x_t}{2} - \frac{y_t^2}{2e^{\phi x_{t-1}}}(1 + \phi x_{t-1} - x_t)\right\}$$

which is clearly Gaussian (and hence easy to simulate from). Note that this linear 'approximation' does not imply that the resulting SMC algorithm is approximate in some way: provided G_t is set to (3), the resulting algorithm targets exactly the filtering distribution of the model, as we have already discussed.

2.6 *Sequential Quasi-Monte Carlo*

2.6.1 QMC Basics

As mentioned in the introduction, we assume that the reader is already familiar with QMC and RQMC (randomised QMC); otherwise see e.g. the books of [21, 22]. We only recall briefly the gist of QMC. Consider an expectation with respect to $\mathscr{U}\left([0, 1]^d\right)$, and its standard Monte Carlo approximation:

$$\frac{1}{N}\sum_{n=1}^{N}\varphi(U^n) \approx \int_{[0,1]^d}\varphi(u)\,\mathrm{d}u$$

where the U^n are IID variables. QMC amounts to replacing the U^n by N deterministic points $u^{n,N}$ that have low discrepancy. The resulting error converges faster than with Monte Carlo under certain conditions, in particular regarding the *regularity* of function φ. This is an important point when it comes to apply QMC in practice: rewriting a given algorithm as a deterministic function of uniforms, and replacing these uniforms by a QMC point set, may not warrant better performance. One has also to make sure that this deterministic function is indeed regular, and maintain low discrepancy in some sense.

2.6.2 SQMC When $d = 1$

We explained in Sect. 2.4 that SMC amounts to a sequence of importance sampling steps, with proposal distribution

$$\sum_{n=1}^{N} W_{t-1}^n \delta_{X_{t-1}^n}(\mathrm{d}x_{t-1}) M_t(x_{t-1}, \mathrm{d}x_t) \tag{4}$$

at time t. To derive a QMC version of this algorithm, we must find a way to generate a low-discrepancy sequence with respect to this distribution. The difficulty lies in the fact that the support of (4) is partly discrete (the choice of the ancestor X_{t-1}^n), partly continuous (the kernel $M_t(x_{t-1}, \mathrm{d}x_t)$). We focus on the discrete part below. For the continuous part, we assume that $\mathfrak{X} \subset \mathbb{R}^d$, and that we know of a function $\Gamma_t : \mathfrak{X} \times [0,1]^d \to \mathfrak{X}$ such that, for any $x_{t-1} \in \mathfrak{X}$, $\Gamma_t(x_{t-1}, U)$, $U \sim \mathscr{U}\left([0,1]^d\right)$, has the same distribution as $M_t(x_{t-1}, \mathrm{d}x_t)$. The choice of Γ_t is model-dependent, and is often easy; the default choice would be the Rosenblatt transform associated to $M_t(x_{t-1}, \mathrm{d}x_t)$ (the multivariate inverse CDF).

Example 6 Consider a state-space model with latent process $X_t = \phi X_{t-1} + V_t$, $V_t \sim N(0, \sigma^2)$. Then one would take typically $\Gamma_t(x_{t-1}, u) = \phi x_{t-1} + \sigma \Phi^{-1}(u)$, where Φ is the CDF of a $N(0, 1)$ distribution. In dimension $d > 1$, such a process would take the form $X_t = AX_{t-1} + V_t$, $V_t \sim N(0, \Sigma)$, where A is a $d \times d$ matrix. Then one would define $\Gamma_t(x_{t-1}, u) = Ax_{t-1} + \Pi_\Sigma(u)$, where the second term may be defined in several ways; e.g. (a) $\Pi_\Sigma(u)$ is the Rosenblatt transform of $N(0, \Sigma)$, i.e. first component of $\Pi_\Sigma(u)$ is $\Sigma_{11}^{1/2}\Phi^{-1}(u_1)$ and so on; or (b) $\Pi_\Sigma(u) = C\boldsymbol{\Phi}^{-1}(u)$, where C is the Cholesky lower triangle of Σ, $CC^T = \Sigma$, and $\boldsymbol{\Phi}^{-1}$ is the function which assigns to vector u the vector $\left(\Phi^{-1}(u(1)), \ldots, \Phi^{-1}(u(d))\right)^T$. In both cases, function Γ_t depends on the order of the components of X_t.

We now focus on the discrete component of (4). The standard approach to sample from such a finite distribution is the inverse CDF method: define $F_{t-1}^N(x) = \sum_{n=1}^N W_{t-1}^n \mathbb{1}\{n \leq x\}$, and set $\hat{X}_{t-1}^n = X_{t-1}^{A_t^n}$ with $A_t^n = \left(F_{t-1}^N\right)^{-1}(U_t^n)$, where $U_t^n \sim \mathscr{U}([0,1])$ and $\left(F_t^N\right)^{-1}$ is the generalised inverse of F_t^N. This is precisely how resampling is implemented in a standard particle filter. See Appendix A for a description of the standard algorithm to evaluate in $\mathscr{O}(N)$ time function $\left(F_{t-1}^N\right)^{-1}$ for N inputs.

A first attempt at introducing a QMC point set would be to set again $A_t^n = \left(F_t^N\right)^{-1}(U_t^n)$, but taking this time for U_t^n the first component of a QMC point set (of dimension $d + 1$). The problem with this approach is that this defines a transformation, from the initial uniforms to the points, which is quite irregular. In fact, since the labels of the N particles are arbitrary, this distribution somehow involves a *random permutation* of the N initial points. In other terms, we add some noise in our transformation, which is not a good idea in any type of QMC procedure.

Now consider the special case $\mathfrak{X} \subset \mathbb{R}$, and let $\sigma_{t-1} = \text{argsort}(X_{t-1}^{1:N})$, i.e. σ_{t-1} is a permutation of the N first integers such that:

$$X_{t-1}^{\sigma_{t-1}(1)} \leq \ldots \leq X_{t-1}^{\sigma_{t-1}(N)}$$

and, for $x \in \mathfrak{X}$, let

$$\hat{F}_{t-1}^N(x) = \sum_{n=1}^N W_{t-1}^n \mathbb{1}\left\{X_{t-1}^n \leq x\right\} = \sum_{n=1}^N W_{t-1}^{\sigma_{t-1}(n)} \mathbb{1}\left\{X_{t-1}^{\sigma_{t-1}(n)} \leq x\right\}.$$

Note that $\hat{F}_{t-1}^N$ does not depend on the *labels* of the N ancestors (like F_{t-1}^N does); for instance, the smallest x such that $\hat{F}_{t-1}^N(x) > 0$ is $X_{t-1}^{\sigma_{t-1}(1)}$, the smallest ancestor (whatever its label).

The first main idea in SQMC is to choose A_t^n such that $X_{t-1}^{A_t^n} = \hat{F}_{t-1}^N(U_t^n)$, where U_t^n is the first component of some QMC or RQMC point set. In this way, the resampled ancestors, i.e. the points $X_{t-1}^{A_t^n}$, may be viewed as a low-discrepancy point set with respect to the marginal distribution of component x_{t-1} in distribution (4). In practice, computing A_t^n amounts to (a) sort the N ancestors; and (b) apply the inverse CDF algorithm of Appendix A to these N sorted ancestors.

2.6.3 SQMC for $d > 1$

When $\mathfrak{X} \subset \mathbb{R}^d$, with $d > 1$, it is less clear how to invert the empirical CDF of the ancestors

$$\hat{F}_{t-1}^N(x) = \sum_{n=1}^N W_{t-1}^n \mathbb{1}\left\{X_t^n \leq x\right\}$$

as this function is $\mathbb{R}^d \to [0, 1]$.

The second main idea in SQMC is to transform the N ancestors X_{t-1}^n into N scalars Z_{t-1}^n, in a certain way that maintains the low discrepancy of the N initial points. Then we may construct a QMC point relative to

$$\hat{F}_{t-1,h}^N(z) = \sum_{n=1}^N W_{t-1}^n \delta_{Z_{t-1}^n}(\mathrm{d}z), \quad z \in [0, 1]$$

in the same way as described in the previous section.

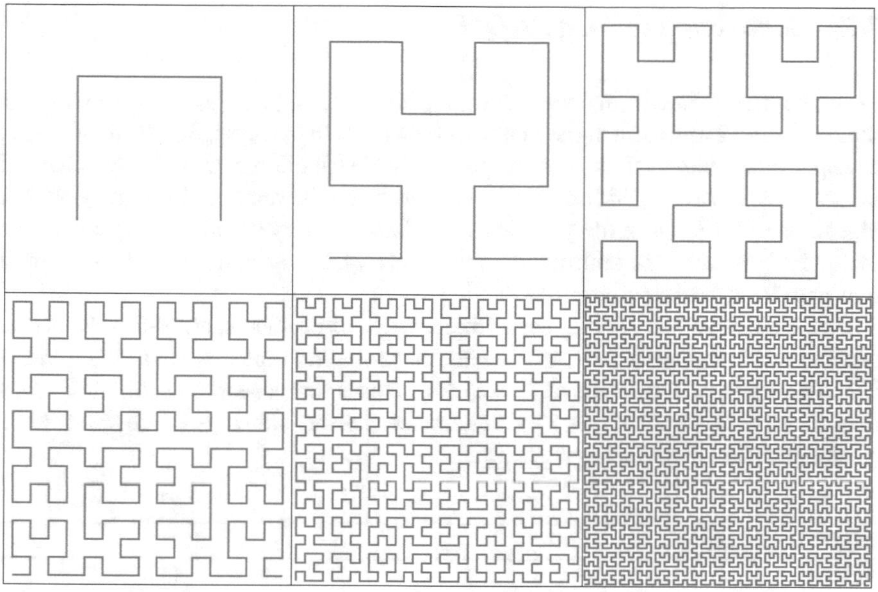

Fig. 1 Sequence of curves of which the Hilbert curve is the limit, for $d = 2$ (Source: Wikipedia)

To do so, we take $Z_{t-1}^n = h \circ \psi(X_{t-1}^n)$, where $h : [0, 1]^d \to [0, 1]$ is the inverse of the Hilbert curve, see below, and $\psi : \mathfrak{X} \to [0, 1]^d$ is model-dependent. (For instance, if $\mathfrak{X} = \mathbb{R}^d$, we may apply a component wise version of the logistic transform.)

The Hilbert curve is a space-filling curve, that is a function $H : [0, 1] \to [0, 1]^d$ with the following properties: it is defined as the limit of the process depicted in Fig. 1; it is Hölder with coefficient $1/d$ (in particular it is continuous); it 'fills' entirely $[0, 1]^d$; the set of points in $[0, 1]^d$ that admit more than one pre-image is of measure 0. Thanks to these properties, it is possible to define a pseudo-inverse $h : [0, 1] \to [0, 1]^d$, such that $H \circ h(u) = u$ for $u \in [0, 1]$.

In addition, the pseudo-inverse h maintains low-discrepancy in the following sense: if the N ancestors X_{t-1}^n are such that $\|\pi^N - \pi\|_E \to 0$ where $\pi^N(\mathrm{d}x) = \sum_{n=1}^N W_{t-1}^n \delta_{X_{t-1}^n}(\mathrm{d}x)$, and π is some limiting probability distribution, then (under appropriate conditions, see Theorem 3 in [16]), $\|\pi_h^N - \pi_h\|_E \to 0$, where π_h^N and π_h are the images of π^N and π through h. The extreme norm $\| \cdot \|_E$ in this theorem is some generalisation of the QMC concept of extreme discrepancy; again see [16] for more details.

We note that other functions $[0, 1]^d \to [0, 1]$ (e.g. pseudo-inverse of other space-filling curves, such as the Lebesgue curve) could be used in lieu of the inverse of the Hilbert curve. However, our impression is that other choices would not necessarily share the same property of "maintaining low discrepancy". At the very least, our proofs in [16] rely on properties that are specific to the Hilbert curve, and would not be easily extended to other functions.

Algorithm 2 summarises the operations performed in SQMC.

2.7 Connection to Array-RQMC

In the Feynman–Kac formalism, taking $G_0(x_0) = 1$, $G_t(x_{t-1}, x_t) = 1$ for all $t \geq 1$, makes $\mathbb{Q}_t$ the distribution of the Markov chain $\{X_t\}$. In that case, SQMC may be used to approximate expectations with respect to the distribution of that Markov chain. In fact, such a SQMC algorithm may be seen as a certain version of the array-RQMC algorithm of [20], where the particles are ordered at every iteration using the inverse of the Hilbert curve. In return, the convergence results established in [16] apply to that particular version of array-RQMC.

Although designed initially for a smaller class of problems, array-RQMC is built on the same insight as SQMC of viewing the problem of interest not a single Monte Carlo exercise, of dimension $d(T+1)$ (e.g. simulating a Markov chain in $\mathfrak{X} \subset R^d$ over $T+1$ time steps), but as $T+1$ exercises of dimension $d+1$. See also [14] for a related idea in the filtering literature.

Algorithm 2 SQMC algorithm

At time 0,

(a) Generate a QMC point set $u_0^{1:N}$ of dimension d.
(b) Compute $X_0^n = \Gamma_0(u_0^n)$ for all $n \in 1:N$.
(b) Compute $W_0^n = G_0(X_0^n)/\sum_{m=1}^N G_0(X_0^m)$ for all $n \in 1:N$.

Recursively, for time $t = 1:T$,

(a) Generate a QMC or RQMC point set $(u_t^{1:N}, v_t^{1:N})$ of dimension $d+1$ (u_t^n being the first component, and v_t^n the vector of the d remaining components, of point n).
(b) Hilbert sort: find permutation σ_t such that $h \circ \psi(X_{t-1}^{\sigma_t(1)}) \leq \ldots \leq h \circ \psi(X_{t-1}^{\sigma_t(N)})$ if $d \geq 2$, or $X_{t-1}^{\sigma(1)} \leq \ldots \leq X_{t-1}^{\sigma(N)}$ if $d = 1$.
(c) Generate $A_t^{1:N}$ using Algorithm 3, with inputs $\text{sort}(u_t^{1:N})$ and $W_t^{\sigma(1:N)}$, and compute $X_t^n = \Gamma_t(X_{t-1}^{\sigma_t(A_t^n)}, v_t^n)$.
(e) Compute $W_t^n = G_t(X_{t-1}^{\sigma(A_t^n)}, X_t^n)/\sum_{m=1}^N G_t(X_{t-1}^{\sigma(A_t^m)}, X_t^m)$ for all $n \in 1:N$.

2.8 Extensions

In state-space modelling, one may be interested in computing other quantities than the filtering distributions: in particular the likelihood of the data up to t, $p_t(y_{0:t})$, and the smoothing distribution, i.e. the joint law of the states $X_{0:T}$, given some complete dataset $Y_{0:T}$.

The likelihood of the data $p_t(y_{0:t})$ equals the normalising constant L_t in any guided Feynman–Kac model. This quantity may be estimated at iteration t as follows:

$$L_t^N = \left(\frac{1}{N}\sum_{n=1}^N G_0(X_0^n)\right)\prod_{s=1}^t\left(\frac{1}{N}\sum_{n=1}^N G_s(X_{s-1}^{A_s^n}, X_s^n)\right).$$

A non-trivial property of SMC algorithms is that this quantity is an unbiased estimate of L_t [8]. This makes it possible to develop MCMC algorithms for parameter estimation of state-space models which (a) runs at each MCMC iteration a particle filter to approximate the likelihood at given value of the parameter; and yet (b) targets the exact posterior distribution of the parameters, despite the fact the likelihood is computed only approximately. The corresponding PMCMC (particle MCMC) algorithms have been proposed in the influential paper of [1]. If we use RQMC (randomised QMC) point steps within SQMC, then L_t^N remains an unbiased estimate of L_t. Thus, SQMC is compatible with PMCMC (meaning that one may use SQMC instead of SMC at every iteration of a PMCMC algorithm), and in fact one may improve the performance of PMCMC in this way; see [16] for more details.

Smoothing is significantly more difficult than filtering. Smoothing algorithms usually amount to (a) run a standard particle filter, forward in time; (b) run a second algorithm, which performs some operations on the output of the first algorithm, backward in time. Such algorithms have complexity $O(N^2)$ in general. We refer the readers to [3, 13] for a general presentation of smoothing algorithms, and to [15] for how to derive QMC smoothing algorithms that offer better performance than standard (Monte Carlo-based) smoothing algorithms.

Finally, we mention that SMC algorithms may also be used in other contexts that the sequential inference of state-space models. Say we wish to approximate expectations with respect to some distribution of interest π, but it is is difficult to sample directly from π (e.g. the density π is strongly multimodal). One may define a geometric bridge between some easy to sample distribution π_0 and π as follows: $\pi_t(x) \propto \pi_0(x)^{1-\gamma_t}\pi(x)^{\gamma_t}$ where $0 = \gamma_0 < \ldots < \gamma_T = 1$. Then one may apply SMC to the sequence (π_t), and use the output of the final iteration to approximate π. Other sequence of distributions may be considered as well. For more background on such applications of SMC see e.g. [6, 10, 23]. The usefulness of SQMC for such problems remains to be explored.

2.9 A Note on the Impact of the Dimension

Reference [16] include a numerical study of the impact of the dimension on the performance of SQMC. It is observed that the extra performance of SQMC (relative to standard SMC) quickly decreases with the dimension.

Three factors may explain this curse of dimensionality:

1. The inherent curse of dimensionality of QMC: the standard discrepancy bounds invoked as a formal justification of QMC deteriorate with the dimension.
2. Regularity of the Hilbert curve: the Hilbert curve is Hölder with coefficient $1/d$. Consequently, the mapping $u_t^n \mapsto X_{t-1}^{\sigma_t(A_{t-1}^n)}$ induced by steps (a) and (b) of Algorithm 2 for time $t \geq 1$ is less and less regular as the dimension increases. (We however believe that this property is not specific to the use of the Hilbert curve but is due to the resampling mechanism itself, where a single point in $u_t^n \in [0, 1]$ is used to select the d-dimensional ancestor $X_{t-1}^{A_t^n}$.)

3. SMC curse of dimensionality: SMC methods also suffer from the curse of dimensionality, for the simple reason that they rely on importance sampling: the larger the dimension, the greater the discrepancy between the proposal distribution and the target distribution. In practice, one observes in high-dimensional filtering problem that, at each iteration, only a small proportion of the particles get a non-negligible weight.

We thought earlier that factor 2 was the 'main culprit'. However, factor 3 seems to play an important part as well. To see this, we compare below the relative performance of SQMC and SMC for the filtering of the following class of linear Gaussian state-space models (as in [17]): $X_0 \sim N_d(0, I_d)$, and

$$X_t = FX_{t-1} + V_t, \qquad V_t \sim N_d(0, I_d),$$
$$Y_t = X_t + W_t, \qquad W_t \sim N_d(0, I_d),$$

with $F = (\alpha^{|i-j|})_{i,j=1:d}$, and $\alpha = 0.4$. For such models, the filtering distribution may be computed exactly using the Kalman filter [18]. We consider two Feynman–Kac formalisms of that problem:

- The bootstrap formalism, where M_t is set to $N_d(FX_{t-1}, I_d)$, the distribution of $X_t|X_{t-1}$ according to the model, and $G_t(x_{t-1}, x_t) = f_t(y_t|x_t) = N_d(y_t; x_t, I_d)$, the probability density at point y_t of distribution $N_d(x_{t-1}, I_d)$.
- The 'optimal' guided formalism where

$$M_t(x_{t-1}, dx_t) \propto P_t(x_{t-1}, dx_t) f_t(y_t|x_t) \sim N_d\left(\frac{Y_t + FX_{t-1}}{2}, \frac{1}{2}I_d\right)$$

  and, by (3),

$$G_t(x_{t-1}, x_t) = N_d(y_t; Fx_{t-1}, 2I_d)$$

  the probability density at point y_t of distribution $N_d(Fx_{t-1}, 2I_d)$.

In both cases, as already explained, we recover the filtering distribution as $\mathbb{Q}_t$. But the latter formalism is chosen so as to minimise the variance of the weights at each iteration.

We simulate $T = 50$ data-points from the model, for $d = 5, 10, 15$ and 20. Figure 2 compares the following four algorithms: SMC-bootstrap, SQMC-bootstrap, SMC-guided, and SQMC-guided. The comparison is in terms of the MSE (mean square error) of the estimate of the filtering expectation of the first component of X_t, i.e. $\mathbb{E}[X_t(1)|Y_{0:t} = y_{0:t}]$. We use SMC-guided as the reference algorithm, and we plot for each of the three other algorithms the variations of the gain (MSE of reference algorithm divided by MSE of considered algorithm) for the T estimates. (We use violin plots, which are similar to box-plots, except that the box is replaced by kernel density estimates.) A gain g means that the considered algorithm would need g times less particles (roughly) to provide an estimate with a similar variance (to that of the reference algorithm). Each algorithm was run with $N = 10^4$.

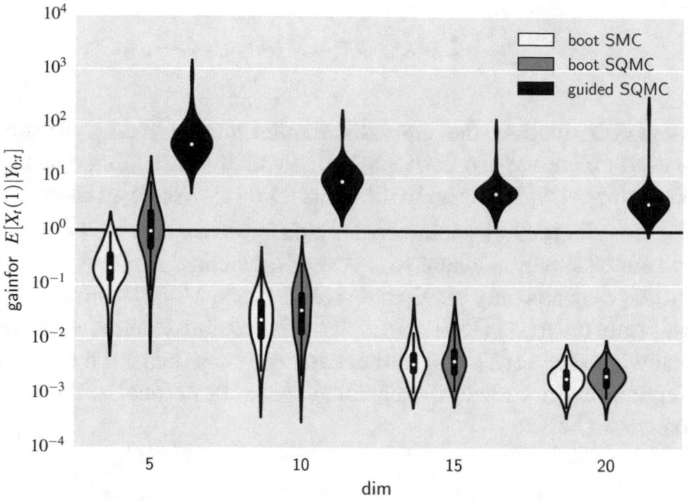

Fig. 2 Violin plots of the gains of the considered algorithms when estimating the filtering expectations $\mathbb{E}[X_t(1)|Y_{0:t}]$ for $t = 0, \ldots, T = 50$. (Each violin plot represents the variability of the T gains for these T estimates.) Gain is MSE (mean square error) of reference algorithm (guided-SMC) divided by MSE of considered algorithm

First, we observe that guided algorithms outperforms bootstrap algorithms more and more as the dimension increases. Second, for bootstrap algorithms, the performance between SMC and SQMC is on par as soon as $d \geq 10$. (In fact, the performance is rather bad in both cases, owning to the aforementioned curse of dimensionality.) On the other hand, for guided formalisms we still observe a gain of order $\mathcal{O}(10^1)$ (resp. $10^{0.5}$) for $d = 10$ (resp. $d = 20$).

The bottom line is that the amount of extra performance brought by SQMC (relative to SMC) depends strongly on the chosen Feynman–Kac formalism. If one is able to construct a Feynman–Kac formalism (for the considered problem) that leads to good performance for the corresponding SMC algorithm (meaning that the variance of the weights is low at each iteration), then one may expect significant extra performance from SQMC, even in high dimension.

3 Application to Diffusions

3.1 Dimension Reduction in SQMC

We start this section by a basic remark, which makes it possible to improve the performance of SQMC when applied to models having a certain structure. We explained in Sect. 2.4 that SMC amounts to performing importance sampling at every step, using as a proposal distribution:

$$\sum_{n=1}^{N} W_{t-1}^n \delta_{X_{t-1}^n}(\mathrm{d}x_{t-1}) M_t(x_{t-1}, \mathrm{d}x_t) \tag{5}$$

and as a target distribution, the same distribution times $G_t(x_{t-1}, x_t)$ (up to a constant). We used this remark to derive SQMC as an algorithm that constructs a low-discrepancy point-set with respect to the distribution above; i.e. to construct N points $(X_{t-1}^{A_t^n}, X_t^n)$, the empirical distribution of which approximates well (5).

Now consider a situation where we know of a function $\Lambda : \mathfrak{X} \to \mathbb{R}^k$, with $k < d$, such that (a) G_t depends only on X_t and $\Lambda(X_{t-1})$; and Markov kernel $M_t(x_{t-1}, \mathrm{d}x_t)$ also depends only on $\Lambda(x_{t-1})$. (In particular, it is possible to simulate X_t conditional on X_{t-1}, knowing only $\Lambda(X_{t-1})$.) In that case, one may define the same importance sampling operation on a lower-dimensional space. In particular, the new proposal distribution would be:

$$\sum_{n=1}^{N} W_{t-1}^n \delta_{\Lambda(X_{t-1}^n)}(\mathrm{d}\lambda_{t-1}) M_t^{\Lambda}(\lambda_{t-1}, \mathrm{d}x_t)$$

where $M_t^{\Lambda}(\lambda_{t-1}, \mathrm{d}x_t)$ is simply the Markov kernel which associates distribution $M_t(x_{t-1}, \mathrm{d}x_t)$ to any x_{t-1} such that $\Lambda(x_{t-1}) = \lambda_{t-1}$. We may use exactly the same ideas as before, i.e. generate a QMC point of dimension $d + 1$, and use the first component to pick the ancestor. However, the Hilbert sorting is now applied to the N points $\Lambda(X_{t-1}^n)$, and therefore operates in a smaller dimension. Thus one may expect better performance, compared to the standard version of SQMC.

This remark is related somehow to the QMC notion of "effective dimension": the performance of QMC may remain good in high-dimensional problems, if one is able to reformulate the problem in such a way that it depends "mostly" (or in our case, "only") on a few dimensions of the state-space.

3.2 Filtering of Diffusion Processes

We now consider the general class of diffusion-driven state-space models:

$$\begin{aligned}
\mathrm{d}\widetilde{X}_t &= \mu_X(\widetilde{X}_t) + \sigma_X(\widetilde{X}_t)\mathrm{d}W_t^X \\
\mathrm{d}\widetilde{Y}_t &= \mu_Y(\widetilde{X}_t) + \sigma_Y(\widetilde{X}_t)\mathrm{d}W_t^Y
\end{aligned}$$

where $(W_t^X)_{t\geq 0}$ and $(W_t^Y)_{t\geq 0}$ are possibly correlated Wiener processes. Functions μ_X, μ_Y, σ_X and σ_Y may also depend on t, and μ_Y, σ_Y may also depend on Y_t, but for the sake of exposition we stick to the simple notations above.

Filtering in continuous time amounts to recover the distribution of $\widetilde{X}_t$ conditional on trajectory $y_{[0:t]}$ (i.e. the observation of process $\{\widetilde{Y}_t\}$ over interval $[0, t]$). However, in most practical situations, one does not observe process $\{\widetilde{Y}_t\}$ continuously, but on a grid. To simplify, we assume henceforth that process $\{\widetilde{Y}_t\}$ is observed at times $t \in \mathbb{N}$ and we rewrite the above model as

$$\begin{aligned} \mathrm{d}\widetilde{X}_t &= \mu_X(\widetilde{X}_t) + \sigma_X(\widetilde{X}_t)\mathrm{d}W_t^X \\ \widetilde{Y}_{t+1} &= \widetilde{Y}_t + \int_t^{t+1} \mu_Y(\widetilde{X}_s)\mathrm{d}s + \int_t^{t+1} \sigma_Y(\widetilde{X}_s)\mathrm{d}W_s^Y. \end{aligned} \tag{6}$$

It is typically too difficult to work directly in continuous time. Thus, as standardly done when dealing with such processes, we replace the initial process $(\widetilde{X}_t)$ by its (Euler-) discretized version $\{X_t\}$, with discretisation step $\delta = 1/M$, $M \geq 1$. That is, $\{X_t\}$ is a $\mathbb{R}^M$−valued process, where X_t is a M-dimensional vector representing the original process at times $t, t + 1/M, \dots, t + 1 - 1/M$, which is defined as:

$$\begin{aligned} X_t(1) &= X_{t-1}(M) + \delta\mu_X\left(X_{t-1}(M)\right) + \sigma_X(X_{t-1}(M))\left\{W_{t+\delta}^X - W_t^X\right\} \\ &\vdots \\ X_t(M) &= X_t(M-1) + \delta\mu_X\left(X_t(M-1)\right) + \sigma_X(X_t(M-1))\left\{W_{t+1}^X - W_{t+1-\delta}^X\right\} \end{aligned} \tag{7}$$

and the resulting dicretization of (6) is given by

$$Y_{t+1} = Y_t + \delta \sum_{m=1}^{M} \mu_Y\left(X_t(m)\right) + \sum_{m=1}^{M} \sigma_Y\left(X_t(m)\right)\left\{W_{t+\delta m}^Y - W_{t+\delta(m-1)}^Y\right\}. \tag{8}$$

SQMC may be applied straightforwardly to the filtering of the discretized model defined by (7) and (8). However, the choice of the $\delta = 1/M$ becomes problematic. We would like to take M large, to reduce the discretization bias. But M is also the dimension of the state-space, so a large M may mean a degradation of performance for SQMC (relative to SMC).

Fortunately, the dimension reduction trick of the previous section applies here. For simplicity, consider the bootstrap Feynman–Kac formalism of this particular state-space model:

- $M_t(x_{t-1}, \mathrm{d}x_t)$ is the distribution of $X_t|X_{t-1}$ defined by (7); observe that it only depends on $X_{t-1}(M)$, the last component of X_{t-1};
- $G_t(x_{t-1}, x_t)$ is the probability density of datapoint y_t given $X_t = x_t$ and $Y_{t-1} = y_{t-1}$, induced by (7) and (8); observe that it does not depend on x_{t-1} when (W_t^X) and (W_t^Y) are uncorrelated and that it depends on x_{t-1} only through $x_{t-1}(M)$ when these two processes are correlated (see the next subsection).

Hence we may define $\Lambda(x_{t-1}) = x_{t-1}(M) \in \mathbb{R}$. The Hilbert ordering step may be applied to the values $Z_t^n = X_{t-1}^n(M)$. In fact, since these values are scalars, there is no need to implement any Hilbert ordering, a standard sorting is enough.

3.3 QMC and Brownian Motion

We now briefly discuss how to choose Γ_t, the deterministic function such that $\Gamma_t(x_{t-1}, v)$, for $x_{t-1} \in \mathfrak{X}$ and $v \in [0, 1]^d$, returns a variate from kernel $M_t(x_{t-1}, dx_t)$.

The distribution of $X_t|X_{t-1}$ defined in the previous section is a simple linear transform of the distribution of a Brownian path on a regular grid. Thus, defining function Γ_t amounts to constructing a certain function $[0, 1]^M \to \mathbb{R}^M$ that transforms $\mathscr{U}\left([0, 1]^M\right)$ into the joint distribution of $(W^X_{t+\delta}, \ldots, W^X_{t+1})$, conditional on W^X_t.

It is well known in the QMC literature (e.g. Sect. 8.2 of [22]) that there is more than one way to write the simulation of a Brownian path as a function of uniforms, and that the most obvious way may perform poorly when applied in conjunction with QMC. More precisely, consider the following two approaches:

1. Forward construction: simulate independently the increments $W^X_{t+\delta m} - W^X_{t+\delta(m-1)}$ from a $N(0, \delta)$ distribution.
2. Brownian bridge construction [4]: Simulate $(W^X_{t+\delta}, \ldots, W^X_{t+1})$ given W^X_t sequentially according to the Van der Corput sequence: $W^X_{t+\delta\lceil M/2\rceil}, W^X_{t+\delta\lceil M/4\rceil}, W^X_{t+\delta\lceil 3M/4\rceil}$ until all the components of vector $(W^X_{t+\delta}, \ldots, W^X_{t+1})$ are simulated. For instance, for $s < t' < u$, we use

$$W^X_{t'}|W^X_s, W^X_u \sim N_1\left(\frac{u-t'}{u-s}W^X_s + \frac{t'-s}{u-s}W^X_u, \frac{(u-t')(t'-s)}{u-s}\right)$$

and the fact that (W^X_t) is a Markov process (i.e. $W^X_t|W^X_s$ does not depend on $W^X_{s'}$ for $s' < s$).

In both cases, it is easy to write the simulation of $(W^X_{t+\delta}, \ldots, W^X_{t+1})$ as a function of M uniform variates. However, in the first case, the obtained function depends in the same way on each of the M variates, while in the second case, the function depends less and less on the successive components. This mitigates the inherent curse of dimensionality of QMC [4].

We shall observe the same phenomenon applies to SQMC; even so for a moderate value of M, interestingly. We also mention briefly the PCA (principal components analysis) construction as another interesting way to construct Brownian paths, and refer again to Sect. 8.2 [22] for a more in-depth discussion of QMC and Brownian paths.

Lastly, although we focus on univariate diffusion processes in this section for the sake of simplicity, the above considerations also hold for multivariate models. Notably, the Brownian bridge construction is easily generalizable to the case where (W^X_t) is a d-dimensional vector of correlated Wiener processes. The dimension of the QMC point set used as input of SQMC is then of size $dM + 1$ and the Hilbert ordering would operate on a d-dimensional space.

3.4 Numerical Experiments

To illustrate the discussion of the previous subsections we consider the following diffusion driven stochastic volatility model (e.g. [5])

$$\mathrm{d}\widetilde{X}_t = \left\{\kappa(\mu^X - e^{\widetilde{X}_t})e^{-\widetilde{X}_t} - 0.5\omega^2 e^{-X_t}\right\}\mathrm{d}t + \omega e^{-\widetilde{X}_t/2}\mathrm{d}W_t^X$$

$$\widetilde{Y}_{t+1} = \widetilde{Y}_t + \int_t^{t+1} \{\mu^Y + \beta e^{\widetilde{X}_z}\}\mathrm{d}z + \int_t^{t+1} e^{\widetilde{X}_s/2}\mathrm{d}W_s^Y$$

where (W_t^X) and (W_t^Y) are Wiener processes with correlation coefficient $\rho \in (-1, 1)$, $\omega > 0$, $\kappa > 0$ while the other parameters μ^Y, β are in $\mathbb{R}$.

To fit this model into the bootstrap Feynman–Kac formalism that we consider in this section, note that, for $t \geq 0$,

$$\widetilde{Y}_{t+1}|\widetilde{Y}_t, \widetilde{X}_{[t,t+1]} \sim N\left(\widetilde{Y}_t + \mu^Y + \beta\sigma_{t+1}^2 + \rho Z_{t+1},\ (1-\rho^2)\sigma_{t+1}^2\right)$$

with $\sigma_{t+1}^2 = \int_t^{t+1} e^{\widetilde{X}_s}\mathrm{d}s$ and $Z_{t+1} = \int_t^{t+1} e^{\widetilde{X}_s/2}\mathrm{d}W_s^X$, and thus, as explained in Sect. 3.2,

$$\begin{aligned} G_t(x_{t-1}, x_t) &= \tilde{G}_t(x_{t-1}(M), x_t) \\ &:= N\left(\widetilde{Y}_{t+1}; \widetilde{Y}_t + \mu^Y + \beta\hat{\sigma}_{t+1}^2(x_t) + \rho\hat{Z}_{t+1}(x_{t-1}(M), x_t),\ (1-\rho^2)\hat{\sigma}_{t+1}^2(x_t)\right) \end{aligned}$$

where

$$\hat{\sigma}_{t+1}^2(x_t) = \frac{1}{M}\sum_{m=1}^{M} e^{x_t(m)},\quad \hat{Z}_{t+1}(x_{t-1}(M), x_t) = \sum_{m=1}^{M} e^{\frac{x_t(m)}{2}}\left(W_{t+m\delta}^X - W_{t+(m-1)\delta}^X\right).$$

Note that $W_{t+m\delta}^X - W_{t+(m-1)\delta}^X$ depends on $(x_{t-1}(M), x_t)$ through (7). To complete the model we take for $\mathbb{M}_0(\mathrm{d}x_0)$, the initial distribution of process $\{X_t\}$, the density of the $N\left(\mu^X, \omega^2/(2\kappa)\right)$ distribution.

We set the parameters of the model to their estimated values for the daily return data on the closing price of the S&P 500 index from 5/5/1995 to 4/14/2003 [5] and simulate observations $\{Y_t\}_{t=0}^T$ using the discretized model (7) and (8) with $M = 20\,000$. The number of observations T is set to 4 000.

Below we compare SMC with SQMC based on the forward construction and on the Brownian bridge construction of Brownian paths. In both cases, SQMC is implemented using as input a nested scrambled [24] Sobol' sequence. The performance of these three algorithms is compared, for $t = 1, \ldots, T$, for the estimation of (1) the filtering expectation $\mathbb{E}[X_t|Y_{0:t}]$ and (2) of the log-likelihood function $\log(L_t)$.

Figure 3 shows the ratio of the SMC variance over the SQMC variance for the two alternative implementations of SQMC. Results are presented for a discretization grid of size $M = 5$ and for different number of particles N. Two observations are worth

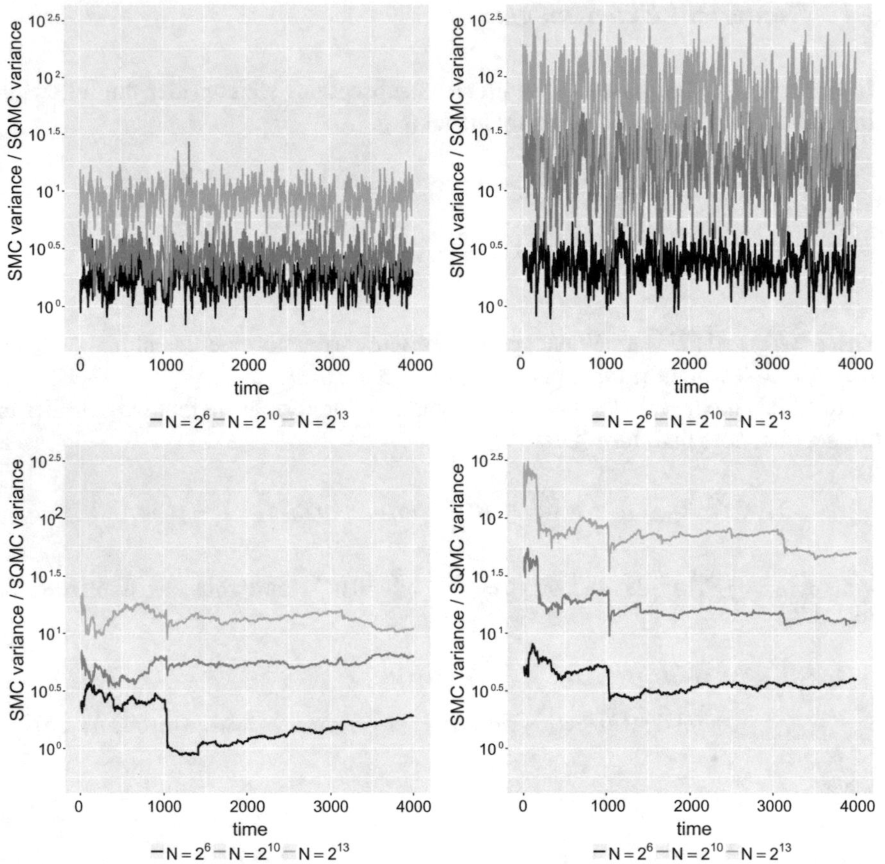

Fig. 3 Estimation of $\mathbb{E}[X_t | Y_{0:t} = y_{0:t}]$ (top plots) and of $\log p(y_{0:t})$ for $t \in \{0, \dots, T\}$ and for different values of N. SQMC is implemented with the forward construction (left plots) and with the Brownian Bridge construction of Brownian paths (right plots), and $M = 5$

noting from this figure. First, the two versions of SQMC outperform SMC in terms of variance. Second, the variance reduction is much larger with the Brownian bridge construction than with the forward construction of Brownian paths, as expected from the discussion of the previous subsection. Note that for both versions of SQMC the ratio of variances increases with the number of particles, showing that SQMC converges faster than the $N^{-1/2}$ Monte Carlo error rate.

In Fig. 4 we perform the same analysis than in Fig. 3 but now with $M = 10$ and $M = 20$ discretization steps. ($M = 10$ is considered as sufficient for parameter estimation by [5].) Results are presented only for the Brownian bridge construction. Despite the large dimension of the QMC point set used as input, we observe that SQMC converges much faster than the $N^{-1/2}$ Monte Carlo error rate. In particular, we observe that the gains in term of variance brought by SQMC are roughly similar

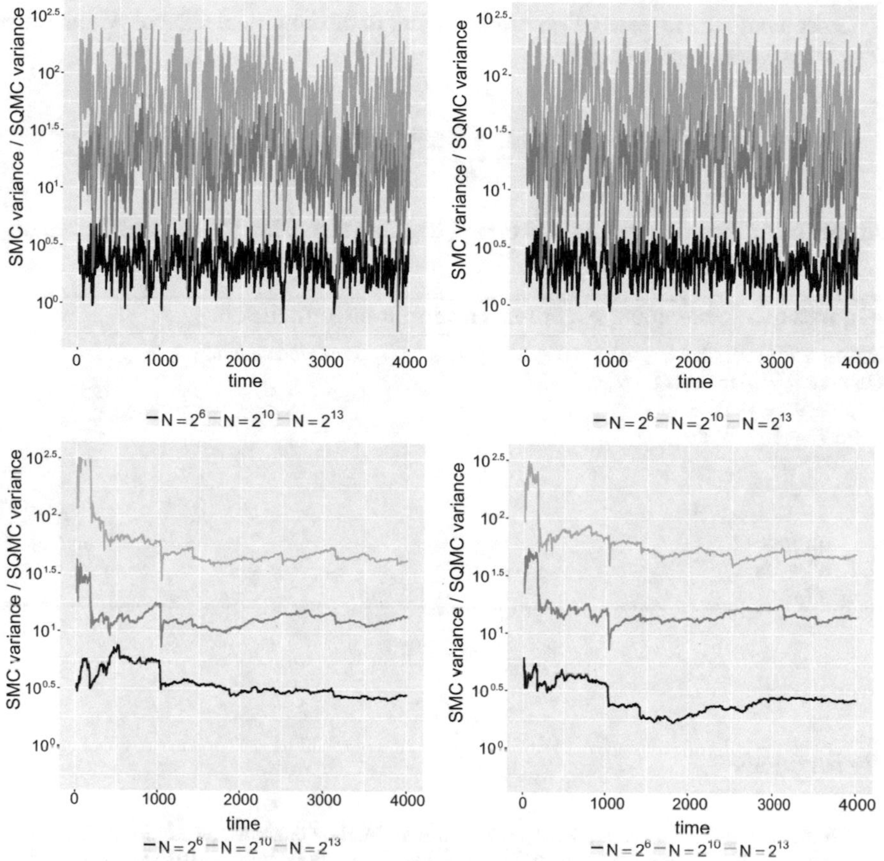

Fig. 4 Estimation of $\mathbb{E}[X_t|Y_{0:t} = y_{0:t}]$ (top) and of $\log p(y_{0:t})$ for $t \in \{0, \ldots, T\}$ and for different values of N. SQMC is implemented with the Brownian Bridge construction of Brownian paths. Results are presented for $M = 10$ (left plots) and for $M = 20$

whatever the choice of M is. As explained above, this observation suggests that the effective dimension of the model remains low (or even constant in the present setting) even when the "true" dimension M increases.

Appendix A: Resampling

Algorithm 3 below takes as input N sorted points $u^1 \leq \ldots \leq u^n$, and N weights W^n, and return as an output the N values $\left(F^N\right)^{-1}(u^n)$, where $\left(F^N\right)^{-1}$ is the inverse CDF relative to CDF $F^N(z) = \sum_{n=1}^{N} W^n \mathbb{1}\{n \leq z\}$, $z \in \mathbb{R}$. Its complexity is $O(N)$.

To compute the inverse CDF corresponding to the empirical CDF of N ancestors (as discussed in Sect. 2.6.2), i.e.

$$F^N(x) = \sum_{n=1}^{N} W^n \mathbb{1}\left\{X^n \leq x\right\}$$

simply order the N ancestors, and apply the same algorithm to the sorted ancestors.

Algorithm 3 Resampling Algorithm (inverse transform method)

Input: $u^{1:N}$ (such that $0 \leq u^1 \leq \ldots \leq u^N \leq 1$, $W^{1:N}$ (normalised weights)
Output: $a^{1:N}$ (labels in $1:N$)

```
s ← 0, m ← 0
for n = 1 → N do
    repeat
        m ← m + 1
        s ← s + W^m
    until s > u^n
    a^n ← m
end for
```

References

1. Andrieu, C., Doucet, A., Holenstein, R.: Particle Markov chain Monte Carlo methods. J. R. Stat. Soc. Ser. B Stat. Methodol. **72**(3), 269–342 (2010). https://doi.org/10.1111/j.1467-9868.2009.00736.x
2. Arulampalam, M.S., Maskell, S., Gordon, N., Clapp, T.: A tutorial on particle filters for online nonlinear/non-Gaussian Bayesian tracking. IEEE Trans. Signal Process. **50**(2), 174–188 (2002)
3. Briers, M., Doucet, A., Maskell, S.: Smoothing algorithms for state-space models. Ann. Inst. Stat. Math. **62**(1), 61–89 (2010). https://doi.org/10.1007/s10463-009-0236-2
4. Caflisch, R.E., Morokoff, W.J., Owen, A.B.: Valuation of mortgage backed securities using Brownian bridges to reduce effective dimension. Department of Mathematics, University of California, Los Angeles (1997)
5. Chib, S., Pitt, M.K., Shephard, N.: Likelihood-based inference for diffusion models. Technical report, Nuffield College, Oxford (2004)
6. Chopin, N.: A sequential particle filter method for static models. Biometrika **89**(3), 539–551 (2002). https://doi.org/10.1093/biomet/89.3.539
7. Chopin, N.: Central limit theorem for sequential Monte Carlo methods and its application to Bayesian inference. Ann. Stat. **32**(6), 2385–2411 (2004). https://doi.org/10.1214/009053604000000698
8. Del Moral, P.: Non-linear filtering: interacting particle resolution. Markov Process. Relat. Fields **2**(4), 555–581 (1996)
9. Del Moral, P., Guionnet, A.: Central limit theorem for nonlinear filtering and interacting particle systems. Ann. Appl. Probab. **9**(2), 275–297 (1999). https://doi.org/10.1214/aoap/1029962742
10. Del Moral, P., Doucet, A., Jasra, A.: Sequential Monte Carlo samplers. J. R. Stat. Soc. Ser. B Stat. Methodol. **68**(3), 411–436 (2006). https://doi.org/10.1111/j.1467-9868.2006.00553.x

11. Doucet, A., Godsill, S., Andrieu, C.: On sequential Monte Carlo sampling methods for Bayesian filtering. Stat. Comput. **10**(3), 197–208 (2000). https://doi.org/10.1023/A:1008935410038
12. Doucet, A., de Freitas, N., Gordon, N.J.: Sequential Monte Carlo Methods in Practice. Springer, New York (2001)
13. Doucet, A., Kantas, N., Singh, S.S., Maciejowski, J.M.: An overview of Sequential Monte Carlo methods for parameter estimation in general state-space models. In: Proceedings IFAC System Identification (SySid) Meeting (2009)
14. Fearnhead, P.: Using random quasi-Monte-Carlo within particle filters, with application to financial time series. J. Comput. Graph. Stat. **14**(4), 751–769 (2005). https://doi.org/10.1198/106186005X77243
15. Gerber, M., Chopin, N.: Convergence of sequential quasi-Monte Carlo smoothing algorithms. arXiv:1506.06117 (2015)
16. Gerber, M., Chopin, N.: Sequential quasi Monte Carlo. J. R. Stat. Soc. Ser. B. Stat. Methodol. **77**(3), 509–579 (2015). https://doi.org/10.1111/rssb.12104
17. Guarniero, P., Johansen, A.M., Lee, A.: The iterated auxiliary particle filter. J. Am. Stat. Assoc. **112**(520), 1636–1647 (2017)
18. Kalman, R.E., Bucy, R.S.: New results in linear filtering and prediction theory. Trans. ASME Ser. D J. Basic Eng. **83**, 95–108 (1961). https://doi.org/10.1115/1.3658902
19. Kim, S., Shephard, N., Chib, S.: Stochastic volatility: likelihood inference and comparison with arch models. Rev. Econ. Stud. **65**(3), 361–393 (1998)
20. L'Ecuyer, P., Lécot, C., Tuffin, B.: A randomized quasi-Monte Carlo simulation method for Markov chain. In: Monte-Carlo and quasi Monte-Carlo methods 2004, pp. 331–342. Springer, Berlin (2006)
21. Lemieux, C.: Monte Carlo and Quasi-Monte Carlo Sampling. Springer Series in Statistics. Springer, Berlin (2009)
22. Leobacher, G., Pillichshammer, F.: Introduction to quasi-Monte Carlo integration and applications. Compact Textbook in Mathematics. Birkhäuser/Springer, Cham (2014). https://doi.org/10.1007/978-3-319-03425-6
23. Neal, R.M.: Annealed importance sampling. Stat. Comput. **11**(2), 125–139 (2001). https://doi.org/10.1023/A:1008923215028
24. Owen, A.B.: Randomly permuted (t, m, s)-nets and (t, s)-sequences. Monte Carlo and Quasi-Monte Carlo Methods in Scientific Computing. Lecture Notes in Statistics, vol. 106, pp. 299–317. Springer, New York (1995)
25. Pitt, M.K., Shephard, N.: Filtering via simulation: auxiliary particle filters. J. Am. Stat. Assoc. **94**(446), 590–599 (1999). https://doi.org/10.2307/2670179

Hot New Directions for Quasi-Monte Carlo Research in Step with Applications

Frances Y. Kuo and Dirk Nuyens

Abstract This article provides an overview of some interfaces between the theory of quasi-Monte Carlo (QMC) methods and applications. We summarize three QMC theoretical settings: first order QMC methods in the unit cube $[0, 1]^s$ and in $\mathbb{R}^s$, and higher order QMC methods in the unit cube. One important feature is that their error bounds can be independent of the dimension s under appropriate conditions on the function spaces. Another important feature is that good parameters for these QMC methods can be obtained by fast efficient algorithms even when s is large. We outline three different applications and explain how they can tap into the different QMC theory. We also discuss three cost saving strategies that can be combined with QMC in these applications. Many of these recent QMC theory and methods are developed not in isolation, but in close connection with applications.

Keywords Quasi-Monte Carlo methods · Uniform · Lognormal · Higher order Randomly shifted lattice rules · Interlaced polynomial lattice rules

1 Introduction

High dimensional computation is a new frontier in scientific computing, with applications ranging from financial mathematics such as option pricing or risk management, to groundwater flow, heat transport, and wave propagation. A tremendous amount of progress has been made in the past two decades on the theory and application of *quasi-Monte Carlo* (*QMC*) *methods* for approximating high dimensional integrals.

F. Y. Kuo (✉)
School of Mathematics and Statistics, University of New South Wales,
Sydney, NSW 2052, Australia
e-mail: f.kuo@unsw.edu.au

D. Nuyens
Department of Computer Science, KU Leuven, Celestijnenlaan 200A,
3001 Leuven, Belgium
e-mail: dirk.nuyens@cs.kuleuven.be

A. B. Owen and P. W. Glynn (eds.), *Monte Carlo and Quasi-Monte Carlo Methods*, Springer Proceedings in Mathematics & Statistics 241,
https://doi.org/10.1007/978-3-319-91436-7_6

See e.g., the classical references [64, 75] and the recent books [8, 60, 61]. One key element is the *fast component-by-component construction* [6, 66–68] which provides parameters for first order or *higher order QMC methods* [8, 12] for sufficiently smooth functions. Another key element is the careful selection of parameters called *weights* [76, 78] to ensure that the worst case errors in an appropriately weighted function space are bounded independently of the dimension. The dependence on dimension is very much the focus of the study on *tractability* [65] of multivariate problems.

We are particularly keen on the idea that new theory and methods for high dimensional computation are developed not in isolation, but in close connection with applications. The theoretical QMC convergence rates depend on the appropriate pairing between the function space and the class of QMC methods. Practitioners are free to choose the theoretical setting or pairing that is most beneficial for their applications, i.e., to achieve the best possible convergence rates under the weakest assumptions on the problems. As QMC researchers we take application problems to be our guide to develop new theory and methods as the needs arise. This article provides an overview of some interfaces between such theory and applications.

We begin in Sect. 2 by summarizing three theoretical settings. The first setting is what we consider to be the standard QMC setting for integrals formulated over the unit cube. Here the integrand is assumed to have square-integrable mixed first derivatives, and it is paired with *randomly shifted lattice rules* [77] to achieve first order convergence. The second setting is for integration over $\mathbb{R}^s$ against a product of univariate densities. Again the integrands have square-integrable mixed first derivatives and we use randomly shifted lattice rules to achieve first order convergence. The third setting returns to the unit cube, but considers integrands with higher order mixed derivatives and pairs them with *interlaced polynomial lattice rules* [30] which achieve higher order convergence. These three settings are discussed in more detail in [50].

Next in Sect. 3 we outline three applications of QMC methods: option pricing, GLMM (generalized linear mixed models) maximum likelihood, PDEs with random coefficients – all with quite different characteristics and requiring different strategies to tackle them. We explain how to match each example application with an appropriate setting from Sect. 2. In the option pricing application, see e.g., [2, 27, 29], none of the settings is applicable due to the presence of a *kink*. We discuss the strategy of *smoothing by preintegration* [38], which is similar to the method known as conditional sampling [1]. In the maximum likelihood application [51], the change of variables plays a crucial role in a similar way to importance sampling for Monte Carlo methods. In the PDE application, see e.g., [5, 39, 50, 73], the *uniform* and the *lognormal* cases correspond to integration over the unit cube and $\mathbb{R}^s$, respectively, and the two cases tap into different QMC settings. For the lognormal case we briefly contrast three ways to generate the random field: *Karhunen–Loève expansion*, *circulant embedding* [18, 32, 33, 35], and *H-matrix technique* [19, 40].

Then in Sect. 4 we discuss three different cost saving strategies that can be applied to all of the above applications. First, *multi-level methods* [26] restructure the required computation as a telescoping sum and tackle different levels separately to improve

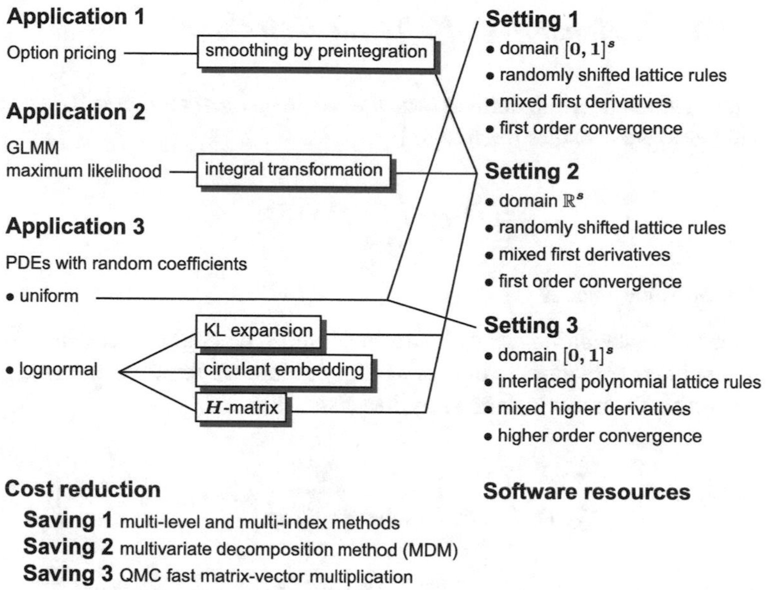
Application 1
Option pricing
smoothing by preintegration
Application 2
GLMM
maximum likelihood
integral transformation
Application 3
PDEs with random coefficients
• uniform
• lognormal
KL expansion
circulant embedding
H-matrix
Setting 1
• domain [0, 1]^s
• randomly shifted lattice rules
• mixed first derivatives
• first order convergence
Setting 2
• domain ℝ^s
• randomly shifted lattice rules
• mixed first derivatives
• first order convergence
Setting 3
• domain [0, 1]^s
• interlaced polynomial lattice rules
• mixed higher derivatives
• higher order convergence
Cost reduction
Saving 1 multi-level and multi-index methods
Saving 2 multivariate decomposition method (MDM)
Saving 3 QMC fast matrix-vector multiplication
Software resources

Fig. 1 The connection between different components of this article

the overall cost versus error balance, while more general *multi-index methods* [41] allow different criteria to be considered simultaneously in a multi-index telescoping sum. Second, the *multivariate decomposition methods* [53, 58, 79] work in a similar way by making an explicit decomposition of the underlying function into functions of only subsets of the variables [54]. The third strategy is *fast QMC matrix-vector multiplication* which carries out the required computation for multiple QMC samples at the same time using an FFT [15].

We provide pointers to some software resources in Sect. 5 and conclude the article in Sect. 6 with a summary and an outlook to future work. An overview of the various components of this article is given in Fig. 1.

2 Three Settings

Here we describe three theoretical function space settings paired with appropriate QMC methods. These three setting are also covered in [50]. Of course these three pairs are not the only possible combinations. We selected them due to our preference for constructive QMC methods that achieve the best possible convergence rates, with the implied constant independent of dimension, under the weakest possible assumptions on the integrands.

2.1 Setting 1: Standard QMC for the Unit Cube

For f a real-valued function defined over the s-dimensional unit cube $[0, 1]^s$, with s finite and fixed, we consider the integral

$$I(f) = \int_{[0,1]^s} f(\boldsymbol{y}) \, \mathrm{d}\boldsymbol{y} \,. \tag{1}$$

Weighted Sobolev Spaces

We assume in this standard setting that the integrand f belongs to a *weighted Sobolev space of smoothness one* in the unit cube $[0, 1]^s$. Here we focus on the *unanchored* variant in which the norm is defined by, see also [78],

$$\|f\|_{\boldsymbol{\gamma}} = \left[\sum_{\mathfrak{u} \subseteq \{1:s\}} \frac{1}{\gamma_{\mathfrak{u}}} \int_{[0,1]^{|\mathfrak{u}|}} \left(\int_{[0,1]^{s-|\mathfrak{u}|}} \frac{\partial^{|\mathfrak{u}|} f}{\partial \boldsymbol{y}_{\mathfrak{u}}}(\boldsymbol{y}) \, \mathrm{d}\boldsymbol{y}_{\{1:s\}\setminus\mathfrak{u}} \right)^2 \mathrm{d}\boldsymbol{y}_{\mathfrak{u}} \right]^{1/2}, \tag{2}$$

where $\{1 : s\}$ is a shorthand notation for the set of indices $\{1, 2, \ldots, s\}$, $(\partial^{|\mathfrak{u}|} f)/(\partial \boldsymbol{y}_{\mathfrak{u}})$ denotes the mixed first derivative of f with respect to the "active" variables $\boldsymbol{y}_{\mathfrak{u}} = (y_j)_{j \in \mathfrak{u}}$, while $\boldsymbol{y}_{\{1:s\}\setminus\mathfrak{u}} = (y_j)_{j \in \{1:s\}\setminus\mathfrak{u}}$ denotes the "inactive" variables.

There is a weight parameter $\gamma_{\mathfrak{u}} \geq 0$ associated with each subset of variables $\boldsymbol{y}_{\mathfrak{u}}$ to model their relative importance. We denote the weights collectively by $\boldsymbol{\gamma}$. Special forms of weights have been considered in the literature. *POD weights* (*product and order dependent weights*), arisen for the first time in [56], take the form

$$\gamma_{\mathfrak{u}} = \Gamma_{|\mathfrak{u}|} \prod_{j \in \mathfrak{u}} \Upsilon_j \,,$$

which is specified by two sequences $\Gamma_0 = \Gamma_1 = 1, \Gamma_2, \Gamma_3, \ldots \geq 0$ and $\Upsilon_1 \geq \Upsilon_2 \geq \cdots > 0$. Here the factor $\Gamma_{|\mathfrak{u}|}$ is said to be order dependent because it is determined solely by the cardinality of $\mathfrak{u}$ and not the precise indices in $\mathfrak{u}$. The dependence of the weight $\gamma_{\mathfrak{u}}$ on the indices $j \in \mathfrak{u}$ is controlled by the product of terms Υ_j. Each term Υ_j in the sequence corresponds to one coordinate direction; the sequence being non-increasing indicates that successive coordinate directions become less important. Taking all $\Gamma_{|\mathfrak{u}|} = 1$ or all $\Upsilon_j = 1$ corresponds to the weights known as *product weights* or *order dependent weights* in the literature [76, 78].

Randomly Shifted Lattice Rules

We pair the weighted Sobolev space with *randomly shifted lattice rules*; the complete theory can be found in [12]. Randomly shifted lattice rules approximate the integral (1) by

$$Q(f) = \frac{1}{n} \sum_{i=1}^{n} f(\boldsymbol{t}_i), \qquad \boldsymbol{t}_i = \left\{ \frac{i\,\boldsymbol{z}}{n} + \boldsymbol{\Delta} \right\}, \tag{3}$$

where $z \in \mathbb{Z}^s$ is known as the *generating vector*, $\boldsymbol{\Delta}$ is a *random shift* drawn from the uniform distribution over $[0,1]^s$, and the braces indicate that we take the fractional parts of each component in a vector.

A randomly shifted lattice rule provides an unbiased estimate of the integral, i.e., $\mathbb{E}[Q(f)] = I(f)$, where the expectation is taken with respect to the random shift $\boldsymbol{\Delta}$. Its quality is determined by the choice of the generating vector z. By analyzing the quantity known as *shift-averaged worst case error*, it is known that good generating vectors can be obtained using a *CBC construction* (*component-by-component construction*), determining the components of z one at a time sequentially, to achieve nearly $\mathcal{O}(n^{-1})$ convergence rate which is optimal in the weighted Sobolev space of smoothness one, and the implied constant in the big $\mathcal{O}$ bound can be independent of s under appropriate conditions on the weights $\boldsymbol{\gamma}$.

More precisely, if n is a power of 2 then we know that the CBC construction yields the root-mean-square error bound, for all $\lambda \in (1/2, 1]$,

$$\sqrt{\mathbb{E}\left[|I(f) - Q(f)|^2\right]} \le \left(\frac{2}{n} \sum_{\emptyset \neq \mathfrak{u} \subseteq \{1:s\}} \gamma_{\mathfrak{u}}^{\lambda} \, [\vartheta(\lambda)]^{|\mathfrak{u}|}\right)^{1/(2\lambda)} \|f\|_{\boldsymbol{\gamma}}, \tag{4}$$

where $\vartheta(\lambda) := 2\zeta(2\lambda)/(2\pi^2)^{\lambda}$, with $\zeta(a) := \sum_{k=1}^{\infty} k^{-a}$ denoting the Riemann zeta function. A similar result holds for general n. The best rate of convergence clearly comes from choosing λ close to $1/2$, but the advantage is offset by the fact that $\zeta(2\lambda) \to \infty$ as $\lambda \to (1/2)_+$.

Choosing the Weights

To apply this abstract theory to a given practical integrand f, we need to first obtain an estimate of the norm $\|f\|_{\boldsymbol{\gamma}}$. Remember that at this stage we do not yet know how to choose the weights $\gamma_{\mathfrak{u}}$. Assuming that bounds on the mixed first derivatives in (2) can be obtained so that

$$\|f\|_{\boldsymbol{\gamma}} \le \left(\sum_{\mathfrak{u} \subseteq \{1:s\}} \frac{B_{\mathfrak{u}}}{\gamma_{\mathfrak{u}}}\right)^{1/2}, \tag{5}$$

we can substitute (5) into (4) and then, with λ fixed but unspecified at this point and $A_{\mathfrak{u}} = [\vartheta(\lambda)]^{|\mathfrak{u}|}$, we choose the weights $\gamma_{\mathfrak{u}}$ to minimizing the product

$$C_{\boldsymbol{\gamma}} := \left(\sum_{\mathfrak{u} \subseteq \{1:s\}} \gamma_{\mathfrak{u}}^{\lambda} A_{\mathfrak{u}}\right)^{1/(2\lambda)} \left(\sum_{\mathfrak{u} \subseteq \{1:s\}} \frac{B_{\mathfrak{u}}}{\gamma_{\mathfrak{u}}}\right)^{1/2}.$$

Elementary calculus leads us to conclude that we should take

$$\gamma_{\mathfrak{u}} := \left(\frac{B_{\mathfrak{u}}}{A_{\mathfrak{u}}}\right)^{1/(1+\lambda)}, \tag{6}$$

which yields

$$C_\gamma = \Bigg(\sum_{\mathfrak{u}\subseteq\{1:s\}} A_{\mathfrak{u}}^{1/(1+\lambda)} B_{\mathfrak{u}}^{\lambda/(1+\lambda)} \Bigg)^{(1+\lambda)/(2\lambda)} .$$

We then specify a value of λ, as close to $1/2$ as possible, to ensure that C_γ can be bounded independently of s. This in turn determines the theoretical convergence rate which is $\mathcal{O}(n^{-1/(2\lambda)})$.

The chosen weights $\gamma_{\mathfrak{u}}$ are then fed into the CBC construction to produce generating vectors for randomly shifted lattice rules that achieve the desired theoretical error bound for this integrand. This strategy for determining weights was first considered in [56].

Fast CBC constructions (using FFT) can produce generating vectors for an n-point rule in s dimensions in $\mathcal{O}(s\,n\log n)$ operations in the case of product weights [66], and in $\mathcal{O}(s\,n\log n + s^2\,n)$ operations in the case of POD weights [55]. Note that these are considered to be pre-computation costs. The actual cost for generating the points on the fly is $\mathcal{O}(s\,n)$ operations, no worse than Monte Carlo simulations. Strategies to improve on the computational cost of approximating the integral are discussed in Sect. 4.

The CBC construction yields a lattice rule which is extensible in dimension s. We can also construct *lattice sequences* which are extensible or embedded in the number of points n, at the expense of increasing the implied constant in the error bound [6, 11, 45, 46].

2.2 Setting 2: QMC Integration over $\mathbb{R}^s$

QMC approximation to an integral which is formulated over the Euclidean space $\mathbb{R}^s$ can be obtained by first mapping the integral to the unit cube as follows:

$$\begin{aligned} I(f) &= \int_{\mathbb{R}^s} f(\boldsymbol{y}) \prod_{j=1}^{s} \phi(y_j)\,\mathrm{d}\boldsymbol{y} = \int_{[0,1]^s} f(\Phi^{-1}(\boldsymbol{w}))\,\mathrm{d}\boldsymbol{w} \\ &\approx \frac{1}{n}\sum_{i=1}^{n} f(\Phi^{-1}(\boldsymbol{t}_i)) = Q(f) . \end{aligned} \qquad (7)$$

(With a slight abuse of notation we have reused $I(f)$ and $Q(f)$ from the previous subsection for integration over $\mathbb{R}^s$ in this subsection.) Here ϕ can be *any general univariate probability density function*, and Φ^{-1} denotes the component-wise application of the inverse of the cumulative distribution function corresponding to ϕ. Note that in many practical applications we need to first apply some clever transformation to convert the integral into the above form; some examples are discussed in Sect. 3. The transformed integrand $f\circ\Phi^{-1}$ arising from practical applications typically does not belong to the Sobolev space defined over the unit cube due to the integrand being unbounded near the boundary of the cube, or because the mixed derivatives of the transformed integrand do not exist or are unbounded. Thus the theory in the preceding subsection generally does not apply in practice. Some theory for QMC on singular integrands is given in [70].

We summarize here a special weighted space setting in $\mathbb{R}^s$ for which randomly shifted lattice rules have been shown to achieve nearly the optimal convergence rate of order one [52, 63]. The norm in this setting is given by

$$\|f\|_{\boldsymbol{\gamma}} = \Bigg[\sum_{\mathfrak{u}\subseteq\{1:s\}} \frac{1}{\gamma_{\mathfrak{u}}} \int_{\mathbb{R}^{|\mathfrak{u}|}} \Bigg(\int_{\mathbb{R}^{s-|\mathfrak{u}|}} \frac{\partial^{|\mathfrak{u}|} f}{\partial \boldsymbol{y}_{\mathfrak{u}}}(\boldsymbol{y}) \Bigg(\prod_{j\in\{1:s\}\setminus\mathfrak{u}} \phi(y_j) \Bigg) \mathrm{d}\boldsymbol{y}_{\{1:s\}\setminus\mathfrak{u}} \Bigg)^2 \\ \times \Bigg(\prod_{j\in\mathfrak{u}} \varpi_j^2(y_j) \Bigg) \mathrm{d}\boldsymbol{y}_{\mathfrak{u}} \Bigg]^{1/2}. \tag{8}$$

Comparing (8) with (2), apart from the difference that the integrals are now over the unbounded domain, there is a probability density function ϕ as well as additional *weight functions* ϖ_j which can be chosen to moderate the tail behavior of the mixed derivatives of f.

The convergence results for the CBC construction of randomly shifted lattice rules in this general setting depend on the choices of ϕ and ω_j. For n a power of 2, the root-mean-square error bound takes the form, for all $\lambda \in (1/(2r), 1]$,

$$\sqrt{\mathbb{E}\big[|I(f) - Q(f)|^2\big]} \le \Bigg(\frac{2}{n} \sum_{\emptyset\neq\mathfrak{u}\subseteq\{1:s\}} \gamma_{\mathfrak{u}}^{\lambda} \prod_{j\in\mathfrak{u}} \vartheta_j(\lambda) \Bigg)^{1/(2\lambda)} \|f\|_{\boldsymbol{\gamma}},$$

with r (appearing in the applicable lower bound on λ) and $\vartheta_j(\lambda)$ depending on ϕ and ϖ_j, see [63, Theorem 8]. Some special cases have been analyzed:

- See [36, Theorem 15] or [50, Theorem 5.2] for $\phi(y) = \phi_{\mathrm{nor}}(y) = \exp(-y^2/2)/\sqrt{2\pi}$ being the standard normal density and $\varpi_j^2(y_j) = \exp(-2\,\alpha_j\,|y_j|)$ with $\alpha_j > 0$.
- See [50, Theorem 5.3] for $\phi = \phi_{\mathrm{nor}}$ and $\varpi_j^2(y_j) = \exp(-\alpha\, y_j^2)$ with $\alpha < 1/2$.
- See [74, Theorem 2] for ϕ being a logistic, normal, or Student density and $\varpi_j = 1$.

To apply this abstract theory to a practical integral over $\mathbb{R}^s$, it is important to realize that the choice of ϕ can be tuned as part of the process of transformation to express the integral in the form (7). (This point will become clearer when we describe the maximum likelihood application in Sect. 3.2.) Then the choice of weight functions ϖ_j arises as part of the process to obtain bounds on the norm of f, as in (5). (This point will become clearer when we describe the PDE application in Sect. 3.3.) Finally we can choose the weights $\gamma_{\mathfrak{u}}$ as in (6) but now with $A_{\mathfrak{u}} = \prod_{j\in\mathfrak{u}} \vartheta_j(\lambda)$ for the appropriate $\vartheta_j(\lambda)$ corresponding to the choice of ϕ and ϖ_j. The choice of density ϕ, weight functions ϖ_j, and weight parameters $\gamma_{\mathfrak{u}}$ then enter the CBC construction to obtain the generating vector of good randomly shifted lattice rules that can achieve the theoretical error bound for this integrand.

In practice, it may well be that the weights $\gamma_{\mathfrak{u}}$ obtained in this way are not sensible because we were working with theoretical upper bounds on the error that may be too pessimistic. It may already be so in the standard setting of the previous subsection, but is more pronounced in the setting for $\mathbb{R}^s$ due to the additional complication associated with the presence of ϕ and ϖ_j.

2.3 Setting 3: Smooth Integrands in the Unit Cube

Now we return to the integration problem over the unit cube (1) and outline a weighted function space setting from [14] for smooth integrands of order α. The norm is given by

$$\|f\|_{\boldsymbol{\gamma}} = \sup_{\mathfrak{u}\subseteq\{1:s\}} \sup_{\boldsymbol{y}_{\mathfrak{v}}\in[0,1]^{|\mathfrak{v}|}} \frac{1}{\gamma_{\mathfrak{u}}} \sum_{\mathfrak{v}\subseteq\mathfrak{u}} \sum_{\boldsymbol{\tau}_{\mathfrak{u}\setminus\mathfrak{v}}\in\{1:\alpha\}^{|\mathfrak{u}\setminus\mathfrak{v}|}} \left| \int_{[0,1]^{s-|\mathfrak{v}|}} (\partial^{(\boldsymbol{\alpha}_{\mathfrak{v}},\boldsymbol{\tau}_{\mathfrak{u}\setminus\mathfrak{v}},\boldsymbol{0})} f)(\boldsymbol{y}) \, \mathrm{d}\boldsymbol{y}_{\{1:s\}\setminus\mathfrak{v}} \right| . \tag{9}$$

Here $(\boldsymbol{\alpha}_{\mathfrak{v}}, \boldsymbol{\tau}_{\mathfrak{u}\setminus\mathfrak{v}}, \boldsymbol{0})$ denotes a multi-index $\boldsymbol{\nu}$ with $\nu_j = \alpha$ for $j \in \mathfrak{v}$, $\nu_j = \tau_j$ for $j \in \mathfrak{u}\setminus\mathfrak{v}$, and $\nu_j = 0$ for $j \notin \mathfrak{u}$. We denote the $\boldsymbol{\nu}$-th partial derivative of f by $\partial^{\boldsymbol{\nu}} f = (\partial^{|\boldsymbol{\nu}|} f)/(\partial_{y_1}^{\nu_1} \partial_{y_2}^{\nu_2} \cdots \partial_{y_s}^{\nu_s})$.

This function space setting can be paired with *interlaced polynomial lattice rules* [30, 31] to achieve higher order convergence rates in the unit cube. A *polynomial lattice rule* [64] is similar to a lattice rule (see (3) without the random shift $\boldsymbol{\Delta}$), but instead of a generating vector of integers we have a generating vector of polynomials, and thus the regular multiplication and division are replaced by their polynomial equivalents. We omit the technical details here. An interlaced polynomial lattice rule with $n = 2^m$ points in s dimensions with interlacing factor α is obtained by taking a polynomial lattice rule in $\alpha\, s$ dimensions and then interlacing the bits from every successive α dimensions to yield one dimension. More explicitly, for $\alpha = 3$, given three coordinates $x = (0.x_1 x_2 \ldots x_m)_2$, $y = (0.y_1 y_2 \ldots y_m)_2$ and $z = (0.z_1 z_2 \ldots z_m)_2$ we interlace their bits to obtain $w = (0.x_1 y_1 z_1 x_2 y_2 z_2 \ldots x_m y_m z_m)_2$.

An interlaced polynomial lattice rule with interlacing factor $\alpha \geq 2$, with irreducible modulus polynomial of degree m, and with $n = 2^m$ points in s dimensions, can be constructed by a CBC algorithm such that, for all $\lambda \in (1/\alpha, 1]$,

$$|I(f) - Q(f)| \leq \left(\frac{2}{n} \sum_{\emptyset\neq\mathfrak{u}\subseteq\{1:s\}} \gamma_{\mathfrak{u}}^{\lambda} \, [\vartheta_\alpha(\lambda)]^{|\mathfrak{u}|} \right)^{1/\lambda} \|f\|_{\boldsymbol{\gamma}} ,$$

where $\vartheta_\alpha(\lambda) := 2^{\alpha\lambda(\alpha-1)/2}([1 + 1/(2^{\alpha\lambda} - 2)]^\alpha - 1)$. This result can be found in [50, Theorem 5.4], which was obtained from minor adjustments of [14, Theorem 3.10].

Given a practical integrand f, if we can estimate the corresponding integrals involving the mixed derivatives in (9), then we can choose the weights $\gamma_{\mathfrak{u}}$ so that every term in the supremum is bounded by a constant, say, c. This strategy in [14] led to a new form of weights called *SPOD weights* (*smoothness-driven product and order dependent weights*); they take the form

$$\gamma_{\mathfrak{u}} = \sum_{\boldsymbol{\nu}_{\mathfrak{u}} \in \{1:\alpha\}^{|\mathfrak{u}|}} \Gamma_{|\boldsymbol{\nu}_{\mathfrak{u}}|} \prod_{j \in \mathfrak{u}} \Upsilon_j(\nu_j) \,.$$

If the weights $\boldsymbol{\gamma}$ are SPOD weights, then the fast CBC construction of the generating vector has cost $\mathcal{O}(\alpha\, s\, n \log n + \alpha^2\, s^2 n)$ operations. If the weights $\boldsymbol{\gamma}$ are product weights, then the CBC algorithm has cost $\mathcal{O}(\alpha\, s\, n \log n)$ operations.

3 Three Applications

Integrals over $\mathbb{R}^s$ often arise from practical applications in the form of multivariate expected values

$$\mathbb{E}_\rho[q] = \int_{\mathbb{R}^s} q(\boldsymbol{y})\, \rho(\boldsymbol{y})\, \mathrm{d}\boldsymbol{y} \,, \tag{10}$$

where q is some quantity of interest which depends on a vector $\boldsymbol{y} = (y_1, \ldots, y_s)$ of parameters or variables in s dimensions, and ρ is some multivariate probability density function describing the distribution of $\boldsymbol{y}$, *not necessarily a product of univariate functions* as we assumed in (7), and so we need to make an appropriate transformation to apply our theory. Below we discuss three motivating applications with quite different characteristics, and we will explain how to make use of the different settings in Sect. 2.

3.1 Application 1: Option Pricing

Following the Black–Scholes model, integrals arising from option pricing problems take the general form (10), with

$$q(\boldsymbol{y}) = \max(\mu(\boldsymbol{y}), 0) \quad \text{and} \quad \rho(\boldsymbol{y}) = \frac{\exp(-\frac{1}{2}\boldsymbol{y}^{\mathrm{T}} \Sigma^{-1} \boldsymbol{y})}{\sqrt{(2\pi)^s \det(\Sigma)}},$$

where the variables $\boldsymbol{y} = (y_1, \ldots, y_s)^{\mathrm{T}}$ correspond to a discretization of the underlying Brownian motion over a time interval $[0, T]$, and the covariance matrix has entries $\Sigma_{ij} = (T/s)\min(i, j)$. For example, in the case of an *arithmetic average Asian call option* [2, 27, 29], the payoff function q depends on the smooth function

$\mu(\boldsymbol{y}) = (1/s)\sum_{j=1}^{s} S_{t_j}(\boldsymbol{y}) - K$ which is the difference between the average of the asset prices S_{t_j} at the discrete times and the strike price K.

The widely accepted strategy to rewrite these option pricing integrals from the form (10) to the form (7) with product densities is to take a factorization $\Sigma = AA^{\mathrm{T}}$ and apply a change of variables $\boldsymbol{y} = A\boldsymbol{y}'$. This yields an integral of the form (7) with

$$f(\boldsymbol{y}') = q(A\boldsymbol{y}') \quad \text{and} \quad \phi = \phi_{\mathrm{nor}}.$$

The choice of factorization therefore determines the function f. For example, A can be obtained through Cholesky factorization (commonly known as the *standard construction*; in this case it is equivalent to generating the Brownian motions sequentially in time), through *Brownian bridge construction* [4], or eigenvalue decomposition sometimes called the *principal component construction* [2]. Note that in practice these factorizations are not carried out explicitly due to the special form of the covariance matrix. In fact, they can be computed in $\mathcal{O}(s)$, $\mathcal{O}(s)$ and $\mathcal{O}(s \ \log s)$ operations, respectively [29].

The success of QMC for option pricing cannot be explained by most existing theory due to the *kink* in the integrand induced by the maximum function. However, for some factorizations it is shown in [37] that all *ANOVA terms* of f are smooth, with the exception of the highest order term. This hints at a *smoothing by preintegration* strategy, where a coordinate with some required property is chosen, say y_k, and we integrate out this one variable (either exactly or numerically with high precision) to obtain a function in $s-1$ variables

$$P_k(f) := \int_{-\infty}^{\infty} f(\boldsymbol{y})\, \phi_{\mathrm{nor}}(y_k)\, \mathrm{d}y_k\,.$$

Under the right conditions (e.g., integrating with respect to y_1 in the case of the principal components construction), this new function is smooth and belongs to the function space setting of Sect. 2.2 (with one less variable) [38]. This strategy is related to the method known as *conditional sampling* [1].

3.2 Application 2: Maximum Likelihood

Another source of integrands which motivated recent developments in the function space setting of Sect. 2.2 is a class of generalized linear mixed models (GLMM) in statistics, as examined in [51, 52, 74]. A specific example of the Poisson likelihood time series model considered in these papers involves an integral of the form (10), with

$$q(\boldsymbol{y}) = \prod_{j=1}^{s} \frac{\exp(\tau_j(\beta + y_j) - e^{\beta + y_j})}{\tau_j!} \quad \text{and} \quad \rho(\boldsymbol{y}) = \frac{\exp(-\frac{1}{2}\boldsymbol{y}^{\mathrm{T}}\Sigma^{-1}\boldsymbol{y})}{\sqrt{(2\pi)^s \det(\Sigma)}}.$$

Here $\beta \in \mathbb{R}$ is a model parameter, $\tau_1, \ldots, \tau_s \in \{0, 1, \ldots\}$ are the count data, and Σ is a Toeplitz covariance matrix with $\Sigma_{ij} = \sigma^2 \varkappa^{|i-j|}/(1-\varkappa^2)$, where σ^2 is the variance and $\varkappa \in (-1, 1)$ is the autoregression coefficient.

An obvious way to rewrite this integral in the form (7) with product densities is to factorize Σ as discussed in the previous subsection for the option pricing applications, but this would yield a very spiky function f. Instead, the strategy developed in [51] recenters and rescales the exponent $T(\boldsymbol{y})$ of the product $q(\boldsymbol{y})\rho(\boldsymbol{y}) =: \exp(T(\boldsymbol{y}))$ as follows:

1. Find the unique stationary point $\boldsymbol{y}^*$ satisfying $\nabla T(\boldsymbol{y}^*) = 0$.
2. Determine the matrix $\Sigma^* = (-\nabla^2 T(\boldsymbol{y}^*))^{-1}$ which describes the convexity of T around the stationary point.
3. Factorise $\Sigma^* = A^* A^{*\mathrm{T}}$.
4. Apply a change of variables $\boldsymbol{y} = A^* \boldsymbol{y}' + \boldsymbol{y}^*$.
5. Multiply and divide the resulting integrand by the product $\prod_{j=1}^s \phi(y_j')$ where ϕ is any univariate density (not necessarily the normal density).

These steps then yield an integral of the form (7) with

$$f(\boldsymbol{y}') = \frac{c \exp(T(A^* \boldsymbol{y}' + \boldsymbol{y}^*))}{\prod_{j=1}^s \phi(y_j')}$$

for some scaling constant $c > 0$. Note that the choice of A^* and ϕ determines f.

The paper [74] provides careful estimates of the norm of the resulting integrand f in the setting of Sect. 2.2 corresponding to three different choices of density ϕ, with the weight functions taken as $\varpi_j = 1$, and gives the formula for the weight parameters $\gamma_{\mathfrak{u}}$ that minimize the overall error bound.

These GLMM problems are extremely challenging not only for QMC but also in general the tools are still lacking. There is still lots of room to develop new QMC methods and theory for these problems.

3.3 Application 3: PDEs with Random Coefficients

Our third application is motivated by fluid flow through a porous medium, typically modelled using Darcy's Law, with random coefficients. A popular toy problem is the elliptic PDE with a random coefficient [5, 39, 73]

$$-\nabla \cdot (a(\boldsymbol{x}, \omega)\, \nabla u(\boldsymbol{x}, \omega)) = \kappa(\boldsymbol{x}) \quad \text{for } \boldsymbol{x} \in D \subset \mathbb{R}^d \text{ and almost all } \omega \in \Omega,$$

with $d \in \{1, 2, 3\}$, subject to homogeneous Dirichlet boundary conditions. The coefficient $a(\boldsymbol{x}, \omega)$ is assumed to be a random field over the spatial domain D (e.g., representing the permeability of a porous material over D), and Ω is the probability space. The goal is to compute the expected values $\mathbb{E}[G(u)]$ of some bounded linear functional G of the solution u over Ω.

For practical reasons it is often assumed that $a(\boldsymbol{x},\omega)$ is a *lognormal random field*, that is, $a(\boldsymbol{x},\omega)=\exp(Z(\boldsymbol{x},\omega))$, where $Z(\boldsymbol{x},\omega)$ is a Gaussian random field with a prescribed mean and covariance function. This is known as the *lognormal case*. However, researchers often analyze a simpler model known as the *uniform case*.

The Uniform Case

In the uniform case, we consider the parametric PDE

$$-\nabla\cdot(a(\boldsymbol{x},\boldsymbol{y})\,\nabla u(\boldsymbol{x},\boldsymbol{y}))=\kappa(\boldsymbol{x})\quad\text{for } \boldsymbol{x}\in D\subset\mathbb{R}^d, \tag{11}$$

together with

$$a(\boldsymbol{x},\boldsymbol{y})\;=\;a_0(\boldsymbol{x})+\sum_{j\ge1} y_j\psi_j(\boldsymbol{x})\,, \tag{12}$$

where the parameters y_j are independently and uniformly distributed on the interval $[-\frac{1}{2},\frac{1}{2}]$, and we assume that $0<a_{\min}\le a(\boldsymbol{x},\boldsymbol{y})\le a_{\max}<\infty$ for all $\boldsymbol{x}$ and $\boldsymbol{y}$.

A (single-level) strategy for approximating $\mathbb{E}[G(u)]$ is as follows:

1. Truncate the infinite sum in (12) to s terms.
2. Solve the PDE using finite element methods with meshwidth h.
3. Approximate the resulting s-dimensional integral using QMC with n points.

So the error is a sum of truncation error, discretization error, and quadrature error.

For the QMC quadrature error in Step 3, we have the integral (1) with

$$f(\boldsymbol{y})\;=\;G\left(u_h^s\left(\cdot,\boldsymbol{y}-\tfrac{1}{2}\right)\right),$$

where u_h^s denotes the finite element solution of the truncated problem, and the subtraction by $\frac{1}{2}$ takes care of the translation from the usual unit cube $[0,1]^s$ to $[-\frac{1}{2},\frac{1}{2}]^s$. By differentiating the PDE (11), we can obtain bounds on the mixed derivatives of the PDE solution with respect to $\boldsymbol{y}$. This leads to bounds on the norm (2) of the integrand f and so we can apply the theoretical setting of Sect. 2.1 to obtain up to first order convergence for QMC. Under appropriate assumptions and with first order finite elements, we can prove that the total error for the above 3-step strategy is of order [56]

$$\mathcal{O}(s^{-2(1/p_0-1)}+h^2+n^{-\min(1/p_0-1/2,1-\delta)}),\quad \delta\in\left(0,\tfrac{1}{2}\right),$$

where $p_0\in(0,1)$ should be as small as possible while satisfying $\sum_{j\ge1}\|\psi_j\|_{L_\infty}^{p_0}<\infty$. This part is presented as a step-by-step tutorial in the article [49] from this volume.

The bounds on the derivatives of the PDE with respect to $\boldsymbol{y}$ also allow us to obtain bounds on the norm (9) and so we can also apply the theoretical setting of Sect. 2.3 to obtain higher order convergence [14]. Specifically, the $\mathcal{O}(n^{-\min(1/p_0-1/2,1-\delta)})$ term can be improved to $\mathcal{O}(n^{-1/p_0})$. Also the $\mathcal{O}(h^2)$ term can be improved by using higher order finite elements. See [49, 50] for more details.

The Lognormal Case with Karhunen–Loève Expansion

In the lognormal case, we have the same parametric PDE (11), but now we use the *Karhunen–Loève expansion* (*KL expansion*) of the Gaussian random field (in the exponent) to write

$$a(\boldsymbol{x}, \boldsymbol{y}) = a_0(\boldsymbol{x}) \exp\left(\sum_{j\ge1} y_j \sqrt{\mu_j}\, \xi_j(\boldsymbol{x})\right),$$

where $a_0(\boldsymbol{x}) > 0$, the μ_j are real, positive and non-increasing in j, the ξ_j are orthonormal in $L_2(D)$, and the parameters $y_j \in \mathbb{R}$ are standard $\mathcal{N}(0, 1)$ random variables. Truncating the infinite series in $a(\boldsymbol{x}, \boldsymbol{y})$ to s terms and solving the PDE with a finite element method as in the uniform case, we have now an integral of the form (7) with

$$f(\boldsymbol{y}) = G(u_h^s(\cdot, \boldsymbol{y})) \quad \text{and} \quad \phi = \phi_{\text{nor}}\,.$$

One crucial step in the analysis of [36] is to choose suitable weight functions ϖ_j so that the function f has a finite and indeed small norm (8), so that the theoretical setting of Sect. 2.2 can be applied. Again see [49, 50] for more details.

In this lognormal case with KL expansion (and also the uniform case), the cost per sample of the random field is $\mathcal{O}(s\,M)$ operations, where M is the number of finite element nodes. This dominates the cost in evaluating the integrand function under the assumption that assembling the stiffness matrix to solve the PDE (which depends on the random field) is higher than the cost of the PDE solve which is $\mathcal{O}(M \log M)$. When s is large the cost of sampling the random field can be prohibitive, and this is why the following alternative strategies emerged.

The Lognormal Case with Circulant Embedding

Since we have a Gaussian random field we can actually sample the random field exactly on any set of M spatial points. This leads to an integral of the form (10) with (assuming the field has zero mean)

$$q(\boldsymbol{y}) = G(u_h^s(\cdot, \boldsymbol{y})) \quad \text{and} \quad \rho(\boldsymbol{y}) = \frac{\exp(-\frac{1}{2}\boldsymbol{y}^{\mathrm{T}} R^{-1} \boldsymbol{y})}{\sqrt{(2\pi)^s \det(R)}},$$

where R is an $M \times M$ covariance matrix, and initially we have $s = M$. (Note the subtle abuse of notation that the second argument in $u_h^s(\boldsymbol{x}, \cdot)$ has a different meaning to the KL case, in the sense that there the covariance is already built in.) This integral can be transformed into the form (7) with a factorization $R = AA^{\mathrm{T}}$ and a change of variables $\boldsymbol{y} = A\boldsymbol{y}'$, as in the option pricing example, to obtain

$$f(\boldsymbol{y}') = G(u_h^s(\cdot, A\boldsymbol{y}')) \quad \text{and} \quad \phi = \phi_{\text{nor}}\,.$$

The advantage of this discrete formulation is that there is no error arising from the truncation of the KL expansion. However, the direct factorization and matrix-vector multiplication require $\mathcal{O}(M^3)$ operations which can be too costly when M is large.

The idea of *circulant embedding* [18, 32, 33, 35] is to sample the random field on a regular grid and to embed the covariance matrix of these points into a larger $s \times s$ matrix which is *nested block circulant with circulant blocks*, so that FFTs can be used to reduce the per sample cost to $\mathcal{O}(s \log s)$ operations. Values of the random field at the finite element quadrature nodes can be obtained by interpolation. Note that this turns the problem into an even higher dimensional integral, and we can have $s \gg M$. For this strategy to work we need to use regular spatial grid points to sample the field and a stationary covariance function (i.e., the covariance depends only on the relative distance between points). An additional difficulty is to ensure positive definiteness of the extended matrix; this is studied in [32].

The Lognormal Case with H-Matrix Technique

Another approach for the discrete matrix formulation of the lognormal case is to first approximate R by an H-matrix [40] and make use of H-matrix techniques to compute the matrix-vector multiplication with the square-root of this H-matrix at essentially linear cost $\mathcal{O}(M)$. Two iterative methods have been proposed in [19] to achieve this (one is based on a variant of the *Lanczos iteration* and the other on the *Schultz iteration*), with full theoretical justification for the error incurred in the H-matrix approximation. An advantage of this approach over circulant embedding is that it does not require the spatial grid to be regular nor that the covariance be stationary.

Other Developments

A different QMC analysis for the lognormal case has been considered in [42]. QMC for holomorphic equations was considered in [17], and for Baysesian inversion in [9, 72]. Recently there is also QMC analysis developed for the situation where the functions in the expansion of $a(\boldsymbol{x}, \boldsymbol{y})$ have local support, see [22, 43, 48].

4 Three Cost Saving Strategies

In this section we discuss the basic ideas of three different kinds of cost saving strategies that can be applied to QMC methods, without going into details. Actually, the circulant embedding and H-matrix technique discussed in the previous section can also be considered as cost saving strategies. These strategies are not mutually exclusive, and it may be possible to mix and match them to benefit from compound savings.

4.1 Saving 1: Multi-level and Multi-index

The *multi-level* idea [26] is easy to explain in the context of numerical integration. Suppose that there is a sequence $(f_\ell)_{\ell\geq 0}$ of approximations to an integrand f, with increasing accuracy and cost as ℓ increases, such that we have the telescoping sum

$$f = \sum_{\ell=0}^{\infty}(f_\ell - f_{\ell-1}), \quad f_{-1} := 0.$$

For example, the different f_ℓ could correspond to different number of time steps in option pricing, different number of mesh points in a finite element solve for PDE, different number of terms in a KL expansion, or a combination of aspects. A multi-level method for approximating the integral of f is

$$A_{\mathrm{ML}}(f) = \sum_{\ell=0}^{L} Q_\ell(f_\ell - f_{\ell-1}),$$

where the parameter L determines the number of *levels*, and for each level we apply a different quadrature rule Q_ℓ to the difference $f_\ell - f_{\ell-1}$.

The integration error (in this simple description with deterministic quadrature rules) satisfies

$$|I(f) - A_{\mathrm{ML}}(f)| \leq \underbrace{|I(f) - I_L(f_L)|}_{\leq \varepsilon/2} + \underbrace{\sum_{\ell=0}^{L} |(I_\ell - Q_\ell)(f_\ell - f_{\ell-1})|}_{\leq \varepsilon/2}.$$

For a given error threshold $\varepsilon > 0$, the idea (as indicated by the underbraces) is that we choose L to ensure that the first term (the truncation error) on the right-hand side is $\leq \varepsilon/2$, and we specify parameters for the quadrature rules Q_ℓ so that the second term (the quadrature error) is also $\leq \varepsilon/2$. The latter can be achieved with a Lagrange multiplier argument to minimize cost subject to the given error threshold. Our hope is that the successive differences $f_\ell - f_{\ell-1}$ will become smaller with increasing ℓ and therefore we would require less quadrature points for the more costly higher levels.

The *multi-index* idea [41] generalizes this from a scalar level index ℓ to a vector index $\boldsymbol{\ell}$ so that we can vary a number of different aspects (e.g., spatial/temporal discretization) simultaneously and independently of each other. It makes use of the sparse grid concept so that the overall cost does not blow up with respect to the dimensionality of $\boldsymbol{\ell}$, i.e., the number of different aspects being considered. A simple example is that we use different finite element meshwidths for different spatial coordinates. This is equivalent to applying sparse finite element methods within a multilevel algorithm, see the article [28] in this volume.

Multi-level and multi-index extensions of QMC methods for the applications from Sect. 3 include e.g., [16, 27, 57, 59, 71].

4.2 Saving 2: Multivariate Decomposition Method

In the context of numerical integration, the *multivariate decomposition method* (*MDM*) [23, 53, 58, 79] makes use of a decomposition of the integrand f of the form

$$f = \sum_{|\mathfrak{u}|<\infty} f_{\mathfrak{u}},$$

where the sum is over all finite subsets $\mathfrak{u} \subset \{1, 2, \ldots\}$ and each function $f_{\mathfrak{u}}$ depends only on the integration variables with indices in the set $\mathfrak{u}$. Then MDM takes the form

$$A_{\text{MDM}}(f) = \sum_{\mathfrak{u}\in\mathscr{A}} Q_{\mathfrak{u}}(f_{\mathfrak{u}})$$

where $\mathscr{A}$ is known as the *active set* of subsets of indices, and for each $\mathfrak{u}$ in the active set we apply a different quadrature rule $Q_{\mathfrak{u}}$ to $f_{\mathfrak{u}}$.

Analogously to the multi-level idea, the error of MDM satisfies

$$|I(f) - A_{\text{MDM}}(f)| \le \underbrace{\sum_{\mathfrak{u}\notin\mathscr{A}} |I_{\mathfrak{u}}(f_{\mathfrak{u}})|}_{\le \varepsilon/2} + \underbrace{\sum_{\mathfrak{u}\in\mathscr{A}} |(I_{\mathfrak{u}} - Q_{\mathfrak{u}})(f_{\mathfrak{u}})|}_{\le \varepsilon/2},$$

where we choose the active set $\mathscr{A}$ to ensure that the truncation error is $\le \varepsilon/2$, and we use a Lagrange multiplier argument to specify parameters for the quadrature rules so that the quadrature error is also $\le \varepsilon/2$. Our hope is that, although the cardinality of the active set $\mathscr{A}$ might be huge (e.g., tens of thousands), the cardinality of the individual subsets $\mathfrak{u} \in \mathscr{A}$ might be relatively small (e.g., at most 8 or 10), and therefore we transfer the problem into that of solving a large number of low dimensional integrals.

There are many important considerations in the implementation of MDM [25]. First, we need to decide on how to decompose the integrand f so that values of the functions $f_{\mathfrak{u}}$ can be computed. One obvious choice is known as the *anchored decomposition* which can be computed via the explicit formula [54]

$$f_{\mathfrak{u}}(\boldsymbol{y}_{\mathfrak{u}}) = \sum_{\mathfrak{v}\subseteq\mathfrak{u}} (-1)^{|\mathfrak{u}|-|\mathfrak{v}|} f(\boldsymbol{y}_{\mathfrak{v}}; \boldsymbol{a}), \tag{13}$$

where $\boldsymbol{a}$ is an *anchor* and $(\boldsymbol{y}_{\mathfrak{v}}; \boldsymbol{a})$ denotes a vector obtained from $\boldsymbol{y}$ by replacing the component y_j with the corresponding component a_j when the index j does not belong to the subset $\mathfrak{v}$. (This is similar to the well-known ANOVA decomposition which, however, involves integrals that cannot be computed in practice.) Second, we need to specify and construct the active set $\mathscr{A}$ and have an efficient data structure to store the sets for later traversing. Third, we need to explore nestedness or embedding in the quadrature rules, taking into account the sum in (13) and develop efficient ways to reuse function evaluations.

4.3 Saving 3: Fast QMC Matrix-Vector Multiplication

There is a certain structure in some QMC methods that can allow for fast matrix-vector multiplication using FFT. This structure has been exploited in the fast CBC construction of lattice rules and polynomial lattice rules [66]. We now explain how this same structure can also be used in more general circumstances [15].

For notational convenience, we denote all QMC points $\boldsymbol{t}_i$ as row vectors in this subsection. Given an arbitrary matrix A, suppose we want to

$$\text{compute} \quad \boldsymbol{y}_i\, A \quad \text{for all} \quad i = 1, \ldots, n\,,$$

with the row vectors $\boldsymbol{y}_i = \chi(\boldsymbol{t}_i)$, where χ denotes an arbitrary univariate function that is applied to every component of the QMC point $\boldsymbol{t}_i$. Typically we have $\boldsymbol{t}_n \equiv \boldsymbol{t}_0 = \boldsymbol{0}$ so we can leave it out. Consider for simplicity the case n is prime and suppose we can write

$$Y := \begin{bmatrix} \boldsymbol{y}_1 \\ \vdots \\ \boldsymbol{y}_{n-1} \end{bmatrix} = C\,P$$

where C is an $(n-1)\times(n-1)$ circulant matrix, while P is a matrix containing a single 1 in each column and 0 everywhere else. Then we can compute $Y\boldsymbol{a}$ in $\mathscr{O}(n \log n)$ operations for any column $\boldsymbol{a}$ of A.

The desired factorization $Y = CP$ is possible if we have deterministic lattice points or deterministic polynomial lattice points, and if we apply the inverse cumulative distribution function mapping or tent transform [7, 13, 44]. However, it does *not* work with random shifting, scrambling [69], or interlacing. This strategy can be used to generate normally distributed points with a general covariance matrix (no need for stationarity as in circulant embedding), solving PDEs with uniform random coefficients, or solving PDEs with lognormal random coefficients involving finite element quadratures.

5 Software Resources

We provide some software resources for the practical application of QMC methods:

- *The Magic Point Shop*: a collection of QMC point generators and generating vectors.
  https://people.cs.kuleuven.be/~dirk.nuyens/qmc-generators/
- *Fast component-by-component constructions*: a collection of software routines for fast CBC constructions of generating vectors.
  https://people.cs.kuleuven.be/~dirk.nuyens/fast-cbc/
- *QMC4PDE*: accompanying software package for the survey [50] on using QMC methods for parametrized PDE problems.
  https://people.cs.kuleuven.be/~dirk.nuyens/qmc4pde/
- *A practical guide to QMC methods*: a non-technical introduction of QMC methods with software demos.
  https://people.cs.kuleuven.be/~dirk.nuyens/taiwan/

6 Summary and Outlook

In this article we summarized three QMC theoretical settings: randomly shifted lattice rules achieving first order convergence in the unit cube and in $\mathbb{R}^s$, and interlaced polynomial lattice rules achieving higher order convergence in the unit cube. One important feature is that the error bound can be independent of the dimension under appropriate conditions on the weights. Another important feature is that these QMC methods can be constructed by fast CBC algorithms.

We outlined three different applications and explained how they can be preprocessed to make use of the different theory. We also discussed three cost saving strategies that can be combined with QMC in these applications.

This paper is not meant to be a comprehensive survey on QMC methods. There are of course many other significant developments on QMC methods and their applications. For example, we did not discuss *tent transformation* (also known as the baker's transform), which can yield second order convergence for randomly shifted rules or first order convergence for deterministic lattice rules [7, 13, 44]. We also did not discuss *scrambling* [69], which is a well-known randomization method that can potentially improve the convergence rates by an extra half order.

For the future we would like to see QMC in new territories, to tackle a significantly wider range of more realistic problems. Some emerging new application areas of QMC include e.g., Bayesian inversion [9, 72], stochastic wave propagation [20, 21], quantum field theory [3, 47], and neutron transport [24, 34].

Looking ahead into future QMC developments, what would be on the top of our wish list? We would very much like to have a *"Setting 4" where we have QMC methods that achieve higher order convergence in $\mathbb{R}^s$, with error bounds that are* *independent of s, and for which fast constructions are possible.* This open problem has seen some partial solutions [10, 62] but there is more to be done!

Acknowledgements The authors acknowledge the financial supports from the Australian Research Council (FT130100655 and DP150101770) and the KU Leuven research fund (OT:3E130287 and C3:3E150478).

References

1. Achtsis, N., Cools, R., Nuyens, D.: Conditional sampling for barrier option pricing under the Heston model. In: Dick, J., Kuo, F.Y., Peters, G., Sloan, I.H. (eds.) Monte Carlo and Quasi-Monte Carlo Methods 2012, pp. 253–269. Springer, Berlin (2013)
2. Acworth, P., Broadie, M., Glasserman, P.: A comparison of some Monte Carlo and quasi-Monte Carlo techniques for option pricing. In: Hellekalek, P., Larcher, G., Niederreiter, H., Zinterhof, P. (eds.) Monte Carlo and quasi-Monte Carlo methods 1996, pp. 1–18. Springer, Berlin (1998)
3. Ammon, A., Hartung, T., Jansen, K., Leövey, H., Vollmer, J.: On the efficient numerical solution of lattice systems with low-order couplings. Comput. Phys. Commun. **198**, 71–81 (2016)
4. Caflisch, R.E., Morokoff, W., Owen, A.B.: Valuation of mortgage backed securities using Brownian bridges to reduce effective dimension. J. Comput. Financ. **1**, 27–46 (1997)
5. Cohen, A., DeVore, R.: Approximation of high-dimensional parametric PDEs. Acta Numer. **24**, 1–159 (2015)
6. Cools, R., Kuo, F.Y., Nuyens, D.: Constructing embedded lattice rules for multivariate integration. SIAM J. Sci. Comput. **28**, 2162–2188 (2006)
7. Cools, R., Kuo, F.Y., Nuyens, D., Suryanarayana, G.: Tent-transformed lattice rules for integration and approximation of multivariate non periodic functions. J. Complex. **36**, 166–181 (2016)
8. Dick, J., Pillichshammer, F.: Digital Nets and Sequences. Cambridge University Press, Cambridge (2010)
9. Dick, J., Gantner, R.N., Le Gia, Q.T., Schwab, Ch.: Higher order Quasi-Monte Carlo integration for Bayesian estimation (in review)
10. Dick, J., Irrgeher, Ch., Leobacher, G., Pillichshammer, F.: On the optimal order of integration in Hermite spaces with finite smoothness. SIAM J. Numer. Anal. (to appear)
11. Dick, J., Pillichshammer, F., Waterhouse, B.J.: The construction of good extensible rank-1 lattices. Math. Comput. **77**, 2345–2374 (2008)
12. Dick, J., Kuo, F.Y., Sloan, I.H.: High-dimensional integration: the Quasi-Monte Carlo way. Acta Numer. **22**, 133–288 (2013)
13. Dick, J., Nuyens, D., Pillichshammer, F.: Lattice rules for nonperiodic smooth integrands. Numer. Math. **126**, 259–291 (2014)
14. Dick, J., Kuo, F.Y., Le Gia, Q.T., Nuyens, D., Schwab, Ch.: Higher order QMC Galerkin discretization for parametric operator equations. SIAM J. Numer. Anal. **52**, 2676–2702 (2014)
15. Dick, J., Kuo, F.Y., Le Gia, Q.T., Schwab, Ch.: Fast QMC matrix-vector multiplication. SIAM J. Sci. Comput. **37**, A1436–A1450 (2015)
16. Dick, J., Kuo, F.Y., Le Gia, Q.T., Schwab, Ch.: Multi-level higher order QMC Galerkin discretization for affine parametric operator equations. SIAM J. Numer. Anal. **54**, 2541–2568 (2016)
17. Dick, J., Le Gia, Q.T., Schwab, Ch.: Higher order Quasi-Monte Carlo integration for holomorphic, parametric operator equations. SIAM/ASA J. Uncertain. Quantif. **4**, 48–79 (2016)
18. Dietrich, C.R., Newsam, G.H.: Fast and exact simulation of stationary Gaussian processes through circulant embedding of the covariance matrix. SIAM J. Sci. Comput. **18**, 1088–1107 (1997)

19. Feischl, M., Kuo, F.Y., Sloan, I.H.: Fast random field generation with H-matrices. Numer. Math. (to appear)
20. Ganesh, M., Kuo, F.Y., Sloan, I.H.: Quasi-Monte Carlo finite element wave propagation in heterogeneous random media (in preparation)
21. Ganesh, M., Hawkins, S.C.: A high performance computing and sensitivity analysis algorithm for stochastic many-particle wave scattering. SIAM J. Sci. Comput. **37**, A1475–A1503 (2015)
22. Gantner, R.N., Herrmann, L., Schwab, Ch.: Quasi-Monte Carlo integration for affine-parametric, elliptic PDEs: local supports and product weights SIAM. J. Numer. Anal. **56**(1), 111–135 (2018)
23. Gilbert, A.D., Wasilkowski, G.W.: Small superposition dimension and active set construction for multivariate integration under modest error demand. J. Complex. **42**, 94–109 (2017)
24. Gilbert, A.D., Graham, I.G., Kuo, F.Y., Scheichl, R., Sloan, I.H.: Analysis of quasi-Monte Carlo methods for elliptic eigenvalue problems with stochastic coefficients (in preparation)
25. Gilbert, A.D., Kuo, F.Y., Nuyens, D., Wasilkowski, G.W.: Efficient implementation of the multivariate decomposition method (in review)
26. Giles, M.B.: Multilevel Monte Carlo methods. Acta Numer. **24**, 259–328 (2015)
27. Giles, M.B., Waterhouse, B.J.: Multilevel quasi-Monte Carlo path simulation. Radon Ser. Comp. Appl. Math. **8**, 1–18 (2009)
28. Giles, M.B., Kuo, F.Y., Sloan, I.H.: Combining sparse grids, multilevel MC and QMC for elliptic PDEs with random coefficients (in this volume)
29. Giles, M.B., Kuo, F.Y., Sloan, I.H., Waterhouse, B.J.: Quasi-Monte Carlo for finance applications. ANZIAM J. **50**, C308–C323 (CTAC2008) (2008)
30. Goda, T.: Good interlaced polynomial lattice rules for numerical integration in weighted Walsh spaces. J. Comput. Appl. Math. **285**, 279–294 (2015)
31. Goda, T., Dick, J.: Construction of interlaced scrambled polynomial lattice rules of arbitrary high order. Found. Comput. Math. **15**, 1245–1278 (2015)
32. Graham, I.G., Kuo, F.Y., Nuyens, D., Scheichl, R., Sloan, I.H.: Analysis of circulant embedding methods for sampling stationary random fields. SIAM J. Numer. Anal. (to appear)
33. Graham, I.G., Kuo, F.Y., Nuyens, D., Scheichl, R., Sloan, I.H.: Circulant embedding with QMC – analysis for elliptic PDE with lognormal coefficients. Numer. Math. (to appear)
34. Graham, I.G., Parkinson, M.J., Scheichl, R.: Modern Monte Carlo variants for uncertainty quantification in neutron transport. In: Dick, J., Kuo. F.Y., Woźniakowski, H. (eds.) Contemporary Computational Mathematics - A Celebration of the 80th Birthday of Ian Sloan pp. 455–481. Springer (2018)
35. Graham, I.G., Kuo, F.Y., Nuyens, D., Scheichl, R., Sloan, I.H.: Quasi-Monte Carlo methods for elliptic PDEs with random coefficients and applications. J. Comput. Phys. **230**, 3668–3694 (2011)
36. Graham, I.G., Kuo, F.Y., Nichols, J.A., Scheichl, R., Schwab, Ch., Sloan, I.H.: QMC FE methods for PDEs with log-normal random coefficients. Numer. Math. **131**, 329–368 (2015)
37. Griebel, M., Kuo, F.Y., Sloan, I.H.: The smoothing effect of integration in $\mathbb{R}^d$ and the ANOVA decomposition. Math. Comput. **82**, 383–400 (2013); and the note in Math. Comput. **86**, 1847–1854 (2017)
38. Griewank, A., Kuo, F.Y., Leövey, H., Sloan, I.H.: High dimensional integration of kinks and jumps – smoothing by preintegration. J. Comput. Appl. Math. (to appear)
39. Gunzburger, M., Webster, C., Zhang, G.: Stochastic finite element methods for partial differential equations with random input data. Acta Numer. **23**, 521–650 (2014)
40. Hackbusch, W.: Hierarchical Matrices: Algorithms and Analysis. Springer, Heidelberg (2015)
41. Haji-Ali, A.L., Nobile, F., Tempone, R.: Multi-index Monte Carlo: when sparsity meets sampling. Numer. Math. **132**, 767–806 (2016)
42. Harbrecht, H., Peters, M., Siebenmorgen, M.: On the quasi-Monte Carlo method with Halton points for elliptic PDEs with log-normal diffusion. Math. Comput. **86**, 771–797 (2017)
43. Herrmann, L., Schwab, Ch.: QMC integration for lognormal-parametric, elliptic PDEs: local supports imply product weights (in review)

44. Hickernell, F.J.: Obtaining $O(N^{-2+\varepsilon})$ convergence for lattice quadrature rules. In: Fang, K.T., Hickernell, F.J., Niederreiter, H. (eds.) Monte Carlo and Quasi-Monte Carlo Methods 2000, pp. 274–289. Springer, Berlin (2002)
45. Hickernell, F.J., Niederreiter, H.: The existence of good extensible rank-1 lattices. J. Complex. **19**, 286–300 (2003)
46. Hickernell, F.J., Hong, H.S., LÉcuyer, P., Lemieux, C.: SIAM J. Sci. Comput. **22**, 1117–1138 (2000)
47. Jansen, K., Leövey, H., Griewank, A., Müller-Preussker, M.: Quasi-Monte Carlo methods for lattice systems: a first look. Comput. Phys. Commun. **185**, 948–959 (2014)
48. Kazashi, Y.: Quasi-Monte Carlo integration with product weights for elliptic PDEs with log-normal coefficients. IMA J. Numer. Anal. (to appear)
49. Kuo, F.Y., Nuyens, D.: Application of quasi-Monte Carlo methods to PDEs with random coefficients – an overview and tutorial (in this volume)
50. Kuo, F.Y., Nuyens, D.: Application of quasi-Monte Carlo methods to elliptic PDEs with random diffusion coefficients - a survey of analysis and implementation. Found. Comput. Math. **16**, 1631–1696 (2016)
51. Kuo, F.Y., Dunsmuir, W.D.M., Sloan, I.H., Wand, M.P., Womersley, R.S.: Quasi-Monte Carlo for highly structured generalised response models. Methodol. Comput. Appl. **10**, 239–275 (2008)
52. Kuo, F.Y., Sloan, I.H., Wasilkowski, G.W., Waterhouse, B.J.: Randomly shifted lattice rules with the optimal rate of convergence for unbounded integrands. J. Complex. **26**, 135–160 (2010)
53. Kuo, F.Y., Sloan, I.H., Wasilkowski, G.W., Woźniakowski, H.: Liberating the dimension. J. Complex. **26**, 422–454 (2010)
54. Kuo, F.Y., Sloan, I.H., Wasilkowski, G.W., Woźniakowski, H.: On decompositions of multi-variate functions. Math. Comput. **79**, 953–966 (2010)
55. Kuo, F.Y., Schwab, Ch., Sloan, I.H.: Quasi-Monte Carlo methods for high dimensional integration: the standard weighted-space setting and beyond. ANZIAM J. **53**, 1–37 (2011)
56. Kuo, F.Y., Schwab, Ch., Sloan, I.H.: Quasi-Monte Carlo finite element methods for a class of elliptic partial differential equations with random coefficient. SIAM J. Numer. Anal. **50**, 3351–3374 (2012)
57. Kuo, F.Y., Schwab, Ch., Sloan, I.H.: Multi-level quasi-Monte Carlo finite element methods for a class of elliptic partial differential equations with random coefficient. Found. Comput. Math. **15**, 411–449 (2015)
58. Kuo, F.Y., Nuyens, D., Plaskota, L., Sloan, I.H., Wasilkowski, G.W.: Infinite-dimensional integration and the multivariate decomposition method. J. Comput. Appl. Math. **326**, 217–234 (2017)
59. Kuo, F.Y., Scheichl, R., Schwab, Ch., Sloan, I.H., Ullmann, E.: Multilevel Quasi-Monte Carlo methods for lognormal diffusion problems. Math. Comput. **86**, 2827–2860 (2017)
60. Lemieux, C.: Monte Carlo and Quasi-Monte Carlo Sampling. Springer, New York (2009)
61. Leobacher, G., Pillichshammer, F.: Introduction to Quasi-Monte Carlo Integration and Applications. Springer, Berlin (2014)
62. Nguyen, D.T.P., Nuyens, D.: Multivariate integration over $\mathbb{R}^s$ with exponential rate of convergence. J. Comput. Appl. Math. **315**, 327–342 (2017)
63. Nichols, J.A., Kuo, F.Y.: Fast CBC construction of randomly shifted lattice rules achieving $\mathscr{O}(N^{-1+\delta})$ convergence for unbounded integrands in $\mathbb{R}^s$ in weighted spaces with POD weights. J. Complex. **30**, 444–468 (2014)
64. Niederreiter, H.: Random Number Generation and Quasi-Monte Carlo Methods. SIAM, Philadelphia (1992)
65. Novak, E., Woźniakowski, H.: Tractability of Multivariate Problems, II: Standard Information for Functionals. European Mathematical Society, Zürich (2010)
66. Nuyens, D.: The construction of good lattice rules and polynomial lattice rules. In: Kritzer, P., Niederreiter, H., Pillichshammer, F., Winterhof, A. (eds.) Uniform Distribution and Quasi-Monte Carlo Methods. Radon Series on Computational and Applied Mathematics, vol. 15, pp. 223–256. De Gruyter, Berlin (2014)

67. Nuyens, D., Cools, R.: Fast algorithms for component-by-component construction of rank-1 lattice rules in shift-invariant reproducing kernel Hilbert spaces. Math. Comput. **75**, 903–920 (2006)
68. Nuyens, D., Cools, R.: Fast component-by-component construction of rank-1 lattice rules with a non-prime number of points. J. Complex. **22**, 4–28 (2006)
69. Owen, A.B.: Scrambled net variance for integrals of smooth functions. Ann. Stat. **25**, 1541–1562 (1997)
70. Owen, A.B.: Halton sequences avoid the origin. SIAM Rev. **48**, 487–503 (2006)
71. Robbe, P., Nuyens, D., Vandewalle, S.: A multi-index quasi-Monte Carlo algorithm for log-normal diffusion problems. SIAM J. Sci. Comput. **39**, S851–S872 (2017)
72. Scheichl, R., Stuart, A., Teckentrup, A.L.: Quasi-Monte Carlo and multilevel Monte Carlo methods for computing posterior expectations in elliptic inverse problems. SIAM/ASA J. Uncertain. Quantif. **5**, 493–518 (2017)
73. Schwab, Ch., Gittelson, C.J.: Sparse tensor discretizations of high-dimensional parametric and stoch astic PDEs. Acta Numer. **20**, 291–467 (2011)
74. Sinescu, V., Kuo, F.Y., Sloan, I.H.: On the choice of weights in a function space for quasi-Monte Carlo methods for a class of generalised response models in statistics. In: Dick, J., Kuo, F.Y., Peters, G., Sloan, I.H. (eds.) Monte Carlo and Quasi-Monte Carlo Methods 2012, pp. 631–647. Springer, Heidelberg (2013)
75. Sloan, I.H., Joe, S.: Lattice Methods for Multiple Integration. Oxford University Press, Oxford (1994)
76. Sloan, I.H., Woźniakowski, H.: When are quasi-Monte Carlo algorithms efficient for high-dimensional integrals? J. Complex. **14**, 1–33 (1998)
77. Sloan, I.H., Kuo, F.Y., Joe, S.: Constructing randomly shifted lattice rules in weighted Sobolev spaces. SIAM J. Numer. Anal. **40**, 1650–1665 (2002)
78. Sloan, I.H., Wang, X., Woźniakowski, H.: Finite-order weights imply tractability of multivariate integration. J. Complex. **20**, 46–74 (2004)
79. Wasilkowski, G.W.: On tractability of linear tensor product problems for ∞-variate classes of functions. J. Complex. **29**, 351–369 (2013)

Stratified Bayesian Optimization

Saul Toscano-Palmerin and Peter I. Frazier

Abstract We consider derivative-free black-box global optimization of expensive noisy functions, when most of the randomness in the objective is produced by a few influential scalar random inputs. We present a new Bayesian global optimization algorithm, called Stratified Bayesian Optimization (SBO), which uses this strong dependence to improve performance. Our algorithm is similar in spirit to stratification, a technique from simulation, which uses strong dependence on a categorical representation of the random input to reduce variance. We demonstrate in numerical experiments that SBO outperforms state-of-the-art Bayesian optimization benchmarks that do not leverage this dependence.

Keywords Bayesian optimization · Gaussian processes · Simulation

1 Introduction

We consider derivative-free black-box global optimization of expensive noisy functions,

$$\max_{x \in A \subset \mathbb{R}^n} \mathbb{E}\left[f(x, w, z)\right], \tag{1}$$

where the expectation is taken over $z \in \mathbb{R}^{d_1}$ and $w \in \mathbb{R}^{d_2}$, which have joint probability density p, A is a simple compact set (e.g., a hyperrectangle, or simplex), and we can directly observe only $f(x, w, z)$ at some collection of chosen or sampled x, w, z, and not its expectation, or the derivative of this expectation. The separation of the random inputs to $f(x, w, z)$ into two variables will be discussed in more detail below. We suppose that f has no special structural properties, e.g., concavity, or linearity, that

S. Toscano-Palmerin (✉) · P. I. Frazier
School of Operations Research and Information Engineering, Cornell University, Ithaca, NY, USA
e-mail: st684@cornell.edu

P. I. Frazier
e-mail: pf98@cornell.edu

A. B. Owen and P. W. Glynn (eds.), *Monte Carlo and Quasi-Monte Carlo Methods*, Springer Proceedings in Mathematics & Statistics 241,
https://doi.org/10.1007/978-3-319-91436-7_7

we can exploit to solve this problem, making it a "black blox." We also suppose that evaluating f is costly or time-consuming, making these evaluations "expensive", severely limiting the number of evaluations we may perform. This typically occurs because each evaluation requires running a complex PDE-based or discrete-event simulation, or requires training a machine learning algorithm on a large dataset. When f comes from a discrete-event simulation, this problem is also called "simulation optimization."

Bayesian optimization is a popular class of techniques for solving this problem, originating with the seminal paper [11], and enjoying early contributions from [12, 13]. This class of techniques was popularized in the 1990s by the introduction in [9] of the most well-known Bayesian optimization method, Efficient Global Optimization (EGO), relying on earlier ideas from [12]. Recently the machine learning community has devoted considerable attention to Bayesian optimization for its applications to tuning computationally intensive machine learning models, as in, e.g., [18]. Textbooks and surveys on Bayesian optimization include [1, 3].

Most work on Bayesian optimization assumes we can observe the objective function directly without noise, but a substantial number of papers, e.g. [1, 8, 17, 20], do allow noise and thus consider (1). These methods all build a statistical model (usually using Gaussian processes) of the function $x \mapsto G(x) := \mathbb{E}[f(x, w, z)]$ using noisy observations, and then use an acquisition criterion, typically expected improvement or probability of improvement [1], to decide where to sample next.

Existing work from Bayesian optimization for solving (1) relies on noisy evaluations in which w and z are drawn iid from their governing joint probability distribution p, and then $f(x, w, z)$ is observed. However, in many applications, we have the ability to choose not just x, but w as well, simulating the remaining components z conditioning on these values. (The choice of which random inputs to include in w instead of z can be made arbitrarily when using existing Bayesian optimization algorithms, but the new algorithm we propose will assume they accommodate easy sampling of z given w.)

This ability to simulate random inputs given the value of some of their values is widely used in stratified sampling to estimate expectations with better precision [6].

For example, in a queuing simulation (we give a detailed example in our numerical experiments), we can simulate the individual arrival times of customers z conditioning on the overall number of arrivals w. In a revenue management simulation, we can simulate individual purchase decisions z conditioned on the overall demand w. In an aerodynamic simulation, we can simulate fine-scale airflows z, conditioned on average wind speed w.

We thus rephrase problem (1) into the equivalent problem

$$\max_{x \in A \subset \mathbb{R}^n} \mathbb{E}\left[F(x, w)\right] \tag{2}$$

where $F(x, w) := \mathbb{E}\left[f(x, w, z) \mid w\right]$, and the problems are equivalent because we have that $\mathbb{E}[F(x, w)] = \mathbb{E}\left[f(x, w, z)\right] = G(x)$.

This equivalent formulation suggests that standard approaches to Bayesian optimization are wasteful from a statistical point of view, as they do not use past observations w to learn $G(x) = \mathbb{E}[F(x, w)]$, treating w only as an unobservable source of noise.

Instead, one can use Bayesian quadrature [15], which builds a Gaussian process model of the function $F(x, w)$ using past observations of x, w, $f(x, w, z)$, and then uses the known relationship $G(x) = \int F(x, w)p(w)\,dw$ (where we assume $p(w) := \int p(w, z)\,dz$ is known in closed form) to imply a second Gaussian process model on $G(x)$.

In this paper, we leverage this ability and develop an algorithm, called stratified Bayesian optimization (SBO), which chooses not just the x at which to evaluate $f(x, w, z)$, but also the w. It chooses these using a one-step Bayes-optimal acquisition function based on a value-of-information [7] analysis. It then samples z from its conditional distribution given w, and uses the resulting observation within a Bayesian quadrature framework to update its Gaussian process posterior on both $(x, w) \mapsto F(x, w)$ and $x \mapsto \mathbb{E}[F(x, w)]$. By using more information, we make our statistical model more powerful, and provide better answers with fewer samples.

This approach is similar in spirit to stratified sampling [6], where our goal is to estimate $G(x) = E[F(x, w)] = E[f(x, w, z)]$ for a fixed x, and we choose which values of w at which to sample rather than sampling them from their marginal distribution, and then compensate for this choice via a known relationship between $F(x, w)$ and $G(x)$ to obtain lower variance estimates.

To choose x and w, SBO uses a decision-theoretic approach that models the utility resulting from solutions to the optimization problem (2). SBO finds the pair of values (x, w) at which to sample that maximizes the expected utility of the final solution, under the assumption, made for tractability, that we may take only one additional sample. Thus, our SBO algorithm is optimal in a decision-theoretic sense, in a one-step setting.

This one-step decision-theoretic approach follows the development of acquisition functions for other settings. In more traditional Bayesian optimization problems, the well-known expected improvement acquisition function [9, 12] has this optimality property when observations are noise-free and the final solution must be taken from previously evaluated solutions [5], and the knowledge-gradient (KG) method [4, 17] has this optimality property when the final solution is not restricted to be a previously evaluated solution, in both the noisy and noise-free setting.

Our approach also builds on, and significantly generalizes, the previous work [19], which developed a similar method, but did not allow for the inclusion of unmodeled random inputs z, instead requiring all inputs to be included and modeled statistically in w. This introduces a heavy computational and statistical burden when dealing with problems in which the combined dimension of w and z is large, which includes many complex stochastic models, significantly limiting its applicability.

This paper is organized as follows: Sect. 2 presents our statistical model. Section 3 presents the SBO algorithm. Section 4 describes the computation of the value of information. Section 5 describes the computation of the gradient of the value of information. Section 6 presents simulation experiments. Section 7 concludes.

2 Statistical Model

The SBO algorithm that we develop relies on a Gaussian process (GP) model of the underlying function F, which then implies (because integration is a linear function) a Gaussian process model over G. This statistical approach mirrors a standard Bayesian quadrature approach, but we summarize it here both to define notation used later, and because its application to Bayesian optimization is new.

We first place a Gaussian process prior distribution over the function F:

$$F(\cdot,\cdot) \sim GP\left(\mu_0(\cdot,\cdot),\, \Sigma_0(\cdot,\cdot,\cdot,\cdot)\right),$$

where μ_0 is a real-valued function taking arguments (x, w), and Σ_0 is a positive semi-definite function taking arguments $\left(x, w, x', w'\right)$. Common choices for μ_0 and Σ_0 from the Gaussian process regression literature [14, 16], e.g., setting μ_0 to a constant and letting Σ_0 be the squared exponential or Màtern kernel, are appropriate here as well.

Our algorithm will take samples sequentially. At each time $n = 1, 2, \ldots, N$, our algorithm will choose x_n and w_n based on previous observations. It will then take M samples of $f(x_n, w_n, z)$ and observe the average response. More precisely, it will sample $z_{n,m} \sim p(z \mid w_n)$ for $m = 1, \ldots, M$ and observe $y_n = \frac{1}{M}\sum_{m=1}^{M} f\left(x_n, w_n, z_{n,m}\right)$. The choice of M is an algorithm parameter, and should be chosen large enough that the central limit theorem may be applied, so that we may reasonably model the (conditional) distribution of y_n as normal. We will then have,

$$y_n|x_n, w_n \sim N\left(F(x_n, w_n), \sigma^2(x_n, w_n)/M\right),$$

where $\sigma^2(w, z) := \mathrm{Var}(f(x, w, z) \mid w)$. We assume that this conditional variance is finite for all x and w. In updating the posterior, we also assume that we observe this value $\sigma^2(x_n, w_n)$, although in practice we estimate it using the empirical variance from our M samples.

Let $H_n = (y_{1:n}, w_{1:n}, x_{1:n})$ be the history observed by time n. Then, the posterior distribution on F at time n is

$$F(\cdot,\cdot) \mid H_n \sim GP\left(\mu_n(\cdot,\cdot),\, \Sigma_n(\cdot,\cdot,\cdot,\cdot)\right),$$

where μ_n and Σ_n can be computed using standard results from Gaussian process regression [16]. To support later analysis, expressions for μ_n and Σ_n are provided in the appendix.

We denote by $\mathbb{E}_n$, Cov_n, and Var_n the conditional expectation, conditional covariance, and conditional variance on F (and thus also on G, since G is specified by F) with respect to the Gaussian process posterior given H_n. By results from Bayesian quadrature [15], which rely on the previously noted fact that $G(x) = \int F(x, w)p(w)\, dw$,

$$\mathbb{E}_n [G(x)] = \int \mu_n(x, w)p(w)\, dw := a_n(x), \tag{3}$$

$$\mathrm{Cov}_n \left(G(x), G(x')\right) = \int \int \Sigma_n \left(x, w, x', w'\right) p(w)\, p\left(w'\right) dwdw.'$$

The first line is derived using interchange of integral and expectation, as in $\mathbb{E}_n [G(x)] = \mathbb{E}_n \left[\int F(x, w)p(w)\, dw\right] = \int \mathbb{E}_n [F(x, w)p(w)]\, dw = \int \mu_n(x, w)\, p(w)\, dw$. The second line is derived similarly, though with more effort, by writing the covariance as an expectation, and interchanging expectation and integration.

3 Stratified Bayesian Optimization (SBO) Algorithm

Our SBO algorithm will choose points to evaluate using a value of information analysis [7], which maximizes the expected gain in the quality of the final solution to (1) that results from a sample.

To support this value of information analysis, we first consider the expected solution quality resulting for a particular set of samples. After n samples, if we were to choose the solution to (1) with the best expected quality with respect to the Bayesian posterior distribution on G, we would choose

$$x_n^* \in \arg\max_x \mathbb{E}_n [G(x)] = \arg\max_x a_n(x).$$

This is the Bayes-optimal solution when we are risk neutral. This solution has expected value (again, with respect to the posterior),

$$\mu_n^* := \max_x \mathbb{E}_n [G(x)] = \max_x a_n(x).$$

The improvement in expected solution quality that results from a sample at (x, w) at time n is

$$V_n(x, w) = \mathbb{E}_n \left[\mu_{n+1}^* - \mu_n^* \mid x_{n+1} = x, w_{n+1} = w\right]. \tag{4}$$

We refer to this quantity as the *value of information*, and if we have one evaluation remaining, then choosing to sample at the point with the largest value of information is optimal from a Bayesian decision-theoretic point of view. If we have more than one evaluation remaining, then it is not necessarily Bayes-optimal, but we argue that it remains a reasonable heuristic.

Thus, our Stratified Bayesian Optimization (SBO) algorithm is defined by

$$(x_{n+1}, w_{n+1}) \in \arg\max_{x,w} V_n(x, w). \tag{5}$$

Detailed computation of this value of information, and its gradient with respect to x and w, is discussed below in Sect. 4. We use this gradient to solve (5) using multi-start gradient ascent or multi-start sequential least squares programming [10].

The SBO algorithm is summarized in Algorithm 1. The complexity of the SBO algorithm is $O(LN^2 + N^4)$ if it is run during N iterations, and L is the number of points in the discretization of the domain of the points x, see Sect. 4.

Algorithm 1 SBO Algorithm

1: **First stage of samples** Evaluate F at n_0 points, chosen uniformly at random from A. Use maximum likelihood or maximum a posteriori estimation to fit the parameters of the GP prior on F, conditioned on these n_0 samples. Let μ_0 and Σ_0 be the mean function and covariance kernel of the resulting GP posterior on F.
2: **Main stage of samples:**
3: **for** $n = 1$ **to** N **do**
4: Update our Gaussian process posterior on F using all samples from the first stage, and samples $x_{1:n}$,$w_{1:n}$, $y_{1:n}$. This allows computation of μ_n and Σ_n as described in the appendix, computation of a_n through (3), and computation of V_n and ∇V_n as described in Sect. 4.
5: Solve $(x_{n+1}, w_{n+1}) \in \arg\max_{x,w} V_n(x, w)$ using multi-start sequential least squares programming or multi-start gradient ascent and the ability to compute ∇V_n. Let (x_{n+1}, w_{n+1}) be the resulting maximizer.
6: Evaluate $y_{n+1} = \frac{1}{M}\sum_{m=1}^{M} f\left(x_{n+1}, w_{n+1}, z_{n+1,m}\right)$ where $z_{n+1,m}$ are iid draws from $p(z \mid w_{n+1})$, and $p(z \mid w) = p(w, z)/p(w)$ is the conditional density of z given w.
7: **end for**
8: Return $x^* = \arg\max_x a_{N+1}(x)$.

Figure 1 illustrates how SBO works, showing one step in the algorithm applied to a simple analytic test problem

$$\max_{x\in[-\frac{1}{2},\frac{1}{2}]} \mathbb{E}[f(x, w, z)] = \max_{x\in[-\frac{1}{2},\frac{1}{2}]} \mathbb{E}\left[zx^2 + w\right] \tag{6}$$

where $w \sim N(0, 1)$ and $z \sim N(-1, 1)$. Direct computation shows $F(x, w) = -x^2 + w$ and $G(x) = -x^2$.

The figure shows the contours of $F(x, w)$, the mean of SBO's posterior on $F(x, w)$ in the first row, and the value of information and SBO's posterior on $G(x)$ in the second row, all after $n = 7$ samples.

SBO's value of information is small near where SBO has already sampled, because it has less uncertainty about $F(x, w)$ in this region. Its value of information is also smaller for w far away from 0 because they have smaller $p(w)$, and thus their $F(x, w)$ have less influence on $G(x)$. SBO's value of information is also small for extreme values of x, because its posterior on G suggests that these x are far from its maximum. SBO's value of information is thus largest for points that are far from previous samples, closer to $x = 0$, and closer to $w = 0$, and SBO samples next at the point with the largest value of information.

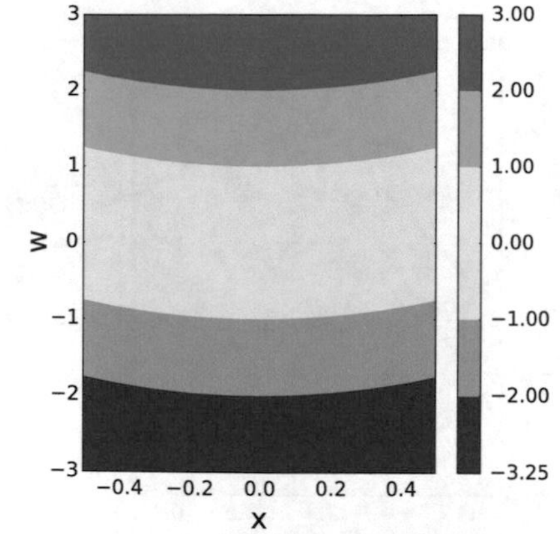

(a) The contours of $F(x,w)$. The objective $G(x)$ is $E[F(x,w)|x]$.

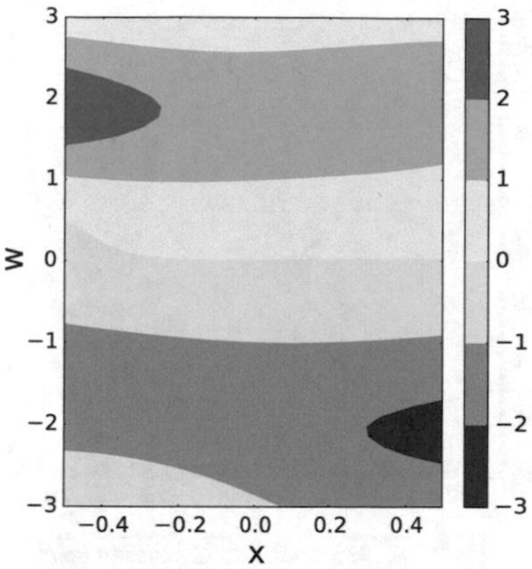

(b) SBO: The contours of SBO's estimate $\mu_n(x,w)$ of $F(x,w)$, at $n=7$.

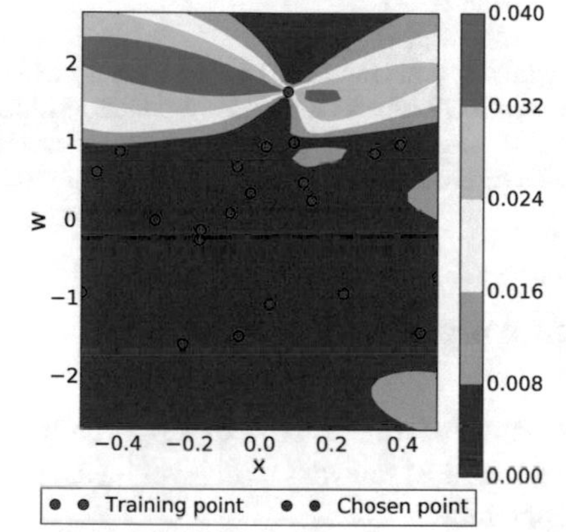

(c) SBO: The contours of the value of information $V_n(x,w)$ under SBO at $n=7$. SBO's value of information depends on both x and w.

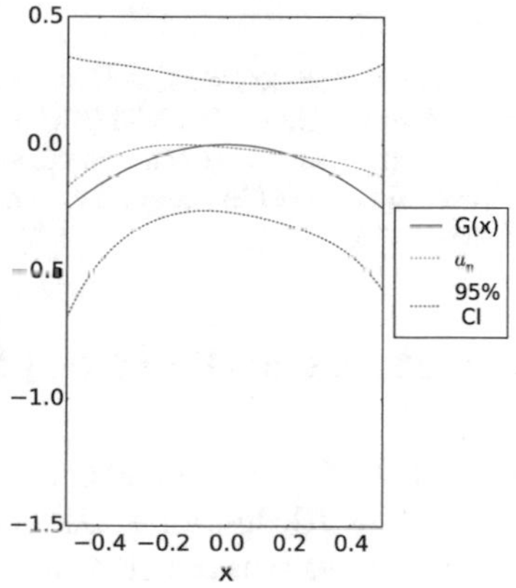

(d) SBO:The objective $G(x)$, and SBO's estimate $a_n(x)$ and 95% credible interval, at $n=7$.

Fig. 1 Illustration of the SBO algorithm on an analytic test problem. SBO models $F(x, w)$ while benchmark methods (KG and EI) model $G(x)$. (**First row**) The contours of $F(x, w)$ (left) and of SBO's estimate of F (right). (**Second row**) Left shows the contours of SBO's value of information, which depends on both x and w, and which SBO uses to choose the pair (x, w) to sample next. Right shows SBO's estimates of $G(x)$, which is based on the estimate of $F(x, w)$ in the first row

Figure 2 shows equivalent quantities for the KG method [4], which, like other Bayesian optimization methods, models $G(x)$ directly, ignoring valuable information from w, and computes a value of information as a function of x only (it believes that observing near $x = 0.1$ or $x = 1$ would be most useful), leaving the choice of w to chance. Furthermore, after $n = 7$ observations, SBO's use of w allows it to have a much more accurate estimate of $G(x)$, and the location of its maximum.

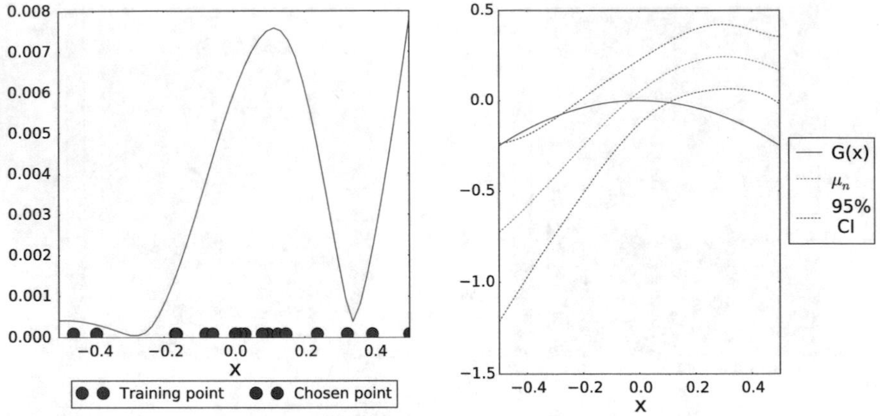

(a) KG: The value of information under KG at $n = 7$. KG's value of information depends only on x.

(b) KG: The objective $G(x)$, and KG's estimate $a_n(x)$ and 95% credible interval, at $n = 7$.

Fig. 2 Illustration of the KG algorithm on an analytic test problem. SBO models $F(x, w)$ while benchmark methods (KG and EI) model $G(x)$. Left shows KG's value of information, which depends only on x, and which KG uses to choose the point x to sample next. Right shows KG's estimates of $G(x)$. This estimate is of lower quality than SBO's estimate above, because it does not use the observed values of w

4 Computation of the Value of Information

In this section we discuss computation of the value information (4), to support implementation of the SBO algorithm. We focus on the most relevant high level ideas, and the detailed computations are given in the appendix.

Table 1 summarizes notation used in this section.

We first rewrite the value of information (4) as

Table 1 Table of notation

$G(x)$	$\triangleq$	$\mathbb{E}[f(x, w, z)]$
V_n	$\triangleq$	Value of Information at time n
$a_n(x)$	$\triangleq$	$\mathbb{E}_n[G(x)]$
H_n	$\triangleq$	History observed by time n
Σ_0	$\triangleq$	Kernel of the Gaussian process prior distribution over the function F
$B(x, i)$	$\triangleq$	$\int \Sigma_0(x, w, x_i, w_i)\, p(w) dw$, for $i = 1, \ldots, n+1$
γ	$\triangleq$	$\begin{bmatrix} \Sigma_0(x_{n+1}, w_{n+1}, x_1, w_1) \\ \vdots \\ \Sigma_0(x_{n+1}, w_{n+1}, x_n, w_n) \end{bmatrix}$
A_n	$\triangleq$	$\left(\Sigma_0\left(x_i, w_i, x_j, w_j\right)\right)_{i,j=1}^{n} + \text{diag}\left(\left(\sigma^2(x_i, w_i)\right)_{i=1}^{n}\right)$

$$V_n\left(x_{n+1}, w_{n+1}\right) = \mathbb{E}_n\left[\max_{x' \in A} a_{n+1}\left(x'\right) \mid x_{n+1}, w_{n+1}\right] - \max_{x' \in A} a_n\left(x'\right). \quad (7)$$

To calculate this expectation, we must find the joint distribution of $a_{n+1}(x)$ across all x conditioned on (x_{n+1}, w_{n+1}) and H_n for any x. This is provided by the following lemma.

Lemma 1 *There exists a standard normal random variable Z_{n+1} such that, for all x,*

$$a_{n+1}(x) = a_n(x) + \tilde{\sigma}_n(x, x_{n+1}, w_{n+1}) Z_{n+1}.$$

where

$$\tilde{\sigma}_n^2(x, x_{n+1}, w_{n+1}) := Var_n\left[G(x)\right] - \mathbb{E}_n\left[Var_{n+1}\left[G(x)\right] \mid x_{n+1}, w_{n+1}\right].$$

To compute the value of information, we then discretize the feasible set A, over which we take the maximum in (7), into $L < \infty$ points. We let A' denote this discrete set of points, so $A' \subseteq A$ and $|A'| = L$. For example, if A is a hyperrectangle, then we may discretize it using a uniform mesh.

Then, we approximate (7) by

$$\begin{aligned} V_n(x_{n+1}, w_{n+1}) &= \mathbb{E}_n\left[\max_{x \in A} a_n(x) + \tilde{\sigma}(x, x_{n+1}, w_{n+1}) Z_{n+1}\right] - \max_{x \in A} a_n(x) \\ &\approx \mathbb{E}_n\left[\max_{x \in A'} a_n(x) + \tilde{\sigma}(x, x_{n+1}, w_{n+1}) Z_{n+1}\right] - \max_{x \in A'} a_n(x) \\ &= h(a_n(A'), \tilde{\sigma}_n(A', x_{n+1}, w_{n+1})), \end{aligned}$$

where $a_n(A') = (a_n(x_i))_{i=1}^L$, $\tilde{\sigma}_n(x, w) = (\tilde{\sigma}_n(x_i, x, w))_{i=1}^L$, and $h : \mathbb{R}^L \times \mathbb{R}^L \to \mathbb{R}$ is a function defined by $h(a, b) = \mathbb{E}\left[\max_i a_i + b_i Z\right] - \max_i a_i$, where a and b are any deterministic vectors, and Z is a one-dimensional standard normal random variable. By convenience, we will denote $a_n(x_i)$ by e_i and $\tilde{\sigma}_n(x_i, x, w)$ by f_i for each i in $\{1, \ldots, L\}$. If $A = A'$, which is possible if A is a finite set, then the approximation in the second line above is exact.

In [4], it is also shown how to compute h. Using the Algorithm 1 in that paper, we can get a subset of indexes $\{j_1, \ldots, j_\ell\}$ from $\{1, \ldots, L\}$, such that

$$\begin{aligned} V_n(x_{n+1}, w_{n+1}) &= h(a_n(A'), \tilde{\sigma}_n(A', x_{n+1}, w_{n+1})) \\ &= \sum_{i=1}^{\ell-1} \left(f_{j_{i+1}} - f_{j_i}\right) f\left(-|c_i|\right) \end{aligned}$$

where

$$\begin{aligned} f(z) &:= \varphi(z) + z\Phi(z), \\ c_i &:= \frac{e_{j_{i+1}} - e_{j_i}}{f_{j_{i+1}} - f_{j_i}}, \; i = 1, \ldots, \ell - 1, \end{aligned}$$

and φ, Φ are the standard normal cdf and pdf, respectively. This shows how to compute the Value of Information V_n.

5 Computation of the Gradient of the Value of Information

In this section we show how to compute the gradient of the Value of Information V_n. Observe that if $\ell = 1$, $V_n(x, w) = 0$ and so $\nabla V_n\left(x, w^{(1)}\right) = 0$. On the other hand, if $\ell > 1$, one can show via direct computation that

$$\nabla V_n(x, w) = \sum_{i=1}^{\ell-1} \left(-\nabla f_{j_{i+1}} + \nabla f_{j_i}\right) \varphi(|c_i|).$$

Consequently, we only need to compute ∇f_{j_i} for each i in $\{1, \ldots, \ell\}$. Another direct computation shows that

$$\nabla \tilde{\sigma}_n(x, x_{n+1}, w_{n+1}) = \beta_1 \beta_3 - \frac{1}{2} \beta_1^3 \beta_2 [\beta_5 - \beta_4]$$

where

$$\begin{aligned}
\beta_1 &= \left[\Sigma_0(x_{n+1}, w_{n+1}, x_{n+1}, w_{n+1}) - \gamma^T A_n^{-1} \gamma\right]^{-1/2}, \\
\beta_2 &= B(x, n+1) - [B(x, 1) \ \cdots \ B(x, n)] A_n^{-1} \gamma, \\
\beta_3 &= \left(\nabla B(x, n+1) - \nabla\left(\gamma^T\right) A_n^{-1} \begin{bmatrix} B(x, 1) \\ \vdots \\ B(x, n) \end{bmatrix}\right), \\
\beta_4 &= 2 \nabla\left(\gamma^T\right) A_n^{-1} \gamma, \\
\beta_5 &= \nabla \Sigma_0(x_{n+1}, w_{n+1}, x_{n+1}, w_{n+1}).
\end{aligned}$$

We now give expressions for a_n to compute the parameters of the posterior distribution of a_{n+1}. First, a_n can be computed using the following formula,

$$\begin{aligned}
a_n(x) &= \mathbb{E}[\mu_n(x, w)] \\
&= \mathbb{E}[\mu_0(x, w)] + [B(x, 1) \ \cdots \ B(x, n)] A_n^{-1} \begin{pmatrix} y_1 - \mu_0(x_1, w_1) \\ \vdots \\ y_n - \mu_0(x_n, w_n) \end{pmatrix}.
\end{aligned}$$

In some cases it is possible to get a closed-form formula for B, e.g. if w follows a normal distribution, the components of w are independent and we use the squared exponential kernel.

Finally, a direct computation detailed in the appendix shows that the formula for $\tilde{\sigma}_n^2(x, x_{n+1}, w_{n+1})$ is

$$\left[\frac{\left(B(x, n+1) - [B(x, 1) \ \cdots \ B(x, n)] A_n^{-1} \gamma\right)}{\sqrt{\left(\Sigma_0(x_{n+1}, w_{n+1}, x_{n+1}, w_{n+1}) - \gamma^T A_n^{-1} \gamma\right)}}\right]^2.$$

6 Numerical Experiments

We now present simulation experiments illustrating how the SBO algorithm can be applied in practice, and comparing its performance against some baseline Bayesian optimization algorithms. We compare on a test problem with a simple analytic form (Sect. 6.1), on a realistic problem arising in the design of the New York City's Citi Bike system (Sect. 6.2), and on a wide variety of problems simulated from Gaussian process priors (Sect. 6.3) designed to provide insight into what problem characteristics allow SBO to provide substantial benefit.

We consider two baseline Bayesian optimization algorithms. We use the Knowledge-Gradient policy of [4] and Expected Improvement criterion [9], which both place the Gaussian process prior directly on $G(x)$, and use a standard sampling procedure, in which w and z are drawn from their joint distribution, and $f(x, w, z)$ is observed. The Knowledge-Gradient policy is equivalent to SBO if all components of w are moved into z. Thus, comparing against KG quantifies the benefit of SBO's core contribution, while holding constant standard aspects of the Bayesian optimization approach.

We also solved the problems from (Sect. 6.1) and (Sect. 6.2) with Probability of Improvement (PI) [1], but we did not include its results in our graphs because both KG and EI substantially outperformed PI. Moreover, according to Brochu [1], "EI's acquisition function is more satisfying than PI's acquisition function".

When implementing the SBO algorithm, we use the squared exponential kernel, which is defined as

$$\Sigma_0\left(x, w, x', w'\right) = \sigma_0^2 \exp\left(-\sum_{k=1}^{n} \alpha_1^{(k)}\left[x_k - x_k'\right]^2 - \sum_{k=1}^{d_1} \alpha_2^{(k)}\left[w_k - w_k'(1)\right]^2\right),$$

where σ_0^2 is the common prior variance and $\alpha_1^{(1)}, \ldots, \alpha_1^{(n)}, \alpha_2^{(1)}, \ldots, \alpha_2^{(d_1)} \in \mathbb{R}_+$ are length scales. These values, σ^2 and the mean μ_0 are calculated using maximum likelihood estimation following the first stage of samples.

6.1 An Analytic Test Problem

In our first example, we consider the problem (6) stated in Sect. 3. Figure 3 compares the performance of SBO, KG and EI on this problem, plotting the number of samples beyond the first stage on the x axis, and the average true quality of the solutions provided, $G(\text{argmax}_x \mathbb{E}_n[G(x)])$, averaging over 3000 independent runs of the three algorithms.

We see that SBO substantially outperforms both benchmark methods. This is possible because SBO reduces the noise in its observations by conditioning on w, allowing it to more swiftly localize the objective's maximum.

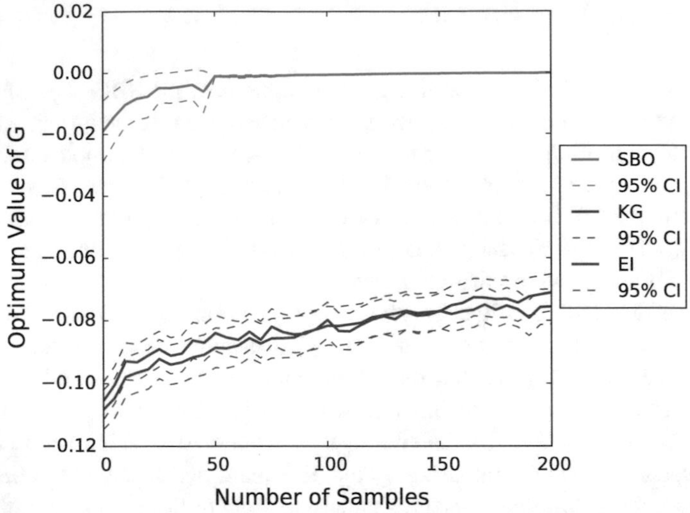

Fig. 3 Performance comparison between SBO and two Bayesian optimization benchmark, the KG and EI methods, on the analytic test problem (6) from Sect. 3. SBO performs significantly better than two benchmarks: knowledge-gradient (KG) and expected improvement (EI)

6.2 New York City's Citi Bike System

We now consider a more realistic problem, using a queuing simulation based on New York City's Citi Bike system, in which system users may remove an available bike from a station at one location within the city, and ride it to a station with an available dock in some other location within the city. The optimization problem that we consider is the allocation of a constrained number of bikes (6000) to available docks within the city at the start of rush hour, so as to minimize, in simulation, the expected number of potential trips in which the rider could not find an available bike at their preferred origination station, or could not find an available dock at their preferred destination station. We call such trips "negatively affected trips."

We simulated in Python the demand of bike trips of a New York City's Bike System on any day from January 1st to December 31st between 7:00 AM and 11:00 AM. We used 329 actual bike stations, locations, and numbers of docks from the Citi Bike system, and estimated demand and average time for trips for every day in a year using publicly available data of the year 2014 from Citi Bike's website [2].

We simulate the demand for trips between each pair of bike stations on a day using an independent Poisson process, and trip times between pairs of stations follows an exponential distribution. If a potential trip's origination station has no available bikes, then that trip does not occur, and we increment our count of negatively affected trips. If a trip does occur, and its preferred destination station does not have an available dock, then we also increment our count of negatively affected trips, and the bike is returned to the closest bike station with available docks.

We divided the bike stations in 4 groups using k-nearest neighbors, and let x be the number of bikes in each group at 7:00 AM. We suppose that bikes are allocated uniformly among stations within a single group.

The random variable w is the total demand of bike trips during the period of our simulation. The random vector z contains all other random quantities within our simulation.

Table 2 provides a concrete mapping of SBO's abstractions onto the CitiBike example.

Figure 4a compares the performance of SBO, KG and EI, plotting the number of samples beyond the first stage on the x axis, and the average true quality of the solutions provided, $G(\text{argmax}_x \mathbb{E}_n[G(x)])$, averaging over 300 independent runs of the three algorithms. We see that SBO was able to quickly find an allocation of bikes to groups that attains a small expected number of negatively affected trips.

Table 2 Table of notation for the citibike problem

$x \in \mathbb{R}^4$	$\triangleq$	Deterministic vector that represents the number of bikes in each group of bike stations at 7:00 AM
$w \in \mathbb{N}$	$\triangleq$	Poisson random variable that represents the total demand of bike trips between 7:00 AM to 11:00 AM
z	$\triangleq$	Random vector that consists of: (i) day of the year where the simulation occurs, (ii) $\binom{329}{2}$ dimensional Poisson random vector that represents the total demand between each pair of bike stations, (iii) exponential random vector that represents the time duration of each bike trip
$-f(x, w, z)$	$\triangleq$	Negatively affected trips between 7:00 AM to 11:00 AM
$G(x)$	$\triangleq$	$\mathbb{E}[f(x, w, z)]$

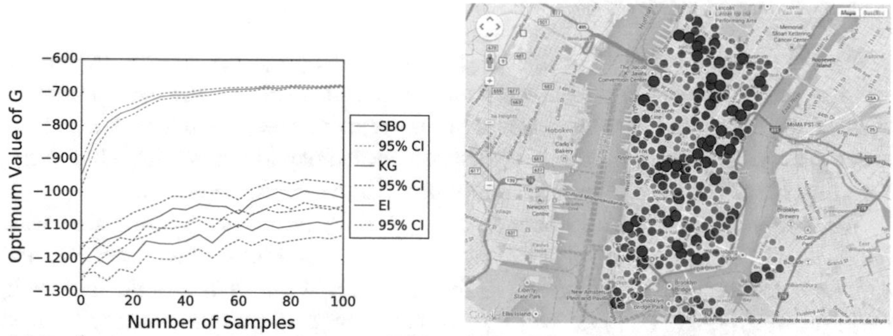

(a) Performance comparison between SBO and two Bayesian optimization benchmark, the KG and EI methods, on the Citi Bike Problem from Sect. 6.2

(b) Location of bike stations (circles) in New York City, where size and color represent the ratio of available bikes to available docks

Fig. 4 Performance results for the Citi Bike problem (plot 1), and a screenshot from our simulation of the Citi Bike problem (plot b)

6.3 Problems Simulated from Gaussian Process Priors

We now compare the performance of SBO against a benchmark Bayesian optimization algorithm on synthetic problems drawn at random from Gaussian process priors. We use the KG algorithm as our benchmark, as it performed as well or better than the other benchmark algorithms (EI and Probability of Improvement) on the test problems in Sects. 6.1 and 6.2. In these experiments, SBO outperforms the benchmark on most problems, in some cases offering an improvement of almost 1000%. On those few problems in which SBO underperforms the benchmark, it underperforms by a much smaller margin of less than 50%.

Our experiments also provide insight into how SBO should be applied in practice. They show that the most important factor in determining SBO's performance over benchmarks is the speed with which the conditional expectation $F(x, w) = \mathbb{E}\left[f\left(x, w, z\right)|w\right]$ varies with w. SBO provides the most value when this variation is large enough to influence performance, and small enough to allow $F(x, w)$ to be modeled with a Gaussian process. Thus, users of SBO should choose a w that plays a big role in overall performance, and whose influence on performance is smooth enough to support predictive modeling.

We now construct these problems in detail. Let $f(x, w, z) = h(x, w) + g(z)$ on $[0, 1]^2 \times \mathbb{R}$, where:

- $g(z)$ is drawn, for each z, independently from a normal distribution with mean 0 and variance α_d (we could have set g to be an Orstein–Uhlenbeck process with large volatility, and obtained an essentially identical result).
- h is drawn from a Gaussian Process with mean 0 and Gaussian covariance function $\Sigma\left((x, w), (x', w')\right) = \alpha_h \exp\left(-\beta \left\|(x, w) - (x', w')\right\|_2^2\right)$.
- w is drawn uniformly from $\{0, 1/49, 2/49, \ldots, 1\}$ and z is drawn uniformly from $[0, 1]$.

We thus have a class of problems parametrized by α_h, α_d, β, the number of samples per iteration n, and an outcome measure determined by the overall number of samples. To reduce the dimensionality of the search space, we first set the number of samples per iteration, n, to 1. (We also performed experiments with other n, not described here, and found the same qualitative behavior described below.)

We reparameterize the dependence on α_h and α_d in a more interpretable way. We first set $\text{Var}[f(x, w, z)|w, z] = \alpha_h + \alpha_d$ to 1, as multiplying both α_h and α_d by a scalar simply scales the problem. Then, the variance reduction ratio $\text{Var}[f(x, w, z)|f, w]/\text{Var}[f(x, w, z)|f]$ achieved by SBO in conditioning on w is approximately $\alpha_h/(\alpha_d + \alpha_h)$, with this estimate becoming exact as β grows large and the values of $h(x, w)$ become uncorrelated across w. We define $A = \alpha_h/(\alpha_d + \alpha_h)$ equal to this approximate variance reduction ratio.

Thus, our problems are parameterized by the approximate variance reduction ratio A, the overall number of samples, and by β, which measures the speed with which the conditional expectation $\mathbb{E}\left[f\left(x, w, z\right)|w\right]$ varies with w.

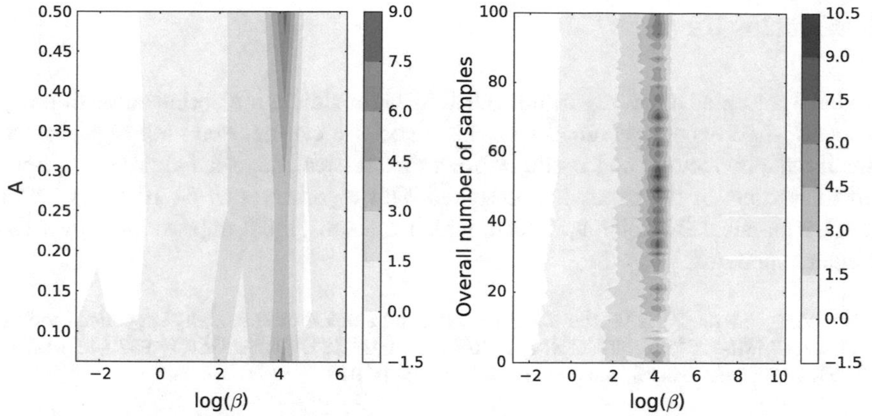

(a) Normalized performance difference as a function of β and A, when the overall number of samples is 50

(b) Normalized performance difference as a function of β and the overall number of iterations, when $A = 1/2$

Fig. 5 Normalized performance difference between SBO and KG in problems simulated from a Gaussian process, as a function of β, which measures how quickly $\mathbb{E}[f(x, w, z)|w]$ varies with w, the approximate variance reduction ratio A, and the overall number of samples. SBO outperforms KG over most of the parameter space, and is approximately 10 times better when β is near exp(4)

Given this parameterization, we sampled problems from Gaussian process priors using all combinations of $A \in \left\{\frac{1}{2}, \frac{1}{4}, \frac{1}{8}, \frac{1}{16}\right\}$ and $\beta \in \left\{2^{-4}, 2^{-3}, \ldots, 2^{9}, 2^{10}\right\}$. We also performed additional simulations at $A = \frac{1}{2}$ for $\beta \in \left\{2^{11}, \ldots, 2^{15}\right\}$.

Figure 5 shows Monte Carlo estimates of the normalized performance difference between SBO and KG for these problems, as a function of $\log(\beta)$ (log is the natural logarithm), A, and the overall number of samples.

The normalized performance difference is estimated for each set of problem parameters by taking a randomly sampled problem generated using those problem parameters, discretizing the domain into 2500 points, running each algorithm independently 500 times on that problem, and averaging $(G(x^*_{\text{SBO}}) - G(x^*_{\text{KG}}))/|G(x^*_{\text{KG}})|$ across these 500 samples, where x^*_{SBO} is the final solution calculated by SBO, and similarly for x^*_{KG}.

We see that the normalized performance difference is robust to A and the overall number of samples, but is strongly influenced by β. We see that SBO is always better than KG whenever $\beta >= 1$. Moreover, it is substantially better than KG when $\log(\beta) \in (3, 5)$, with SBO outperforming KG by as much as a factor of 10. For larger β, SBO remains better than KG, but by a smaller margin. This unimodal dependence of the normalized performance difference on β can be understood as follows: SBO provides value by modeling the dependence of $F(x, w)$ on w. Modeling this dependence is most useful when β takes moderate values because it is here where observations of $F(x, w)$ at one value of w are most useful in predicting the value of $F(x, w)$ at other values of w. When F varies very quickly with w (large β), it is more difficult to generalize, and when F varies very slowly with w (β close to 0), then modeling dependence on w is comparable with modeling F as constant.

7 Conclusion

We have presented a new algorithm called SBO for simulation optimization of noisy derivative-free expensive functions. This algorithm can be used with high dimensional random vectors, and it outperforms the classical Bayesian approach to optimize functions in the examples presented. Our algorithm can be 10 times better than the classical Bayesian approach, which is a substantial improvement over the standard approach.

Acknowledgements Peter Frazier and Saul Toscano-Palmerin were partially supported by NSF CAREER CMMI-1254298, NSF CMMI-1536895, NSF IIS-1247696, AFOSR FA9550-12-1-0200, AFOSR FA9550-15-1-0038, and AFOSR FA9550-16-1-0046.

Appendix

Detailed Computations of the Equations Used in the Statistical Model

In this section we compute the parameters of the posterior distribution of F and G.

Parameters of the Posterior Distribution of F

In this section we are going to calculate the posterior distribution of $F(\cdot,\cdot)$ given that we have placed a Gaussian process (GP) prior distribution over the function F:

$$F(\cdot,\cdot) \sim GP\left(\mu_0(\cdot,\cdot), \Sigma_0(\cdot,\cdot,\cdot,\cdot)\right)$$

where

$$\mu_0 : (x, w) \to \mathbb{R},$$
$$\Sigma_0 : \left(x, w, x', w'\right) \to \mathbb{R},$$

and Σ_0 is a positive semi-definite function. We choose Σ_0 such that closer arguments are more likely to correspond to similar values, i.e. $\Sigma_0\left(x, w, x', w'\right)$ is a decreasing function of the distance between (x, w) and $\left(x', w'\right)$. Specifically, we can use the squared exponential covariance function:

$$\Sigma_0\left(x, w^{(1)}, x', w'^{(1)}\right) = \sigma_0^2 \exp\left(-\sum_{k=1}^{n} \alpha_{1,k}\left[x_k - x_k'\right]^2 - \sum_{k=1}^{d_1} \alpha_{2,k}\left[w_k - w_k'\right]^2\right)$$

where σ_0^2 is the common prior variance, and $\alpha_{1,1}, \ldots, \alpha_{1,n}, \alpha_{2,1}, \ldots, \alpha_{2,d_1} \in \mathbb{R}_+$ are the length scales. These values are calculated using likelihood estimation from the observations of F.

First, observe that standard results from Gaussian process regression provide the following expressions for μ_n and Σ_n (the parameters of the posterior distribution of F),

$$\mu_n(x,w) = \mu_0(x,w) + [\Sigma_0(x,w,x_1,w_1) \cdots \Sigma_0(x,w,x_n,w_n)]\, A_n^{-1} \times \begin{pmatrix} y_1 - \mu_0(x_1,w_1) \\ \vdots \\ y_n - \mu_0(x_n,w_n) \end{pmatrix}$$

$$\Sigma_n\left(x,w,x',w'\right) = \Sigma_0\left(x,w,x',w'\right) - [\Sigma_0(x,w,x_1,w_1) \cdots \Sigma_0(x,w,x_n,w_n)]\, A_n^{-1} \begin{pmatrix} \Sigma_0\left(x',w',x_1,w_1\right) \\ \vdots \\ \Sigma_0\left(x',w',x_n,w_n\right) \end{pmatrix}$$

where

$$A_n = \begin{bmatrix} \Sigma_0(x_1,w_1,x_1,w_1) & \cdots & \Sigma_0(x_1,w_1,x_n,w_n) \\ \vdots & \ddots & \vdots \\ \Sigma_0(x_n,w_n,x_1,w_n) & \cdots & \Sigma_0(x_n,w_n,x_n,w_n) \end{bmatrix} + \mathrm{diag}\left(\sigma^2(x_1,w_1), \ldots, \sigma^2(x_n,w_n)\right),$$

and $\sigma^2(x,w) = \mathrm{Var}(f(x,w,z) \mid w)$.

Parameters of the Posterior Distribution of G

In this section, we compute the parameters of the posterior distribution of G, $\tilde{\sigma}_n(x, x_{n+1}, w_{n+1})$ and $a_n(x)$. We give close formulas for these parameters when we use the squared exponential kernel, and w follows a normal distribution ($w_i \sim N\left(\mu_i, \sigma_i^2\right)$) and its components are independent.

We first compute $\tilde{\sigma}_n(x, x_{n+1}, w_{n+1})$,

$$\begin{aligned}
\tilde{\sigma}_n^2(x, x_{n+1}, w_{n+1}) &= \mathrm{Var}_n[G(x)] - \mathbb{E}_n\left[\mathrm{Var}_{n+1}[G(x)] \mid x_{n+1}, w_{n+1}\right] \\
&= \mathrm{Var}_n\left[G(x) \mid x_{n+1}, w_{n+1}\right] - \mathrm{Var}_{n+1}\left[G(x) \mid x_{n+1}, w_{n+1}\right] \\
&= \int\int \Sigma_n\left(x,w,x,w'\right) p(w)\, p\left(w'\right) dw dw' \\
&\quad - \int\int \Sigma_{n+1}\left(x,w,x,w'\right) p(w)\, p\left(w'\right) dw^{(1)} dw'^{(1)}
\end{aligned}$$

$$= \int \int \Sigma_n (x, w, x_{n+1}, w_{n+1}) \frac{\Sigma_n \left(x, w', x_{n+1}, w_{n+1}\right)}{\Sigma_n (x_{n+1}, w_{n+1}, x_{n+1}, w_{n+1})}$$
$$\times p (w) p \left(w'\right) dwdw'$$
$$= \left[\frac{\int \Sigma_n (x, w, x_{n+1}, w_{n+1})}{\sqrt{\Sigma_n (x_{n+1}, w_{n+1}, x_{n+1}, w_{n+1})}} p (w) dw \right]^2$$
$$= \left[\frac{\int \Sigma_n (x, w, x_{n+1}, w_{n+1})}{\sqrt{\Sigma_n (x_{n+1}, w_{n+1}, x_{n+1}, w_{n+1})}} p (w) dw \right]^2$$
$$= \left[\frac{\left(B (x, n+1) - [B (x, 1) \ \cdots \ B (x, n)] A_n^{-1} \gamma\right)}{\sqrt{\left(\Sigma_0 (x_{n+1}, w_{n+1}, x_{n+1}, w_{n+1}) - \gamma^T A_n^{-1} \gamma\right)}} \right]^2.$$

We now compute $a_n (x)$,

$$a_n (x) = \mathbb{E} [\mu_n (x, w)]$$
$$= \mathbb{E} [\mu_0 (x, w)] + [B (x, 1) \ \cdots \ B (x, n)] A_n^{-1} \begin{pmatrix} y_1 - \mu_0 (x_1, w) \\ \vdots \\ y_n - \mu_0 (x_n, w) \end{pmatrix}.$$

In the particular case that we use the squared exponential kernel, and w follows a normal distribution ($w_i \sim N\left(\mu_i, \sigma_i^2\right)$) and its components are independent, we have that

$$B (x, i) = \int \Sigma_0 (x, w, x_i, w_i) \, dw$$
$$= \sigma_0^2 \exp\left(- \sum_{k=1}^{n} \alpha_{1,k} \left[x_k - x_{i,k}\right]^2 \right) \prod_{k=1}^{d_1} \int \exp\left(-\alpha_{2,k} \left[w_k - w_{i,k}\right]^2\right) p (w_k) \, dw_k$$

for $i = 1, \ldots, n$.

We can also compute $\int \exp\left(-\alpha_{2,k} \left[w_k - w_{i,k}\right]^2\right) p (w_k) \, d (w_k)$ for any k and i,

$$\int \exp\left(-\alpha_{2,k} \left[w_k - w_{i,k}\right]^2\right) p (w_k) \, d (w_k)$$
$$= \frac{1}{\sqrt{2\pi} \sigma_k} \int \exp\left(-\alpha_{2,k} \left[z - w_{i,k}\right]^2 - \frac{[z - \mu_k]^2}{2\sigma_k^2}\right) dz$$
$$= \frac{1}{\sqrt{2\pi} \sigma_k} \exp\left(- \frac{\mu_k^2}{2\sigma_k^2} - \alpha_{2,k} w_{i,k}^2 - \frac{\left(\frac{\mu_k}{\sigma_k^2} + 2\alpha_{2,k} w_{i,k}\right)^2}{4\left(-\alpha_{2,k} - \frac{1}{2\sigma_k^2}\right)} \right)$$

$$\times \int \exp\left(-\left(\alpha_{2,k}+\frac{1}{2\sigma_k^2}\right)\left[z-\frac{\frac{\mu_k}{\sigma_k^2}+2\alpha_{2,k}w_{i,k}}{2\left(b+\frac{1}{2\sigma_k^2}\right)}\right]^2\right)dz$$

$$=\frac{1}{\sqrt{2}\sigma_k}\frac{1}{\sqrt{\alpha_{2,k}+\frac{1}{2\sigma_k^2}}}\exp\left(-\frac{\mu_k^2}{2\sigma_k^2}-\alpha_{2,k}w_{i,k}^2-\frac{\left(\frac{\mu_k}{\sigma_k^2}+2\alpha_{2,k}w_{i,k}\right)^2}{4\left(-\alpha_{2,k}-\frac{1}{2\sigma_k^2}\right)}\right).$$

Detailed Computations of the Equations Used in the Computation of the Value of Information and Its Gradient

In this section, we prove the Lemma 1 of the paper.

Proposition 1 *We have that*

$$a_{n+1}(x) \mid \mathscr{F}_n, (x_{n+1}, w_{n+1}) \sim N\left(a_n(x), \tilde{\sigma}_n^2(x, x_{n+1}, w_{n+1})\right)$$

where

$$\tilde{\sigma}_n^2(x, x_{n+1}, w_{n+1}) = Var_n[G(x)] - \mathbb{E}_n\left[Var_{n+1}[G(x)] \mid x_{n+1}, w_{n+1}\right].$$

Proof By Eq. (8),

$$a_{n+1}(x) = \mathbb{E}\left[\mu_{n+1}(x,w)\right] \tag{8}$$
$$= \mathbb{E}\left[\mu_0(x,w)\right] \tag{9}$$
$$+\left[B(1) \;\cdots\; B(n+1)\right] A_{n+1}^{-1}\begin{pmatrix} y_1-\mu_0(x_1,w_1) \\ \vdots \\ y_{n+1}-\mu_0(x_{n+1},w_{n+1})\end{pmatrix}. \tag{10}$$

Since y_{n+1} conditioned on $\mathscr{F}_n, x_{n+1}, w_{n+1}$ is normally distributed, then $a_{n+1}(x) \mid \mathscr{F}_n, x_{n+1}, w_{n+1}$ is also normally distributed. By the tower property,

$$\begin{aligned}\mathbb{E}_n\left[a_{n+1}(x) \mid x_{n+1}, w_{n+1}\right] &= \mathbb{E}_n\left[\mathbb{E}_{n+1}[G(x)] \mid x_{n+1}, w_{n+1}\right]\\ &= \mathbb{E}_n[G(x)]\\ &= a_n(x)\end{aligned}$$

and

$$\begin{aligned}\tilde{\sigma}_n^2 (x, x_{n+1}, w_{n+1}) &= \mathrm{Var}_n \left[\mathbb{E}_{n+1} [G (x)] \mid x_{n+1}, w_{n+1}\right] \\ &= \mathrm{Var}_n [G (x)] - \mathbb{E}_n \left[\mathrm{Var}_{n+1} [G (x)] \mid x_{n+1}, w_{n+1}\right].\end{aligned}$$

This proves the proposition.

Proof of Lemma 1 Using the Eq. (8) and the previous proposition, we get the following formula for a_{n+1},

$$a_{n+1} = a_n + \tilde{\sigma}_n (x, x_{n+1}, w_{n+1}) Z$$

where $Z \sim N(0, 1)$, which is the Lemma 1 of the paper.

We now compute the gradient of the value of information.

First, we compute the gradient in the general case,

$$\begin{aligned}\nabla V_n (x_{n+1}, w_{n+1}) &= \nabla h \left(a_n \left(A'\right), \tilde{\sigma}_n \left(A', x_{n+1}, w_{n+1}\right)\right) \\ &= \sum_{i=1}^{l-1} \left(f_{j_{i+1}} - f_{j_i}\right) (-\Phi (-|c_i|)) \nabla (|c_i|) \\ &\quad - \left(\nabla f_{j_{i+1}} - \nabla f_{j_i}\right) f (-|c_i|) \\ &= \sum_{i=1}^{l-1} \left(\nabla f_{j_{i+1}} - \nabla f_{j_i}\right) (-\Phi (-|c_i|) |c_i| - f (-|c_i|)) \\ &= \sum_{i=1}^{l-1} \left(-\nabla f_{j_{i+1}} + \nabla f_{j_i}\right) \varphi (|c_i|).\end{aligned}$$

We only need to compute ∇f_{j_i} for all i,

$$\nabla \tilde{\sigma}_n (x, x_{n+1}, w_{n+1}) = \nabla \left(\sqrt{\left(\mathrm{Var}_n [G (x)] - \mathbb{E}_n \left[\mathrm{Var}_{n+1} [G (x)] \mid x_{n+1}, w_{n+1}\right]\right)} \right)$$

$$= \beta_1 \left(\nabla B (x, n+1) - \nabla \left(\gamma^T\right) A_n^{-1} \begin{bmatrix} B (x, 1) \\ \vdots \\ B (x, n) \end{bmatrix} \right) \tag{11}$$

$$-\frac{1}{2} \beta_1^3 \beta_2 \left[\nabla \Sigma_0 (x_{n+1}, w_{n+1}, x_{n+1}, w_{n+1}) \right. \tag{12}$$

$$\left. -2 \nabla \left(\gamma^T\right) A_n^{-1} \gamma\right] \tag{13}$$

where

$$\beta_1 = \left[\Sigma_0 (x_{n+1}, w_{n+1}, x_{n+1}, w_{n+1}) - \gamma^T A_n^{-1} \gamma\right]^{-1/2}$$
$$\beta_2 = B(x, n+1) - [B(x, 1) \cdots B(x, n)] A_n^{-1} \gamma.$$

Now, we give a closed formula for this gradient when we use the squared exponential kernel, and w follows a normal distribution ($w_i \sim N\left(\mu_i, \sigma_i^2\right)$) and its components are independent. Observe that we can compute (11) explicitly by plugging in:

$$\nabla_{x_{n+1},j} \Sigma_0 (x_{n+1}, w_{n+1}, x_i, w_i) = \begin{cases} 0, & i = n+1 \\ -2\alpha_{1,j} \left[x_{n+1,j} - x_{i,j}\right] \Sigma_0 (x_{n+1}, w_{n+1}, x_i, w_i), & i < n+1 \end{cases}$$

$$\nabla_{w_{n+1},j} \Sigma_0 (x_{n+1}, w_{n+1}, x_i, w_i) = \begin{cases} 0, & i = n+1 \\ -2\alpha_{2,j} \left[w_{n+1,j} - w_{i,j}\right] \Sigma_0 (x_{n+1}, w_{n+1}, x_i, w_i), & i < n+1 \end{cases}$$

where $\nabla_{x_{n+1},j}$ is the derivative respect to the jth entry of x_{n+1}. Finally, we only need to compute

$$\nabla_{x_{n+1,j}} B(x, n+1) = -2\alpha_1^{(j)} \left(x_j - x_{n+1,j}\right) B(x, n+1)$$

$$\begin{aligned} \nabla_{w_{n+1,k}} B(x, n+1) = \sigma_0^2 \exp\left(-\sum_{i=1}^{n} \alpha_1^{(i)} \left[x_i - x_{n+1,i}\right]^2\right) \\ \times \prod_{j \neq k} \int \exp\left(-\alpha_2^{(j)} \left[w_j - w_{n+1,j}\right]^2\right) p(w_j)\, d(w_j) \\ \times \int \left[\left(-2\alpha_2^{(k)} \left(w_k - w_{n+1,k}\right)\right) \exp\left(-\alpha_2^{(k)} \left[w_k - w_{n+1,k}\right]^2\right) \right. \\ \left. \times p(w_k) \right] d(w_k). \end{aligned}$$

References

1. Brochu, E., Cora, V.M., De Freitas, N.: A tutorial on bayesian optimization of expensive cost functions, with application to active user modeling and hierarchical reinforcement learning (2010). arXiv preprint arXiv:1012.2599
2. Citi: Citi bike website (2015). https://www.citibikenyc.com/. Accessed May 2015
3. Forrester, A., Sobester, A., Keane, A.: Engineering Design Via Surrogate Modelling: A Practical Guide. Wiley, New York (2008)
4. Frazier, P., Powell, W., Dayanik, S.: The knowledge-gradient policy for correlated normal beliefs. INFORMS J. Comput. **21**(4), 599–613 (2009)
5. Frazier, P.I., Wang, J.: Bayesian optimization for materials design (2015). arXiv:1506.01349
6. Glasserman, P.: Monte Carlo Methods in Financial Engineering, vol. 53. Springer Science and Business Media, Berlin (2003)

7. Howard, R.: Information value theory. IEEE Trans. Syst. Sci. Cybern. **2**(1), 22–26 (1966)
8. Huang, D., Allen, T.T., Notz, W.I., Zeng, N.: Global optimization of stochastic black-box systems via sequential kriging meta-models. J. Glob. Optim. **34**(3), 441–466 (2006)
9. Jones, D.R., Schonlau, M., Welch, W.J.: Efficient global optimization of expensive black-box functions. J. Glob. Optim. **13**(4), 455–492 (1998)
10. Kraft, D., et al.: A Software Package for Sequential Quadratic Programming. DFVLR Obersfaffeuhofen, Germany (1988)
11. Kushner, H.J.: A new method of locating the maximum of an arbitrary multi-peak curve in the presence of noise. J. Basic Eng. **86**, 97–106 (1964)
12. Mockus, J.: Bayesian Approach to Global Optimization: Theory and Applications. Kluwer Academic, Dordrecht (1989)
13. Mockus, J., Tiesis, V., Zilinskas, A.: The application of bayesian methods for seeking the extremum. Towards Glob. Optim. **2**(117–129), 2 (1978)
14. Murphy, K.P.: Machine Learning: A Probabilistic Perspective. MIT Press, Cambridge (2012)
15. O'Hagan, A.: Bayes-hermite quadrature. J. Stat. Plan. Inference **29**(3), 245–260 (1991)
16. Rasmussen, C., Williams, C.: Gaussian Processes for Machine Learning. MIT Press, Cambridge (2006)
17. Scott, W., Frazier, P., Powell, W.: The correlated knowledge gradient for simulation optimization of continuous parameters using gaussian process regression. SIAM J. Optim. **21**(3), 996–1026 (2011)
18. Snoek, J., Larochelle, H., Adams, R.P.: Practical bayesian optimization of machine learning algorithms. Adv. Neural Inf. Process. Syst. 2951–2959 (2012)
19. Xie, J., Frazier, P., Sankaran, S., Marsden, A., Elmohamed, S.: Optimization of computationally expensive simulations with gaussian processes and parameter uncertainty: application to cardiovascular surgery. In: 50th Annual Allerton Conference on Communication, Control, and Computing (2012)
20. Villemonteix, J., Vazquez, E., Walter, E.: An informational approach to the global optimization of expensive-to-evaluate functions. J. Glob. Optim. **44**(4), 509–534 (2009)

Part III
Regular Talks

Tusnády's Problem, the Transference Principle, and Non-uniform QMC Sampling

Christoph Aistleitner, Dmitriy Bilyk and Aleksandar Nikolov

Abstract It is well-known that for every $N \geq 1$ and $d \geq 1$ there exist point sets $x_1, \ldots, x_N \in [0, 1]^d$ whose discrepancy with respect to the Lebesgue measure is of order at most $(\log N)^{d-1} N^{-1}$. In a more general setting, the first author proved together with Josef Dick that for any normalized measure μ on $[0, 1]^d$ there exist points $x_1, \ldots, x_N$ whose discrepancy with respect to μ is of order at most $(\log N)^{(3d+1)/2} N^{-1}$. The proof used methods from combinatorial mathematics, and in particular a result of Banaszczyk on balancings of vectors. In the present note we use a version of the so-called transference principle together with recent results on the discrepancy of red-blue colorings to show that for any μ there even exist points having discrepancy of order at most $(\log N)^{d-\frac{1}{2}} N^{-1}$, which is almost as good as the discrepancy bound in the case of the Lebesgue measure.

Keywords Low-discrepancy sequences · Non-uniform sampling · combinatorial discrepancy · Tusnády's problem · Gates of Hell

1 Introduction and Statement of Results

Many problems from applied mathematics require the calculation or estimation of the expected value of a function depending on several random variables; such problems include, for example, the calculation of the fair price of a financial derivative

C. Aistleitner (✉)
Institute of Analysis and Number Theory, TU Graz, Graz, Austria
e-mail: aistleitner@math.tugraz.at

D. Bilyk
School of Mathematics, University of Minnesota, Minneapolis, MN, USA
e-mail: dbilyk@math.umn.edu

A. Nikolov
Department of Computer Science, University of Toronto, Toronto, ON, Canada
e-mail: anikolov@cs.toronto.edu

A. B. Owen and P. W. Glynn (eds.), *Monte Carlo and Quasi-Monte Carlo Methods*, Springer Proceedings in Mathematics & Statistics 241,
https://doi.org/10.1007/978-3-319-91436-7_8

and the calculation of the expected loss of an insurance risk. The expected value $\mathbb{E}\big(g(Y^{(1)}, \ldots, Y^{(d)})\big)$ can be written as

$$\int_{\mathbb{R}^d} g\big(y^{(1)}, \ldots, y^{(d)}\big)\, d\nu\big(y^{(1)}, \ldots, y^{(d)}\big), \tag{1}$$

where ν is an appropriate probability measure describing the joint distribution of the random variables $Y^{(1)}, \ldots, Y^{(d)}$. Since a precise calculation of the value of such an integral is usually not possible, one looks for a numerical approximation instead. Two numerical methods for such problems are the *Monte Carlo method* (MC, using random sampling points) and the *Quasi-Monte Carlo method* (QMC, using cleverly chosen deterministic sampling points). The QMC method is often preferred due to a faster convergence rate and deterministic error bounds. However, in the QMC literature it is often assumed that the problem asking for the value of (1) has at the outset already been transformed into the problem asking for the value of

$$\int_{[0,1]^d} f\big(x^{(1)}, \ldots, x^{(d)}\big)\, dx^{(1)} \cdots x^{(d)}. \tag{2}$$

Thus it is assumed that the integration domain is shrunk from $\mathbb{R}^d$ to $[0, 1]^d$, and that the integration measure is changed from ν to the Lebesgue measure. In principle, such a transformation always exists – this is similar to the way how general multivariate random sampling is reduced to sampling from the multivariate uniform distribution, using for example the Rosenblatt transform [28] (which employs sequential conditional inverse functions).

The shrinking procedure is less critical, albeit also non-trivial – if one wishes to avoid this shrinking process and carry out QMC integration directly on $\mathbb{R}^d$ instead, then one way of doing so is to "lift" a set of sampling points from $[0, 1]^d$ to $\mathbb{R}^d$ rather than shrinking the domain of the function. See for example [15] and the references there.

However, the change from a general measure to the uniform measure is highly problematic, in particular because QMC error bounds depend strongly on the regularity of the integrand. Note that the change from ν to the Lebesgue measure induces a transformation of the original function g to a new function f, and this transformation may totally ruin all "nice" properties of the initial function g such as the existence of derivatives or the property of having bounded variation. This is not the place to discuss this topic in detail; we just note that the main problem is not that each of $Y^{(1)}, \ldots, Y^{(d)}$ may have a non-uniform marginal distribution, but rather that there may be a strong dependence in their joint distribution (which by Sklar's theorem may be encoded in a so-called *copula* – see [26] for details). We refer to [12] for a discussion of these issues from a practitioner's point of view.

There are two possible ways to deal with the problems mentioned in the previous paragraph, which can be seen as two sides of the same coin.

Firstly, one may try to transform the points of a classical QMC point set in such a way that they can be used for integration with respect to a different, general measure, and such that one obtains a bound for the integration error in terms of the discrepancy (with respect to the uniform measure) of the original point set – this is the approach of Hlawka and Mück [18, 19], which has been taken up by several authors.

On the other hand, one may try to sample QMC points directly in such a way that they have small discrepancy with respect to a given measure μ, and use error bounds which apply in this situation. This is the approach discussed in [1, 2], where it is shown that both key ingredients for QMC integration are given also in the setting of a general measure μ: there exist (almost) low-discrepancy point sets, and there exists a Koksma–Hlawka inequality giving bounds for the integration error in terms of the discrepancy (with respect to μ) of the sampling points and the degree of regularity of the integrand function. More precisely, let μ be a normalized Borel measure on $[0, 1]^d$, and let

$$D_N^*(x_1, \ldots, x_N; \mu) = \sup_{A \in \mathscr{A}} \left| \frac{1}{N} \sum_{n=1}^{N} \mathbf{1}_A(x_n) - \mu(A) \right| \tag{3}$$

be the star-discrepancy of $x_1, \ldots, x_N \in [0, 1]^d$ with respect to μ; here $\mathbf{1}_A$ denotes the indicator function of A, and the supremum is extended over the class $\mathscr{A}$ of all axis-parallel boxes contained in $[0, 1]^d$ which have one vertex at the origin. Then in [1] it is shown that for arbitrary μ there exist points $x_1, \ldots, x_N \in [0, 1]^d$ such that

$$D_N^*(x_1, \ldots, x_N; \boldsymbol{\mu}) \le 63\sqrt{d}\,\frac{(2 + \log_2 N)^{(3d+1)/2}}{N}. \tag{4}$$

This improves an earlier result of Beck [7], where the exponent of the logarithmic term was $2d$. Note the amazing fact that the right-hand side of (4) does not depend on μ (whereas the choice of the points $x_1, \ldots, x_N$ clearly does). The estimate in (4) should be compared with corresponding results for the case of the uniform measure, where it is known that there exist so-called *low-discrepancy point sets* whose discrepancy is of order at most $(\log N)^{d-1} N^{-1}$.

(*Remark*: While preparing the final version of this manuscript we learned that already in 1989, József Beck [8, Theorem 1.2] proved a version of (4) with the stronger upper bound $\mathscr{O}\left(N^{-1}(\log N)^{d+2}\right)$. However, the implied constant was not specified in his result.)

A Koksma–Hlawka inequality for general measures was first proved by Götz [17] (see also [2]). For any points $x_1, \ldots, x_N$, any normalized measure μ on $[0, 1]^d$ and any d-variate function f whose variation Var f on $[0, 1]^d$ (in the sense of Hardy–Krause) is bounded, we have

$$\left| \int_{[0,1]^d} f(x) d\mu(x) - \frac{1}{N} \sum_{n=1}^{N} f(x_n) \right| \le D_N^*(x_1, \ldots, x_N; \mu) \operatorname{Var} f. \tag{5}$$

This is a perfect analogue of the original Koksma–Hlawka inequality for the uniform measure. Combining (4) and (5) we see that in principle QMC integration is possible for the numerical approximation of integrals of the form

$$\int_{[0,1]^d} f(x^{(1)}, \ldots, x^{(d)}) d\mu(x^{(1)} \ldots x^{(d)}),$$

and that the convergence rate is almost as good as in the classical setting for the uniform measure. However, it should be noted that for the case of the uniform measure many explicit constructions of low-discrepancy point sets are known (see [16]), whereas the proof of (4) is a pure existence result and it is not clear how (and if) such point sets can be constructed with reasonable effort.

The purpose of the present note is to show that for any μ there actually exist points $x_1, \ldots, x_N$ whose discrepancy with respect to μ is of order at most $(\log N)^{d-\frac{1}{2}} N^{-1}$. This is quite remarkable, since it exceeds the corresponding bound for the case of the uniform measure only by a factor $(\log N)^{\frac{1}{2}}$.

Theorem 1 *For every $d \geq 1$ there exists a constant c_d (depending only on d) such that the following holds. For every $N \geq 2$ and every normalized Borel measure μ on $[0, 1]^d$ there exits points $x_1, \ldots, x_N \in [0, 1]^d$ such that*

$$D_N^*(x_1, \ldots, x_N; \mu) \leq c_d \frac{(\log N)^{d-\frac{1}{2}}}{N}.$$

The proof of this theorem uses a version of the so-called transference principle, which connects the combinatorial and geometric discrepancies, see Theorem 2. The novelty and the main observation of this paper lies in the fact that this principle is still valid for general measures μ. This observation was made earlier by Matoušek in [23], without providing any details, but otherwise it has been largely overlooked. In addition we shall use new upper bounds for Tusnády's problem due to the third author (for the discussion of Tusnády's problem, see Sect. 2.2). If one wants to refrain from the application of unpublished results, one can use Larsen's [20] upper bounds for Tusnády's problem instead, which yield Theorem 1 with the exponent $d + 1/2$ instead of $d - 1/2$ for the logarithmic term (see also [25]). An exposition of the connection between geometric and combinatorial discrepancies, together with the proof of Theorem 1, is given in the following section.

Before turning to combinatorial discrepancy, we want to make several remarks concerning Theorem 1.

Firstly, whereas the conclusion of Theorem 1 is stated for finite point sets, one can use a well-known method to find an infinite sequence $(x_n)_{n \geq 1}$ whose discrepancy is of order at most $(\log N)^{d+1/2} N^{-1}$ for all $N \geq 1$. A proof can be modeled after the proof of Theorem 2 in [1].

Secondly, while the upper bound in Theorem 1 is already very close to the corresponding upper bound in the case of the classical discrepancy for the uniform measure, there is still a gap. One wonders whether this gap is a consequence of a deficiency of our method of proof, or whether the discrepancy bound in the case of general measures really has to be larger than that in the case of the uniform measure.

In other words, we have the following open problem, which we state in a slightly sloppy formulation.

Open problem: Is the largest upper bound for the smallest possible discrepancy the one for the case of the Lebesgue measure? In other words, is there any measure which is more difficult to approximate by finite atomic measures with equal weights than the Lebesgue measure?

We think that this is a problem of significant theoretical interest. Note, however, that even in the classical case of the Lebesgue measure the problem asking for the minimal order of the discrepancy of point sets is famously still open; while in the upper bounds for the best known constructions the logarithmic term has exponent $d - 1$, in the best known lower bounds the exponent is $(d - 1)/2 + \varepsilon_d$ for some small $\varepsilon_d > 0$ (see [11] for the latter result, and [10] for a survey). This is known as the *Great Open Problem* of discrepancy theory, see, e.g., [9, p. 8], and [24, p. 178].

It is clear that some measures are much easier to approximate than Lebesgue measure – think of the measure having full mass at a single point, which can be trivially approximated with discrepancy zero. On the other hand, if the measure has a non-vanishing continuous component then one can carry over the orthogonal functions method of Roth [29] – this is done in [14]. However, the lower bounds for the discrepancy which one can obtain in this way are the same as those for the uniform case (and not larger ones). Intuitively, it is tempting to assume that the Lebesgue measure is essentially extremal – simply because intuition suggests that a measure which is difficult to approximate should be "spread out everywhere", and should be "equally large" throughout the cube.

2 Combinatorial Discrepancy and the Transference Principle

2.1 Combinatorial Discrepancy

Let $V = \{1, \ldots, n\}$ be a ground set and $\mathscr{C} = \{C_1, \ldots, C_m\}$ be a system of subsets of V. Then the (combinatorial) discrepancy of $\mathscr{C}$ is defined as

$$\operatorname{disc}\mathscr{C} = \min_{y \in \{-1,1\}^n} \operatorname{disc}(\mathscr{C}, y),$$

where the vector y is called a *(red-blue) coloring* of V and

$$\operatorname{disc}(\mathscr{C}, y) = \max_{i \in \{1,\ldots,m\}} \left| \sum_{j \in C_i} y_j \right|$$

is the discrepancy of the coloring y. We may visualize the entries of the vector y as representing two different colors (usually red and blue). Then $\operatorname{disc}(\mathscr{C}, y)$ is the maximal difference between the number of red and blue elements of V contained in a test set C_i, and disc$\mathscr{C}$ is the minimal value of disc$(\mathscr{C}, y)$ over all possible colorings y. For more information on this combinatorial notion of discrepancy, see [13, 24].

We will only be concerned with a geometric variation of this notion, i.e. the case when V is a point set in $[0, 1]^d$, and when the set system $\mathscr{C}$ is the collection of all sets of the form $G \cap V$, where $G \in \mathscr{G}$ and $\mathscr{G}$ is some collection of geometric subsets of $[0, 1]^d$: standard choices include, e.g., balls, convex sets, boxes (axis-parallel or arbitrarily rotated) etc. Let disc$(N, \mathscr{G}, d)$ denote the maximal possible discrepancy in this setting; that is, set

$$\mathrm{disc}(N, \mathscr{G}, d) = \max_P (\mathrm{disc}\mathscr{C}),$$

where the maximum is taken over all sets P of N points in $[0, 1]^d$ and where $\mathscr{C} = \mathscr{G}|_P = \{G \cap P : G \in \mathscr{G}\}$.

2.2 *Tusnády's Problem*

We denote by $\mathscr{A}$ (as in Sect. 1) the class of all axis-parallel boxes having one vertex at the origin. The problem of finding sharp upper and lower bounds for disc$(N, \mathscr{A}, d)$ as a function of N and d is known as *Tusnády's problem*. For the history and background of the problem, see [6, 23, 25].

We will use the following result, which has been recently announced by the third author [27]. A slightly weaker result [5] (with exponent d) by Bansal and Garg has been presented at MCQMC 2016 (and is also still unpublished). A yet weaker, but already published result (with exponent $d + 1/2$) is contained in [20].

Proposition 1 *For every $d \geq 1$ there exists a constant c_d (depending only on d) such that for all $N \geq 2$*

$$\mathrm{disc}(N, \mathscr{A}, d) \leq c_d (\log N)^{d-\frac{1}{2}}.$$

Finally, we want to note that Tusnády's problem is still unsolved; the known (and conjecturally optimal) lower bounds are of the order $(\log N)^{d-1}$, see [25]. As will become clear from the next subsection, further improvements of the upper bounds for Tusnády's problem would directly imply improved upper bounds in Theorem 1. Note that actually Tusnády's problem also falls within the framework of "discrepancy with respect to a general measure μ". Given N points, let μ be the discrete measure that assigns mass $1/N$ to each of these points. Then Tusnády's problem asks for a set P of roughly $N/2$ points such that the discrepancy of P with respect to μ is small – under the additional requirement that the elements of P are chosen from the original set of N points. This additional requirement also explains why lower bounds for Tusnády's problem do not imply lower bounds for the problem discussed in the present paper.

2.3 Transference Principle

It is known that upper bounds for the combinatorial discrepancy of red-blue colorings can be turned into upper bounds for the smallest possible discrepancy of a point set in the unit cube. This relation is called the *transference principle*.

For a system $\mathscr{G}$ of measurable subsets of $[0, 1]^d$, its (unnormalized) geometric discrepancy is defined as

$$D_N(\mathscr{G}) = \inf_{P:\#P=N} \sup_{G\in\mathscr{G}} \left| \sum_{p\in P} \mathbf{1}_G(p) - N\lambda(G) \right|, \tag{6}$$

where the infimum is taken over all N-point sets $P \subset [0, 1]^d$ and $\lambda(G)$ is the Lebesgue measure of G. The *transference principle*, roughly speaking, says that (under some mild assumptions on the collection $\mathscr{G}$) ***the geometric discrepancy is bounded above by the combinatorial discrepancy***, i.e. $D_N(\mathscr{G}) \lesssim \mathrm{disc}(N, \mathscr{G}, d)$ with the symbol "$\lesssim$" interpreted loosely. Thus upper bounds on combinatorial discrepancy yield upper bounds for its geometric counterpart. This relation, in general, cannot be reversed: in the case when $\mathscr{G}$ is the collection of all convex subsets of the unit cube, we have that $D_N(\mathscr{G})$ is of the order $N^{1-\frac{2}{d+1}}$, while $\mathrm{disc}(N, \mathscr{G}, d)$ is of the order N as $N \to \infty$ (see, e.g., [24]).

In [24, 25] it is mentioned that M. Simonovits attributes the idea of this principle to Vera T. Sós. It was used in the context of Tusnády's problem by Beck [6] in 1981, and is stated in a rather general form in [22]. It can be found also in Matoušek's book [24, p. 20] and in [25]. In all these instances it is formulated in a version which bounds the geometric discrepancy of point sets *with respect to the Lebesgue measure* (as defined in (6)) in terms of the combinatorial discrepancy of red-blue colorings.

However, upon examination of the proof, it turns out that the argument carries over to the case of the discrepancy with respect to an arbitrary measure (the only significant requirement is that the measure allows an ε-approximation for arbitrary ε; see below). Similar to (3) and (6), we define the geometric discrepancy of $\mathscr{G}$ with respect to a Borel measure μ as

$$D_N(\mathscr{G}, \mu) = \inf_{P:\#P=N} D(P, \mathscr{G}, \mu), \tag{7}$$

where

$$D(P, \mathscr{G}, \mu) = \sup_{G\in\mathscr{G}} \left| \sum_{p\in P} \mathbf{1}_G(p) - N\mu(G) \right|. \tag{8}$$

Note that with this definition $D_N^*(x_1, \ldots, x_N; \mu)$, as defined in (3), satisfies

$$D_N^*(x_1, \ldots, x_N; \mu) = \frac{1}{N} D\big(\{x_1, \ldots, x_N\}, \mathscr{A}, \mu\big).$$

Below we state and prove the transference principle in a rather general form for arbitrary measures. The statement is similar to that in [24, 25] in the case of Lebesgue measure, and the proof follows along the lines of [24].

Theorem 2 (Transference principle for general measures) *Let μ be a Borel probability measure on $[0, 1]^d$ and let $\mathcal{G}$ be a class of Borel subsets of $[0, 1]^d$ such that $[0, 1]^d \in \mathcal{G}$. Suppose that*

$$D_N(\mathcal{G}, \mu) = o(N) \quad \text{as} \quad N \to \infty. \tag{9}$$

Assume furthermore that the combinatorial discrepancy of $\mathcal{G}$ satisfies

$$\operatorname{disc}(N, \mathcal{G}, d) \leq h(N), \qquad N \geq 1, \tag{10}$$

for some function h with the property that $h(2N) \leq (2 - \delta)h(N)$ for all $N \geq 1$ for some fixed $\delta > 0$. Then

$$D_N(\mathcal{G}, \mu) = \mathcal{O}(h(N)) \qquad \text{as} \quad N \to \infty, \tag{11}$$

i.e., there is a constant $C = C(\delta)$ such that for every $N \geq 1$ there exist points $x_1, \ldots, x_N \in [0, 1]^d$ so that $D\big(\{x_1, \ldots, x_N\}, \mathcal{G}, \mu\big) \leq Ch(N)$.

Proof Set $\varepsilon = h(N)/N$, and using (9) choose a positive integer k so large that there exists a set P_0 of $2^k N$ points in $[0, 1]^d$ so that $\frac{D(P_0, \mathcal{G}, \mu)}{2^k N} \leq \varepsilon$. By (10) we can find a red-blue coloring of the set P_0 with discrepancy at most $h(2^k N)$. The difference between the total number of red and blue points is also at most $h(2^k N)$, since the full unit cube is an element of our class of test sets. Without loss of generality we may assume that there are no more red than blue points (otherwise switch the roles of the red and the blue points). We keep all the red points and only so many of the blue points as to make sure that in total we have half the number of the original points, while we dispose of all the other blue points. Write P_1 for the new set. The cardinality of P_1 is $2^{k-1}N$. Furthermore,

$$\frac{D(P_1, \mathcal{G}, \mu)}{2^{k-1}N} \leq \varepsilon + \frac{h(2^k N)}{2^{k-1}N}.$$

To see why this is the case, note that an arbitrary set $G \in \mathcal{G}$ contains between $2^k N\mu(G) - \varepsilon 2^k N$ and $2^k N\mu(G) + \varepsilon 2^k N$ elements of P_0. Thus it contains between

$$\frac{1}{2}\big(2^k N\mu(G) - \varepsilon 2^k N - h(2^k N)\big) \quad \text{and} \quad \frac{1}{2}\big(2^k N\mu(G) + \varepsilon 2^k N + h(2^k N)\big)$$

red elements of P_0, and consequently between

$$\frac{1}{2}\big(2^k N\mu(G) - \varepsilon 2^k N - h(2^k N)\big) \quad \text{and} \quad \frac{1}{2}\big(2^k N\mu(G) + \varepsilon 2^k N + 2h(2^k N)\big)$$

elements of P_1, where the upper bound is increased by $h(2^k N)/2$ since we have to add at most so many blue points in order to make sure that P_1 has the desired cardinality. Repeating this procedure, we obtain a point set P_2 of cardinality $2^{k-2}N$ whose discrepancy with respect to μ is at most

$$\frac{D(P_2, \mathscr{G}, \mu)}{2^{k-2}N} \leq \varepsilon + \frac{h(2^k N)}{2^{k-1}N} + \frac{h(2^{k-1}N)}{2^{k-2}N}.$$

We repeat this procedure over and over again, until we arrive at a point set P_k which has cardinality N, and whose discrepancy with respect to μ satisfies

$$D(P_k, \mathscr{G}, \mu) \leq \underbrace{\varepsilon N}_{=h(N)} + \sum_{j=0}^{k} \frac{h(2^{k-j}N)}{2^{k-j-1}} \leq Ch(N),$$

where we have used the condition that $h(2N) \leq (2-\delta)h(N)$. Note that the value of C does not depend on the measure μ. This finishes the proof.

2.4 *Proof of Theorem 1*

Theorem 1 now easily follows from the combinatorial discrepancy estimate in Tusnády's problem (Proposition 1) and the transference principle for general measures (Theorem 2) applied in the case $\mathscr{G} = \mathscr{A}$. The only point that needs checking is whether μ satisfies the approximability condition (9) with respect to the collection $\mathscr{A}$ of axis-parallel boxes with one vertex at the origin. But (9) follows trivially from the prior result (4).

A slightly more direct way to prove (9) is to observe that the collection $\mathscr{A}$ is a VC class (its VC-dimension is d), see e.g. [13] for definitions. This implies (see [30]) that it is a uniform Glivenko–Cantelli class, i.e.

$$\sup_{\mu} \mathbf{E} \sup_{A \in \mathscr{A}} \left| \frac{1}{N} \sum_{n=1}^{N} \mathbf{1}_A(x_n) - \mu(A) \right| \to 0 \text{ as } N \to \infty,$$

where the expectation is taken over independent random points $x_1,\ldots,x_N$ with distribution μ, and the supremum is taken over all probability measures μ on $[0, 1]^d$. This immediately yields (9).

In conclusion we want to make some remarks on possible algorithmic ways of finding a point set satisfying the conclusion of Theorem 1. Following the proof of the transference principle, two steps are necessary. First one has to find the ε-approximation of μ. The existence of such a point set is guaranteed in the proof as a consequence of (4). However, this is not of much practical use since the point set for (4) cannot be constructed explicitly. However, in Corollary 1 of [2] it is proved that a set of $2^{26}d\varepsilon^{-2}$ random points which are sampled randomly and independently from the distribution μ will have, with positive probability, a discrepancy with respect to μ which is less than a given ε. This result is deduced from large deviation inequalities, and thus for a larger value of the constant (say 2^{35} instead of 2^{26}) the probability that the random point set has discrepancy at most ε with respect to μ will be extremely close to 1. We can think of ε as roughly $1/N$; for the cardinality of the random point set, this would give roughly $2^{35}dN^2$. Note, however, that in each iteration of the coloring procedure the number of points is halved; accordingly, starting with $2^{35}dN^2$ points leads to a point set of cardinality N after a rather small number of steps. Admittedly, by using random points for the ε-approximation the whole problem is only shifted rather than solved; it is typically rather difficult to draw independent random samples from a general multivariate distribution μ.

The second part of the proof (the coloring procedure) is less of a problem from an algorithmic point of view, since in recent years much work has been done on algorithms for actually finding balanced colorings in combinatorial discrepancy theory. In particular, the recent bound [5] for Tusnády's problem due to Bansal and Garg mentioned in Sect. 2 is algorithmic in the following sense: they describe an efficient randomized algorithm that, given a set V of N points in $[0, 1]^d$, finds a red-blue coloring of V which has discrepancy at most $c_d(\log N)^d$ with probability arbitrarily close to 1. "Efficient" here means that the running time of the algorithm is bounded by a polynomial in N. In another recent preprint, Levy, Ramadas and Rothvoss [21] describe an efficient deterministic algorithm that achieves the same guarantees as the randomized algorithm from [4] for the Komlós problem. Since the techniques used in [5] are very closely related to those of [4], it seems likely that an efficient deterministic algorithm to find colorings for Tusnády's problem with discrepancy bounded by $c_d(\log N)^d$ can be constructed via the methods of [21]. Unfortunately, the stronger bound of $c_d(\log N)^{d-\frac{1}{2}}$ proved in [27] relies on an existential result of Banaszczyk [3], and no efficient algorithm is currently known that achieves this bound.

Acknowledgements This paper was conceived while the authors were taking a walk in the vicinity of Rodin's *Gates of Hell* sculpture on Stanford University campus during the MCQMC 2016 conference. We want to thank the MCQMC organizers for bringing us together. Based on this episode, we like to call the open problem stated in the first section of this paper the *Gates of Hell Problem*.

The first author is supported by the Austrian Science Fund (FWF), project Y-901. The third author is supported by an NSERC Discovery Grant.

References

1. Aistleitner, C., Dick, J.: Low-discrepancy point sets for non-uniform measures. Acta Arith. **163**(4), 345–369 (2014). https://doi.org/10.4064/aa163-4-4
2. Aistleitner, C., Dick, J.: Functions of bounded variation, signed measures, and a general Koksma–Hlawka inequality. Acta Arith. **167**(2), 143–171 (2015). https://doi.org/10.4064/aa167-2-4

3. Banaszczyk, W.: On series of signed vectors and their rearrangements. Random Struct. Algorithms **40**(3), 301–316 (2012). https://doi.org/10.1002/rsa.20373
4. Bansal, N., Dadush, D., Garg, S.: An algorithm for Komlós conjecture matching Banaszczyk's bound. In: Dinur, I. (ed.) IEEE 57th Annual Symposium on Foundations of Computer Science, FOCS 2016, 9–11 October 2016, Hyatt Regency, New Brunswick, New Jersey, USA, pp. 788–799. IEEE Computer Society (2016). https://doi.org/10.1109/FOCS.2016.89
5. Bansal, N., Garg, S.: Algorithmic discrepancy beyond partial coloring. http://arxiv.org/abs/1611.01805
6. Beck, J.: Balanced two-colorings of finite sets in the square. I. Combinatorica **1**(4), 327–335 (1981). https://doi.org/10.1007/BF02579453
7. Beck, J.: Some upper bounds in the theory of irregularities of distribution. Acta Arith. **43**(2), 115–130 (1984)
8. Beck, J.: Balanced two-colorings of finite sets in the cube. In: Proceedings of the Oberwolfach Meeting "Kombinatorik" (1986), vol. 73, pp. 13–25. (1989). https://doi.org/10.1016/0012-365X(88)90129-X
9. Beck, J., Chen, W.W.L.: Irregularities of Distribution. Cambridge Tracts in Mathematics, vol. 89. Cambridge University Press, Cambridge (1987). https://doi.org/10.1017/CBO9780511565984
10. Bilyk, D.: Roth's orthogonal function method in discrepancy theory and some new connections. A Panorama of Discrepancy Theory. Lecture Notes in Mathematics, vol. 2107, pp. 71–158. Springer, Cham (2014). https://doi.org/10.1007/978-3-319-04696-9_2
11. Bilyk, D., Lacey, M.T., Vagharshakyan, A.: On the small ball inequality in all dimensions. J. Funct. Anal. **254**(9), 2470–2502 (2008). https://doi.org/10.1016/j.jfa.2007.09.010
12. Cambou, M., Hofert, M., Lemieux, C.: Quasi-random numbers for copula models. http://arxiv.org/abs/1508.03483
13. Chazelle, B.: The Discrepancy Method. Cambridge University Press, Cambridge (2000). https://doi.org/10.1017/CBO9780511626371
14. Chen, W.W.L.: On irregularities of distribution and approximate evaluation of certain functions. Q. J. Math. Oxf. Ser. (2) **36**(142), 173–182 (1985). https://doi.org/10.1093/qmath/36.2.173
15. Dick, J., Irrgeher, C., Leobacher, G., Pillichshammer, F.: On the optimal order of integration in Hermite spaces with finite smoothness. http://arxiv.org/abs/1608.06061
16. Dick, J., Pillichshammer, F.: Digital Nets and Sequences. Cambridge University Press, Cambridge (2010). https://doi.org/10.1017/CBO9780511761188
17. Götz, M.: Discrepancy and the error in integration. Monatsh. Math. **136**(2), 99–121 (2002). https://doi.org/10.1007/s006050200037
18. Hlawka, E., Mück, R.: A transformation of equidistributed sequences. In: Applications of Number Theory to Numerical Analysis (Proc. Sympos., Univ. Montréal, Montreal, Que., 1971), pp. 371–388. Academic Press, New York (1972)
19. Hlawka, E., Mück, R.: Über eine Transformation von gleichverteilten Folgen. II. Computing (Arch. Elektron. Rechnen) **9**, 127–138 (1972)
20. Larsen, K.G.: On range searching in the group model and combinatorial discrepancy. SIAM J. Comput. **43**(2), 673–686 (2014)
21. Levy, A., Ramadas, H., Rothvoss, T.: Deterministic discrepancy minimization via the multiplicative weight update method. http://arxiv.org/abs/1611.08752
22. Lovász, L., Spencer, J., Vesztergombi, K.: Discrepancy of set-systems and matrices. Eur. J. Comb. **7**(2), 151–160 (1986). https://doi.org/10.1016/S0195-6698(86)80041-5
23. Matoušek, J.: On the discrepancy for boxes and polytopes. Monatsh. Math. **127**(4), 325–336 (1999). https://doi.org/10.1007/s006050050044
24. Matoušek, J.: Geometric Discrepancy. Algorithms and Combinatorics, vol. 18. Springer, Berlin (2010). https://doi.org/10.1007/978-3-642-03942-3
25. Matoušek, J., Nikolov, A., Talwar, K.: Factorization norms and hereditary discrepancy. https://arxiv.org/abs/1408.1376
26. Nelsen, R.B.: An Introduction to Copulas. Springer Series in Statistics, 2nd edn. Springer, New York (2006)

27. Nikolov, A.: Tighter bounds for the discrepancy of boxes and polytopes. Mathematika **63** (3), 1091Â–1113 (2017)
28. Rosenblatt, M.: Remarks on a multivariate transformation. Ann. Math. Stat. **23**, 470–472 (1952)
29. Roth, K.F.: On irregularities of distribution. Mathematika **1**, 73–79 (1954)
30. Vapnik, V.N., Chervonenkis, A.Y.: On the uniform convergence of relative frequencies of events to their probabilities. In: Measures of Complexity, pp. 11–30. Springer, Cham (2015). (Reprint of Theor. Probability Appl. **1**6 (1971), 264–280)

Learning Light Transport the Reinforced Way

Ken Dahm and Alexander Keller

Abstract We show that the equations of reinforcement learning and light transport simulation are related integral equations. Based on this correspondence, a scheme to learn importance while sampling path space is derived. The new approach is demonstrated in a consistent light transport simulation algorithm that uses reinforcement learning to progressively learn where light comes from. As using this information for importance sampling includes information about visibility, too, the number of light transport paths with zero contribution is dramatically reduced, resulting in much less noisy images within a fixed time budget.

Keywords Rendering · Path tracing · Reinforcement learning · Integral equations · Monte Carlo and quasi-Monte Carlo methods

1 Introduction

One application of light transport simulation is the computational synthesis of images that cannot be distinguished from real photographs. In such simulation algorithms [24], light transport is modeled by a Fredholm integral equation of the second kind and pixel colors are determined by estimating functionals of the solution of the Fredholm integral equation. The estimators are averages of contributions of sampled light transport paths that connect light sources and camera sensors.

Compared to reality, where photons and their trajectories are abundant, a computer may only consider a tiny fraction of path space, which is one of the dominant reasons for noisy images. It is therefore crucial to efficiently find light transport paths that have an important contribution to the image. While a lot of research in computer graphics has been focussing on importance sampling [1, 3, 4, 18, 23], for long there

K. Dahm · A. Keller (✉)
NVIDIA, Fasanenstr. 81, 10623 Berlin, Germany
e-mail: keller.alexander@gmail.com

K. Dahm
e-mail: ken.dahm@gmail.com

A. B. Owen and P. W. Glynn (eds.), *Monte Carlo and Quasi-Monte Carlo Methods*, Springer Proceedings in Mathematics & Statistics 241,
https://doi.org/10.1007/978-3-319-91436-7_9

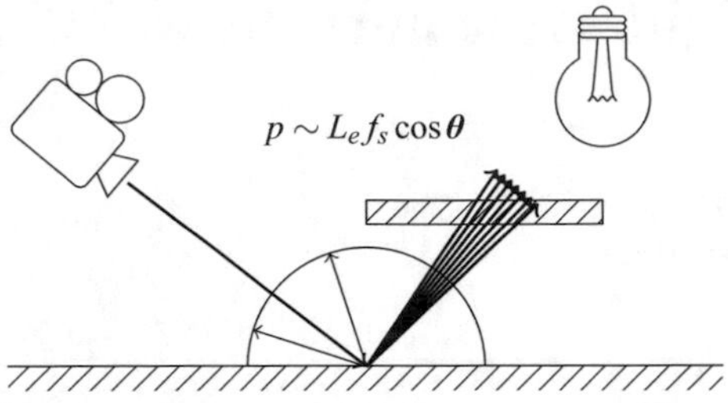

Fig. 1 In the illustration, radiance is integrated by sampling proportional to the product of emitted radiance L_e and the bidirectional scattering distribution function f_s representing the physical surface properties taking into account the fraction of radiance that is incident perpendicular to the surface, which is the cosine of the angle θ between the surface normal and the direction of incidence. As such importance sampling does not consider blockers, light transport paths with zero contributions cannot be avoided unless visibility is considered

has not been a simple and efficient online method that can substantially reduce the number of light transport paths with zero contribution [33].

The majority of zero contributions are caused by unsuitable local importance sampling using only a factor instead of the complete integrand (see Fig. 1) or by trying to connect vertices of light transport path segments that are occluded, for example shooting shadow rays to light sources or connecting path segments starting both from the light sources and the camera sensors. An example for this inefficiency has been investigated early on in computer graphics [30, 31]: The visible part of the synthetic scene shown in Fig. 4 is lit through a door. By closing the door more and more the problem can be made arbitrarily more difficult to solve.

We therefore propose a method that is based on reinforcement learning [27] and allows one to sample light transport paths that are much more likely to connect lights and sensors. Complementary to first approaches of applying machine learning to image synthesis [33], in Sect. 2 we show that light transport and reinforcement learning can be modeled by the same integral equation. As a consequence, importance in light transport can be learned using any light transport algorithm.

Deriving a relationship between reinforcement learning and light transport simulation, we establish an automatic importance sampling scheme as introduced in Sect. 3. Our approach allows for controlling the memory footprint, for suitable representations of importance does not require preprocessing, and can be applied during image synthesis and/or across frames, because it is able to track distributions over time. A second parallel between temporal difference learning and next event estimation is pointed out in Sect. 4.

As demonstrated in Sect. 5 and shown in Fig. 8, already a simple implementation can dramatically improve light transport simulation. The efficiency of the scheme is based on two facts: Instead of shooting towards the light sources, we are guiding light transport paths to where the light comes from, which effectively shortens path length, and we learn importance from a smoothed approximation instead from higher variance path space samples [9, 18, 22].

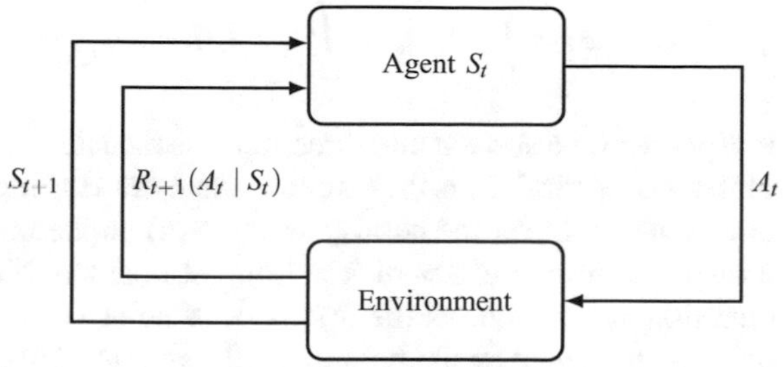

Fig. 2 The setting for reinforcement learning: At time t, an agent is in state S_t and takes an action A_t, which after interaction with the environment brings him to the next state S_{t+1} with a scalar reward R_{t+1}

2 Identifying Q-Learning and Light Transport

The setting of reinforcement learning [27] is depicted in Fig. 2: An agent takes an action thereby transitioning to the resulting next state and receiving a reward. In order to maximize the reward, the agent has to learn which action to choose in what state. This process very much resembles how humans learn.

Q-learning [39] is a model free reinforcement learning technique. Given a set of states S and a set of actions A, it determines a function $Q(s, a)$ that for any $s \in S$ values taking the action $a \in A$. Thus given a state s, the action a with the highest value may be selected next and

$$Q(s,a) = (1-\alpha) \cdot Q(s,a) + \alpha \cdot \left(r(s,a) + \gamma \cdot \max_{a' \in A} Q(s',a') \right) \tag{1}$$

may be updated by a fraction of $\alpha \in [0, 1]$, where $r(s, a)$ is the reward for taking the action resulting in a transition to a state s'. In addition, the maximum Q-value of possible actions in s' is considered and discounted by a factor of $\gamma \in [0, 1)$.

Instead of taking into account only the best valued action,

$$Q(s,a) = (1-\alpha) \cdot Q(s,a) + \alpha \cdot \left(r(s,a) + \gamma \cdot \sum_{a' \in A} \pi(s',a') Q(s',a') \right)$$

averages all possible actions in s' and weighs their values $Q(s', a')$ by a transition kernel $\pi(s', a')$, which is a strategy called *expected SARSA* [27, Sect. 6.6]. This is especially interesting, as later it will turn out that always selecting the "best" action does not perform as well as considering all options (see Fig. 4). For a continuous space A of actions, we then have

$$Q(s,a) = (1-\alpha) \cdot Q(s,a) + \alpha \cdot \left(r(s,a) + \gamma \cdot \int_A \pi(s',a') Q(s',a') da' \right). \tag{2}$$

On the other hand, the radiance

$$L(x, \omega) = L_e(x, \omega) + \int_{\mathscr{S}^+(x)} L(h(x, \omega_i), -\omega_i) f_s(\omega_i, x, \omega) \cos \theta_i d\omega_i \qquad (3)$$

in a point x on a surface into direction ω is modeled by a Fredholm integral equation of the second kind. L_e is the source radiance and the integral accounts for all radiance that is incident over the hemisphere $\mathscr{S}^+(x)$ aligned by the surface normal in x and transported into direction ω. The hitpoint function $h(x, \omega_i)$ traces a ray from x into direction ω_i and returns the first surface point intersected. The radiance from this point is attenuated by the bidirectional scattering distribution function f_s, where the cosine term of the angle θ_i between surface normal and ω_i accounts for only the fraction that is perpendicular to the surface.

A comparison of Eq. 2 for $\alpha = 1$ and Eq. 3 reveals structural similarities of the formulation of reinforcement learning and the light transport integral equation, respectively, which lend themselves to matching terms: Interpreting the state s as a location $x \in \mathbb{R}^3$ and an action a as tracing a ray from location x into direction ω resulting in the point $y := h(x, \omega)$ corresponding to the state s', the reward term $r(s, a)$ can be linked to the emitted radiance $L_e(y, -\omega) = L_e(h(x, \omega), -\omega)$ as observed from x. Similarly, the integral operator can be applied to the value Q, yielding

$$Q(x, \omega) = L_e(y, -\omega) + \int_{\mathscr{S}^+(y)} Q(y, \omega_i) f_s(\omega_i, y, -\omega) \cos \theta_i d\omega_i, \qquad (4)$$

where we identified the discount factor γ multiplied by the policy π and the bidirectional scattering distribution function f_s. Taking a look at the geometry and the physical meaning of the terms, it becomes obvious that Q in fact must be the radiance $L_i(x, \omega)$ incident in x from direction ω and in fact is described by a Fredholm integral equation of the second kind - like the light transport equation 3.

3 *Q*-Learning while Path Tracing

In order to synthesize images, we need to compute functionals of the radiance equation 3, i.e. project the radiance onto the image plane. For the purpose of this article, we start with a simple forward path tracer [16, 24]: From a virtual camera, rays are traced through the pixels of the screen. Upon their first intersection with the scene geometry, the light transport path is continued into a scattering direction determined according to the optical surface properties. Scattering and ray tracing are repeated until a light source is hit. The contribution of this complete light transport path is added to the pixel pierced by the initial ray of this light transport path when started at the camera.

In this simple form, the algorithm exposes quite some variance as can be seen in the images on the left in Fig. 3. This noise may be reduced by importance sampling. We therefore progressively approximate Eq. 4 using reinforcement learning: Once a direction has been selected and a ray has been traced by the path tracer,

Fig. 3 Comparison of simple path tracing without (left) and with (right) reinforcement learning importance sampling. The top row is using 8 paths per pixel, while 32 are used for the bottom row. The challenge of the scene is the area light source on the left indirectly illuminating the right part of the scene. The enlarged insets illustrate the reduction of noise level

$$Q'(x,\omega) = (1-\alpha)\cdot Q(x,\omega) + \alpha \cdot \left(L_e(y,-\omega) + \int_{\mathscr{S}^+(y)} Q(y,\omega_i) f_s(\omega_i, y, -\omega)\cos\theta_i d\omega_i \right) \quad (5)$$

is updated using a learning rate α. The probability density function resulting from normalizing Q in turn is used for importance sampling a direction to continue the path. As a consequence more and more light transport paths are sampled that contribute to the image. Computing a global solution to Q in a preprocess would not allow for focussing computations on light transport paths that contribute to the image.

3.1 Implementation

Often, approximations to Q are tabulated for each pair of state and action. In computer graphics, there are multiple choices to represent radiance and for the purpose of this article, we chose the data structure as used for irradiance volumes [6] to approximate Q. Figure 5 shows an exemplary visualization of such a discretization during rendering: For selected points y in space, the hemisphere is stratified and one value $Q_k(y)$ is stored per sector, i.e. stratum k. Figure 4f illustrates the placement of probe centers y, which results from mapping a two-dimensional low discrepancy sequence onto the scene surface (Fig. 5).

Now the integral

$$\int_{\mathscr{S}^+(y)} Q(y,\omega_i) f_s(\omega_i, y, -\omega)\cos\theta_i d\omega_i \approx \frac{2\pi}{n}\sum_{k=0}^{n-1} Q_k(y) f_s(\omega_k, y, -\omega)\cos\theta_k$$

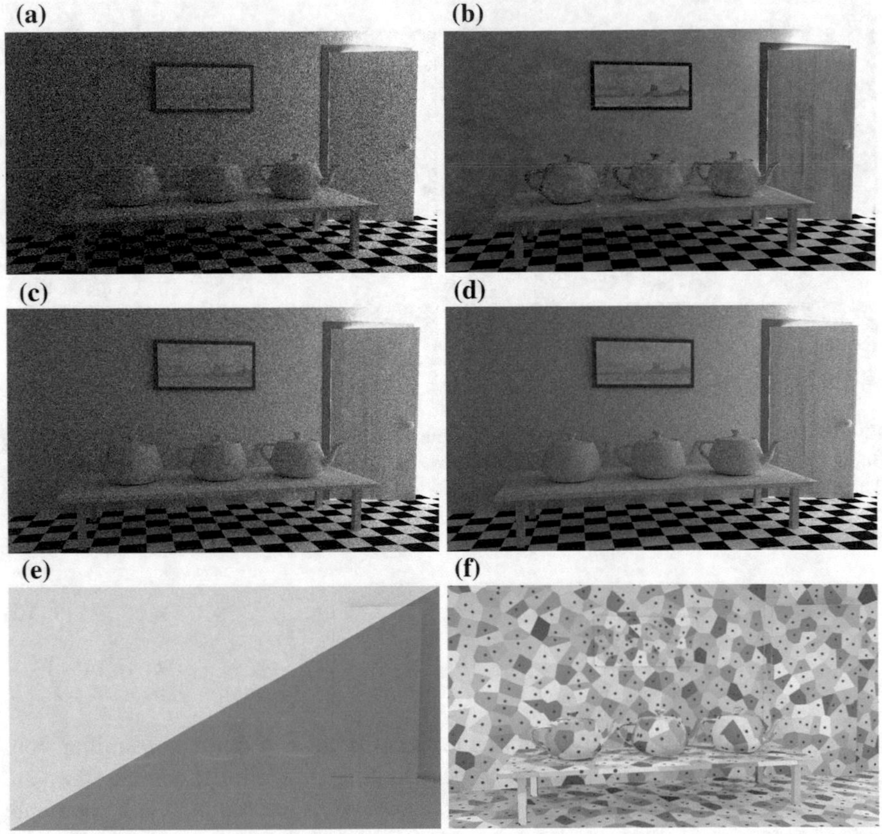

Fig. 4 Comparison at 1024 paths per pixel (the room behind the door is shown in Fig. 5): **a** A simple path tracer with cosine importance sampling, **b** the Kelemen variant of the Metropolis light transport algorithm, **c** scattering proportional to Q, while updating Q with the maximum as in Eq. 1 and **d** scattering proportional to Q weighted by the bidirectional scattering distribution function and updating accordingly by Eq. 5. The predominant reinforcement approach of always taking the best next action is inferior to selecting the next action proportional to Q, i.e. considering all alternatives. A comparison to the Metropolis algorithm reveals much more uniform lighting, especially much more uniform noise and the lack of the typical splotches. **e** The average path length of path tracing (above image diagonal) is about 215, while with reinforcement learning it amounts to an average of 134. The average path length thus is reduced by 40% in this scene. **f** Discretized hemispheres to approximate Q are stored in points on the scene surfaces determined by samples of the Hammersley low discrepancy point set. Retrieving Q for a query point results in searching for the nearest sample of Q that has a similar normal to the one in the query point (see especially the teapot handles). The red points indicate where in the scene hemispheres to hold the Q_k are stored. The colored areas indicate their corresponding Voronoi cells. Storing the Q_k in this example requires about 2 MB of memory. Scene courtesy (cc) 2013 Miika Aittala, Samuli Laine, and Jaakko Lehtinen (https://mediatech.aalto.fi/publications/graphics/GMLT/)

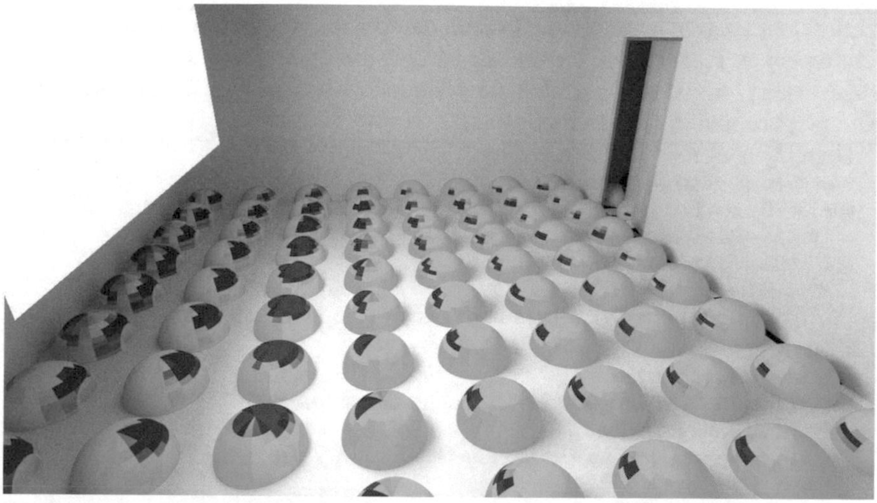

Fig. 5 The image shows parts of an example discretization of Q by a grid, where hemispheres are uniformly distributed across the ground plane. The false colors indicate magnitude, where small values are green and large values are red. The large values on each hemisphere point towards the part of the scene, where the light is coming from. For example, under the big area light source on the left, most radiance is incident as reflected radiance from the wall opposite to the light source

in Eq. 5 can be estimated by using each one uniform random direction ω_k in each stratum k, where θ_k is the angle between the surface normal in y and ω_k.

The method has been implemented in an importance driven forward path tracer as shown in Algorithm 1: Only two routines for updating Q and selecting a scattering direction proportional to Q need to be added. Normalizing the Q in a point y then results in a probability density that is used for importance sampling during scattering by inverting the cumulative distribution function. In order to guarantee ergodicity, meaning that every light transport path remains possible, all $Q(y)$ are initialized to be positive, for example by a uniform probability density or proportional to a factor of the integrand (see Fig. 1). When building the cumulative distribution functions in parallel every accumulated frame, values below a small positive threshold are replaced by the threshold.

The parameters exposed by our implementation are the resolution of the discretization and the learning rate α.

3.2 Consistency

It is desirable to craft consistent rendering algorithms [16], because then all renderer introduced artifacts, like for example noise, are guaranteed to vanish over time. This

Algorithm 1 Given a camera and scene description, augmenting a path tracer by reinforcement learning for importance sampling requires only two additions: The importance Q needs to be updated along the path and scattering directions are sampled proportional to Q as learned so far.

```
throughput ← 1
ray ← setupPrimaryRay(camera)
for i ← 0 to ∞ do
  (y, n) ← intersect(scene, ray) // corresponds to y := h(x, ω)
  // addition 1: update Q
  if i > 0 then
    Q(x, ω) ← (1 − α)Q(x, ω) + α (L_e(y, −ω) + ∫_{S^2_+(y)} f_s(ω_i, y, −ω) cos θ_i Q(y, ω_i)dω_i)
  end if
  if isEnvironment(y) then
    return throughput· getRadianceFromEnvironment(ray, y)
  else if isAreaLight(y) then
    return throughput· getRadianceFromAreaLight(ray, y)
  end if
  // addition 2: scatter proportional to Q
  (ω, p_ω, f_s) ← sampleScatteringDirectionProportionalTo(Q, y)
  throughput ← throughput · f_s · cos(n, ω)/p_ω
  ray ← (y, ω)
end for
```

requires the $Q_k(y)$ to converge, which may be accomplished by a vanishing learning rate α.

In reinforcement learning [27], a typical approach is to count the number of visits to each pair of state s and action a and using

$$\alpha(s, a) = \frac{1}{1 + \text{visits}(s, a)}.$$

The method resembles the one used to make progressive photon mapping consistent [7], where consistency has been achieved by decreasing the search radius around a query point every time a photon hits sufficiently close. Similarly, the learning rate may also depend on the total number of visits to a state s alone, or even may be chosen to vanish independently of state and action. Again, such approaches have been explored in consistent photon mapping [12].

While the $Q_k(y)$ converge, they do not necessarily converge to the incident radiance in Eq. 4. First, as they are projections onto a basis, the $Q_k(y)$ at best only are an approximation of Q in realistic settings. Second, as the coefficients $Q_k(y)$ are learned during path tracing, i.e. image synthesis, and used for importance sampling, it may well happen that they are not updated everywhere at the same rate. Nevertheless, since all operators are linear, the number of visits will be proportional to the number of light transport paths [12] and consequently as long as $Q_k(y) > 0$ whenever $L_i(y, \omega_i) > 0$ all $Q_k(y)$ will be updated eventually.

3.3 Learning While Light Tracing

For guiding light transport paths from the light sources towards the camera, the transported weight W of a measurement (see [29]), i.e. the characteristic function of the image plane, has to be learned instead of the incident radiance Q. As W is the adjoint of Q, the same data structures may be used for its storage. Learning both Q and W allows one to implement bidirectional path tracing [29] with reinforcement learning for importance sampling to guide both light and camera path segments including visibility information for the first time. Note that guiding light transport paths this way may reach efficiency levels that even can make bidirectional path tracing and multiple importance sampling redundant [33] in many common cases.

4 Temporal Difference Learning and Next Event Estimation

Besides the known shortcomings of (bidirectional) path tracing [17, Sect. 2.4 Problem of insufficient techniques], the efficiency may be restricted by the approximation quality of Q: For example, the smaller the light sources, the finer the required resolution of Q to reliably guide rays to hit a light source. This is where next event estimation may help [5, 13, 32].

Already in [38] the contribution of light sources has been "learned": A probability per light source has been determined by the number of successful shadow rays divided by the total number of shadow rays shot. This idea has been refined subsequently [2, 14, 36, 37].

For reinforcement learning, the state space may be chosen as a regular grid over the scene, where in each grid cell c for each light source l a value $V_{c,l}$ is stored that is initialized with zero. Whenever a sample on a light source l is visible to a point x to be illuminated in the cell c upon next event estimation, its value

$$V'_{c,l} = (1 - \alpha)V_{c,l} + \alpha \cdot \|C_l(x)\|_\infty \tag{6}$$

is updated using the norm of the contribution $C_l(x)$. Building a cumulative distribution function from all values $V_{c,l}$ within a cell c, light may be selected by importance sampling. Figure 6 shows the efficiency gain of this reinforcement learning method over uniform light source selection for 16 paths per pixel.

It is interesting to see that this is another relation to reinforcement learning: While the Q-learning equation 5 takes into account the values of the next, non-terminal state, the next state in event estimation is always a terminal state and Q-learning coincides with plain temporal difference learning [26] as in Eq. 6.

Fig. 6 Two split-image comparisons of uniformly selecting area light sources and selection using temporal difference learning, both at 16 paths per pixel. The scene on the left has 5000 area light sources, whereas the scene on the right has about 15000 (San Miguel scene courtesy Guillermo M. Leal Llaguno (http://www.evvisual.com/))

4.1 Learning Virtual Point Light Sources

The vertices generated by tracing photon trajectories (see Sect. 3.3) can be considered a photon map [10] and may be used in the same way. Furthermore, they may be used as a set of virtual point light sources for example the instant radiosity [15] algorithm.

Continuously updating and learning the measurement contribution function W [29] across frames and using the same seed for the pseudo- or quasi-random sequences allows for generating virtual point light sources that expose a certain coherency over time, which reduces temporal artifacts when rendering animations with global illumination.

4.2 Learning Environment Lighting

Rendering sun and sky is usually done by distributing samples proportional to the brightness of pixels in the environment texture. More samples should end up in brighter regions, which is achieved by constructing and sampling from a cumulative distribution function, for example using the alias method [35]. Furthermore, the sun may be separated from the sky and simulated separately. The efficiency of such importance sampling is highly dependent on occlusion, i.e. what part of the environment can be seen from the point to be shaded (see Fig. 1).

Similar to Sect. 3.1 and in order to consider the actual contribution including occlusion, an action space is defined by partitioning the environment map into tiles and learning the importance per tile. Figure 7 shows the improvement for an example setting.

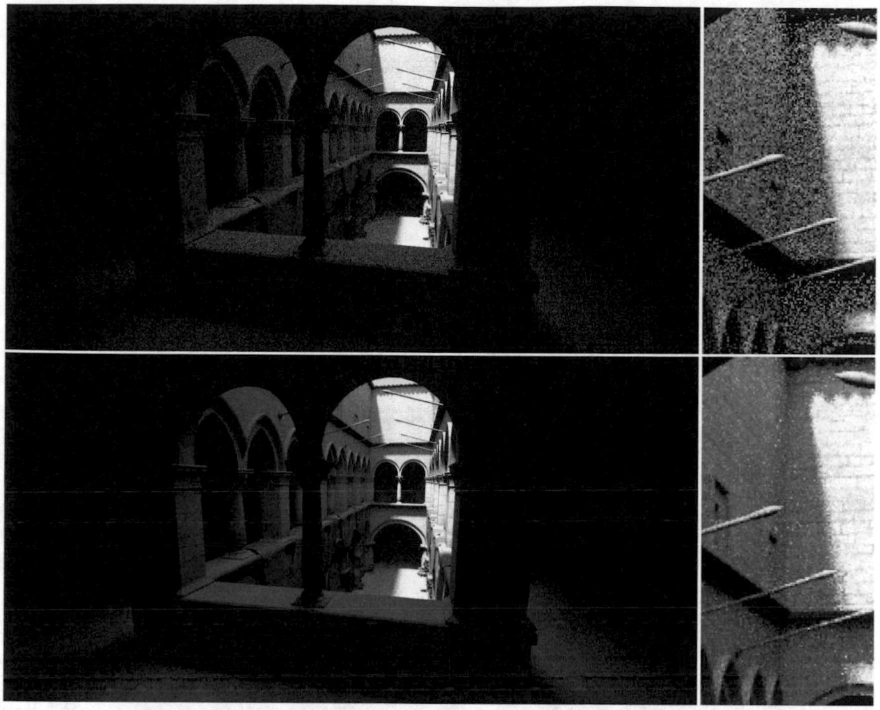

Fig. 7 Sun and sky illumination at 32 paths per pixel. Top: simple importance sampling considering only the environment map as a light source. Bottom: Importance sampling with reinforcement learned importance. The enlargements on the right illustrate the improved noise reduction. Scene courtesy Frank Meinl, Crytek (http://graphics.cs.williams.edu/data/meshes/crytek-sponza-copyright.html)

5 Results and Discussion

Figure 4 compares the new reinforcement learning algorithm to common algorithms: For the same budget of light transport paths, the superiority over path tracing with importance sampling according to the reflection properties is obvious. A comparison with the Metropolis algorithm for importance sampling [11, 30] reveals much more uniform noise lacking the typical splotchy structure inherent with the local space exploration of Metropolis samplers. Note, however, that the new reinforcement learning importance sampling scheme could as well be combined with Metropolis sampling. Finally, updating Q by Eq. 1, i.e. the "best possible action" strategy is inferior to using the weighted average of all possible next actions according to Eq. 5. In light transport simulation this is not surprising, as the deviation of the integrand from its estimated maximum very often is much larger than from a piecewise constant approximation.

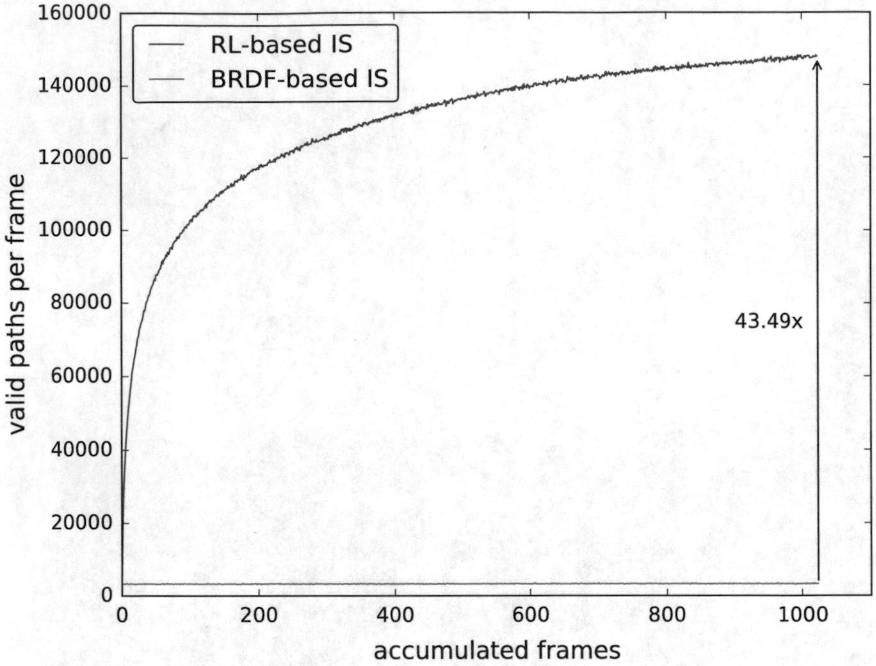

Fig. 8 Using reinforcement learning (RL), the number of paths actually connecting to a light source is dramatically improved over classic importance sampling (IS) using only the bidirectional scattering distribution function for importance sampling. As a result, more non-zero contributions are accumulated for the same number of paths, see also Fig. 4

The big gain in quality is due to the dramatic reduction of zero contribution light transport paths (see Fig. 8), even under complex lighting. In Fig. 4a–d, the same number of paths has been used. In each iteration, for path tracing with and without reinforcement learning one path has been started per pixel, while for the Metropolis variant the number of Markov chains equals the number of pixels of the image. Rendering the image at 1280×720 pixels, each iteration takes 41ms for path tracing, 49ms for Metropolis light transport [11, 30], and 51ms for the algorithm with reinforcement learned importance sampling. Hence the 20% overhead is well paid off by the level of noise reduction.

Shooting towards where the radiance comes from naturally shortens the average path length as can be seen in Fig. 4e. Based on the approach to guide light paths using a pre-trained Gaussian mixture model [33] to represent probabilities, in [34] in addition the density of light transport paths is controlled across the scene using splitting and Russian roulette. These ideas have the potential to further improve the efficiency of our approach.

While the memory requirements for storing our data structure for Q are small, the data structure is not adaptive. An alternative is an adaptive hierarchical approximation to Q as used in [18, 22]. Yet, another variant would be learning parameters for lobes to guide light transport paths [1]. In principle any data structure that has been used in graphics to approximate irradiance or radiance is a candidate. Which data structure and what parameters are best, may depend on the scene to be rendered. For example, using discretized hemispheres limits the resolution with respect to solid angle. If the resolution is chosen too fine, learning is slow, if the resolution is to coarse, convergence is slow.

Given that Q asymptotically approximates the incident radiance L_i, it is worthwhile to investigate how it can be used for the separation of the main part as explored in [18, 23] to further speed up light transport simulation or even as an alternative to importance sampling.

Beyond what we explore, path guiding has been extended to consider product importance sampling [8] and reinforcement learning [27] offers more policy evaluation strategies to consider.

6 Conclusion

Guiding light transport paths has been explored in [1, 4, 9, 18, 22, 23, 33]. However, key to our approach is that by using a representation of Q in Eq. 5 instead of solving the equation by recursion, i.e. a Neumann series, Q may be learned much faster and in fact during sampling light transport paths without any preprocess. This results in a new algorithm to increase the efficiency of path tracing by approximating importance using reinforcement learning during image synthesis. Identifying Q learning and light transport, heuristics have been replaced by physically based functions, and the only parameters that the user may control are the learning rate and the discretization of Q.

The combination of reinforcement learning and deep neural networks [19–21, 28] is an obvious avenue of future research: Representing the radiance on hemispheres already has been successfully explored [25] and the interesting question is how well Q can be represented by neural networks.

Acknowledgements The authors would like to thank Jaroslav Křivánek, Tero Karras, Toshiya Hachisuka, and Adrien Gruson for profound discussions and comments.

References

1. Bashford-Rogers, T., Debattista, K., Chalmers, A.: A significance cache for accelerating global illumination. Comput. Graph. Forum **31**(6), 1837–1851 (2012)
2. Benthin, C., Wald, I., Slusallek, P.: A scalable approach to interactive global illumination. Computer Graphics Forum (Proc. Eurographics 2003) **22**(3), 621–629 (2003)
3. Cline, D., Adams, D., Egbert, P.: Table-driven adaptive importance sampling. Comput. Graph. Forum **27**(4), 1115–1123 (2008)

4. Dutré, P., Willems, Y.: Potential-driven Monte Carlo particle tracing for diffuse environments with adaptive probability functions. In: Rendering Techniques 1995 (Proceedings of the 6th Eurographics Workshop on Rendering), pp. 306–315. Springer, Berlin (1995)
5. Estevez, C., Kulla, C.: Importance sampling of many lights with adaptive tree splitting. In: ACM SIGGRAPH 2017 Talks, SIGGRAPH '17, pp. 33:1–33:2. ACM (2017)
6. Greger, G., Shirley, P., Hubbard, P., Greenberg, D.: The irradiance volume. IEEE Comput. Graph. Appl. **18**(2), 32–43 (1998)
7. Hachisuka, T., Ogaki, S., Jensen, H.: Progressive photon mapping. ACM Trans. Graph. **27**(5), 130:1–130:8 (2008)
8. Herholz, S., Elek, O., Vorba, J., Lensch, H., Křivánek, J.: Product importance sampling for light transport path guiding. In: Proceedings of the Eurographics Symposium on Rendering, EGSR '16, pp. 67–77. Eurographics Association (2016). https://doi.org/10.1111/cgf.12950
9. Jensen, H.: Importance driven path tracing using the photon map. In: Hanrahan P., Purgathofer W. (eds.) Rendering Techniques 1995 (Proceedings of the 6th Eurographics Workshop on Rendering), pp. 326–335. Springer, Berlin (1995)
10. Jensen, H.: Realistic Image Synthesis Using Photon Mapping. AK Peters (2001)
11. Kelemen, C., Szirmay-Kalos, L., Antal, G., Csonka, F.: A simple and robust mutation strategy for the metropolis light transport algorithm. Comput. Graph. Forum **21**(3), 531–540 (2002)
12. Keller, A., Binder, N.: Deterministic consistent density estimation for light transport simulation. In: Dick J., Kuo F., Peters G., Sloan I. (eds.) Monte Carlo and Quasi-Monte Carlo Methods 2012, pp. 467–480. Springer, Berlin (2013)
13. Keller, A., Wächter, C., Raab, M., Seibert, D., Antwerpen, D., Korndörfer, J., Kettner, L.: The Iray light transport simulation and rendering system (2017). CoRR arXiv:abs/1705.01263
14. Keller, A., Wald, I.: Efficient importance sampling techniques for the photon map. In: Proceedings of Vision, Modeling, and Visualization, pp. 271–279. IOS Press (2000)
15. Keller, A.: Instant radiosity. In: SIGGRAPH '97: Proceedings of the 24th Annual Conference on Computer Graphics and Interactive Techniques, pp. 49–56 (1997)
16. Keller, A.: Quasi-Monte Carlo image synthesis in a nutshell. In: Dick J., Kuo F., Peters G., Sloan I. (eds.) Monte Carlo and Quasi-Monte Carlo Methods 2012, pp. 203–238. Springer, Berlin (2013)
17. Kollig, T., Keller, A.: Efficient bidirectional path tracing by randomized quasi-Monte Carlo integration. In: Niederreiter H., Fang K, Hickernell F. (eds.) Monte Carlo and Quasi-Monte Carlo Methods 2000, pp. 290–305. Springer, Berlin (2002)
18. Lafortune, E., Willems, Y.: A 5D tree to reduce the variance of Monte Carlo ray tracing. In: Hanrahan P., Purgathofer W. (eds.) Rendering Techniques 1995 (Proceedings of the 6th Eurographics Workshop on Rendering), pp. 11–20. Springer, Berlin (1995)
19. Lillicrap, T., Hunt, J., Pritzel, A., Heess, N., Erez, T., Tassa, Y., Silver, D., Wierstra, D.: Continuous control with deep reinforcement learning (2015). CoRR arXiv:abs/1509.02971
20. Mnih, V., Badia, A., Mirza, M., Graves, A., Lillicrap, T., Harley, T., Silver, D., Kavukcuoglu, K.: Asynchronous methods for deep reinforcement learning (2016). CoRR arXiv:abs/1602.01783
21. Mnih, V., Kavukcuoglu, K., Silver, D., Graves, A., Antonoglou, I., Wierstra, D., Riedmiller, M.: Playing Atari with deep reinforcement learning (2013). CoRR arXiv:abs/1312.5602
22. Müller, T., Gross, M., Novák, J.: Practical path guiding for efficient light-transport simulation. In: Proceedings of the Eurographics Symposium on Rendering (2017)
23. Pegoraro, V., Brownlee, C., Shirley, P., Parker, S.: Towards interactive global illumination effects via sequential Monte Carlo adaptation. In: Proceedings of the 3rd IEEE Symposium on Interactive Ray Tracing, pp. 107–114 (2008)
24. Pharr, M., Jacob, W., Humphreys, G.: Physically Based Rendering - From Theory to Implementation, 3rd edn. Morgan Kaufmann, San Francisco (2016)
25. Satilmis, P., Bashford-Rogers, T., Chalmers, A., Debattista, K.: A machine learning driven sky model. IEEE Comput. Graph. Appl. 1–9 (2016)
26. Sutton, R.: Learning to predict by the methods of temporal differences. Mach. Learn. **3**(1), 9–44 (1988)

27. Sutton, R., Barto, A.: Introduction to Reinforcement Learning, 2nd edn. MIT Press, Cambridge, USA (2017)
28. van Hasselt, H., Guez, A., Silver, D.: Deep reinforcement learning with double Q-learning. CoRR arXiv:abs/1509.06461 (2015)
29. Veach, E., Guibas, L.: Bidirectional estimators for light transport. In: Proceedings of the 5th Eurographics Workshop on Rendering, pp. 147–161. Darmstadt, Germany (1994)
30. Veach, E., Guibas, L.: Metropolis light transport. In: Whitted, T. (ed.) Proceedings of the SIGGRAPH 1997, Annual Conference Series, pp. 65–76. ACM SIGGRAPH, Addison Wesley (1997)
31. Veach, E., Guibas, L.: Optimally combining sampling techniques for Monte Carlo rendering. In: SIGGRAPH '95 Proceedings of the 22nd Annual Conference on Computer Graphics and Interactive Techniques, pp. 419–428 (1995)
32. Vévoda, P., Křivánek, J.: Adaptive direct illumination sampling. In: SIGGRAPH ASIA 2016 Posters, pp. 43:1–43:2. ACM, New York, USA (2016)
33. Vorba, J., Karlík, O., Šik, M., Ritschel, T., Křivánek, J.: On-line learning of parametric mixture models for light transport simulation. ACM Trans. Graph. (Proceedings of SIGGRAPH 2014) **33**(4) (2014)
34. Vorba, J., Křivánek, J.: Adjoint-driven Russian roulette and splitting in light transport simulation. ACM Trans. Graph. (Proceedings of SIGGRAPH 2016) **35**(4), 1–11 (2016)
35. Vose, M.: A linear algorithm for generating random numbers with a given distribution. IEEE Trans. Softw. Eng. **17**(9), 972–975 (1991)
36. Wald, I., Benthin, C., Slusallek, P.: Interactive global illumination in complex and highly occluded environments. In: Christensen P., Cohen-Or D. (eds.) Rendering Techniques 2003 (Proceedings of the 14th Eurographics Workshop on Rendering), pp. 74–81 (2003)
37. Wald, I., Kollig, T., Benthin, C., Keller, A., Slusallek, P.: Interactive global illumination using fast ray tracing. In: Debevec P., Gibson S. (eds.) Rendering Techniques 2002 (Proc. 13th Eurographics Workshop on Rendering), pp. 15–24 (2002)
38. Ward, G.: Adaptive shadow testing for ray tracing. In: 2nd Eurographics Workshop on Rendering. Barcelona, Spain (1991)
39. Watkins, C., Dayan, P.: Q-learning. Mach. Learn. **8**(3), 279–292 (1992)

Successive Coordinate Search and Component-by-Component Construction of Rank-1 Lattice Rules

Adrian Ebert, Hernan Leövey and Dirk Nuyens

Abstract The (fast) component-by-component (CBC) algorithm is an efficient tool for the construction of generating vectors for quasi-Monte Carlo rank-1 lattice rules in weighted reproducing kernel Hilbert spaces. We consider product weights, which assign a weight to each dimension. These weights encode the effect a certain variable (or a group of variables by the product of the individual weights) has. Smaller weights indicate less importance. Kuo (J Complex 19:301–320, 2003 [3]) proved that CBC constructions achieve the optimal rate of convergence in the respective function spaces, but this does not imply the algorithm will find the generating vector with the smallest worst-case error. In fact it does not. We investigate a generalization of the component-by-component construction that allows for a general successive coordinate search (SCS), based on an initial generating vector, and with the aim of getting closer to the smallest worst-case error. The proposed method admits the same type of worst-case error bounds as the CBC algorithm, independent of the choice of the initial vector. Under the same summability conditions on the weights as in (Kuo J Complex 19:301–320, 2003 [3]) the error bound of the algorithm can be made independent of the dimension d and we achieve the same optimal order of convergence for the function spaces from (Kuo, J Complex 19:301–320, 2003 [3]). Moreover, a fast version of our method, based on the fast CBC algorithm as in Nuyens and Cools (Math Comput 75:903–920, 2006, [5]), is available, reducing the computational cost of the algorithm to $O(d\,n\log(n))$ operations, where n denotes the number of function evaluations. Numerical experiments seeded by a Korobov-type generating vector show that the new SCS algorithm will find better choices than the CBC algorithm and the effect is better for slowly decaying weights.

A. Ebert (✉) · D. Nuyens
KU Leuven, Celestijnenlaan 200A, Leuven, Belgium
e-mail: adrian.ebert@cs.kuleuven.be

D. Nuyens
e-mail: dirk.nuyens@cs.kuleuven.be

H. Leövey
Structured Energy Management Team, Axpo, Baden, Switzerland
e-mail: hernaneugenio.leoevey@axpo.com

A. B. Owen and P. W. Glynn (eds.), *Monte Carlo and Quasi-Monte Carlo Methods*, Springer Proceedings in Mathematics & Statistics 241,
https://doi.org/10.1007/978-3-319-91436-7_10

Keywords Numerical Integration · Quasi-Monte Carlo methods · Lattice points · Component-by-component construction · Successive coordinate search algorithm

1 Introduction

In this article we study the numerical approximation of integrals of the form

$$I(f) = \int_{[0,1]^d} f(\boldsymbol{x}) \, \mathrm{d}\boldsymbol{x}$$

for d-variate functions f via quasi-Monte Carlo quadrature rules. Quasi-Monte Carlo rules are equal-weight quadrature rules of the form

$$Q_{n,d}(f) = \frac{1}{n} \sum_{k=0}^{n-1} f(\boldsymbol{x}_k),$$

where the quadrature points $\boldsymbol{x}_0, \ldots, \boldsymbol{x}_{n-1} \in [0, 1]^d$ are chosen deterministically. Here, we consider integrands $f : [0, 1]^d \to \mathbb{R}$ which belong to some normed function space $(H, \|\cdot\|_H)$. In order to assess the quality of a particular QMC rule $Q_{n,d}$ with underlying point set $P_n = \{\boldsymbol{x}_0, \ldots, \boldsymbol{x}_{n-1}\}$, we introduce the notion of the so-called worst-case error, see, e.g., [1], defined by

$$e_{n,d}(P_n, H) = \sup_{\|f\|_H \le 1} \left| \int_{[0,1]^d} f(\boldsymbol{x}) \, \mathrm{d}\boldsymbol{x} - \frac{1}{n} \sum_{k=0}^{n-1} f(\boldsymbol{x}_k) \right| .$$

In other words, $e_{n,d}(P_n, H)$ is the worst error that is attained over all functions in the unit ball of H using the quasi-Monte Carlo rule with quadrature points in P_n. It is often possible to obtain explicit expressions to calculate $e_{n,d}(P_n, H)$, see, e.g., [4]. In particular, we consider weighted Korobov and weighted shift-averaged Sobolev spaces, which are both reproducing kernel Hilbert spaces, for details see, e.g., [1, 3, 8, 9]. In this paper we will limit ourselves to the original choice of "product weights". In essence, the idea is to quantify the varying importance of the coordinate directions x_j with $j = 1, \ldots, d$ w.r.t. the function values by a sequence $\boldsymbol{\gamma} = \{\gamma_j\}_{j=1}^d$ of positive weights.

There are many ways to choose the underlying point set P_n of a QMC rule, ranging from lattice rules and sequences, digital nets and sequences and more recent constructions such as interlaced polynomial lattice rules. In this paper, however, we will restrict ourselves to rank-1 lattice rules. This type of QMC rules has an underlying point set $P_n \subseteq [0, 1]^d$ of the form

$$P_n = \left\{ \left\{ \frac{k\,\boldsymbol{z}}{n} \right\} \,\middle|\, 0 \le k < n \right\},$$

where $z \in \mathbb{Z}^d$ is the *generating vector* of the rank-1 lattice rule and $\{\cdot\}$ denotes the fractional part, componentwise if applied to a vector. It is clear that any vector congruent modulo n is equivalent and so we only consider values modulo n. The components of z are often restricted to the set of integers in $\{1, \ldots, n-1\}$ that are relatively prime to n, see, e.g., [3], such that one obtains n distinct points for all one-dimensional projections, and as such for any projection. In this article we consider generating vectors $z \in \mathbb{Z}_n^d$ with n prime and $\mathbb{Z}_n = \{0, 1, 2, \ldots, n-1\}$. For rank-1 lattice rules in a weighted shift-invariant tensor-product reproducing kernel Hilbert space $H(K)$ with reproducing kernel $K(\boldsymbol{x}, \boldsymbol{y}) = \prod_{j=1}^d (\beta_j + \gamma_j\, \omega(x_j - y_j))$ the squared worst-case error can be written as

$$e_{n,d}^2(P_n, H) = e_{n,d}^2(z) = -\prod_{j=1}^d \beta_j + \frac{1}{n}\sum_{k=0}^{n-1}\prod_{j=1}^d \left(\beta_j + \gamma_j\, \omega\left(\left\{\frac{k\, z_j}{n}\right\}\right)\right),$$

with positive weights $\boldsymbol{\gamma} = \{\gamma_j\}_{j=1}^d$ and $\boldsymbol{\beta} = \{\beta_j\}_{j=1}^d$, and where $\int_0^1 \omega(t)\,\mathrm{d}t = 0$, see, e.g., [5]. We note that the weights β_j are to easily accommodate for some types of shift-averaged Sobolev spaces. Moreover, the initial squared worst-case error in this function space, i.e., using $n = 0$ samples and with the convention that $Q_{0,d}(f) = 0$, is given by

$$e_{0,d}^2(0, H) = e_{0,d}^2 = \prod_{j=1}^d \beta_j.$$

Remark 1 It is always possible to consider the normalized worst-case error by dividing by the initial worst-case error for the zero-algorithm. The squared normalized worst-case error then takes the form

$$\frac{e_{n,d}^2(z)}{e_{0,d}^2} = -1 + \frac{1}{n}\sum_{k=0}^{n-1}\prod_{j=1}^d \left(1 + \frac{\gamma_j}{\beta_j}\,\omega\left(\left\{\frac{k\, z_j}{n}\right\}\right)\right), \tag{1}$$

with $e_{0,d}^2 = \prod_{j=1}^d \beta_j$. This is equivalent to considering the worst-case error $e_{n,d}(z)$ with modified weight sequences $\hat{\beta}_j = 1$ and $\hat{\gamma}_j = \gamma_j / \beta_j$.

One of the most commonly considered methods to construct good rank-1 lattice rules is the component-by-component (CBC) construction, see, e.g., [4], which extends the generating vector one component at a time by selecting the next components z_s which minimizes the worst-case error of the s-dimensional rule. The pseudo-code of the CBC algorithm is given below.

Algorithm 1 Component-by-component construction (CBC)

Output: $z \in \mathbb{Z}_n^d$
for $s = 1$ **to** d **do**
  for all $z_s \in \mathbb{Z}_n$ **do**

$$e_{n,s}^2(z_1, \ldots, z_{s-1}, z_s) = -\prod_{j=1}^{s} \beta_j + \frac{1}{n} \sum_{k=0}^{n-1} \prod_{j=1}^{s} \left(\beta_j + \gamma_j \, \omega \left(\left\{ \frac{k \, z_j}{n} \right\} \right) \right)$$

  end for

$$z_s = \underset{z \in \mathbb{Z}_n}{\operatorname{argmin}} \; e_{n,s}^2(z_1, \ldots, z_{s-1}, z)$$

end for

It was shown in [3] that the component-by-component construction generates lattice rules which achieve optimal rates of convergence in weighted Korobov and Sobolev function spaces. Additionally, a fast construction method is available, see [5, 6], that reduces the construction cost to $O(d \, n \, \log(n))$ operations.

Even though the CBC algorithm constructs generating vectors z which exhibit the optimal error asymptotics, the constructed vector is not necessarily the one minimizing the worst-case error $e_{n,d}(z)$. We will therefore introduce and investigate a different construction method which can generate lattice rules with a smaller worst-case error than the CBC construction.

The article is structured as follows. In Sect. 2 we introduce the successive coordinate search (SCS) algorithm and analyse some properties. In Sect. 3 we prove that the SCS construction achieves optimal rates of convergence in the weighted Korobov and weighted shift-averaged Sobolev space. To get dimension-independent bounds, i.e., achieve tractability, we show that the summability condition on the weights is the same as for the normal CBC construction. Finally we report on various numerical experiments in Sect. 4.

2 Formulation of the Successive Coordinate Search Algorithm

In this section we introduce an algorithm of similar nature to the component-by-component construction. One advantage of the component-by-component construction is that the algorithm is extensible in the dimension d, i.e., to find the $(d+1)$-dimensional generating vector, the algorithm does not need to restart but just starts from the generating vector of dimension d. In our setting this also implies that a d-dimensional vector, with d large enough, could be constructed and used for all problems with less than d dimensions. This allows us to fix the maximum number of dimensions to some large enough d and successively try to find the best s-th component of a d-dimensional generating vector, keeping all other $d-1$ choices fixed. The pseudocode of the successive coordinate search (SCS) algorithm is given below.

Algorithm 2 Successive coordinate search algorithm (SCS)

Input: $z^0 \in \mathbb{Z}_n^d$
Output: $z \in \mathbb{Z}_n^d$
for $s = 1$ **to** d **do**
  for all $z_s \in \mathbb{Z}_n$ **do**
    $e^2_{n,d}(z_1, \ldots, z_{s-1}, z_s, z_{s+1}=z^0_{s+1}, \ldots, z_d = z^0_d) = -\prod_{j=1}^{d} \beta_j + \frac{1}{n} \sum_{k=0}^{n-1} \prod_{j=1}^{d} \left(\beta_j + \gamma_j\, \omega\left(\left\{\frac{k z_j}{n}\right\}\right)\right)$
  end for
  $z_s = \underset{z \in \mathbb{Z}_n}{\operatorname{argmin}}\; e^2_{n,d}(z_1, \ldots, z_{s-1}, z, z^0_{s+1}, \ldots, z^0_d)$
end for

Instead of increasing the dimension in each step of the algorithm, we keep d fixed during all calculations. Based on a starting vector $z^0 \in \mathbb{Z}_n^d = \{0, 1, \ldots, n-1\}^d$, the algorithm successively selects the coordinate $z_s \in \mathbb{Z}_n$ which minimizes the squared worst-case error $e^2_{n,d}(z_1, \ldots, z_{s-1}, z_s, z^0_{s+1}, \ldots, z^0_d)$ while keeping all other coordinates of z fixed. Thus, in the process of the SCS algorithm the coordinates of the starting vector z^0 are altered in each step of the algorithm. Our construction is very similar to the component-by-component construction, with the only difference being that an initial vector z^0 is required as input for the algorithm. In fact, we can prove that the successive coordinate search algorithm is a generalized version of the CBC algorithm as the following theorem shows by starting from an initial vector $z^0 = (0, \ldots, 0) \in \mathbb{Z}_n^d$. We note that this is a degenerate vector as it generates only a 1-point rule, and thus is in some sense the worst possible choice for any $n \geq 1$.

Theorem 1 *The component-by-component (CBC) algorithm and the successive coordinate search (SCS) algorithm with starting vector $z^0 = (0, \ldots, 0)$ both yield the same generating vector as outcome (with equivalent choices selected in the same way in both algorithms).*

Proof Denote by 0^r the r-dimensional zero vector, where $1 \leq r \leq d$. For an arbitrary $z \in \mathbb{Z}_n^s$ with $1 \leq s \leq d$ and with $\tilde{z} = (z, 0^{d-s}) \in \mathbb{Z}_n^d$, the squared worst-case error equals

$$
\begin{aligned}
e^2_{n,d}(\tilde{z}) = e^2_{n,d}(z, 0^{d-s}) &= -\prod_{j=1}^{d} \beta_j + \frac{1}{n} \sum_{k=0}^{n-1} \prod_{j=1}^{d} \left(\beta_j + \gamma_j\, \omega\left(\left\{\frac{k\, \tilde{z}_j}{n}\right\}\right)\right) \\
&= -\prod_{j=1}^{d} \beta_j + \frac{1}{n} \sum_{k=0}^{n-1} \prod_{j=1}^{s} \left(\beta_j + \gamma_j\, \omega\left(\left\{\frac{k\, z_j}{n}\right\}\right)\right) \prod_{j=s+1}^{d} (\beta_j + \gamma_j\, \omega(0)) \\
&= -\prod_{j=1}^{d} \beta_j + \frac{C_s}{n} \sum_{k=0}^{n-1} \prod_{j=1}^{s} \left(\beta_j + \gamma_j\, \omega\left(\left\{\frac{k\, z_j}{n}\right\}\right)\right) \\
&= -\prod_{j=1}^{d} \beta_j + C_s \left(e^2_{n,s}(z) + \prod_{j=1}^{s} \beta_j \right),
\end{aligned}
$$

where $C_s = \prod_{j=s+1}^{d} (\beta_j + \gamma_j\, \omega(0))$. Note that due to the non-negativity of the squared worst-case error $e_{n,d}^2$ the function ω is such that $\omega(0) \geq 0$ and so the constants C_s are positive for all $s = 1, \ldots, d$.

Now, in each step $1 \leq s \leq d$ of the SCS algorithm with initial vector $z^0 = (0, \ldots, 0)$, we search for the $z_s \in \mathbb{Z}_n$ that minimizes $e_{n,d}^2(z_1, \ldots, z_{s-1}, z_s, 0^{d-s})$, where $z_1, \ldots, z_{s-1}$ have been determined in the previous steps of the algorithm. By the above identity we have that

$$e_{n,d}^2(z_1, \ldots, z_{s-1}, z_s, 0^{d-s}) = -\prod_{j=1}^{d} \beta_j + C_s \left(e_{n,s}^2(z_1, \ldots, z_{s-1}, z_s) + \prod_{j=1}^{s} \beta_j \right),$$

and so, since the remaining terms on the right-hand side are independent of z_s, this is equivalent to finding $z_s \in \mathbb{Z}_n$ such that $e_{n,s}^2(z_1, \ldots, z_{s-1}, z_s)$ is minimized. As this is exactly the same quantity which is minimized in each step of the component-by-component construction algorithm, we see that the CBC algorithm and the SCS algorithm with starting vector $z^0 = (0, \ldots, 0)$ yield exactly the same outcome under the assumption that both algorithms select the same minimizer whenever multiple choices occur in a minimization step.

Furthermore, the formulation of the successive coordinate search construction guarantees that the generating vector z obtained by the SCS algorithm with initial vector z^0 is never worse than the input vector z^0.

Proposition 1 *Let $z^0 \in \mathbb{Z}_n^d$ be an arbitrary generating vector for a rank-1 lattice rule and denote by $z^1 \in \mathbb{Z}_n^d$ the generating vector constructed by the SCS algorithm with starting vector z^0. Then we have that $e_{n,d}(z^1) \leq e_{n,d}(z^0)$, i.e., the SCS method constructs a generating vector with worst-case error smaller than or equal to the worst-case error of the initial vector.*

Proof The statement follows directly from the formulation of the algorithm. □

Similar to the case of the component-by-component construction there is a fast version available that allows for the construction of generating vectors with time complexity $O(d\, n \log(n))$. In case n is a prime number this results in the following algorithm.

Algorithm 3 Fast version of the SCS algorithm for prime n

Input: $z^0 \in \mathbb{Z}_n^d$
$\mathbf{q}_0 = \mathbf{1}_{n\times 1} \in \mathbb{R}^n$
for $s = 1$ **to** d **do**
  $\mathbf{q}_s = \left(\beta_s \mathbf{1}_{1\times n} + \gamma_s\, \Omega_n^{\langle g\rangle}(z_s^0, :)\right) .* \mathbf{q}_{s-1}$ ▷ **initialize q**
end for
for $s = 1$ **to** d **do**
  $\mathbf{q}_d = \mathbf{q}_d \;./\; \left(\beta_s \mathbf{1}_{1\times n} + \gamma_s\, \Omega_n^{\langle g\rangle}(z_s^0, :)\right)$ ▷ **divide out initial choice** z_s^0
  $\mathbf{E}_s^2 = -\overline{\beta}_d \mathbf{1}_{n\times 1} + \frac{1}{n}\left(\beta_s \mathbf{1}_{n\times n} + \gamma_s\, \Omega_n^{\langle g\rangle}\right) \mathbf{q}_d$ ▷ **use FFT for matrix-vector product**
  $z_s = \operatorname{argmin}_{z\in\mathbb{Z}_n} E_s^2(z)$ ▷ **select component**
  $\mathbf{q}_d = \left(\beta_s \mathbf{1}_{1\times n} + \gamma_s\, \Omega_n^{\langle g\rangle}(z_s, :)\right) .* \mathbf{q}_d$ ▷ **update** q **with new choice** z_s
end for

Here we used the notations

$$\overline{\beta}_s = \prod_{j=1}^{s} \beta_j, \qquad \Omega_n = \left[\omega\left(\left\{\frac{k\,z}{n}\right\}\right)\right]_{\substack{z=0,\ldots,n-1\\ k=0,\ldots,n-1}}$$

and $\Omega_n^{\langle g\rangle}$ denotes the reordering of Ω_n w.r.t. a generator g for the cyclic group of $\mathbb{Z}_n$. For more details see [4, 5]. The symbols $.*$ and $./$ denote componentwise vector multiplication and division, respectively, and $\Omega_n^{\langle g\rangle}(j, :)$ stands for the j-th row of $\Omega_n^{\langle g\rangle}$. Note that the computation is slightly more expensive than the fast CBC algorithm since $\mathbf{q}$ has to be initialized and updated using z^0, but the computational complexity is still $O(d\, n \log(n))$.

3 Error Bounds for the SCS Algorithm

In this section we derive worst-case error bounds and show that the previously introduced successive coordinate search construction achieves optimal convergence rates in the respective function space. Here, we consider two of the most common function spaces in QMC theory, the weighted Korobov space and the weighted shift-averaged (anchored) Sobolev space.

3.1 The Weighted Korobov Space

Let $\boldsymbol{\gamma} = \{\gamma_j\}$ and $\boldsymbol{\beta} = \{\beta_j\}$ be two weight sequences. The reproducing kernel of the corresponding d-dimensional weighted Korobov space is then given by

$$K_{d,\boldsymbol{\gamma},\boldsymbol{\beta}}(\boldsymbol{x},\boldsymbol{y}) = \prod_{j=1}^{d} \left(\beta_j + \gamma_j \sum_{0 \neq h = -\infty}^{\infty} \frac{\mathrm{e}^{2\pi \mathrm{i}\, h(x_j - y_j)}}{r_\alpha(h)} \right),$$

where $\alpha > \frac{1}{2}$ is referred to as the smoothness parameter and we define

$$r_\alpha(\boldsymbol{h}) = \prod_{j=1}^{d} r_\alpha(h_j), \qquad r_\alpha(h) = \begin{cases} |h|^{2\alpha}, & \text{if}, h \neq 0, \\ 1, & \text{otherwise.} \end{cases}$$

For integer α the smoothness can be interpreted as the number of mixed partial derivatives $f^{(\tau_1,\ldots,\tau_d)}$ with $(\tau_1, \ldots, \tau_d) \leq (\alpha, \ldots, \alpha)$ that exist and are square-integrable. The space consists of functions which can be represented as absolutely summable Fourier series with norm

$$\|f\|^2_{K_{d,\boldsymbol{\gamma},\boldsymbol{\beta}}} = \sum_{\boldsymbol{h} \in \mathbb{Z}^d} |\hat{f}_{\boldsymbol{h}}|^2\, r_\alpha(\boldsymbol{h}) \prod_{\substack{j=1 \\ h_j \neq 0}}^{d} \gamma_j \prod_{\substack{j=1 \\ h_j = 0}}^{d} \beta_j,$$

where the $\hat{f}_{\boldsymbol{h}}$ denote the Fourier coefficients of f.

We prove that the successive coordinate search (SCS) algorithm achieves the optimal rate of convergence for multivariate integration in the weighted Korobov space for any initial vector. As is usual practice, we restrict ourselves to a prime number of points to simplify the needed proof techniques. We need the following lemma in the proof of the theorem.

Lemma 1 *For $s \in \{1, \ldots, d\}$, n prime, arbitrary integers z_j, $j \in \{1{:}d\} \setminus \{s\}$, and $r(\boldsymbol{h}) = \prod_{j=1}^{d} r(h_j)$ with $r(h) > 0$ such that for $h \neq 0$ we have $r(nh) \geq n^c\, r(h)$ for $c \geq 1$, then*

$$0 \leq \frac{1}{n} \sum_{z_s=0}^{n-1} \sum_{\substack{\boldsymbol{h} \in (\mathbb{Z}\setminus\{0\})^d \\ \boldsymbol{h}\cdot\boldsymbol{z} \equiv 0 \pmod{n}}} r^{-1}(\boldsymbol{h}) \leq \frac{2}{n} \sum_{\boldsymbol{h} \in (\mathbb{Z}\setminus\{0\})^d} r^{-1}(\boldsymbol{h}).$$

Proof The condition $\boldsymbol{h} \cdot \boldsymbol{z} \equiv 0 \pmod{n}$ can be written equivalently by

$$\frac{1}{n} \sum_{z_s=0}^{n-1} \sum_{\substack{\boldsymbol{h} \in (\mathbb{Z}\setminus\{0\})^d \\ \boldsymbol{h}\cdot\boldsymbol{z} \equiv 0 \pmod{n}}} r^{-1}(\boldsymbol{h}) = \frac{1}{n} \sum_{z_s=0}^{n-1} \sum_{\boldsymbol{h} \in (\mathbb{Z}\setminus\{0\})^d} r^{-1}(\boldsymbol{h}) \left[\frac{1}{n} \sum_{k=0}^{n-1} \prod_{j=1}^{d} \mathrm{e}^{2\pi \mathrm{i}\, k\, h_j z_j / n} \right].$$

Further, for $0 \le k < n$ and n prime

$$\frac{1}{n}\sum_{z=0}^{n-1} \mathrm{e}^{2\pi \mathrm{i} khz/n} = \begin{cases} 1, & \text{if } k = 0 \text{ or } h \equiv 0 \pmod{n}, \\ 0, & \text{otherwise,} \end{cases}$$

and hence for $h \in \mathbb{Z}$

$$\frac{1}{n}\sum_{k=0}^{n-1} \frac{1}{n}\sum_{z=0}^{n-1} \mathrm{e}^{2\pi \mathrm{i} khz/n} = \begin{cases} 1, & \text{if } h \equiv 0 \pmod{n}, \\ 1/n, & \text{if } h \not\equiv 0 \pmod{n}. \end{cases}$$

Thus

$$\begin{aligned}
&\sum_{\boldsymbol{h}\in(\mathbb{Z}\setminus\{0\})^d} r^{-1}(\boldsymbol{h}) \frac{1}{n}\sum_{k=0}^{n-1}\left(\prod_{s\neq j=1}^{d} \mathrm{e}^{2\pi \mathrm{i}\, k\, h_j z_j/n}\right)\frac{1}{n}\sum_{z_s=0}^{n-1} \mathrm{e}^{2\pi \mathrm{i}\, k\, h_s z_s/n} \\
&\le \sum_{\boldsymbol{h}\in(\mathbb{Z}\setminus\{0\})^d} r^{-1}(\boldsymbol{h}) \frac{1}{n}\sum_{k=0}^{n-1}\left(\prod_{s\neq j=1}^{d} \left|\mathrm{e}^{2\pi \mathrm{i}\, k\, h_j z_j/n}\right|\right)\left|\frac{1}{n}\sum_{z_s=0}^{n-1} \mathrm{e}^{2\pi \mathrm{i}\, k\, h_s z_s/n}\right| \\
&\le \sum_{\boldsymbol{h}\in(\mathbb{Z}\setminus\{0\})^d} r^{-1}(\boldsymbol{h}) \frac{1}{n}\sum_{k=0}^{n-1} \frac{1}{n}\sum_{z_s=0}^{n-1} \mathrm{e}^{2\pi \mathrm{i}\, k\, h_s z_s/n} \\
&= \sum_{\substack{\boldsymbol{h}\in(\mathbb{Z}\setminus\{0\})^d \\ h_s\equiv 0 \pmod{n}}} r^{-1}(\boldsymbol{h}) + \frac{1}{n} \sum_{\substack{\boldsymbol{h}\in(\mathbb{Z}\setminus\{0\})^d \\ h_s\not\equiv 0 \pmod{n}}} r^{-1}(\boldsymbol{h}) \\
&\le \frac{1}{n}\sum_{\boldsymbol{h}\in(\mathbb{Z}\setminus\{0\})^d} r^{-1}(\boldsymbol{h}) + \frac{1}{n} \sum_{\substack{\boldsymbol{h}\in(\mathbb{Z}\setminus\{0\})^d \\ h_s\not\equiv 0 \pmod{n}}} r^{-1}(\boldsymbol{h}) \le \frac{2}{n}\sum_{\boldsymbol{h}\in(\mathbb{Z}\setminus\{0\})^d} r^{-1}(\boldsymbol{h})\,.
\end{aligned}$$

Which completes the proof.

Theorem 2 *Let n be a prime number and $\boldsymbol{z}^0 = (z_1^0, \ldots, z_d^0) \in \mathbb{Z}_n^d$ be an arbitrary initial vector. Furthermore, denote by $\boldsymbol{z}^* = (z_1^*, \ldots, z_d^*) \in \mathbb{Z}_n^d$ the generating vector constructed by the successive coordinate search method with initial vector $\boldsymbol{z}^0$. Then the squared worst-case error $e_{n,d}^2(\boldsymbol{z}^*)$ in the Korobov space with kernel $K_{d,\boldsymbol{\gamma},\boldsymbol{\beta}}$ satisfies*

$$e_{n,d}^2(\boldsymbol{z}^*) \le C_{d,\lambda}\, n^{-\lambda} \quad \textit{for all} \quad 1 \le \lambda < 2\alpha\,,$$

where the constant $C_{d,\lambda}$ is given by

$$C_{d,\lambda} = 2^\lambda \left(\sum_{j=1}^{d} \frac{\gamma_j^{1/\lambda}}{\beta_j^{1/\lambda}}\right)^{\lambda} \prod_{j=1}^{d}\left(\beta_j^{1/\lambda} + \gamma_j^{1/\lambda}\mu_{\alpha,\lambda}\right)^{\lambda} \mu_{\alpha,\lambda}^{\lambda} \max_{s=1,\ldots,d}\left(1 + \frac{\gamma_s^{1/\lambda}}{\beta_s^{1/\lambda}}\mu_{\alpha,\lambda}\right)^{-\lambda}$$

with $\mu_{\alpha,\lambda} = 2\zeta(2\alpha/\lambda)$ *. Additionally, if the weights satisfy the summability condition*

$$\sum_{j=1}^{\infty} \frac{\gamma_j^{1/\lambda}}{\beta_j^{1/\lambda}} < \infty$$

then $C_{d,\lambda} \leq C_\lambda < \infty$*, and the constant* C_λ *is bounded independent of the dimension* d*. Hence, the worst-case error* $e_{n,d}(z^*)$ *is arbitrarily close to* $O(n^{-\alpha})$*, with the implied constant independent of* n*, and independent of* d *if the summability condition holds.*

Proof We use the notation from [4]: for a subset $\mathfrak{u} \subseteq \{1{:}d\} = \{1, \ldots, d\}$, we set $\mathbb{Z}_\mathfrak{u} = \{\boldsymbol{h} \in \mathbb{Z}^d : h_j \neq 0 \text{ for all } j \in \mathfrak{u} \text{ and } h_j = 0 \text{ for } j \notin \mathfrak{u}\}$, and define the dual lattice $L_\mathfrak{u}^\perp(z, n) = L_\mathfrak{u}^\perp(z_\mathfrak{u}, n) = \{\boldsymbol{h} \in \mathbb{Z}_\mathfrak{u} : \boldsymbol{h}_\mathfrak{u} \cdot z_\mathfrak{u} \equiv 0 \pmod{n}\}$ where we write $z_\mathfrak{u}$ and $\boldsymbol{h}_\mathfrak{u}$ to refer only to those components in z and $\boldsymbol{h}$. For $\boldsymbol{h} \in \mathbb{Z}_\mathfrak{u}$ we will write $\boldsymbol{h}_\mathfrak{u} \in \mathbb{Z}_\mathfrak{u}$ and $r_\alpha(\boldsymbol{h}_\mathfrak{u})$ to explicitly denote the dependence on the dimensions in $\mathfrak{u}$ only. We also write $\gamma_u = \prod_{j \in u} \gamma_j$ and set $\gamma_\emptyset = 1$. Now, without loss of generality, we consider the case where $\beta_j = 1$ for all j and correct the final expression afterwards, see Remark 1. Then from (1), or, see, e.g., [4, p. 5, Eq. (6)] with $q = 2$ and $\varphi(\boldsymbol{x}_k) = \mathrm{e}^{2\pi \mathrm{i}\, k\, \boldsymbol{h} \cdot z/n}$, we have

$$e_{n,d}^2(z) = \sum_{\boldsymbol{0} \neq \boldsymbol{h} \in \mathbb{Z}^d} \left| \frac{1}{n} \sum_{k=0}^{n-1} \mathrm{e}^{2\pi \mathrm{i}\, k\, \boldsymbol{h} \cdot z/n} \right|^2 r_\alpha^{-1}(\boldsymbol{h}) \prod_{\substack{j=1 \\ h_j \neq 0}}^{d} \gamma_j = \sum_{\emptyset \neq \mathfrak{u} \subseteq \{1:d\}} \gamma_\mathfrak{u} \sum_{\boldsymbol{h}_\mathfrak{u} \in L_\mathfrak{u}^\perp(z_\mathfrak{u}, n)} r_\alpha^{-1}(\boldsymbol{h}_\mathfrak{u}) \,.$$

Now define

$$g_\mathfrak{u}(z_\mathfrak{u}) = \gamma_\mathfrak{u} \sum_{\boldsymbol{h}_\mathfrak{u} \in L_\mathfrak{u}^\perp(z_\mathfrak{u}, n)} r_\alpha^{-1}(\boldsymbol{h}_\mathfrak{u}) \quad \text{and} \quad T_s(z_1, \ldots, z_s) = \sum_{s \in \mathfrak{u} \subseteq \{1:s\}} g_\mathfrak{u}(z_\mathfrak{u}) \,,$$

which gives

$$e_{n,d}^2(z) = \sum_{\emptyset \neq \mathfrak{u} \subseteq \{1:d\}} g_\mathfrak{u}(z_\mathfrak{u}) = \sum_{s=1}^{d} \sum_{s \in \mathfrak{u} \subseteq \{1:s\}} g_\mathfrak{u}(z_\mathfrak{u}) = \sum_{s=1}^{d} T_s(z_1, \ldots, z_s) \,.$$

Minimizing $e_{n,d}(z)$ over $z_s \in \mathbb{Z}_n$ is equivalent to minimizing only those parts which depend on z_s, resulting in the auxiliary target function

$$\theta_s(z) = \sum_{s \in \mathfrak{u} \subseteq \{1:d\}} g_\mathfrak{u}(z_\mathfrak{u}) = \sum_{s \in \mathfrak{u} \subseteq \{1:d\}} \gamma_\mathfrak{u} \sum_{\boldsymbol{h}_\mathfrak{u} \in L_\mathfrak{u}^\perp(z_\mathfrak{u}, n)} r_\alpha^{-1}(\boldsymbol{h}_\mathfrak{u}) \,.$$

We note that in the standard CBC proofs this quantity only depends on the dimensions up to s while here it depends on all d dimensions. Obviously

$$e_{n,d}^2(z) = \sum_{s=1}^{d} T_s(z_1,\ldots,z_s) = \sum_{s=1}^{d} \sum_{s\in\mathfrak{u}\subseteq\{1:s\}} g_{\mathfrak{u}}(z_{\mathfrak{u}}) \le \sum_{s=1}^{d} \sum_{s\in\mathfrak{u}\subseteq\{1:d\}} g_{\mathfrak{u}}(\tilde{z}_{\mathfrak{u}}) = \sum_{s=1}^{d} \theta_s(\tilde{z}),$$

where the tilde on top of z means that in replacing the sum over $s \in \mathfrak{u} \subseteq \{1\colon s\}$ by the sum over $s \in \mathfrak{u} \subseteq \{1\colon d\}$ we choose arbitrary z_j for $j > s$. We are free to do so since we are just adding positive quantities. Furthermore, for $1 \le \lambda < \infty$, using the so-called Jensen's inequality, we obtain

$$\begin{aligned}\left(e_{n,d}^2(z)\right)^{1/\lambda} &= \left(\sum_{s=1}^{d} \sum_{s\in\mathfrak{u}\subseteq\{1:s\}} g_{\mathfrak{u}}(z_{\mathfrak{u}})\right)^{1/\lambda} \le \left(\sum_{s=1}^{d} \sum_{s\in\mathfrak{u}\subseteq\{1:d\}} g_{\mathfrak{u}}(\tilde{z}_{\mathfrak{u}})\right)^{1/\lambda} \\ &= \left(\sum_{s=1}^{d} \theta_s(z_1,\ldots,z_{s-1},z_s,w_{s+1},\ldots,w_d)\right)^{1/\lambda} \\ &\le \sum_{s=1}^{d} \theta_s^{1/\lambda}(z_1,\ldots,z_{s-1},z_s,w_{s+1},\ldots,w_d),\end{aligned}$$

which holds for all choices of $\boldsymbol{w}$.

Since in minimizing $e_{n,d}^2(z_1^*,\ldots,z_{s-1}^*,z_s,z_{s+1}^0,\ldots,z_d^0)$ we are in fact minimizing $\theta_s(z_1^*,\ldots,z_{s-1}^*,z_s,z_{s+1}^0,\ldots,z_d^0)$ we now use the standard reasoning that the best choice $z_s = z_s^*$ makes θ_s at least as small as the average over all choices, and the same reasoning holds if we raise θ_s to the power $1/\lambda$. Therefore we obtain

$$\begin{aligned}\theta_s^{1/\lambda}(z_1^*,\ldots,z_{s-1}^*,z_s^*,z_{s+1}^0,\ldots,z_d^0) &\le \frac{1}{n}\sum_{z_s\in\mathbb{Z}_n} \theta_s^{1/\lambda}(z_1^*,\ldots,z_{s-1}^*,z_s,z_{s+1}^0,\ldots,z_d^0) \\ \le \sum_{s\in\mathfrak{u}\subseteq\{1:d\}} \gamma_{\mathfrak{u}}^{1/\lambda} \frac{1}{n} \sum_{z_s\in\mathbb{Z}_n} \sum_{\boldsymbol{h}_{\mathfrak{u}}\in L_{\mathfrak{u}}^{\perp}(\bar{z}_{\mathfrak{u}},n)} r_\alpha^{-1/\lambda}(\boldsymbol{h}_{\mathfrak{u}}) &\le \frac{2}{n} \sum_{s\in\mathfrak{u}\subseteq\{1:d\}} \gamma_{\mathfrak{u}}^{1/\lambda} \sum_{\boldsymbol{h}_{\mathfrak{u}}\in\mathbb{Z}_{\mathfrak{u}}} r_\alpha^{-1/\lambda}(\boldsymbol{h}_{\mathfrak{u}}),\end{aligned}$$

where we used Jensen's inequality to obtain the second line (and where $\bar{z}_{\mathfrak{u}}$ means we take $\bar{z} = (z_1^*,\ldots,z_{s-1}^*,z_s,z_{s+1}^0,\ldots,z_{s+1}^d)$) and Lemma 1, relabeling the set $\mathfrak{u}$ to be $\{1,\ldots,|\mathfrak{u}|\}$, $d = |\mathfrak{u}|$, and with $r(h) = |h|^{2\alpha/\lambda}$ and $c = 2\alpha/\lambda \ge 1$, to obtain the last line. For convenience we define

$$\mu_{\alpha,\lambda} = \sum_{0\ne h\in\mathbb{Z}} r_\alpha^{-1/\lambda}(h) = 2\sum_{h=1}^{\infty} h^{-2\alpha/\lambda} = 2\zeta(2\alpha/\lambda) < \infty$$

from which it follows that $2\alpha/\lambda > 1$ and we thus need $\lambda < 2\alpha$. Since for $\boldsymbol{h}_{\mathfrak{u}} \in \mathbb{Z}_{\mathfrak{u}}$ we have $r_\alpha^{-1/\lambda}(\boldsymbol{h}_{\mathfrak{u}}) = \prod_{j\in\mathfrak{u}} r_\alpha^{-1/\lambda}(h_j)$ we find $\sum_{\boldsymbol{h}_{\mathfrak{u}}\in\mathbb{Z}_{\mathfrak{u}}} r_\alpha^{-1/\lambda}(\boldsymbol{h}_{\mathfrak{u}}) = \mu_{\alpha,\lambda}^{|\mathfrak{u}|}$.

In each step of the SCS algorithm we now have a bound on $\theta_s^{1/\lambda}$ which we insert in our bound for the worst-case error, each time choosing the components of z^0 for $\boldsymbol{w}$, to obtain

$$
\begin{aligned}
\left(e_{n,d}^2(z^*)\right)^{1/\lambda} &\le \sum_{s=1}^{d} \theta_s^{1/\lambda}(z_1^*, \ldots, z_s^*, z_{s+1}^0, \ldots, z_d^0) \le \frac{2}{n} \sum_{s=1}^{d} \sum_{s \in \mathfrak{u} \subseteq \{1:d\}} \gamma_{\mathfrak{u}}^{1/\lambda} \mu_{\alpha,\lambda}^{|\mathfrak{u}|} \\
&= \frac{2}{n} \sum_{s=1}^{d} \left(\sum_{\mathfrak{u} \subseteq \{1:d\} \setminus \{s\}} \gamma_{\mathfrak{u}}^{1/\lambda} \mu_{\alpha,\lambda}^{|\mathfrak{u}|} \right) (\gamma_s^{1/\lambda} \mu_{\alpha,\lambda}) \\
&\le \frac{2}{n} \left(\sum_{s=1}^{d} \gamma_s^{1/\lambda} \right) \mu_{\alpha,\lambda} \max_{s=1,\ldots,d} \left(\sum_{\mathfrak{u} \subseteq \{1:d\} \setminus \{s\}} \gamma_{\mathfrak{u}}^{1/\lambda} \mu_{\alpha,\lambda}^{|\mathfrak{u}|} \right) \\
&= \frac{2}{n} \left(\sum_{s=1}^{d} \gamma_s^{1/\lambda} \right) \mu_{\alpha,\lambda} \max_{s=1,\ldots,d} \left(\prod_{s \neq j=1}^{d} \left(1 + \gamma_j^{1/\lambda} \mu_{\alpha,\lambda}\right) \right) \\
&= \frac{2}{n} \left(\sum_{s=1}^{d} \gamma_s^{1/\lambda} \right) \mu_{\alpha,\lambda} \max_{s=1,\ldots,d} \left(\frac{\prod_{j=1}^{d} \left(1 + \gamma_j^{1/\lambda} \mu_{\alpha,\lambda}\right)}{1 + \gamma_s^{1/\lambda} \mu_{\alpha,\lambda}} \right) \\
&= \frac{2}{n} \left(\sum_{s=1}^{d} \gamma_s^{1/\lambda} \right) \prod_{j=1}^{d} \left(1 + \gamma_j^{1/\lambda} \mu_{\alpha,\lambda}\right) \mu_{\alpha,\lambda} \max_{s=1,\ldots,d} \left(1 + \gamma_s^{1/\lambda} \mu_{\alpha,\lambda}\right)^{-1} .
\end{aligned}
$$

To show that the summability condition $\sum_{j=1}^{\infty} \gamma_j^{1/\lambda} < \infty$ gives a bound independent of d we note that

$$
\prod_{j=1}^{d} \left(1 + \gamma_j^{1/\lambda} \mu_{\alpha,\lambda}\right) < \infty \quad \text{if and only if} \quad \log \left(\prod_{j=1}^{d} \left(1 + \gamma_j^{1/\lambda} \mu_{\alpha,\lambda}\right) \right) < \infty .
$$

Now using that $\log(1 + x) \le x$ for $x > -1$, we find that

$$
\log \left(\prod_{j=1}^{d} \left(1 + \gamma_j^{1/\lambda} \mu_{\alpha,\lambda}\right) \right) = \sum_{j=1}^{d} \log \left(1 + \gamma_j^{1/\lambda} \mu_{\alpha,\lambda}\right) \le \mu_{\alpha,\lambda} \sum_{j=1}^{d} \gamma_j^{1/\lambda} \le \mu_{\alpha,\lambda} \sum_{j=1}^{\infty} \gamma_j^{1/\lambda},
$$

which implies the result.

3.2 The Weighted Sobolev Space

Again, let $\boldsymbol{\gamma} = \{\gamma_j\}$ and $\boldsymbol{\beta} = \{\beta_j\}$ be two weight sequences. There is a close relationship between the weighted Korobov space with smoothness parameter $\alpha = 1$ and the shift-averaged weighted Sobolev space. The shift-invariant kernel of the weighted Sobolev space with anchor $\boldsymbol{a} = (a_1, \ldots, a_d)$ of d-variate functions is given by

$$K^*_{d,\boldsymbol{\gamma},\boldsymbol{\beta}}(\boldsymbol{x},\boldsymbol{y}) = \prod_{j=1}^{d}\left(\hat{\beta}_j + \hat{\gamma}_j \sum_{0\neq h=-\infty}^{\infty} \frac{\mathrm{e}^{2\pi\mathrm{i}\,h(x_j-y_j)}}{|h|^2}\right),$$

where $\hat{\beta}_j = \beta_j + \gamma_j\left(a_j^2 - a_j + \frac{1}{3}\right)$ and $\hat{\gamma}_j = \frac{\gamma_j}{2\pi^2}$ for anchor values a_j. Furthermore, for $c_j = a_j^2 - a_j + \frac{1}{3}$ the shift-averaged squared worst-case error $\hat{e}^2_{n,d}(\boldsymbol{z})$ with generating vector $\boldsymbol{z}$ takes the following form

$$\begin{aligned}\hat{e}^2_{n,d}(\boldsymbol{z}) &= -\prod_{j=1}^{d}(\beta_j + \gamma_j c_j) + \frac{1}{n}\sum_{k=0}^{n-1}\prod_{j=1}^{d}\left(\beta_j + \gamma_j\left[B_2\left(\left\{\frac{k z_j}{n}\right\}\right) + c_j\right]\right)\\ &= -\prod_{j=1}^{d}\hat{\beta}_j + \frac{1}{n}\sum_{k=0}^{n-1}\prod_{j=1}^{d}\left(\hat{\beta}_j + \hat{\gamma}_j \sum_{0\neq h=-\infty}^{\infty}\frac{\mathrm{e}^{2\pi\mathrm{i}\,khz_j/n}}{h^2}\right).\end{aligned}$$

Additionally, the initial worst-case error in the weighted Sobolev space is given by

$$\hat{e}_{0,d}(0, K_{d,\boldsymbol{\gamma},\boldsymbol{\beta}}) = \prod_{j=1}^{d}\hat{\beta}_j^{1/2} = \prod_{j=1}^{d}\left(\beta_j + \gamma_j\left(a_j^2 - a_j + \frac{1}{3}\right)\right)^{1/2}.$$

Since these are precisely the worst-case error expressions as for the weighted Korobov space with $\alpha = 1$ and weights $\hat{\beta}_j$ and $\hat{\gamma}_j$, we obtain similar error bounds as before.

Theorem 3 *Let n be a prime number and $\boldsymbol{z}^0 = (z_1^0, \ldots, z_d^0) \in \mathbb{Z}_n^d$ be an arbitrary initial vector. Furthermore, denote by $\boldsymbol{z}^* = (z_1^*, \ldots, z_d^*) \in \mathbb{Z}_n^d$ the generating vector constructed by the successive coordinate search method with initial vector $\boldsymbol{z}^0$. Then the squared worst-case error $\hat{e}^2_{n,d}(\boldsymbol{z}^*)$ in the shift-averaged (anchored) Sobolev space with kernel $K^*_{d,\boldsymbol{\gamma},\boldsymbol{\beta}}$ satisfies*

$$\hat{e}^2_{n,d}(\boldsymbol{z}^*) \le \hat{C}_{d,\lambda}\, n^{-\lambda} \quad \textit{for all} \quad 1 \le \lambda < 2,$$

where the constant $\hat{C}_{d,\lambda}$ is given by the expression for $C_{d,\lambda}$ from Theorem 2 with $\alpha = 1$ and weights $\hat{\beta}_j = \beta_j + \gamma_j\left(a_j^2 - a_j + \frac{1}{3}\right)$ and $\hat{\gamma}_j = \frac{\gamma_j}{2\pi^2}$.

Additionally, if the weights satisfies the summability condition

$$\sum_{j=1}^{\infty}\frac{\hat{\gamma}_j^{1/\lambda}}{\hat{\beta}_j^{1/\lambda}} < \infty$$

then $\hat{C}_{d,\lambda} \le \hat{C}_\lambda < \infty$, and the constant C_λ is bounded independent of the dimension d. Hence, the worst-case error $\hat{e}_{n,d}(\boldsymbol{z}^*)$ can be taken arbitrarily close to $O(n^{-1})$, *with the implied constant independent of n, and independent of d if the summability condition holds.*

Proof The theorem follows directly from the previous result in Theorem 2. □

4 Numerical Results and Experiments

The idea regarding the SCS algorithm is to obtain generating vectors with smaller error values than obtained by the CBC algorithm, provided we choose a suitable initial vector $z^0 \in \mathbb{Z}_n^d$. The formulation of the algorithm suggests that the performance of the SCS construction strongly depends on the starting vector z^0 which we select beforehand. In this section we conduct some numerical experiments in the same setting as for the CBC algorithm in order to assess the performance of the SCS algorithm. For the experiments we prefer to have n distinct points in each dimension and so restrict our generating vector choices to exclude the choice $z_s = 0$ for the components of z for prime n, i.e., $z \in \{1, \ldots, n-1\}^d$. Allowing the choice $z_s = 0$ has effect on the results which depend on the weights since the CBC algorithm can now pick a zero component when the weights γ_j decay too slow.

4.1 Construction Methods

As we do not know how to best choose the initial vectors for the SCS algorithm, we propose to start from randomly selected initial vectors. This is different from the randomized CBC construction, see, e.g., [7], where in each minimization step the number of possible candidates z_s is restricted to r random integers in $\{1, \ldots, n-1\}$. We consider the following two methods.

1. Uniform random vectors + SCS algorithm: Choose q initial vectors $z^0 \in \mathbb{Z}_n^d$ at random, apply the fast SCS algorithm to them and then select the one with the smallest worst-case error $e_{n,d}(z)$.

2. Korobov-type generating vector + SCS algorithm: Take q randomly chosen Korobov-type generating vectors $z^0 = z(a) \equiv (a^0, a^1, \ldots, a^{d-1}) \pmod n$, with $a \in \{1, \ldots, n-1\}$, as initial vectors, apply the fast SCS algorithm to them and then select the one with the smallest worst-case error $e_{n,d}(z)$.

As the successive coordinate search algorithm has time complexity $O(d\,n\,\log(n))$, both proposed construction methods have time complexity $O(q\,d\,n\,\log(n))$.

Remark 2 The obvious candidate for the initial vector z^0 would of course be the generating vector constructed by the CBC method since by Proposition 1 one would construct z^1 such that $e_{n,d}(z^1) \le e_{n,d}(z^0)$. However, experiments show that in most cases the CBC vector is a fixed point with respect to the SCS method, i.e., applying the SCS algorithm to the CBC vector z^0 leaves the coordinates of z^0 unchanged. Thus, this approach yields usually no further improvement.

4.2 Exhaustive Search in Low Dimensions

In order to test the effectivity of our method we perform some numerical experiments in low dimensions and for a low number of points. Here, we can compute the best generating vector for the respective function space via an exhaustive search over the full set $\mathbb{Z}_n^d$ and then compare its worst-case error to the error values of the generating vectors obtained by our method.

For the weighted unanchored Sobolev space the squared worst-case error is given by

$$e_{n,d}^2(z) = -\prod_{j=1}^{d} \beta_j + \frac{1}{n} \sum_{k=0}^{n-1} \prod_{j=1}^{d} \left(\beta_j + \gamma_j \, B_2 \left(\left\{ \frac{k \, z_j}{n} \right\} \right) \right),$$

where $B_2(x) = x^2 - x + \frac{1}{6}$ denotes the Bernoulli polynomial of degree 2. Furthermore, z_{full} denotes the generating vector obtained by the full exhaustive search, z_{cbc} denotes the generating vector obtained via the CBC construction and z_{rand}^* and z_{kor}^* are the best generating vectors obtained out of $q = 100$ initial random choices by the above construction methods 1 and 2, respectively. For two different weight sequences γ_j and a selection of prime n we obtain the results in Tables 1 and 2, where $\gamma_j = (0.95)^j$ and $\gamma_j = (0.7)^j$, respectively. To be able to find the global minimum z_{full} over the whole set we limited the dimensionality to $d = 5$ and the number of points to $n \leq 199$. This leads to exhaustive searches over about 6 to 96 million possible choices for z, where we used the symmetry of the kernel and the fact that we only need to consider generating vectors with $z_1 = 1$ since multiplication by the multiplicative inverse of the first component normalizes any generating vector to have $z_1 = 1$.

Table 1 Weighted unanchored Sobolev space: $d = 5, \beta_j = 1, \gamma_j = (0.95)^j, q = 100$

n	$e_{n,d}(z_{\text{kor}}^*)$	$e_{n,d}(z_{\text{rand}}^*)$	$e_{n,d}(z_{\text{cbc}})$	$e_{n,d}(z_{\text{full}})$
101	2.6003e-02	2.6000e-02	2.6022e-02	2.6000e-02
127	2.1794e-02	2.1834e-02	2.2180e-02	2.1751e-02
151	1.8886e-02	1.8893e-02	1.9175e-02	1.8843e-02
181	1.5963e-02	1.5937e-02	1.6453e-02	1.5928e-02
199	1.4813e-02	1.4808e-02	1.5368e-02	1.4802e-02

Table 2 Weighted unanchored Sobolev space: $d = 5, \beta_j = 1, \gamma_j = (0.7)^j, q = 100$

n	$e_{n,d}(z^*_{\text{kor}})$	$e_{n,d}(z^*_{\text{rand}})$	$e_{n,d}(z_{\text{cbc}})$	$e_{n,d}(z_{\text{full}})$
101	1.0721e-02	1.0695e-02	1.0878e-02	1.0695e-02
127	8.7079e-03	8.6296e-03	8.6700e-03	8.6275e-03
151	7.4913e-03	7.4913e-03	7.5295e-03	7.4913e-03
181	6.2679e-03	6.2594e-03	6.3898e-03	6.2421e-03
199	5.7456e-03	5.7682e-03	5.8758e-03	5.7352e-03

The results in Tables 1 and 2 show that, even for a moderate value of q, the randomized SCS method generates lattice rules which have a smaller worst-case error than the one obtained via the CBC construction. Additionally, we see that our method generates worst-case errors that lie in the region of the smallest worst-case error $e_{n,d}(z_{\text{full}})$ and sometimes even constructs the best possible lattice rule. Although we only show two small tables here, similar results were observed for other test cases as well. In particular, we considered weight sequences of the form $\gamma_j = q^j$ with $0 < q < 1$ and $\gamma_j = j^{-k}$ with $k \in \{2, 3, 4\}$, for additional results see [2]. The experiments showed that the SCS algorithm outperforms the CBC construction when the decay of the weight sequence γ_j is slow.

4.3 Numerical Experiments for Higher Dimensions

In higher dimensions and/or for higher number of points it is not possible to perform an exhaustive search in order to obtain a reference value to measure the quality of the constructed generating vectors. Thus, we compare the outcome of the SCS method with the generating vector constructed by the CBC algorithm. Additionally, the empirical numerical results suggested that the use of Korobov-type initial vectors is to be preferred over uniform random vectors and we will therefore only consider Korobov-type initial vectors in this section. We denote by $\overline{e_{n,d}(z_{\text{kor}})}$ the average over the q random choices of the worst-case errors of the SCS constructed vectors z_{kor} and with $e_{n,d}(z^*_{\text{kor}})$ the best over the q random choices.

The numerical results presented in Tables 3 and 4 are for a Korobov space with dimension $d = 100$, $\alpha = 1$ and two different choices of weights, being $\beta_j = \frac{2}{3}$ and $\gamma_j = \frac{2}{3}(0.95)^j$, and $\beta_j = 1$ and $\gamma_j = (0.7)^j$, respectively, both with $q = 100$ random initial Korobov-type vectors. Our experiments show that the SCS method can construct good lattice rules for high dimensions and large n. For our choice of parameters, the SCS algorithm performs moderately better than the CBC construction when the weight sequence $\boldsymbol{\gamma} = \{\gamma_j\}_{j=1}^d$ is slowly decaying, as can be seen by comparing the relative difference between $e_{n,d}(z^*_{\text{kor}})$ and $e_{n,d}(z_{\text{cbc}})$ for the two different weight sequences in Tables 3 and 4. For a more extensive analysis of this behaviour we refer again to [2] where a wider range of weight sequences is considered.

Table 3 Weighted Korobov space: $d = 100, \alpha = 1, \beta_j = \frac{2}{3}, \gamma_j = \frac{2}{3}(0.95)^j, q = 100$

n	$\overline{e_{n,d}(z_{\text{kor}})}$	$e_{n,d}(z^*_{\text{kor}})$	$e_{n,d}(z_{\text{cbc}})$
1009	1.6554e-02	1.6221e-02	1.6566e-02
2003	1.1759e-02	1.1474e-02	1.1719e-02
4001	8.3025e-03	8.1204e-03	8.2869e-03
8009	5.8655e-03	5.7730e-03	5.8500e-03
32003	2.9320e-03	2.8874e-03	2.9301e-03

Table 4 Weighted Korobov space: $d = 100, \alpha = 1, \beta_j = 1, \gamma_j = (0.7)^j, q = 100$

n	$\overline{e_{n,d}(z_{\text{kor}})}$	$e_{n,d}(z^*_{\text{kor}})$	$e_{n,d}(z_{\text{cbc}})$
1009	3.1185e-01	3.0834e-01	3.0931e-01
2003	2.0902e-01	2.0661e-01	2.0708e-01
4001	1.3894e-01	1.3713e-01	1.3658e-01
8009	9.1757e-02	9.0445e-02	8.9611e-02
32003	3.9467e-02	3.8763e-02	3.8528e-02

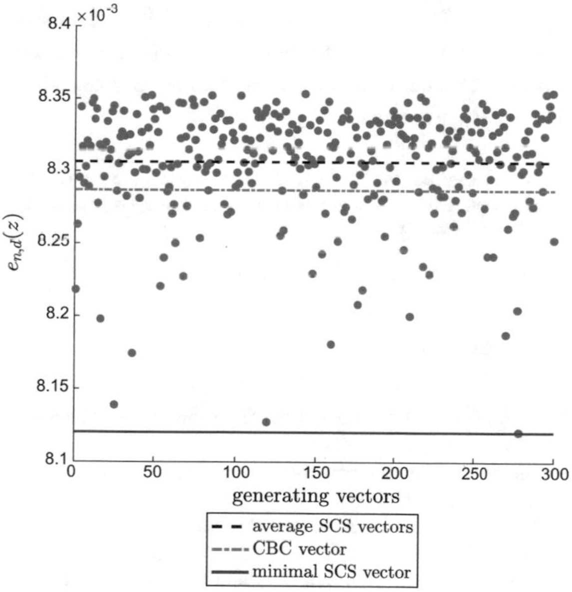

Fig. 1 Numerical results of the SCS method in the weighted Korobov space with $d = 100$, $\alpha = 1$, $\beta_j = \frac{2}{3}, \gamma_j = \frac{2}{3}(0.95)^j$ where $n = 4001$ and $q = 300$ in comparison to the CBC algorithm

Figure 1 illustrates the performance of the SCS method compared to the CBC method. The blue dots represent the worst-case error values of lattice rules with $n = 4001$ points constructed by the SCS method with $q = 300$ Korobov-type initial vectors. The minimal error amongst the constructed lattice rules and the average over the q random seed choices is indicated by the red or black line, respectively.

The error corresponding to the CBC method is indicated by the green line. From the figure it becomes evident that the CBC algorithm outperforms the average of the SCS algorithm applied to randomly selected Korobov-type rules, but the best SCS results clearly win over the generating vector constructed by the CBC method.

5 Conclusion

The results and experiments in the previous section, see [2] for additional results, showed that it is possible to use the successive coordinate search algorithm to construct good generating vectors for rank-1 lattice rules. They also confirmed that randomized methods based on the SCS construction can provide generating vectors with smaller worst-case errors than the CBC vector. However, the computational cost of the SCS method can be several times higher while the gained improvement depends on the weight sequence $\boldsymbol{\gamma}$. Future research could help to find a selection criterion for the starting vector z^0 in order to reduce the construction cost of the SCS algorithm. The SCS algorithm should further be regarded as a generalization of the existing component-by-component construction rather than a completely new algorithm. Due to the formulation of the successive coordinate search method it can also be used to improve existing lattice rules. Numerical experiments show that the improvements of the SCS method are higher when the decay of the weights $\boldsymbol{\gamma} = \{\gamma_j\}_{j=1}^d$ is slow.

Acknowledgements We thank Peter Kritzer for some useful comments and discussions about the manuscript and we acknowledge financial support from the KU Leuven research fund (OT:3E130287 and C3:3E150478).

References

1. Dick, J., Kuo, F.Y., Sloan, I.H.: High-dimensional integration: the quasi-Monte Carlo way. Acta Numer. **22**, 133–288 (2013)
2. Ebert, A.: The component-by-component construction in weighted reproducing kernel Hilbert spaces - an optimization approach. Available on the document repository Lirias of the KU Leuven, Master's thesis at the Humboldt University of Berlin (2015)
3. Kuo, F.Y.: Component-by-component constructions achieve the optimal rate of convergence for multivariate integration in weighted Korobov and Sobolev spaces. J Complex. **19**(3), 301–320 (2003)
4. Nuyens, D.: The construction of good lattice rules and polynomial lattice rules. In: Kritzer, P., Niederreiter, H., Pillichshammer, F., Winterhof, A. (eds.) Uniform Distribution and Quasi-Monte Carlo Methods: Discrepancy, Integration and Applications, vol. 15, Radon Series on Computational and Applied Mathematics, pp. 223–256. De Gruyter (2014)
5. Nuyens, D., Cools, R.: Fast algorithms for component-by-component construction of rank-1 lattice rules in shift-invariant reproducing kernel Hilbert spaces. Math. Comput. **75**, 903–920 (2006)

6. Nuyens, D., Cools, R.: Fast component-by-component construction of rank-1 lattice rules with a non-prime number of points. J. Complex. **22**(1), 4–28 (2006)
7. Sinescu, V., L'Ecuyer, P.: On the behavior of weighted star discrepancy bounds for shifted lattice rules. In: L'Ecuyer, P., Owen, A.B (eds.) Monte Carlo and Quasi-Monte Carlo Methods 2008, pp. 603–616 (2009)
8. Sloan, I.H., Woźniakowski, H.: When are quasi-Monte Carlo algorithms efficient for high dimensional integrals? J. Complex. **14**, 1–33 (1998)
9. Sloan, I.H., Woźniakowski, H.: Tractability of multivariate integration for weighted Korobov classes. J. Complex. **17**, 697–721 (2001)

Adaptive Euler–Maruyama Method for SDEs with Non-globally Lipschitz Drift

Wei Fang and Michael B. Giles

Abstract This paper, based on two main papers Fang and Giles (Adaptive Euler–Maruyama method for SDEs with non-globally Lipschitz drift: Part I, finite time interval, 2016, [2]), Fang and Giles (Adaptive Euler–Maruyama method for SDEs with non-globally Lipschitz drift: Part II, infinite time interval, 2017, [3]) which contains the full details of the literature review, numerical analysis and numerical experiments, aims to give an overview of the adaptive Euler–Maruyama method for SDEs with non-globally Lipschitz drift in a concise structure without any proof. It shows that if the timestep is bounded appropriately, then over a finite time interval the numerical approximation is stable, and the expected number of timesteps is finite. Furthermore, the order of strong convergence is the same as usual, i.e. order $\frac{1}{2}$ for SDEs with a non-uniform globally Lipschitz volatility, and order 1 for Langevin SDEs with unit volatility and a drift with sufficient smoothness. For a class of ergodic SDEs, we also show that the bound for the moments and the strong error of the numerical solution are uniform in T, which allow us to introduce the adaptive multilevel Monte Carlo method to compute the expectations with respect to the invariant measure. The analysis is supported by numerical experiments.

Keywords SDE · Euler–Maruyama · Strong convergence · Adaptive timestep · Ergodicity · MLMC

1 Introduction

In this paper we consider an m-dimensional stochastic differential equation (SDE) driven by a d-dimensional Brownian motion:

W. Fang (✉) · M. B. Giles
Mathematical Institute, University of Oxford, Oxford OX2 6GG, UK
e-mail: wei.fang@maths.ox.ac.uk

M. B. Giles
e-mail: mike.giles@maths.ox.ac.uk

A. B. Owen and P. W. Glynn (eds.), *Monte Carlo and Quasi-Monte Carlo Methods*, Springer Proceedings in Mathematics & Statistics 241,
https://doi.org/10.1007/978-3-319-91436-7_11

$$\mathrm{d}X_t = f(X_t)\,\mathrm{d}t + g(X_t)\,\mathrm{d}W_t, \tag{1}$$

with a fixed initial value x_0. The standard theory assumes the drift $f : \mathbb{R}^m \to \mathbb{R}^m$ and the volatility $g : \mathbb{R}^m \to \mathbb{R}^{m\times d}$ are both globally Lipschitz. Under this assumption, there is well-established theory on the existence and uniqueness of strong solutions, and the numerical approximation $\widehat{X}_t$ obtained from the Euler–Maruyama discretization

$$\widehat{X}_{(n+1)h} = \widehat{X}_{nh} + f(\widehat{X}_{nh})\,h + g(\widehat{X}_{nh})\,\Delta W_n$$

using a uniform timestep of size h with Brownian increments ΔW_n, plus a suitable interpolation within each timestep, is known [16] to have a strong error which is $O(h^{1/2})$ so that for any $T, p > 0$,

$$\mathbb{E}\left[\sup_{0\le t\le T} \|\widehat{X}_t - X_t\|^p\right] = O(h^{p/2}).$$

The interest in this paper is in other cases in which g is again globally Lipschitz, but f is only locally Lipschitz. If, for some $\alpha, \beta \ge 0$, f also satisfies the one-sided growth condition

$$\langle x, f(x)\rangle \le \alpha\|x\|^2 + \beta,$$

where $\langle\cdot,\cdot\rangle$ denotes an inner product, then it is again possible to prove the existence and uniqueness of strong solutions (see Theorems 2.3.5 and 2.4.1 in [20]). Furthermore (see Lemma 3.2 in [10]), these solutions are stable in the sense that for any $T, p > 0$, $\mathbb{E}\left[\sup_{0\le t\le T} \|X_t\|^p\right] < \infty$. The problem is that the numerical approximation given by the uniform timestep Euler–Maruyama discretization may not be stable. Indeed, for the SDE

$$\mathrm{d}X_t = -X_t^3\,\mathrm{d}t + \mathrm{d}W_t, \tag{2}$$

it has been proved [13] that for any $T > 0$ and $p \ge 2$, $\lim_{h\to 0} \mathbb{E}\left[\|\widehat{X}_T\|^p\right] = \infty$.

This behaviour has led to research on numerical methods which achieve strong convergence for these SDEs with a non-globally Lipschitz drift, see [2, 10, 12, 14, 21, 22, 26, 32] and the references therein.

The other motivation for this paper is the analysis of a class of ergodic SDEs which exponentially converge to some invariant measure π, for example, the FENE model in [1]. Evaluating the expectation of some function $\varphi(x)$ with respect to that invariant measure π is of great interest in mathematical biology, physics and Bayesian inference in statistics:

$$\pi(\varphi) \triangleq \int \varphi(x)\,\mathrm{d}\pi(x) = \lim_{t\to\infty} \mathbb{E}\left[\varphi(X_t)\right],$$

which drives us to consider the stability and strong convergence of the algorithm in the infinite time interval. Different approaches to computing the expectation include numerical solution of the Fokker–Planck equation, see [30] and the reference therein, and ergodic numerical solutions, see [9, 19, 23, 25, 27, 29, 31]. We assume that the SDEs have a locally Lipschitz drift $f:\mathbb{R}^m\to\mathbb{R}^m$ satisfying the dissipative condition: for some $\alpha,\beta>0$,

$$\langle x, f(x)\rangle \le -\alpha\|x\|^2+\beta, \tag{3}$$

and a bounded and non-degenerate volatility $g:\mathbb{R}^m\to\mathbb{R}^{m\times d}$.

In this paper, we propose instead to use the standard explicit Euler–Maruyama method, but with an adaptive timestep h_n which is a function of the current approximate solution $\widehat{X}_{t_n}$. Adaptive timesteps have been used in previous research to improve the accuracy of numerical approximations, see [4, 11, 15, 17–19, 24, 28] and the references therein. The idea of using an adaptive timestep in this paper comes from considering the divergence of the uniform timestep method for the SDE (2). When there is no noise, the requirement for the explicit Euler approximation of the corresponding ODE to have a stable monotonic decay is that its timestep satisfies $h<\widehat{X}_{t_n}^{-2}$. An intuitive explanation for the instability of the uniform timestep Euler–Maruyama approximation of the SDE is that there is always a very small probability of a large Brownian increment ΔW_n which pushes the approximation $\widehat{X}_{t_{n+1}}$ into the region $h>2\,\widehat{X}_{t_{n+1}}^{-2}$ leading to an oscillatory super-exponential growth. Using an adaptive timestep avoids this problem.

For the ergodic SDEs, by setting a suitable condition for h, we can show that, instead of an exponential bound, the numerical solution has a uniform bound with respect to T for both moments and the strong error. Then, multi-level Monte Carlo (MLMC) methodology [5, 6] is employed and non-nested timestepping is used to construct an adaptive MLMC [7]. Following the idea of Glynn and Rhee [8] to estimate the invariant measure of some Markov chains, we introduce an adaptive MLMC algorithm for the infinite time interval, in which each level ℓ has a different time interval length T_ℓ, to achieve a better computational performance.

The rest of the paper is organized as follows. The adaptive algorithm is presented and the main theorems both in finite time interval and infinite time interval are stated in Sect. 2. Section 3 introduces the MLMC schemes, and the relevant numerical experiments are provided in Sect. 4. Finally, Sect. 5 concludes.

In this paper we consider both the finite time interval $[0,T]$ with $T>0$ be a fixed positive real number and the infinite time interval $[0,\infty)$. Let $(\Omega,\mathscr{F},\mathbb{P})$ be a probability space with normal filtration $(\mathscr{F}_t)_{t\in[0,\infty)}$ for Sect. 2 and $(\mathscr{F}_t)_{t\in(-\infty,0]}$ for Sect. 3 corresponding to a d-dimensional standard Brownian motion $W_t=(W^{(1)},W^{(2)},\ldots,W^{(d)})^T_t$. We denote the vector norm by $\|v\|\triangleq(|v_1|^2+|v_2|^2+\cdots+|v_m|^2)^{\frac{1}{2}}$, the inner product of vectors v and w by $\langle v,w\rangle\triangleq v_1w_1+v_2w_2+\cdots+v_mw_m$, for any $v,w\in\mathbb{R}^m$ and the Frobenius matrix norm by $\|A\|\triangleq\sqrt{\sum_{i,j}A_{i,j}^2}$ for all $A\in\mathbb{R}^{m\times d}$.

2 Adaptive Algorithm and Theoretical Results

2.1 Adaptive Euler–Maruyama Method

The adaptive Euler–Maruyama discretization is

$$t_{n+1} = t_n + h_n, \quad \widehat{X}_{t_{n+1}} = \widehat{X}_{t_n} + f(\widehat{X}_{t_n})\, h_n + g(\widehat{X}_{t_n})\, \Delta W_n,$$

where $h_n \triangleq h(\widehat{X}_{t_n})$ and $\Delta W_n \triangleq W_{t_{n+1}} - W_{t_n}$, and there is fixed initial data $t_0 = 0$, $\widehat{X}_0 = x_0$.

One key point in the analysis is to prove that t_n increases without bound as n increases. More specifically, the analysis proves that for any $T > 0$, almost surely for each path there is an N such that $t_N \geq T$.

We use the notation $\underline{t} \triangleq \max\{t_n : t_n \leq t\}$, $n_t \triangleq \max\{n : t_n \leq t\}$ for the nearest time point before time t, and its index.

We define the piecewise constant interpolant process $\bar{X}_t = \widehat{X}_{\underline{t}}$ and also define the standard continuous interpolant [16] as

$$\widehat{X}_t = \widehat{X}_{\underline{t}} + f(\widehat{X}_{\underline{t}})(t - \underline{t}) + g(\widehat{X}_{\underline{t}})(W_t - W_{\underline{t}}),$$

so that $\widehat{X}_t$ is the solution of the SDE

$$\mathrm{d}\widehat{X}_t = f(\widehat{X}_{\underline{t}})\, \mathrm{d}t + g(\widehat{X}_{\underline{t}})\, \mathrm{d}W_t = f(\bar{X}_t)\, \mathrm{d}t + g(\bar{X}_t)\, \mathrm{d}W_t. \tag{4}$$

In the following two subsections, we state the key results on stability and strong convergence in both finite and infinite time intervals, and related results on the number of timesteps, introducing various assumptions as required for each. All the proofs are in [2, 3].

2.2 Finite Time Interval

2.2.1 Stability

Assumption 1 *(Local Lipschitz and linear growth)* f and g are both locally Lipschitz, so that for any $R > 0$ there is a constant C_R such that

$$\|f(x) - f(y)\| + \|g(x) - g(y)\| \leq C_R\, \|x - y\|$$

for all $x, y \in \mathbb{R}^m$ with $\|x\|, \|y\| \leq R$. Furthermore, there exist constants $\alpha, \beta \geq 0$ such that for all $x \in \mathbb{R}^m$, f satisfies the one-sided linear growth condition:

$$\langle x, f(x)\rangle \leq \alpha \|x\|^2 + \beta, \tag{5}$$

and g satisfies the linear growth condition:

$$\|g(x)\|^2 \leq \alpha \|x\|^2 + \beta. \tag{6}$$

Together, (5) and (6) imply the monotone condition $\langle x, f(x)\rangle + \frac{1}{2}\|g(x)\|^2 \leq \frac{3}{2}(\alpha\|x\|^2+\beta)$, which is a key assumption in the analysis of Mao and Szpruch [22] and Mao [21] for SDEs with volatilities which are not globally Lipschitz. However, in our analysis we choose to use this slightly stronger assumption, which provides the basis for the following lemma on the stability of the SDE solution.

Lemma 1 (SDE stability) *If the SDE satisfies Assumption 1, then for all $p>0$*

$$\mathbb{E}\left[\sup_{0\leq t\leq T} \|X_t\|^p\right] < \infty.$$

We now specify the critical assumption about the adaptive timestep.

Assumption 2 *(Adaptive timestep)* The adaptive timestep function $h : \mathbb{R}^m \to \mathbb{R}^+$ is continuous and strictly positive, and there exist constants $\alpha, \beta > 0$ such that for all $x \in \mathbb{R}^m$, $h(x)$ satisfies the inequality

$$\langle x, f(x)\rangle + \tfrac{1}{2}\, h(x)\, \|f(x)\|^2 \leq \alpha \|x\|^2 + \beta. \tag{7}$$

Note that if another timestep function $h^\delta(x)$ is smaller than $h(x)$, then $h^\delta(x)$ also satisfies the Assumption 2. Note also that the form of (7), which is motivated by the requirements of the proof of the next theorem, is very similar to (5). Indeed, if (7) is satisfied then (5) is also true for the same values of α and β.

Theorem 1 (Finite time stability) *If the SDE satisfies Assumption 1, and the timestep function h satisfies Assumption 2, then T is almost surely attainable (i.e. for $\omega \in \Omega$, $\mathbb{P}(\exists N(\omega) < \infty$ s.t. $t_{N(\omega)} \geq T) = 1$) and for all $p>0$ there exists a constant $C_{p,T}$ which depends solely on p, T and the constants α, β in Assumption 2, such that*

$$\mathbb{E}\left[\sup_{0\leq t\leq T} \|\widehat{X}_t\|^p\right] < C_{p,T}.$$

2.2.2 Strong Convergence

Standard strong convergence analysis for an approximation with a uniform timestep h considers the limit $h \to 0$. This clearly needs to be modified when using an adaptive timestep, and we will instead consider a timestep function $h^\delta(x)$ controlled by a scalar parameter $0 < \delta \leq 1$, and consider the limit $\delta \to 0$.

Given a timestep function $h(x)$ which satisfies Assumption 2, ensuring stability as analysed in the previous section, there are two quite natural ways in which we might introduce δ to define $h^\delta(x)$:

$$h^\delta(x) = \delta \, \min(T, h(x)), \quad h^\delta(x) = \min(\delta\, T, h(x)).$$

The first refines the timestep everywhere, while the latter concentrates the computational effort on reducing the maximum timestep, with $h(x)$ introduced to ensure stability when $\|\widehat{X}_t\|$ is large.

In our analysis, we will cover both possibilities by making the following assumption.

Assumption 3 The timestep function h^δ, satisfies the inequality

$$\delta \, \min(T, h(x)) \leq h^\delta(x) \leq \min(\delta\, T, h(x)), \tag{8}$$

and h satisfies Assumption 2.

Given this assumption, we obtain the following theorem:

Theorem 2 (Strong convergence) *If the SDE satisfies Assumption 1, and the timestep function h^δ satisfies Assumption 3, then for all $p>0$*

$$\lim_{\delta \to 0} \mathbb{E}\left[\sup_{0 \leq t \leq T} \|\widehat{X}_t - X_t\|^p \right] = 0.$$

To prove an order of strong convergence requires new assumptions on f and g:

Assumption 4 *(Lipschitz properties)* There exists a constant $\alpha > 0$ such that for all $x, y \in \mathbb{R}^m$, f satisfies the one-sided Lipschitz condition:

$$\langle x-y, f(x)-f(y) \rangle \leq \tfrac{1}{2}\alpha \|x-y\|^2, \tag{9}$$

and g satisfies the Lipschitz condition:

$$\|g(x)-g(y)\|^2 \leq \tfrac{1}{2}\alpha \|x-y\|^2. \tag{10}$$

In addition, f satisfies the polynomial growth Lipschitz condition

$$\|f(x)-f(y)\| \leq \big(\gamma \, (\|x\|^q + \|y\|^q) + \mu\big) \, \|x-y\|, \tag{11}$$

for some $\gamma, \mu, q > 0$.

Note that setting $y=0$ gives

$$\langle x, f(x) \rangle \;\leq\; \tfrac{1}{2}\alpha \|x\|^2 + \langle x, f(0) \rangle \;\leq\; \alpha \|x\|^2 + \tfrac{1}{2}\alpha^{-1} \|f(0)\|^2,$$

$$\|g(x)\|^2 \;\leq\; 2\|g(x)-g(0)\|^2 + 2\|g(0)\|^2 \;\leq\; \alpha \|x\|^2 + 2\|g(0)\|^2.$$

Hence, Assumption 4 implies Assumption 1, with the same α and an appropriate β.

Theorem 3 (Strong convergence order) *If the SDE satisfies Assumption 4, and the timestep function h^δ satisfies Assumption 3, then for all $p>0$ there exists a constant $C_{p,T}$ such that*

$$\mathbb{E}\left[\sup_{0\le t\le T} \|\widehat{X}_t - X_t\|^p\right] \le C_{p,T}\,\delta^{p/2}.$$

To bound the expected number of timesteps, we require an assumption on how quickly $h(x)$ can approach zero as $\|x\| \to \infty$.

Assumption 5 *(Timestep lower bound)* There exist constants ξ, ζ, $q>0$, such that the adaptive timestep function satisfies the inequality

$$h(x) \ge \left(\xi\,\|x\|^q + \zeta\right)^{-1}.$$

Lemma 2 (Number of timesteps) *If the SDE satisfies Assumption 1, and the timestep function $h^\delta(x)$ satisfies Assumption 3, with $h(x)$ satisfying Assumptions 2 and Assumption 5, then for all $p>0$ there exists a constant $c_{p,T}$ such that*

$$\mathbb{E}\left[(N_T - 1)^p\right] \le c_{p,T}\,\delta^{-p}.$$

where N_T is again the number of timesteps required by a path approximation.

The conclusion from Theorem 3 and Lemma 2 is that

$$\mathbb{E}\left[\sup_{0\le t\le T} \|\widehat{X}_t - X_t\|^p\right]^{1/p} \le C_{p,T}^{1/p}\, c_{1,T}^{1/2}\, (\mathbb{E}\,[N_T])^{-1/2},$$

which corresponds to order $\frac{1}{2}$ strong convergence when comparing the accuracy to the expected cost.

First order strong convergence is achievable for Langevin SDEs in which $m=d$ and g is the identity matrix I_m, but this requires stronger assumptions on the drift f.

Assumption 6 *(Enhanced Lipschitz properties)* f satisfies the Assumption 4 and in addition, f is differentiable, and f and ∇f satisfy the polynomial growth Lipschitz condition

$$\|f(x)-f(y)\| + \|\nabla f(x)-\nabla f(y)\| \le \left(\gamma\,(\|x\|^q+\|y\|^q) + \mu\right)\|x-y\|, \tag{12}$$

for some γ, μ, $q > 0$.

We now state the theorem on improved strong convergence.

Theorem 4 (Strong convergence for Langevin SDEs) *If $m=d$, $g \equiv I_m$, f satisfies Assumption 6, and the timestep function h^δ satisfies Assumption 3, then for all T, $p \in (0,\infty)$ there exists a constant $C_{p,T}$ such that*

$$\mathbb{E}\left[\sup_{0\le t\le T} \|\widehat{X}_t - X_t\|^p\right] \le C_{p,T}\,\delta^p.$$

Comment: first order strong convergence can also be achieved for a general $g(x)$ by using an adaptive timestep Milstein discretization, provided ∇g satisfies an additional Lipschitz condition. However, this numerical approach is only practical in cases in which the commutativity condition is satisfied and therefore there is no need to simulate the Lévy areas which the Milstein method otherwise requires [16].

2.3 *Infinite Time Interval*

Now, we focus on a class of ergodic SDEs and show that the moment bounds and strong error bound is uniform in T which is a stronger result than for the finite time interval.

2.3.1 Stability

Assumption 7 *(Dissipative condition)* f and g satisfy the Assumption 1 and there exist constants $\alpha, \beta > 0$ such that for all $x \in \mathbb{R}^m$, f satisfies the dissipative one-sided linear growth condition:

$$\langle x, f(x)\rangle \le -\alpha\|x\|^2 + \beta, \tag{13}$$

and g is globally bounded and non-degenerate:

$$\|g(x)\|^2 \le \beta. \tag{14}$$

Theorem 4.4 in [23] and Theorem 6.1 in [25] show that this Assumption ensures the existence and uniqueness of the invariant measure. We can also prove the following uniform moment bound for the SDE solution.

Lemma 3 (SDE stability in infinite time interval) *If the SDE satisfies Assumption 7 with $X_0 = x_0$, then for all $p \in (0, \infty)$, there is a constant C_p which only depends on x_0 and p such that, $\forall\, t \ge 0$,*

$$\mathbb{E}\left[\|X_t\|^p\right] \le C_p.$$

We now specify the critical assumption about the adaptive timestep for infinite time interval.

Assumption 8 *(Adaptive timestep for infinite time interval)* The adaptive timestep function $h : \mathbb{R}^m \to (0, h_{max}]$ is continuous and bounded, with $0 < h_{max} < \infty$, and there exist constants $\alpha, \beta > 0$ such that for all $x \in \mathbb{R}^m$, h satisfies the inequality

$$\langle x, f(x)\rangle + \tfrac{1}{2}\, h(x)\, \|f(x)\|^2 \le -\alpha \|x\|^2 + \beta. \tag{15}$$

Note that if another timestep function $h^\delta(x)$ is smaller than $h(x)$, then $h^\delta(x)$ also satisfies this Assumption. Note also that the form of (15), which is motivated by the requirements of the proof of the next theorem, is very similar to (7). Indeed, if (15) is satisfied then (7) is also true for the same values of α and β. Compared with the condition in the finite time analysis, we need additionally to bound h properly to achieve the uniform bound.

Theorem 5 (Stability in infinite interval) *If the SDE satisfies Assumption 7, and the timestep function h satisfies Assumption 8, then for all $p \in (0, \infty)$ there exists a constant C_p which depends solely on p, x_0, h_{max} and the constants α, β in Assumption 8 such that, $\forall t \ge 0$,*

$$\mathbb{E}\left[\|\widehat{X}_t\|^p\right] < C_p, \;\; \mathbb{E}\left[\|\bar{X}_t\|^p\right] < C_p.$$

2.3.2 Strong Convergence

To prove an order of strong convergence requires new assumptions on f and g:

Assumption 9 *(Contractive Lipschitz properties)* f and g satisfy Assumption 4 and for some fixed $p^* \in (1, \infty)$, there exist constants $\lambda > 0$ such that for all $x, y \in \mathbb{R}^m$, f and g satisfy the contractive Lipschitz condition:

$$\langle x-y,\, f(x)-f(y)\rangle + \frac{p^*-1}{2}\, \|g(x)-g(y)\|^2 \le -\lambda\, \|x-y\|^2, \tag{16}$$

Note that this Assumption ensures that two solutions to this SDE starting from different places but driven by the same Brownian increment, will come together exponentially, as shown in the following lemma.

Lemma 4 (SDE contractivity) *If the SDE satisfies Assumption 9 and for some fixed $p^* \in (1, \infty)$, then for $p \in (0, p^*]$ any two solutions to the SDE: X_t and Y_t, driven by the same Brownian motion but starting from x_0 and y_0, where $x_0 \neq y_0$, satisfy that, $\forall\, t > 0$,*

$$\mathbb{E}\left[\|X_t - Y_t\|^p\right] \le e^{-\lambda p t}\, \mathbb{E}\left[\|X_0 - Y_0\|^p\right].$$

This lemma means the error made on previous time steps will decay exponentially and then we can prove a uniform bound for the strong error.

Theorem 6 (Strong convergence order in infinite time interval) *If the SDE satisfies Assumption 9, and the timestep function h^δ satisfies Assumption 3 with h satisfying Assumption 8, then for all $p \in (0, p^*]$ there exists a constant C_p such that, $\forall t \ge 0$,*

$$\mathbb{E}\left[\, \|\widehat{X}_t - X_t\|^p\right] \le C_p\, \delta^{p/2}.$$

For the infinite time interval, we can show that the expected number of timesteps per path is linear in T, which is the same as for uniform timesteps.

Lemma 5 (Number of timesteps) *If the SDE satisfies Assumption 9, and the timestep function h^δ satisfies Assumption 3, with $h(x)$ satisfying Assumption 5 and Assumption 8, then for all T, $p \in (0, \infty)$ there exists a constant c_p such that*

$$\mathbb{E}\left[(N_T - 1)^p\right] \le c_p \, T^p \, \delta^{-p}.$$

where N_T is again the number of timesteps required by a path approximation.

First order strong convergence is also achievable for Langevin SDEs in which $m = d$ and g is the identity matrix I_m, but this requires stronger assumptions on the drift f.

Assumption 10 *(Enhanced contractive Lipschitz properties)* f satisfies Assumption 9 and in addition, f is differentiable, and f and ∇f satisfy the polynomial growth Lipschitz condition (12).

Theorem 7 (Strong convergence for Langevin SDEs in infinite time interval) *If $m = d$, $g \equiv I_m$, f satisfies Assumption 10, and the timestep function h^δ satisfies Assumptions 3 and 8, then for all $p \in (0, \infty)$ there exists a constant C_p such that, $\forall\, t \ge 0$,*

$$\mathbb{E}\left[\|\widehat{X}_t - X_t\|^p\right] \le C_p \, \delta^p.$$

3 Multi-level Monte Carlo in Infinite Time Interval

We are interested in the problem of approximating:

$$\pi(\varphi) := \mathbb{E}_\pi \varphi = \int_{\mathbb{R}^m} \varphi(x) \pi(\mathrm{d}x), \quad \varphi \in L^1(\pi),$$

where π is the invariant measure of the SDE (1). Numerically, we can approximate this quantity by simulating $\mathbb{E}\left[\varphi(X_T)\right]$ for a sufficiently large T. In the following subsections, we will introduce our adaptive multilevel Monte Carlo algorithm and its numerical analysis.

3.1 Algorithm

To estimate $\mathbb{E}\left[\varphi(X_T)\right]$, the simplest Monte Carlo estimator is

$$\frac{1}{N} \sum_{n=1}^{N} \varphi(\widehat{X}_T^{(n)}),$$

where $\widehat{X}_T^{(n)}$ is the terminal value of the nth numerical path in the time interval $[0, T]$ using a suitable adaptive function h^δ. It can be extended to Multilevel Monte Carlo by using non-nested timesteps [7]. Consider the identity

$$\mathbb{E}[\varphi_L] = \mathbb{E}[\varphi_0] + \sum_{\ell=1}^{L} \mathbb{E}[\varphi_\ell - \varphi_{\ell-1}], \tag{17}$$

where $\varphi_\ell := \varphi(\widehat{X}_T^\ell)$ with $\widehat{X}_T^\ell$ being the numerical estimator of X_T, which uses adaptive function h^δ with $\delta = M^{-\ell}$ for some positive integer $M > 1$. Then the standard MLMC estimator is the following telescoping sum:

$$\frac{1}{N_0}\sum_{n=1}^{N_0} \varphi(\widehat{X}_T^{(n,0)}) + \sum_{\ell=1}^{L}\left\{\frac{1}{N_\ell}\sum_{n=1}^{N_\ell}\left(\varphi(\widehat{X}_T^{(n,\ell)}) - \varphi(\widehat{X}_T^{(n,\ell-1)})\right)\right\},$$

where $\widehat{X}_T^{(n,\ell)}$ is the terminal value of the nth numerical path in the time interval $[0, T]$ using a suitable adaptive function h^δ with $\delta = M^{-\ell}$.

Unlike the standard MLMC with fixed time interval $[0, T]$, we now allow different levels to have a different length of time interval T_ℓ, satisfying $0 < T_0 < T_1 < \cdots < T_\ell < \cdots < T_L = T$, which means that as level ℓ increases, we obtain a better approximation not only by using smaller timesteps but also by simulating a longer time interval. However, the difficulty is how to construct a good coupling on each level ℓ since the fine path and coarse path have different lengths of time interval T_ℓ and $T_{\ell-1}$.

Following the idea of Glynn and Rhee [8] to estimate the invariant measure of some Markov chains, we perform the coupling by starting a level ℓ fine path simulation at time $t_0^f = -T_\ell$ and a coarse path simulation at time $t_0^c = -T_{\ell-1}$ and terminating both paths at $t = 0$. Since the drift f and volatility g do not depend explicitly on time t, the distribution of the numerical solution simulated on the time interval $[-T_\ell, 0]$ is the same as one simulated on $[0, T_\ell]$. The key point here is that the fine path and coarse path share the same driving Brownian motion during the overlap time interval $[-T_{\ell-1}, 0]$. Owing to the result of Lemma 4, two solutions to the SDE satisfying Assumption 9, starting from different initial points and driven by the same Brownian motion will converge exponentially. Therefore, the fact that different levels terminate at the same time is crucial to the variance reduction of the multilevel scheme.

Our new multilevel scheme still has the identity (17) but with $\varphi_\ell = \varphi(\widehat{X}_0^\ell)$ with $\widehat{X}_0^\ell$ being the terminal value of the numerical path approximation on the time interval $[-T_\ell, 0]$ using adaptive function h^δ with $\delta = M^{-\ell}$. The corresponding new MLMC estimator is

$$\widehat{Y} \triangleq \frac{1}{N_0}\sum_{n=1}^{N_0} \varphi(\widehat{X}_0^{(n,0)}) + \sum_{\ell=1}^{L}\left\{\frac{1}{N_\ell}\sum_{n=1}^{N_\ell}\left(\varphi(\widehat{X}_0^{(n,\ell)}) - \varphi(\widehat{X}_0^{(n,\ell-1)})\right)\right\}, \tag{18}$$

where $\widehat{X}_0^{(n,\ell)}$ is the terminal value of the nth numerical path through time interval $[-T_\ell, 0]$ using adaptive function h^δ with $\delta = M^{-\ell}$. Algorithm 1 outlines the detailed implementation of a single adaptive MLMC sample using a non-nested adaptive timestep on level ℓ with $M = 2$.

Algorithm 1 Outline of the algorithm for a single adaptive MLMC sample for scalar SDE on level ℓ in time interval $[-T_\ell, 0]$.

$t := -T_\ell;\ t^c := -T_{\ell-1};\ t^f := -T_\ell$
$h^c := 0;\ h^f := 0$
$\Delta W^c := 0;\ \Delta W^f := 0$
$\widehat{X}^c = x_0;\ \widehat{X}^f = x_0$
while $t < 0$ **do**
  $t_{old} := t$
  $t := \min(t^c, t^f)$
  $\Delta W := N(0, t - t_{old})$
  $\Delta W^c := \Delta W^c + \Delta W$
  if $t = -T_{\ell-1}$ **then**
    $\Delta W^c := 0$
  end if
  $\Delta W^f := \Delta W^f + \Delta W$
  if $t = t^c$ **then**
    update coarse path $\widehat{X}^c$ using h^c and ΔW^c
    compute new adapted coarse path timestep $h^c = h^{2\delta}(\widehat{X}^c)$
    $h^c := \min(h^c, -t^c)$
    $t^c := t^c + h^c$
    $\Delta W^c := 0$
  end if
  if $t = t^f$ **then**
    update fine path $\widehat{X}^f$ using h^f and ΔW^f
    compute new adapted fine path timestep $h^f = h^\delta(\widehat{X}^f)$
    $h^f := \min(h^f, -t^f)$
    $t^f := t^f + h^f$
    $\Delta W^f := 0$
  end if
end while
return $\widehat{X}^f - \widehat{X}^c$

3.2 Numerical Analysis

First, we state the exponential convergence to the invariant measure of the original SDEs, which can help us to measure the approximation error caused by truncating the infinite time interval.

Lemma 6 (Exponential convergence) *If the SDE satisfies Assumptions 7 and 9, and φ satisfies the Lipschitz condition: there exists a constant $\kappa > 0$ such that*

$$\|\varphi(x) - \varphi(y)\| \le \kappa \|x - y\|, \tag{19}$$

then there exists a constant $\mu > 0$ *depending on* x_0, κ *and* C_1 *in Lemma 3 such that*

$$|\mathbb{E}\left[\varphi(X_t) - \pi(\varphi)\right]| \le \mu\, e^{-\lambda t}. \tag{20}$$

With this, we can bound the variance of the MLMC correction for each level.

Lemma 7 (Variance of MLMC corrections for bounded volatility) *If* φ *satisfies the Lipschitz condition* (19)*, the SDE satisfies Assumption 9 and the timestep function* h^δ *satisfies Assumption 3 with* $\delta = M^{-\ell}$ *for each level, then for each level* ℓ*, there exist constants* c_1 *and* c_2 *such that the variance of correction* $V_\ell := \mathbb{V}\left[\varphi(\widehat{X}_0^\ell) - \varphi(\widehat{X}_0^{\ell-1})\right]$ *satisfies*

$$V_\ell \le c_1\, M^{-\ell} + c_2\, e^{-2\lambda T_{\ell-1}}. \tag{21}$$

Note that if we set $T_\ell = \frac{\log M}{2\lambda}(\ell + 1)$, then $V_\ell \le (c_1 + c_2) M^{-\ell}$, which has the same magnitude order as the standard MLMC. In some cases, λ needs to be estimated numerically through Lemma 6. N_ℓ can be optimized following the same approach in the MLMC theorem in [6].

Theorem 8 (MLMC for infinite time interval) *If* φ *satisfies the Lipschitz condition* (19)*, the SDE satisfies Assumption 9 and the timestep function* h^δ *satisfies Assumption 3 with* $\delta = M^{-\ell}$ *for each level, then by choosing suitable* T_ℓ, N_ℓ *for each level* ℓ*, there exists a constant* c_3 *such that the MLMC estimator (18) has a mean square error (MSE) with bound*

$$\mathbb{E}\left[(\widehat{Y} - \pi(\varphi))^2\right] \le \varepsilon^2,$$

and a computational cost $\boldsymbol{C}$ *with bound*

$$\mathbb{E}\left[\boldsymbol{C}\right] \le c_3\, \varepsilon^{-2} |\log \varepsilon|^3.$$

For Langevin SDEs, the computational cost can be reduced to $O(\varepsilon^{-2})$.

Theorem 9 (Langevin SDEs) *If* φ *satisfies the Lipschitz condition* (19)*, and for the SDE,* $m = d$, $g \equiv I_m$, f *satisfies Assumption 10, and the timestep function* h^δ *satisfies Assumption 3 with* $\delta = M^{-\ell}$ *for each level, then for each level* ℓ*, there exist constants* c_1 *and* c_2 *such that*

$$V_l \le c_1\, M^{-2\ell} + c_2\, e^{-2\lambda T_{\ell-1}}. \tag{22}$$

By choosing suitable $T_\ell = \frac{\log M}{\lambda}(\ell + 1)$ *and* N_ℓ *for each level* ℓ *in the MLMC estimator (18) such that it achieves the MSE bound* ε^2*, there exists a constant* c_3 *such that*

$$\mathbb{E}[C] \le c_3\, \varepsilon^{-2}.$$

Note that the choice of T_ℓ for Langevin equation is different from the one for SDEs with bounded volatility. In other words, the strong convergence result and the contractive convergence rate λ determine T_ℓ.

4 Examples and Numerical Results

In this section we first discuss some example SDEs with non-globally Lipschitz drift, then present the numerical result for finite time interval and its extension to infinite time interval.

For scalar SDEs, the drift is often of the form

$$f(x) \approx -c\, \mathrm{sign}(x)\, |x|^q, \quad \text{as } |x| \to \infty \tag{23}$$

for some constants $c>0, q>1$. Therefore, as $|x|\to\infty$, the maximum stable timestep satisfying Assumption 2 corresponds to $\langle x, f(x)\rangle + \frac{1}{2}h(x)\,|f(x)|^2 \approx 0$ and hence $h(x) \approx 2|x|/|f(x)| \approx 2\,c^{-1}|x|^{1-q}$. A suitable choice for $h(x)$ and $h^\delta(x)$ is therefore

$$h(x) = \min\left(T, c^{-1}|x|^{1-q}\right), \quad h^\delta(x) = \delta\, h(x). \tag{24}$$

For example, the Ginzburg–Landau equation, which describes a phase transition from the theory of superconductivity [13, 16], is

$$\mathrm{d}X_t = \left((\eta + \tfrac{1}{2}\sigma^2)X_t - \lambda X_t^3\right) \mathrm{d}t + \sigma X_t\, \mathrm{d}W_t,$$

where $\eta \ge 0$, $\lambda, \sigma > 0$. The drift and volatility satisfy Assumptions 1 and 4, and therefore all of the theory is applicable, with a suitable choice for $h^\delta(x)$, based on (23) and (24), being

$$h^\delta(x) = \delta\, \min\left(T, \lambda^{-1}x^{-2}\right).$$

For multi-dimensional SDEs, there are two cases of particular interest. For SDEs with a drift which, for some $\beta > 0$ and sufficiently large $\|x\|$, satisfies the condition

$$\langle x, f(x)\rangle \le -\beta\, \|x\|\, \|f(x)\|,$$

one can take $\langle x, f(x)\rangle + \frac{1}{2}h(x)\,|f(x)|^2 \approx 0$ and therefore a suitable definition of $h(x)$ for large $\|x\|$ is

$$h(x) = \min(T, \|x\|/\|f(x)\|).$$

For SDEs with a drift which does not satisfy the condition, but for which $\|f(x)\| \to \infty$ as $\|x\| \to \infty$, an alternative choice for large $\|x\|$ is to use

$$h(x) = \min(T, \gamma \, \|x\|^2 / \|f(x)\|^2), \tag{25}$$

for some $\gamma > 0$. For example, the Stochastic Lorenz equation, which is a three-dimensional system modelling convection rolls in the atmosphere [12], is

$$\begin{aligned} \mathrm{d}X_t^{(1)} &= \left(\alpha_1 X_t^{(2)} - \alpha_1 X_t^{(1)}\right) \mathrm{d}t + \beta_1 X_t^{(1)} \mathrm{d}W_t^{(1)} \\ \mathrm{d}X_t^{(2)} &= \left(\alpha_2 X_t^{(1)} - X_t^{(2)} - X_t^{(1)} X_t^{(3)}\right) \mathrm{d}t + \beta_2 X_t^{(2)} \mathrm{d}W_t^{(2)} \\ \mathrm{d}X_t^{(3)} &= \left(X_t^{(1)} X_t^{(2)} - \alpha_3 X_t^{(3)}\right) \mathrm{d}t + \beta_3 X_t^{(3)} \mathrm{d}W_t^{(3)} \end{aligned}$$

where $\alpha_1, \alpha_2, \alpha_3, \beta_1, \beta_2, \beta_3 > 0$. The diffusion coefficient is globally Lipschitz, and since $\langle x, f(x) \rangle$ consists solely of quadratic terms, the drift satisfies the one-sided linear growth condition. Noting that $\|f\|^2 \approx x_1^2(x_2^2 + x_3^2) < \|x\|^4$ as $\|x\| \to \infty$, an appropriate maximum timestep is $h(x) = \min(T, \gamma \|x\|^{-2})$, for any $\gamma > 0$. However, the drift does not satisfy the one-sided Lipschitz condition, and therefore the theory on the order of strong convergence is not applicable.

All the adaptive functions above satisfy the Assumptions 2 and 5. Other example applications include the stochastic Verhulst equation and a large class of Langevin equations.

The testcase taken from [14] is

$$\mathrm{d}X_t = -X_t - X_t^3 \, \mathrm{d}t + \mathrm{d}W_t, \quad x_0 = 1,$$

with $T = 1$. The three methods tested are the Tamed Euler scheme, the implicit Euler scheme, and the new Euler scheme with adaptive timestep. We can set $h_{max} = 1$, $M = 2$ and choose the adaptive function h, h^δ to be

$$h(x) = \frac{\max(1, |x|)}{\max(1, |x + x^3|)}, \quad h^\delta(x) = 2^{-\ell} h(x).$$

Figure 1 shows the the root-mean-square error plotted against the average timestep. The plot on the left shows the error in the terminal time, while the plot on the right shows the error in the maximum magnitude of the solution. The error in each case is computed by comparing the numerical solution to a second solution with a timestep, or δ, which is 2 times smaller.

When looking at the error in the final solution, all 3 methods have similar accuracy with $\frac{1}{2}$ order strong convergence. However, as reported in [14], the cost of the implicit method per timestep is much higher. The plot of the error in the maximum magnitude shows that the new method is slightly more accurate, presumably because it uses smaller timesteps when the solution is large. The plot was included to show that comparisons between numerical methods depend on the choice of accuracy measure being used.

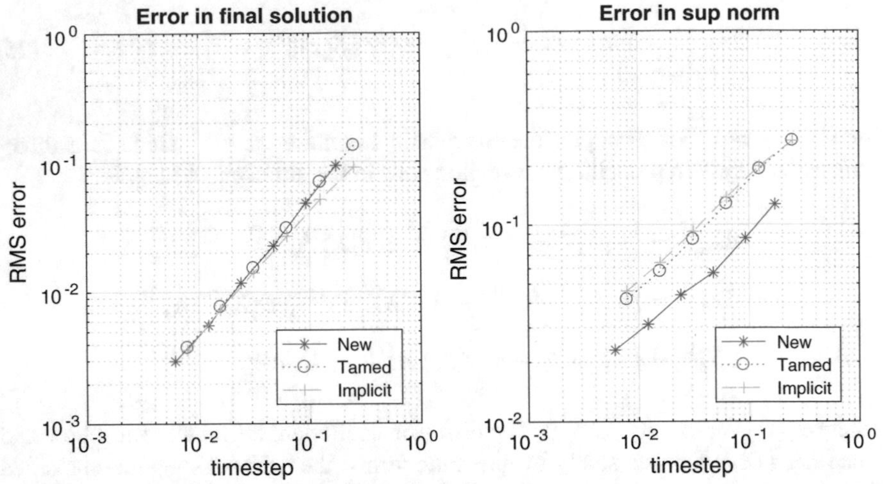

Fig. 1 Numerical results for finite time interval

Fig. 2 Variance of corrections on each level ℓ

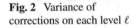

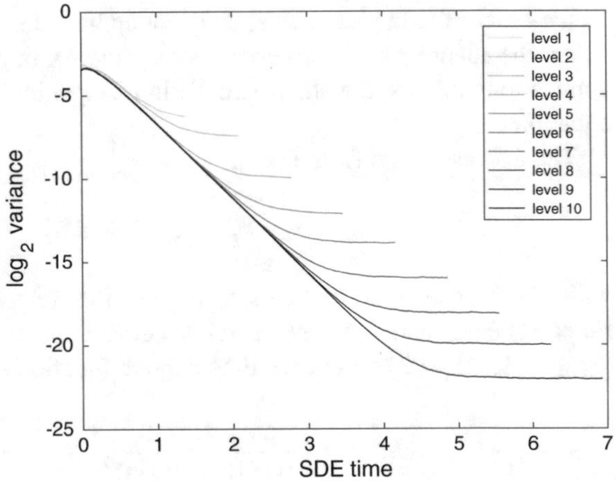

Next, we extend it to adaptive MLMC for the infinite time interval, since it also satisfies the dissipative condition (5) and the contractive condition (16). Our interest is to compute $\pi(\varphi)$ where $\varphi(x) = \|x\|$ satisfies a Lipschitz condition.

First we need to determine T_ℓ for each level. By differentiating drift f we know $\lambda \geq 1$ and choose λ to be 1 in our numerical scheme to simulate a sufficiently long time interval and control the truncation error. Then we choose

$$T_\ell = \log 2\,(\ell + 1).$$

The variance result (22) for the Langevin equation is illustrated in Fig. 2. The exponential part dominates the variance at the beginning, so the variance decays exponentially. As time increase, the $M^{-2\ell}$ term becomes the major part of the variance and the variance stops decreasing.

For level 10, we have $T_{10} = 7.62$ and the variance already stopped decreasing since $T = 5$ as shown in the Fig. 2, which shows that the setting of T_ℓ is sufficient. Then, all the convergence results are the same as the standard MLMC and our algorithm works well. For more detail, see [3].

5 Conclusion

The central conclusion from this paper is that by using an adaptive timestep it is possible to make the Euler–Maruyama approximation stable for SDEs with a globally Lipschitz volatility and a drift which is not globally Lipschitz but is locally Lipschitz and satisfies a one-sided linear growth condition. If the drift also satisfies a one-sided Lipschitz condition then the order of strong convergence is $\frac{1}{2}$, when looking at the accuracy versus the expected cost of each path. For the important class of Langevin equations with unit volatility, the order of strong convergence is 1. For ergodic SDEs satisfying the dissipative and contractive condition, we have shown that the moments and strong error of the numerical solutions are bounded and independent of time T. Moreover, we extend this adaptive scheme to MLMC for the infinite time interval by allowing different lengths of time intervals and carefully coupling the fine path and coarse path in each level ℓ. All the schemes work well and numerical experiments support the theoretical results.

References

1. Barrett, J.W., Süli, E.: Existence of global weak solutions to some regularized kinetic models for dilute polymers. SIAM Multiscale Model. Simul. **6**(2), 506–546 (2007)
2. Fang, W., Giles, M.B.: Adaptive Euler–Maruyama method for SDEs with non-globally Lipschitz drift: Part I, finite time interval (2016). arXiv preprint arXiv:1609.08101
3. Fang, W., Giles, M.B.: Adaptive Euler–Maruyama method for SDEs with non-globally Lipschitz drift: Part II, infinite time interval (2017). arXiv preprint arXiv:1703.06743
4. Gaines, J.G., Lyons, T.J.: Variable step size control in the numerical solution of stochastic differential equations. SIAM J. Appl. Math. **57**(5), 1455–1484 (1997)
5. Giles, M.B.: Multilevel Monte Carlo path simulation. Oper. Res. **56**(3), 607–617 (2008)
6. Giles, M.B.: Multilevel Monte Carlo methods. Acta Numerica **24**, 259 (2015)
7. Giles, M.B, Lester, C., Whittle, J.: Non-nested adaptive timesteps in multilevel Monte Carlo computations (2016)
8. Glynn, P.W., Rhee, C., et al.: Exact estimation for Markov chain equilibrium expectations. J. Appl. Probab. **51**, 377–389 (2014)
9. Hansen, N.R.: Geometric ergodicity of discrete-time approximations to multivariate diffusions. Bernoulli **9**, 725–743 (2003)

10. Higham, D.J., Mao, X., Stuart, A.M.: Strong convergence of Euler-type methods for nonlinear stochastic differential equations. SIAM J. Numer. Anal. **40**(3), 1041–1063 (2002)
11. Hofmann, N., Müller-Gronbach, T., Ritter, K.: The optimal discretization of stochastic differential equations. J. Complex. **17**(1), 117–153 (2001)
12. Hutzenthaler, M., Jentzen, A.: Numerical Approximations of Stochastic Differential Equations with Non-globally Lipschitz Continuous Coefficients, vol. 236. American Mathematical Society, Providence (2015)
13. Hutzenthaler, M., Jentzen, A., Kloeden, P.E.: Strong and weak divergence in finite time of Euler's method for stochastic differential equations with non-globally Lipschitz continuous coefficients. Proc. R. Soc. Lond. A Math. Phys. Eng. Sci. **467**(2130), 1563–1576 (2011)
14. Hutzenthaler, M., Jentzen, A., Kloeden, P.E.: Strong convergence of an explicit numerical method for SDEs with nonglobally Lipschitz continuous coefficients. Ann. Appl. Probab. **22**(4), 1611–1641 (2012)
15. Kelly, C., Lord, G.J.: Adaptive timestepping strategies for nonlinear stochastic systems (2016). arXiv preprint arXiv:1610.04003
16. Kloeden, P.E., Platen, E.: Numerical Solution of Stochastic Differential Equations. Springer, Berlin (1992)
17. Lamba, H.: An adaptive timestepping algorithm for stochastic differential equations. J. Comput. Appl. Math. **161**(2), 417–430 (2003)
18. Lamba, H., Mattingly, J.C., Stuart, A.M.: An adaptive Euler–Maruyama scheme for SDEs: convergence and stability. J. Numer. Anal. **27**, 479–506 (2007)
19. Lemaire, V.: An adaptive scheme for the approximation of dissipative systems. Stoch. Process. Appl. **117**(10), 1491–1518 (2007)
20. Mao, X.: Stochastic Differential Equations and Applications. Horwood Publishers Ltd., Chichester (1997)
21. Mao, X.: The truncated Euler–Maruyama method for stochastic differential equations. J. Comput. Appl. Math. **290**, 370–384 (2015)
22. Mao, X., Szpruch, L.: Strong convergence and stability of implicit numerical methods for stochastic differential equations with non-globally Lipschitz continuous coefficients. J. Comput. Appl. Math. **238**, 14–28 (2013)
23. Mattingly, J.C., Stuart, A.M., Higham, D.J.: Ergodicity for SDEs and approximations: locally Lipschitz vector fields and degenerate noise. Stoch. Process. Appl. **101**(2), 185–232 (2002)
24. Mauthner, S.: Step size control in the numerical solution of stochastic differential equations. J. Comput. Appl. Math. **100**(1), 93–109 (1998)
25. Meyn, S.P., Tweedie, R.L.: Stability of Markovian processes III: Foster–Lyapunov criteria for continuous-time processes. Adv. Appl. Probab. **25**, 518–548 (1993)
26. Milstein, G.N., Tretyakov, M.V.: Numerical integration of stochastic differential equations with nonglobally Lipschitz coefficients. SIAM J. Numer. Anal. **43**(3), 1139–1154 (2005)
27. Milstein, G.N., Tretyakov, M.V.: Computing ergodic limits for Langevin equations. Phys. D Nonlinear Phenom. **229**(1), 81–95 (2007)
28. Müller-Gronbach, T.: Optimal pointwise approximation of SDEs based on Brownian motion at discrete points. Ann. Appl. Probab. **14**, 1605–1642 (2004)
29. Roberts, G.O., Tweedie, R.L.: Exponential convergence of Langevin distributions and their discrete approximations. Bernoulli **2**, 341–363 (1996)
30. Soize, C.: The Fokker–Planck Equation for Stochastic Dynamical Systems and its Explicit Steady State Solutions, vol. 17. World Scientific, Singapore (1994)
31. Talay, D.: Second-order discretization schemes of stochastic differential systems for the computation of the invariant law. Stoch. Int. J. Probab. Stoch. Process. **29**(1), 13–36 (1990)
32. Wang, X., Gan, S.: The tamed Milstein method for commutative stochastic differential equations with non-globally Lipschitz continuous coefficients. J. Differ. Equ. Appl. **19**(3), 466–490 (2013)

Monte Carlo with User-Specified Relative Error

J. Feng, M. Huber and Y. Ruan

Abstract Consider an estimate $\hat{a}$ for a with the property that the distribution of the relative error $\hat{a}/a - 1$ does not depend upon a, but can be chosen by the user ahead of time. Such an estimate will be said to have user-specified relative error (USRE). USRE estimates for continuous distributions such as the exponential have long been known, but only recently have unbiased USRE estimates for Bernoulli and Poisson data been discovered. In this work, biased USRE estimates are examined, and it is shown how to precisely choose the bias in order make the chance that the absolute relative error lies above a threshold decay as quickly as possible. In fact, for Poisson data this decay (on average) is slightly faster than if the CLT approximation is used.

Keywords Discrete scalable distribution · Gamma Bernoulli approximation scheme · Gamma Poisson approximation scheme · Randomized approximation scheme · Tootsie pop algorithm

1 Introduction

Consider the problem of generating an estimate $\hat{a}$ for a such that the relative error $(\hat{a}/a) - 1$ is bounded by user given ε, with user given failure rate δ.

Definition 1 Call an estimate $\hat{a}$ for a an (ε, δ)- *randomized approximation scheme* or (ε, δ)- *ras* for nonnegative ε and δ if

J. Feng
Penn State University, Old Main, State College, PA, USA
e-mail: jpf5265@psu.edu

M. Huber (✉)
Claremont McKenna College, 850 Columbia AV, Claremont, CA, USA
e-mail: mhuber@cmc.edu

Y. Ruan
Pitzer College, 1050 N. Mills AV, Claremont, CA, USA
e-mail: soruan@students.pitzer.edu

A. B. Owen and P. W. Glynn (eds.), *Monte Carlo and Quasi-Monte Carlo Methods*, Springer Proceedings in Mathematics & Statistics 241,
https://doi.org/10.1007/978-3-319-91436-7_12

$$\mathbb{P}\left(\left|\frac{\hat{a}}{a} - 1\right| > \varepsilon\right) < \delta.$$

A stronger form is that the user actually knows precisely the distribution of the relative error.

Definition 2 Say that an estimate $\hat{a}$ for a has *user-specified relative error* or *USRE* if the distribution of $\hat{a}/a$ does not depend on a, but only on parameters specified by the user in constructing $\hat{a}$.

Until recently, the only data distributions with user-specified relative error estimates were continuous and scalable.

Example 1 Say that X has an exponential distribution with rate λ (and mean $1/\lambda$) if the density of X is $f_X(s) = \lambda \exp(-\lambda s)\mathbb{1}(s \geq 0)$. Write $X \sim \mathsf{Exp}(\lambda)$. (Here $\mathbb{1}(\cdot)$ is the usual indicator function that is 1 if the argument is true and 0 if the argument is false.) Given $X_1, X_2, \ldots, X_k$ independent identically distributed (iid) data $\mathsf{Exp}(\lambda)$, an unbiased estimate for λ is

$$\hat{\lambda} = \frac{k-1}{X_1 + \cdots + X_k}.$$

Say Y has a gamma distribution with shape parameter k and rate λ (write $Y \sim \mathsf{Gamma}(k, \lambda)$) if Y has density $f_Y(s) = \lambda^k s^{k-1} \exp(-\lambda s)\mathbb{1}(s \geq 0)/\Gamma(k)$. Then it is well known that $\lambda/\hat{\lambda}$ has a gamma distribution with shape parameter k and rate parameter $k-1$. Therefore $\hat{\lambda}$ is a USRE estimate.

Example 2 Say that X is uniform over $[0, \theta]$ (write $X \sim \mathsf{Unif}([0, \theta])$) if X has density $f_X(s) = \theta^{-1}\mathbb{1}(s \in [0, \theta])$. Suppose $X_1, X_2, \ldots, X_n$ are iid $\mathsf{Unif}([0, \theta])$. Then

$$\hat{\theta} = \frac{n+1}{n} \max_i \{X_i\}$$

is an unbiased USRE estimate of θ. This is because

$$\frac{\hat{\theta}}{\theta} = \frac{n+1}{n} \max_i \left\{\frac{X_i}{\theta}\right\},$$

and it is well known that $X_i/\theta \sim \mathsf{Unif}([0, 1])$. Therefore the maximum of the X_i/θ, which is a beta distributed random variable with parameters n and 1, does not depend on $\hat{\theta}$ in any way. Such a variable has mean $n/(n+1)$, so multiplying by $(n+1)/n$ makes the estimate unbiased.

Remark 1 Throughout this work, we will always use k to denote the number of exponential random variables used in constructing our estimate. The variable n will used more generally to denote the number of samples drawn from any other distribution.

1.1 Discrete Scalable Distributions

The output of Monte Carlo algorithms often come from discrete rather than continuous distributions, and so the creation of user-specified relative error estimates seemed out of reach for many problems. One feature of Monte Carlo data, however, it the ability to generate as much data as needed for the estimate. That is, unlike fixed length experiments where the data output is $X_1, \ldots, X_n$, it is typically easy with Monte Carlo output to have a stream of data and use $X_1, \ldots, X_T$ for some stopping time T as the final set of data.

By carefully using this advantage and exploiting connections between discrete and continuous distributions, it was shown how to build unbiased user-specified relative error estimates for the means of Bernoulli [2] and Poisson [3] iid data.

We open here with a new estimate for the "German tank problem", that is, estimation of the integer θ where $X_1, X_2, \ldots$ are independent $\mathsf{Unif}(\{1, 2, \ldots, \theta\})$ random variables.

Example 3 Let $X_1, X_2, \ldots$ be iid $\mathsf{Unif}(\{1, 2, \ldots, \theta\})$. Then it is well known that for $U_1, U_2, \ldots$ iid $\mathsf{Unif}([0, 1])$ and independent of the X_i, that $Y_i = X_i - U_i$ are iid $\mathsf{Unif}([0, \theta])$. Therefore, from Example 2, the estimate

$$\hat{\theta}_{\text{USRE}} = \frac{n+1}{n} \max_i \{X_i - U_i\}$$

is a user-specified relative error unbiased estimate of θ for the $\{X_i\}$.

The new estimate smooths the data slightly in order to obtain our USRE for θ. What do we lose by doing this? The answer is: a little, but not much. Consider the classic minimum variance unbiased estimator for θ. Given $X_1, \ldots, X_n \sim \mathsf{Unif}(\{1, 2, \ldots, \theta\})$,

$$\hat{\theta}_{\text{mvue}} = \frac{1}{1 + 1/n} \max_i \{X_i\} - 1.$$

The variance of this estimate is

$$\mathbb{V}(\hat{\theta}_{\text{mvue}}) = \frac{(\theta - n)(\theta + 1)}{n(n+2)}.$$

Compare with the USRE, where

$$\mathbb{V}(\hat{\theta}_{\text{USRE}}) = \frac{\theta^2}{n(n+2)}$$

When $n \ll \theta$, the variances are very close together, but it always holds that the variance of the mvue is smaller than that of the USRE.

So what is lost is a small amount of variance, What is gained is the ability to give exact confidence intervals that depend very simply on the data. For instance, for $n = 35$, it holds that a beta distributed random variable with parameters n and

Table 1 Discrete distributions

Distribution	Density $f_X(s)$	Notation
Bernoulli	$p\mathbb{1}(s = 1) + (1 - p)\mathbb{1}(s = 0)$	$\mathsf{Bern}(p)$
Geometric	$p(1 - p)^{s-1}\mathbb{1}(s \in \{1, 2, \ldots\})$	$\mathsf{Geo}(p)$
Poisson	$[\exp(-\mu)\mu^s/s!]\mathbb{1}(s \in \{0, 1, 2, \ldots\})$	$\mathsf{Pois}(\mu)$

1 is within 10% of its maximum value with probability $1 - 0.02503$. Therefore, the same holds for $\hat{\theta}_{\text{usre}}$, regardless of the true value of θ. Hence an exact 95% confidence interval for θ is $[\hat{\theta}_{\text{usre}}[n/(n+1)], \hat{\theta}_{\text{usre}}[n/(n+1)]/(1-0.1)]$.

Now consider data which is either geometric, Bernoulli, or Poisson. Table 1 gives the densities for these distributions.

Example 4 Consider $G_1, G_2, \ldots, G_n \sim \mathsf{Geo}(p)$, so $\mathbb{P}(G_i = i) = p(1-p)^{i-1}$ for $i \in \{1, 2, \ldots\}$. The method of moments estimator for p is

$$\hat{p}_{\text{mom}} = \frac{n}{G_1 + \cdots + G_n}$$

While biased, this does converge to p with probability 1 as k goes to infinity.

As noted in [2], a USRE is obtained for geometric random variables using the following well known fact.

Lemma 1 *If $G \sim \mathsf{Geo}(p)$ and $[A|G] \sim \mathsf{Gamma}(G, 1)$, then $A \sim \mathsf{Exp}(p)$.*

For each G_i, generate $[A_i|G_i] \sim \mathsf{Gamma}(G_i, 1)$. By Lemma 1, each $A_i \sim \mathsf{Exp}(p)$, and then use $\hat{p}$ for p from Example 1 to obtain the USRE for p.

Example 5 For $B_1, B_2, \ldots$ iid $\mathsf{Bern}(p)$, first use the $\{B_i\}$ to generate $\{G_i\}$.

$$G_1 = \inf\{t : B_t = 1\}, \quad G_i = \inf\{t : t > G_{i-1}, B_t = 1\} - G_{i-1}.$$

Then use the $\{G_i\}$ to give $\hat{p}$ from the previous example.

Because this uses Bernoulli random variables together with gamma random variables to give the estimate, this is known as the Gamma Bernoulli Approximation Scheme (GBAS). Each geometric requires (on average) $1/p$ Bernoulli random draws to generate, so the expected number of Bernoulli random variables used by this algorithm is k/p.

The final distribution considered here, Poisson, generates a random number of exponential random variables with each Poisson by using the following well known fact about Poisson point processes.

Lemma 2 *Let $P_1, P_2, \ldots$ be iid $\mathsf{Pois}(\mu)$. Then for each interval $[i, i+1]$ for $i \in \{0, 1, \ldots\}$, let C_i be a set of P_i values drawn independently and uniformly over $[i, i+1]$. Let $D_1 \leq D_2 \leq \cdots$ be the sorted values of $\cup_i C_i$. Then $D_1, D_2 - D_1, D_3 - D_2, \ldots$ form an iid sequence of $\mathsf{Exp}(\mu)$ random variables.*

Example 6 For $P_1, P_2, \ldots$ iid $\textsf{Pois}(\mu)$ and fixed k, use Lemma 2 to generate $A_1, A_2, A_3, \ldots, A_k$ iid $\textsf{Exp}(\mu)$ and then proceed as in Example 1. This estimate is called the Gamma Poisson Approximation Scheme, or GPAS for short.

Each draw of the Poisson generates (on average) μ exponential random variables, and so between k/μ and $k/\mu + 1$ Poisson draws are needed (on average) to generate the exponential random variables.

1.2 Main Results

Let a denote the mean of the exponential, Bernoulli, geometric, or Poisson data used to generate a random variable $R \sim \textsf{Gamma}(k, a)$, where k is chosen by the user. Then it is simple matter to check that $\hat{a} = (k-1)/R$ is unbiased.

Since the gamma distribution is skewed, this $\hat{a}$ estimate is more likely to be too large than too small in the relative error sense. So a better estimate is

$$\hat{a}_c = \frac{k-1}{cR},$$

where c is a fixed constant. When $c = 1$, the estimate is just $\hat{a}$ which is unbiased. By choosing $c > 1$, it is possible to balance the upper and lower tails and return an estimate where the relative error is at most ε with failure probability that decays at the fastest possible rate.

The main result is the following.

Theorem 1 *Let*

$$c = \frac{2\varepsilon}{(1-\varepsilon^2)\ln(1 + 2\varepsilon/(1-\varepsilon))},$$

and $\hat{a}_c = (k-1)/[cR]$ *where* $R \sim \textsf{Gamma}(k, a)$. *Then define*

$$c_1 = \frac{1}{c(1-\varepsilon)}, \quad c_2 = \frac{1}{c(1+\varepsilon)}, \quad b(t) = te^{1-t}. \tag{1}$$

Note that $b(t) < 1$ *for* $t \neq 1$ *and for this choice of* c_1 *and* c_2, $b(c_1) = b(c_2)$, *so let* b *equal this common value. Then*

$$\begin{aligned}
\mathbb{P}\left(\left|\frac{\hat{a}_c}{a} - 1\right| > \varepsilon\right) &\leq \frac{1}{\sqrt{2\pi(k-1)}}\left[\left|\frac{c_1}{c_1 - 1}\right| b^{k-1} + \left|\frac{c_2}{c_2 - 1}\right| b^{k-1}\right] \\
&\leq \sqrt{\frac{2}{\pi\varepsilon^2(k-1)}} \exp\left(-(k-1)\left(\frac{\varepsilon^2}{2} + \frac{11\varepsilon^4}{36}\right)\right).
\end{aligned}$$

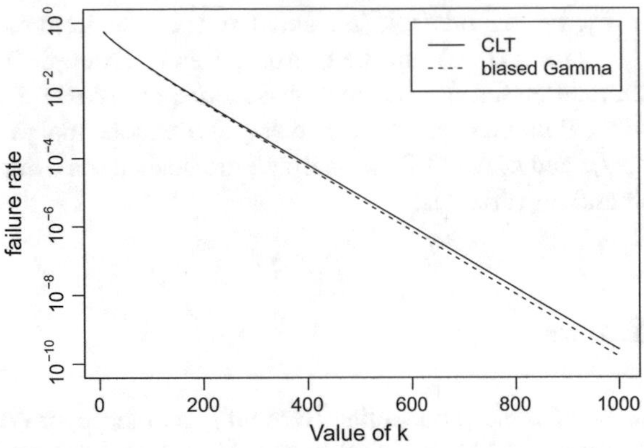

Fig. 1 Given exponential random variables with rate a, consider an estimate of a a failure if the relative error of the estimate is greater than ε. Both the problem of estimating the mean of a Bernoulli and the mean of a Poisson can be converted into this exponential problem. This plot compares the use of k exponential draws to form the estimate of a. The solid line treats the sample average of the exponentials as a normal random variable, while the dotted line uses a biased Gamma estimator. For the same k, the biased Gamma is a better estimator in this sense than the CLT. These particular failure rates use $\varepsilon = 0.2$. The CLT line has asymptotic slope against the log failure rate (to second order in ε) equal to $-\varepsilon^2/2$. The biased gamma line has asymptotic slope against the log failure rate (to the fourth order in ε) equal to $-\varepsilon^2/2 - (11/36)\varepsilon^4$

By using this choice of c, it is often possible to generate an estimate with bounded relative error using fewer samples on average than a CLT analysis. For example, consider $P_1, P_2, P_3, \ldots$ iid $\mathsf{Pois}(\mu)$. The mean and variance of the $\{P_i\}$ is both μ, so consider estimating μ for $W_1, W_2, \ldots$ iid normal with mean and variance μ. The GPAS algorithm uses on average k/μ samples to generate $R \sim \mathsf{Gamma}(k, \mu)$.

So setting $n = \lfloor k/\mu \rfloor$, the sample average $\hat{\mu}_n = (W_1 + \cdots + W_n) \sim \mathsf{N}(\mu, \mu/n)$, and

$$\mathbb{P}(|(\hat{\mu}_n/\mu) - 1| > \varepsilon) > \mathbb{P}(|Z|/\sqrt{k} > \varepsilon),$$

where Z is a standard normal random variable. As shown in Sect. 2.1,

$$\mathbb{P}(|Z| > \varepsilon\sqrt{k}) \approx \sqrt{\frac{2}{\pi\varepsilon^2 k}} \exp\left(-k\frac{\varepsilon^2}{2}\right),$$

so when k is large, the probability for the biased Gamma concentrates slightly faster than for a normal.

For example, when $\varepsilon = 0.1$, to get $\mathbb{P}(|Z|/\sqrt{k} > \varepsilon) < 0.01$ requires $k \geq 663.4897$. But using the value of c from Theorem 1, the value of k needed using GPAS is 661. So GPAS requires on average at most $661/\mu + 1$ samples, while the normal requires at least $663/\mu$. For small μ then, the biased estimator requires fewer samples on average than the CLT approach. See Fig. 1 for the failure rates as a function of k for $\varepsilon = 0.2$.

The remainder of this work is organized as follows. The next section reviews relevant bounds on the tails of gamma and normal distributions, and proves Theorem 1. Finally, Sect. 3 looks at several applications of these results in Monte Carlo integration.

2 Biased Estimates for Minimizing the Failure Probability

For both GBAS and GPAS, the first step is generating a random variable $R \sim \mathsf{Gamma}(k, a)$, where a is the quantity to be estimated. Then $\hat{a}_c = (k-1)/(cR)$ becomes the estimate. The goal is to make

$$\begin{aligned}\mathbb{P}\left(\left|\frac{\hat{a}_c}{a} - 1\right| > \varepsilon\right) &= \mathbb{P}\left(\frac{(k-1)}{acR} > 1 + \varepsilon \text{ or } \frac{(k-1)}{acR} < 1 - \varepsilon\right)\\ &= \mathbb{P}\left(\frac{k-1}{aR} > c(1+\varepsilon)\right) + \mathbb{P}\left(\frac{k-1}{aR} < c(1-\varepsilon)\right)\end{aligned}$$

as small as possible. Since $(aR)/(k-1) \sim \mathsf{Gamma}(k, k-1)$, our work will focus on developing good bounds for the upper and lower tails of this distribution.

Lemma 3 *Let $f_X(s) = \alpha^\beta s^{\alpha-1}\exp(-\beta s)\mathbb{1}(s \geq 0)/\Gamma(k)$ be the density of $X \sim \mathsf{Gamma}(\alpha, \beta)$. Then*

$$f_X(t)\frac{1}{\beta} \leq \mathbb{P}(X \in A) \leq f_X(t)\frac{t}{|\beta t - (\alpha - 1)|}.$$

for $A = [0, t]$ where $t < (\alpha-1)/\beta$ or $A = [t, \infty)$ where $t > (\alpha-1)/\beta$.

Proof Consider for $s > 0$,

$$f_X'(s) = f_X(s)\beta\left[\frac{\alpha-1}{\beta s} - 1\right].$$

For $s \geq t > (a-1)/\beta$, this gives

$$-\beta f_X(s) \leq f_X'(s) \leq f_X(s)\beta[(\alpha-1)/(\beta t) - 1]$$

and

$$f_X'(s)t/(\beta t - (\alpha-1)) \geq f_X(s) \geq f_X'(s)/(-\beta).$$

Integrating these inequalities for s running from t to infinity and 0 to t gives the upper and lower bounds.

The $s \leq t < (\alpha - 1)/\beta$ case is similar. □

Now to understand how $f_X(s)$ behaves.

Lemma 4 *For* $\alpha = k$ *and* $\beta = k - 1$,

$$\exp(-1/[12(k-1)])\sqrt{\frac{k-1}{2\pi}}\left(te^{1-t}\right)^{k-1} \leq f_X(t) \leq \sqrt{\frac{k-1}{2\pi}}\left(te^{1-t}\right)^{k-1}$$

Proof Let $f_1(k-1) = \sqrt{2\pi(k-1)}((k-1)/e)^{k-1}$. Then Stirling's bound can be written

$$f_1(k-1) \leq \Gamma(k) \leq f_1(k-1)\exp(1/[12(k-1)]).$$

The density of a $\mathsf{Gamma}(k, k-1)$ at a is

$$f_X(a) = (k-1)^k t^{k-1}\exp(-(k-1)t)/\Gamma(k).$$

Using Stirling's bound on $\Gamma(k)$ and simplifying gives the result. □

Let $g(t)$ denote $\ln(\mathbb{P}((k-1)/(aR) > t))$ for $t > 1$ and $\ln(\mathbb{P}((k-1)/(aR) < t))$ for $t < 1$. From the previous lemma $g(t) = (k-1)[1 - t + \ln(t)]$ plus lower order terms. Setting $w = 1 - t$ gives $g(1-w) = (k-1)[w + \ln(1-w)]$. The Taylor series expansion of $g(1-w)/(k-1)$ with respect to w is

$$w + \ln(1-w) = -\frac{w^2}{2} - \frac{w^3}{3} - \frac{w^4}{4} - \cdots.$$

It is of course no surprise that the leading term of the logarithm of the tail probability is $-w^2/2$, as a $\mathsf{Gamma}(k, k-1)$ is the sum of k independent $\mathsf{Exp}(k-1)$ random variables, and therefore the CLT gives that the result is approximately normally distributed.

In the rest of this section it helps to define two values based on c and ε, as well as a function that encapsulates our rate. Recall that

$$c_1 = \frac{1}{c(1-\varepsilon)}, \quad c_2 = \frac{1}{c(1+\varepsilon)}, \quad b(t) = te^{1-t}$$

Lemma 5 *For* $\hat{a}_c = (k-1)/(acR)$, *let* c_1, c_2, *and* b *be as in* (1). *Then* $\mathbb{P}(|(\hat{a}_c/a) - 1| > \varepsilon)$ *is in*

$$\frac{1}{\sqrt{2\pi(k-1)}}\left[b(c_1)^{k-1} + b(c_2)^{k-1}, \left|\frac{c_1}{c_1 - 1}\right| b(c_1)^{k-1} + \left|\frac{c_2}{c_2 - 1}\right| b(c_2)^{k-1}\right]$$

Proof For $\hat{a}_c = (k-1)/(cR)$,

$$\mathbb{P}\left(\left|\frac{\hat{a}_c}{a} - 1\right| > \varepsilon\right) = \mathbb{P}\left(\frac{aR}{k-1} > \frac{1}{c(1-\varepsilon)}\right) + \mathbb{P}\left(\frac{aR}{k-1} < \frac{1}{c(1+\varepsilon)}\right)$$

Since $aR/(k-1) \sim \mathsf{Gamma}(k, k-1)$, the rest follows from the previous two lemmas. □

Since $b(t)$ is a unimodal function with maximum at $t = 1$ that goes to 0 as t goes to 0 and infinity, the log of the probability in the tail is minimized when $b(c_1) = b(c_2)$.

Lemma 6 *When*

$$c = \frac{2\varepsilon}{(1-\varepsilon^2)\ln(1+2\varepsilon/(1-\varepsilon))}, \tag{2}$$

and $\hat{a}_c = (k-1)/(cR)$, *then* $b(1/(c(1-\varepsilon))) = b(1/(c(1+\varepsilon))) = b$ *and*

$$\mathbb{P}\left(\left|\frac{\hat{a}_c}{a} - 1\right| > \varepsilon\right) \leq \frac{1}{\sqrt{2\pi(k-1)}}\left[\frac{c_1}{c_1-1} + \frac{c_2}{1-c_2}\right]b^{k-1}.$$

Proof It is easy to verify that $b(c_1) = b(c_2)$ for this choice of c. This choice makes $c_1 > 1$ and $c_2 < 1$. Applying the previous lemma then finishes the proof. □

It helps to have an idea of how good this bound is in terms of ε. Recall that c_1, c_2, and $b = b(c_1) = b(c_2)$ are all functions of ε.

Lemma 7 *For* $\varepsilon > 0$,

$$\frac{c_1}{c_1-1} + \frac{c_2}{1-c_2} \leq \frac{2}{\varepsilon}$$

and

$$b \leq \exp\left(-\frac{1}{2}\varepsilon^2 - \frac{11}{36}\varepsilon^4\right).$$

Proof This follows directly from the Taylor series expansions of these functions in terms of ε, and the continuity of all higher derivatives for $\varepsilon > 0$. □

Combining this with the previous lemma gives the following.

Corollary 1 *For c as in* (2),

$$\mathbb{P}\left(\left|\frac{k-1}{acR} - 1\right| > \varepsilon\right) \leq \sqrt{\frac{2}{\pi\varepsilon^2(k-1)}}\exp\left(-\frac{\varepsilon^2(k-1)}{2} - \frac{11\varepsilon^4(k-1)}{36}\right).$$

Therefore the log failure rate is asymptotically at most $-(k-1)(\varepsilon^2/2 + (11/36)\varepsilon^4)$. This is smaller than the asymptotic log failure rate of $-k\varepsilon^2/2$ for a normally distributed random variable.

2.1 Comparison to Normal Random Variables

A Poisson random variable with mean μ also has variance μ. So consider $X_1, \ldots, X_n$ random variables that are normal with mean and variance μ. In Sect. 1 it was noted that for such random variables the sample average $\hat{\mu}_n = \sum_i X_i/n$ satisfies

$$\mathbb{P}(|(\hat{\mu}_n/\mu) - 1| > \varepsilon) = \mathbb{P}(|Z| > \varepsilon\sqrt{n\mu})$$

where Z is a standard normal random variable.

Well known bounds connect the tail probabilities of a standard normal with the density of a standard normal. For instance, Gordon [1] showed that for all $s > 0$

$$\frac{1}{s + 1/s}\frac{1}{\sqrt{2\pi}}\exp(-s^2/2) \leq \mathbb{P}(Z > s) \leq \frac{1}{s}\frac{1}{\sqrt{2\pi}}\exp(-s^2/2) \tag{3}$$

For $s = \varepsilon\sqrt{n\mu}$, this says

$$\mathbb{P}(|\hat{\mu}_n/\mu - 1| > \varepsilon) = \Omega(\varepsilon^{-1}(n\mu)^{-1/2}\exp(-\varepsilon^2 n\mu/2)), \tag{4}$$

(Recall that we write $f(n) = \Omega(g(n))$ if $\limsup_{n\to\infty} f(n)/g(n) > 0$.) To compare this to the failure probabilities for the Poisson random variable, note that the average number of draws of the Poisson is k/μ where k is the parameter set by the user. So if $n \approx k/\mu$, then the failure probability for the normal random variables will be

$$\Omega(\varepsilon^{-1}k^{-1/2}\exp(-\varepsilon^2 k/2),$$

while for the gamma based estimate,

$$\mathbb{P}(|\hat{p}/p - 1| > \varepsilon) = O(\varepsilon^{-1}(k-1)^{-1/2}\exp(-[\varepsilon^2/2 + 11\varepsilon^4/36](k-1)). \tag{5}$$

So for fixed ε, as $k \to \infty$, eventually the failure probability will fall below that for the normals.

As seen in Sect. 1, this is not some far-off asymptotic range: for $\varepsilon = 0.1$ and $\delta = 0.01$, the gamma based method sets $k = 661$ but the normals require $k > 663$ to achieve the same level of accuracy. This fact that gammas are more highly concentrated than normals about their center is to be expected, as gamma random variables are always positive while for normals both tails are unbounded.

2.2 Biased Beta Estimates

Now consider the problem of estimating θ when $X_1, X_2, \ldots, X_n$ are iid $\mathsf{Unif}(\{1, 2, \ldots, \theta\})$. The unbiased smoothing method generated $U_1, \ldots, U_n$ independent of $X_1, \ldots, X_n$, and set $X_i' = X_i - U_i$. This makes X_i' uniform over $[0, \theta]$. Now an unbiased USRE estimate of θ is $\hat{\theta}_{\text{USRE}} = [(n+1)/n] \max_i (X_i - U_i)$ (see Example 2.)

As earlier, given $\varepsilon > 0$, the failure probability of an estimate $\hat{\theta}$ for θ is $\mathbb{P}(|\hat{\theta}/\theta - 1| > \varepsilon)$. However, the unbiased estimate does not minimize the failure probability.

Instead, note that $\max_i (X_i - U_i) \leq \theta$, so $\hat{\theta} = (1+\varepsilon) \max_i (X_i - U_i)$ can never have relative error greater than ε. The only way the relative error can be less than $-\varepsilon$ is if $(1+\varepsilon) \max_i (X_i - U_i) < (1-\varepsilon)\theta$, or equivalently, $\max_i (X_i - U_i)/\theta < (1+\varepsilon)/(1-\varepsilon)$. Recalling that each $(X_i - U_i)/\theta \sim \mathsf{Unif}([0,1])$, this gives the following lemma.

Lemma 8 *Given $X_1, \ldots, X_n$ iid uniform over $\{1, 2, \ldots, \theta\}$, and $U_1, \ldots, U_n$ iid uniform over $[0, 1]$ (and independent of the $\{X_i\}$), let*

$$\hat{\theta} = (1+\varepsilon) \max_i (X_i - U_i)$$

Then

$$\mathbb{P}(|(\hat{\theta}/\theta) - 1| > \varepsilon) = \left(\frac{1-\varepsilon}{1+\varepsilon} \right)^n.$$

Since $\ln((1-\varepsilon)/(1+\varepsilon)) = -2\varepsilon - (2/3)\varepsilon^3 - \cdots$, to first order the number of samples n necessary for an (ε, δ)-ras is $(1/2)\varepsilon^{-1} \ln(\delta^{-1})$, which is very much smaller than in the exponential or normal cases.

3 Applications

This section considers applications of the GBAS and GPAS algorithms. Suppose our goal is to approximate the value of an integral of dimenson m:

$$I = \int_{x \in \mathbb{R}^m} f(x) \, \mathrm{d}x.$$

Here $f(x) \geq 0$ and m is typically very large. For instance, $f(x)$ could be the unnormalized posterior distribution of a Bayesian model (so prior density times the likelihood of data) or the solution to some #P complete problem.

Our approach is to build three sets, $C \subseteq B \subseteq A$. Set A will have Lebesgue measure equal to the integral I. Set C will have Lebesgue measure that can be computed exactly. Then, random samples will be used to estimate the ratio of the measure of A to that of B, and the ratio of the measure of B to that of C. The product then estimates that ratio of the measure of A to that of C, and then multiply by the known measure of C to estimate the measure of A which is just I.

3.1 Acceptance Rejection Integration

Using acceptance rejection to approximately integrate functions goes back to at least Von Neumann [7].

For a measure ν, say that $X \sim \nu$ over B, if for all measurable $F \subseteq B$, $\mathbb{P}(X \in F) = \nu(F)/\nu(B)$.

Given a region A, and a region B that contains A from which is possible to sample $X \sim \nu$ over B, $\mathbb{P}(X \in A) = \nu(A)/\nu(B)$. Usually it is possible to compute either $\nu(B)$ or $\nu(A)$ easily. Let $\hat{p}$ be an estimate for $\mathbb{P}(X \in A)$ obtained using biased GBAS.

If $\nu(B)$ is known, then $\hat{p}\nu(B)$ is an estimate for $\nu(A)$. If $\nu(A)$ is known then $\nu(A)/\hat{p}$ is an estimate for $\nu(A)$. Either way, to obtain $\nu(A)$ (or $\nu(B)$) within a fixed relative error requires that $\hat{p}$ estimate p within a fixed relative error.

Now consider how this idea can be turned into an algorithm for estimating I. Suppose that $f(x)$ is known through either analysis or numerical experiments to have a local maximum at x^*, and $f(x) \leq f(x^*)$ for all $x : ||x^* - x||_2 \leq \alpha$. Consider three sets,

$$\begin{aligned} A &= \{(x, y) : x \in \mathbb{R}^n, 0 \leq y \leq f(x)\} \\ B &= \{(x, y) : ||x - x^*|| \leq \alpha, 0 \leq y \leq f(x)\} \\ C &= \{(x, y_: ||x - x^*|| \leq \alpha, 0 \leq y \leq f(x^*)\}. \end{aligned}$$

For ν Lebesgue measure, $\nu(A) = I$, the value of the integral that we are looking for.

It is easy to sample from C: just generate x uniformly from the hypersphere about x^* of radius α, and then generate y uniformly from $[0, f(x^*)]$.

Sampling from A is usually (approximately) accomplished using Markov chain Monte Carlo, or in some instances using perfect simulation (see [4, 6]) methods.

Then $B \subseteq A$ and $B \subseteq C$. For ν Lebesgue measure, $\nu(C) = f(x^*)\alpha^n V_n$, where V_n is the volume of an n dimensional hypersphere under $|| \cdot ||_2$.

So the strategy is to use two steps: estimate $\nu(B)/\nu(C)$ with $\hat{p}_1$, and $\nu(B)/\nu(A)$ with $\hat{p}_2$ using biased GBAS. Then $\nu(C)\hat{p}_1/\hat{p}_2 \approx \nu(A) = I$, and the relative error bounds for $\hat{p}_1$ and $\hat{p}_2$ can be used to find a relative error bound for the estimate of $\nu(A)$.

Of course, it is not necessary to know the value of x^* exactly. As an example, consider the function $f(x) = \exp(-x^2/2) + 1.5\exp(-(x-4)^2/2)$. Let $x^* = 0$, and $\alpha = 1$. For $x \in [-1, 1]$, $f(x) \leq 1.1$. Then $A = \{(x, y) : 0 \leq y \leq f(x)\}$, $B = \{x \in [-1, 1], 0 \leq y \leq f(x)\}$, and $C = \{x \in [-1, 1], 0 \leq y \leq 1.1\}$. Then $\nu(C) = 2.2$, so $\nu(A) = 2.2(\nu(B)/\nu(C))/(\nu(B)/\nu(A))$.

The value of $\nu(B)/\nu(C)$ can be estimated by sampling points uniformly from C, and letting the Bernoulli variables be the indicator that the points fall into B. Similarly, the value of $\nu(B)/\nu(A)$ can be estimated by drawing samples from A and letting the Bernoulli random variables be the indicator that the points fall into B.

In this example $\nu(B)/\nu(C) \approx 0.7801$ and $\nu(B)/\nu(A) = 0.273886$. So for a given choice of k, on average $k/0.7801$ samples from C are needed to get $\hat{p}_1$ an estimate for

$\nu(B)/\nu(C)$, and on average $k/0.273886$ samples from A are needed to get $\hat{p}_2$ an estimate for $\nu(B)/\nu(A)$. Recall $k = 661$ gives $\varepsilon = 0.1$ and $\delta = 0.01$. Therefore, using the union bound, $2.2\hat{p}_1/\hat{p}_2$ lies in $[(0.9/1.1)\nu(A), (1.1/0.9)\nu(A)]$ with probability at least 98%.

Suppose we use $\alpha = 0.1$. Then $\nu(B)/\nu(A) \approx 0.03187$, while $\nu(B)/\nu(C) \approx 0.908047$. The number of samples needed grows dramatically to get the $\hat{p}_2$ estimate as α becomes smaller.

3.2 TPA Integration

Generally, as α becomes smaller $\nu(B)/\nu(C)$ typically moves to 1 while $\nu(B)/\nu(A)$ becomes smaller. Therefore, it is helpful to have an alternate way to estimate $\nu(B)/\nu(A)$ when B is small relative to A. In fact, usually $\nu(B)$ is exponentially smaller than $\nu(A)$ in the dimension of the problem.

A solution to this issue is to use the Tootsie Pop Algorithm (TPA) [5]. which in this context operates as follows. Let $A_0 = A$, and draw a sample X_0 from ν over A_0. Let $A_1 = \{(x, y) : ||x - x^*|| \leq ||X_0 - x^*||\}$. Draw X_1 from ν over A_1 in the same way to get A_2, and continue into this fashion until $X_{T-1} \notin B$ and $X_T \in B$. That is, $T = \inf\{i : X_i \in B\}$.

Then Theorem 1 of [5] implies that $T - 1 \sim \mathsf{Pois}(\ln(\nu(B)/\nu(A)))$. GPAS gives us an estimate $\hat{a}$ for $a = \ln(\nu(B)/\nu(A))$, along with exact confidence intervals. These in turn gives exact confidence intervals for $\exp(\hat{a})$ which estimates $\nu(B)/\nu(A)$. Combined with the exact confidence intervals for $\nu(C)/\nu(B)$, the result is an exact confidence interval for the estimate $\nu(C)\hat{p}_1 \exp(-\hat{a})$ of $\nu(A)$.

Consider again our problem from earlier of estimating $\nu(B)/\nu(A)$ when the true answer is 0.0318787. Recall using $k = 661$ and directly drawing from A and forming Bernoullis from the indicator that the points fall in B used on average $k/0.0318787$ to get an estimate within relative error 0.1 with probability at least 99%.

By using TPA with $k = 661$, we obtain an estimate for $-\ln(0.0318787)$ by drawing $-661/\ln(0.0318787)$ Poisson random variables, each of which requires $-\ln(0.0318787) + 1$ draws from various subsets of A. Note $(-\ln(0.0318787) + 1)/(-\ln(0.0318787)) \approx 1.290$, much smaller than $1/0.0318787 \approx 31.37$.

However, the error bounds have changed. The estimate must be exponentiated to get back to the original problem. Letting $a = -\ln(0.0318787)$, we will find $\hat{a}$ such that $\hat{a} = a\xi$ where $\xi \in [0.9, 1.1]$ Hence $\exp(-a) \in [\exp(-\hat{a}/0.9), \exp(-\hat{a}/1.1)]$.

For instance, if $\hat{a} = 3.723$ (off from the true value of $a = -\ln(0.0318787) = 3.445817$) then we could say with 99% confidence that $\exp(a) = \nu(B)/\nu(A) \in [0.01597, 0.03390]$.

This is an exact confidence interval, but does not have relative error of 0.1 as desired. Using the geometric mean of the endpoints at the best estimate, the relative error could be up to 0.46. So we obtain an exact confidence interval, but not at the level of relative accuracy that we desired.

At this point, by knowing a lower bound on $\nu(B)/\nu(A)$, a second run of TPA could be undertaken that would guarantee our desired level of accuracy. Details of this two-phase procedure are given in [5].

Acknowledgements This material is based upon work supported by the National Science Foundation under Grant No. DMS 1418495. We also wish to thank the anonymous referee for several important corrections and suggestions.

References

1. Gordon, R.D.: Values of Mills' ratio of area to bounding ordinate of the normal probability integral for large values of the argument. Ann. Math. Stat. **12**, 364–366 (1941)
2. Huber, M.: A Bernoulli mean estimate with known relative error distribution. Random Struct. Algorithms (2016). arXiv:1309.5413. To appear
3. Huber, M.: An estimator for Poisson means whose relative error distribution is known (2016). arXiv:1605.09445. Submitted
4. Huber, M.L.: Perfect Simulation. Chapman & Hall/CRC Monographs on Statistics and Applied Probability, vol. 148. CRC Press, Boca Raton (2015)
5. Huber, M.L., Schott, S.: Random construction of interpolating sets for high dimensional integration. J. Appl. Probab. **51**(1), 92–105 (2014). arXiv:1112.3692
6. Propp, J.G., Wilson, D.B.: Exact sampling with coupled Markov chains and applications to statistical mechanics. Random Struct. Algorithms **9**(1–2), 223–252 (1996)
7. von Neumann, J.: Various techniques used in connection with random digits. Monte Carlo Method. Applied Mathematics Series, vol. 12. National Bureau of Standards, Washington (1951)

Dimension Truncation in QMC for Affine-Parametric Operator Equations

Robert N. Gantner

Abstract An application of quasi-Monte Carlo methods of significant recent interest in the MCQMC community is the quantification of uncertainties in partial differential equation models. Uncertainty quantification for both forward problems and Bayesian inverse problems leads to high-dimensional integrals that are well-suited for QMC approximation. One of the approximations required in a general formulation as an affine-parametric operator equation is the truncation of the formally infinite-parametric operator to a finite number of dimensions. To date, a numerical study of the available theoretical convergence rates for this error have to the author's knowledge not been published. We present novel results for a selection of model problems, the computation of which has been enabled by recently developed, higher-order QMC methods based on interlaced polynomial lattice rules. Surprisingly, the observed rates are one order better in the case of integration over the parameters than the commonly cited theory suggests; a proof of this higher rate is included, resulting in a theoretical statement consistent with the observed numerics.

Keywords Quasi Monte Carlo · QMC · Dimension truncation · Interlaced polynomial lattice rules

1 Introduction

An important application of quasi-Monte Carlo methods that has been of interest to the MCQMC community in recent years is the quantification of uncertainties in partial differential equation (PDE) models which depend on uncertain inputs, see e.g. [2, 3, 6, 7, 14–16, 18] to name but a few. The goal of computation is usually the mathematical expectation of a goal functional which depends on the solution to the PDE, corresponding to an integral over the uncertain inputs. Especially in the case where distributed uncertain inputs are considered, the problems often involve

---

R. N. Gantner (✉)
Seminar for Applied Mathematics, ETH Zürich, Rämistrasse 101, Zürich, Switzerland
e-mail: robert.gantner@sam.math.ethz.ch

A. B. Owen and P. W. Glynn (eds.), *Monte Carlo and Quasi-Monte Carlo Methods*, Springer Proceedings in Mathematics & Statistics 241,
https://doi.org/10.1007/978-3-319-91436-7_13

high-dimensional input parameter vectors, with the corresponding expectations being integrals over high-dimensional spaces. This fits naturally into a quasi-Monte Carlo framework, and various advances have been achieved in this field in recent years.

A large class of such problems can be formulated as so-called affine-parametric operator equations, for which many general theoretical results are available. These equations are formulated based on infinite parameter sequences, each corresponding to a realization of the uncertain input. In order to make computations feasible, a truncation to finitely many parameters is inevitable, and introduces an error into the computation. This error is called the *dimension truncation error*, and its study is the subject of this article.

Bounds on the dimension truncation error in this context are known [7, 15, 16], but to the author's knowledge, no numerical evidence has been published to support their sharpness. One reason for this may be that obtaining conclusive measurements is computationally very intensive, requiring approximations of integrals in a high number of dimensions to possibly very high accuracy, where each evaluation additionally involves an approximation of the solution to the operator equation by some numerical method, also with high precision. We fill this gap by providing measurements of this error for selected PDE test problems, where we apply a recently introduced higher-order quasi-Monte Carlo method based on interlaced polynomial lattice (IPL) rules [7, 10, 13] to attain the required accuracy in the approximation of the involved integral at reasonable cost. Combined with evaluation on a massively parallel computer system, approximations with sufficient accuracy are obtained in reasonable time.

Remarkably, the measured convergence rate of the error of an integral over the parameters in terms of the truncation dimension s is found to be one order higher than the current theoretical results as stated in e.g. [3, 7, 15, 16, 18]. This prompted a more detailed investigation into this convergence rate, and a proof of this higher rate is given below in Sect. 3 under some minor additional assumptions on the probability measure which are often fulfilled in practice. The higher rate shown here is due to a sharper analysis of the error, which was prompted by the reported numerical results.

We continue now by stating the setting of affine-parametric operator equations and present in Sect. 3 our main result, a novel estimate of the dimension truncation error which improves the known convergence rate by one order. In order to measure this error and verify the predicted rate, the higher-order QMC method used in the experiments is briefly mentioned in Sect. 4. Results supporting sharpness of the derived rate are then given in Sect. 5.

2 Affine-Parametric Operator Equations

Let $\mathscr{X}$, $\mathscr{Y}$ denote two separable Banach spaces with norms $\|\cdot\|_{\mathscr{X}}$, $\|\cdot\|_{\mathscr{Y}}$ and duals $\mathscr{X}'$, $\mathscr{Y}'$, respectively. We denote by $\boldsymbol{y} = (y_1, y_2, \ldots)$ a sequence of parameters taking values in $U = [-1/2, 1/2]^{\mathbb{N}}$, i.e. the set of sequences with entries $y_j \in [-1/2, 1/2]$. For each $\boldsymbol{y} \in U$, we denote by $A(\boldsymbol{y})$ a bounded linear operator from $\mathscr{X}$ to $\mathscr{Y}'$, i.e. $A(\boldsymbol{y}) \in \mathscr{L}(\mathscr{X}, \mathscr{Y}')$. In the following, we denote by ${}_{\mathscr{Y}'}\langle \cdot, \cdot \rangle_{\mathscr{Y}}$ the duality pairing in $\mathscr{Y}$. Then, for a given deterministic forcing function $f \in \mathscr{Y}'$ we seek for $\boldsymbol{y} \in U$ a solution $q(\boldsymbol{y}) \in \mathscr{X}$ to the problem

$$A(\boldsymbol{y})q(\boldsymbol{y}) = f \quad \text{in} \quad \mathscr{Y}' . \tag{1}$$

In the following, we will assume the operator $A(\boldsymbol{y})$ to depend on the y_j in an affine manner. More specifically, for a *nominal operator* A_0 and a sequence of *fluctuation operators* $(A_j)_{j\geq 1}$ we assume $A(\boldsymbol{y})$ to be of the form

$$A(\boldsymbol{y}) = A_0 + \sum_{j\geq 1} y_j A_j . \tag{2}$$

We now state some assumptions on $(A_j)_{j\geq 0}$ that are required for the well-posedness of (1) with $A(\boldsymbol{y})$ given by (2), or for the dimension truncation statements in Sect. 3.

Assumption 1 Assume that the nominal operator $A_0 \in \mathscr{L}(\mathscr{X}, \mathscr{Y}')$ is boundedly invertible. Additionally, assume that the fluctuation operators $(A_j)_{j\geq 1}$ are small wrt. A_0, i.e. there exists a $\kappa < 2$ such that for the sequence $\boldsymbol{b} = (b_1, b_2, \ldots)$, defined by $b_j := \|A_0^{-1} A_j\|_{\mathscr{L}(\mathscr{X})}$ it holds that $\|\boldsymbol{b}\|_{\ell^1(\mathbb{N})} := \sum_{j\geq 1} b_j \leq \kappa < 2$, cp. [18, Assumption 2].

Assumption 2 Assume that there exists $0 < p < 1$ such that for $\boldsymbol{b} = (b_j)_{j\geq 1}$ from Assumption 1 it holds that $\boldsymbol{b} \in \ell^p(\mathbb{N})$, i.e. $\sum_{j\geq 1} b_j^p < \infty$.

Assumption 3 Assume the fluctuation operators $(A_j)_{j\geq 1}$ to be arranged such that $\boldsymbol{b} = (b_j)_{j\geq 1}$ from Assumption 1 is non-increasing.

Proposition 1 ([18, Theorem 2]) *Under Assumption 1, for every parameter sequence $\boldsymbol{y} \in U = [-1/2, 1/2]^{\mathbb{N}}$ the parametric operator $A(\boldsymbol{y})$ is boundedly invertible. Furthermore, for any $\boldsymbol{y} \in U$ and any $f \in \mathscr{Y}'$, the weak parametric equation*

$${}_{\mathscr{Y}'}\langle A(\boldsymbol{y})q(\boldsymbol{y}), v \rangle_{\mathscr{Y}} = {}_{\mathscr{Y}'}\langle f, v \rangle_{\mathscr{Y}} , \quad \forall v \in \mathscr{Y}$$

admits a unique solution $q(\boldsymbol{y})$ and there holds the a-priori estimate

$$\sup_{\boldsymbol{y}\in U} \|q(\boldsymbol{y})\|_{\mathscr{X}} \leq C \|f\|_{\mathscr{Y}'} ,$$

where $C > 0$ is a constant independent of f.

Often, a quantity of interest (QoI) depending on the solution $q(\boldsymbol{y})$ is to be computed. We consider here as QoI a linear goal functional $G \in \mathscr{X}'$, and assume given a product probability measure $\boldsymbol{\mu}(\boldsymbol{y}) = \prod_{j\geq 1} \mu_j(y_j)$ on U. The goal of computation is the mathematical expectation

$$\mathbb{E}[G(q)] = \int_U G(q(\boldsymbol{y}))\, \boldsymbol{\mu}(\mathrm{d}\boldsymbol{y}) . \tag{3}$$

Examples of the goal functional G are point evaluation of the solution, or an average over (a subset of) the spatial domain. The following statement on the parametric regularity of the solution, i.e. a bound on the partial derivatives of $q(\boldsymbol{y})$ with respect to the y_j, will be required for the higher-order quasi-Monte Carlo method to be presented in Sect. 4.

Proposition 2 ([7, Theorem 2.2]) *Let* $\mathscr{F} = \{\boldsymbol{\nu} \in \mathbb{N}_0^{\mathbb{N}} : |\boldsymbol{\nu}| := \sum_{j\geq 1} \nu_j < \infty\}$ *denote the set of finitely supported multiindices and denote by* $\partial_{\boldsymbol{y}}^{\boldsymbol{\nu}}$ *partial derivatives of order* ν_j *with respect to coordinate* y_j *and let* $q(\boldsymbol{y})$ *be the solution to* (1) *with* $A(\boldsymbol{y})$ *as in* (2) *satisfying Assumption 1. Then, there exists a constant* $C > 0$ *and a sequence* $\boldsymbol{\beta}$ *satisfying Assumption 2 such that for all* $f \in \mathscr{Y}'$ *and every* $\boldsymbol{y} \in U$ *it holds that*

$$\forall \boldsymbol{\nu} \in \mathscr{F} : \quad \|\partial_{\boldsymbol{y}}^{\boldsymbol{\nu}} q(\boldsymbol{y})\|_{\mathscr{X}} \leq C|\boldsymbol{\nu}|!\boldsymbol{\beta}^{\boldsymbol{\nu}} := C\Big(\sum_{j\geq 1} \nu_j\Big)! \prod_{j\geq 1} \beta_j^{\nu_j} .$$

2.1 Approximation

In order to obtain a computable approximation to (3), three approximations are required: (i) dimension truncation of the affine-parametric operator from (2), (ii) Petrov–Galerkin discretization of the Eq. (1) based on the dimensionally truncated operator, and (iii) quasi-Monte Carlo approximation of the integral over $\boldsymbol{y} \in U$.

We denote by $A^s(\boldsymbol{y}) = A(y_1, \ldots, y_s, 0, \ldots)$ the dimensionally truncated operator, and by $q^s(\boldsymbol{y})$ the solution to (1) based on $A^s(\boldsymbol{y})$. Petrov–Galerkin discretization yields for fixed $\boldsymbol{y} \in U$ a discrete solution $q_h^s(\boldsymbol{y})$ approximating $q^s(\boldsymbol{y})$, where the discretization parameter h usually signifies the maximal meshwidth when using the finite element method. The third and final approximation is replacing the integral over U by an N-point QMC quadrature rule with point set $\mathscr{P}_N = \{\boldsymbol{y}^{(0)}, \ldots, \boldsymbol{y}^{(N-1)}\} \subset [0, 1]^s$, yielding the full approximation

$$\mathbb{E}[G(q)] = \int_U G(q(\boldsymbol{y}))\, \boldsymbol{\mu}(\mathrm{d}\boldsymbol{y}) \quad \approx \quad \frac{1}{N} \sum_{n=0}^{N-1} G(q_h^s(\boldsymbol{y}^{(n)} - \mathbf{1/2})) . \tag{4}$$

By the triangle inequality, we can write the total error $E_{s,h,N}$ as

$$\begin{aligned} E_{s,h,N} &= \Big| \int_U G(q(\boldsymbol{y}))\, \boldsymbol{\mu}(\mathrm{d}\boldsymbol{y}) - \frac{1}{N} \sum_{n=0}^{N-1} G(q_h^s(\boldsymbol{y}^{(n)} - \mathbf{1/2})) \Big| \\ &\leq \Big| \int_U G(q_h^s(\boldsymbol{y}))\, \boldsymbol{\mu}(\mathrm{d}\boldsymbol{y}) - \frac{1}{N} \sum_{n=0}^{N-1} G(q_h^s(\boldsymbol{y}^{(n)} - \mathbf{1/2})) \Big| \\ &\quad + \Big| \int_U G(q^s(\boldsymbol{y})) - G(q_h^s(\boldsymbol{y}))\, \boldsymbol{\mu}(\mathrm{d}\boldsymbol{y}) \Big| + \Big| \int_U G(q(\boldsymbol{y})) - G(q^s(\boldsymbol{y}))\, \boldsymbol{\mu}(\mathrm{d}\boldsymbol{y}) \Big| . \end{aligned} \tag{5}$$

The dimension truncation error is the last term in (5), which we will bound in the following section and approximate computationally in Sect. 5. In order to do the latter, we must still rely on an approximation of the form (4), in principle choosing N and h large and small enough, respectively, to ensure that the first two errors are negligible. We comment more on the choice of these values in Sect. 5 below.

3 Dimension Truncation Error Estimates

We begin by recalling existing estimates on the dimension truncation error pointwise in $\boldsymbol{y}$, i.e. $\|q(\boldsymbol{y}) - q^s(\boldsymbol{y})\|_{\mathscr{X}}$. Then, we detail in Theorem 1 a novel result which gives a statement on the convergence of the error of the integral $\left|\int_U G(q(\boldsymbol{y}) - q^s(\boldsymbol{y}))\,\mathrm{d}\boldsymbol{y}\right|$ in the truncation dimension s, improving upon known bounds.

Proposition 3 *For every $f \in \mathscr{Y}'$, $\boldsymbol{y} \in U$, $s \in \mathbb{N}$, denote by $q^s(\boldsymbol{y})$ the solution to a problem of the form $A^s(\boldsymbol{y})q^s(\boldsymbol{y}) = f$ with $A^s(\boldsymbol{y}) = A(y_1, \ldots, y_s, 0, \ldots)$. Let Assumptions 1 and 2 hold, and assume additionally that μ_j is such that $\int_U y_j\,\mu_j(\mathrm{d}y_j) = 0$ for all $j \geq 1$. Then, for a constant $C > 0$ which is independent of s and f it holds that*

$$\forall \boldsymbol{y} \in U: \quad \left\|q(\boldsymbol{y}) - q^s(\boldsymbol{y})\right\|_{\mathscr{X}} \leq C\|f\|_{\mathscr{Y}'} s^{-1/p+1}. \tag{6}$$

Proof See e.g. [1, 7, 15]. □

We now make the following additional assumption on the measure $\boldsymbol{\mu}$, noting that the first part holds in particular for all symmetric distributions.

Assumption 4 Assume that $\boldsymbol{\mu}(\boldsymbol{y}) = \prod_{j\geq 1} \mu_j(y_j)$ is a product probability measure and that the factor measures μ_j on the parameters y_j are such that for all $j \geq 1$ it holds that $\int_{-1/2}^{1/2} y_j\,\mu_j(\mathrm{d}y_j) = 0$ and $\int_{-1/2}^{1/2} y_j^k\,\mu_j(\mathrm{d}y_j) \leq C_k < \infty$ for all integers $k \geq 2$.

In [15], the bound $\left|\int_U q(\boldsymbol{y}) - q^s(\boldsymbol{y})\,\mathrm{d}\boldsymbol{y}\right| \leq Cs^{-2(1/p-1)}$ was shown for equations of the type considered here under Assumption 2 and $\int_{-1/2}^{1/2} y_j\,\mu(\mathrm{d}y_j) = 0$, which we improve here to $\mathscr{O}(s^{-2/p+1})$, which is one order better. We begin by proving the following Lemma.

Lemma 1 *Let $A^s(\boldsymbol{y})$ denote the operator $A(\boldsymbol{y})$ of the form (2) truncated after dimension s, i.e. $A^s(\boldsymbol{y}) = A(y_1, \ldots, y_s, 0, \ldots)$. Assume Assumptions 1 and 3. Then, for sufficiently large s it holds that*

$$\sup_{\boldsymbol{y}\in U} \|(A^s(\boldsymbol{y}))^{-1}(A(\boldsymbol{y}) - A^s(\boldsymbol{y}))\|_{\mathscr{L}(\mathscr{X})} \leq \frac{1}{2-\kappa}\sum_{j>s} b_j < 1. \tag{7}$$

Proof We have $\|(A^s)^{-1}(A - A^s)\|_{\mathscr{L}(\mathscr{X})} \le \|(A^s)^{-1}A_0\|_{\mathscr{L}(\mathscr{X})}\|A_0^{-1}(A - A^s)\|_{\mathscr{L}(\mathscr{X})}$, which we bound individually. For the first factor, Assumption 1 implies that for all $\boldsymbol{y} \in U$ it holds that $\|A_0^{-1}A^s\|_{\mathscr{L}(\mathscr{X})} \le \sum_{j=1}^s y_j\|A_0^{-1}A_j\|_{\mathscr{L}(\mathscr{X})} < 1$, implying with the Neumann series the bound

$$\|(A^s)^{-1}A_0\|_{\mathscr{L}(\mathscr{X})} = \|(A_0^{-1}A^s)^{-1}\|_{\mathscr{L}(\mathscr{X})} \le \frac{1}{1 - \frac{1}{2}\sum_{j=1}^s b_j} \le \frac{1}{1 - \frac{1}{2}\|\boldsymbol{b}\|_{\ell^1(\mathbb{N})}}. \tag{8}$$

For the second factor, we have

$$\|A_0^{-1}(A - A^s)\|_{\mathscr{L}(\mathscr{X})} = \|A_0^{-1}\sum_{j>s} y_jA_j\|_{\mathscr{L}(\mathscr{X})} \le \frac{1}{2}\sum_{j>s}\|A_0^{-1}A_j\|_{\mathscr{L}(\mathscr{X})} = \frac{1}{2}\sum_{j>s} b_j.$$

Combining these two bounds and recalling $\|\boldsymbol{b}\|_{\ell^1(\mathbb{N})} \le \kappa$ yields the first inequality. The bound is less than 1 for sufficiently large s since $\boldsymbol{b} \in \ell^1(\mathbb{N})$ and the b_j are assumed in Assumption 3 to be non-increasing. □

Theorem 1 *For every $s \in \mathbb{N}$, denote by $q^s(\boldsymbol{y})$ the solution to a problem of the form $A^s q^s = f$ with A^s as in (2) where $\boldsymbol{y} = (y_1, \dots, y_s, 0, \dots)$. Let Assumptions 1–4 hold. Then, for any $s \in \mathbb{N}$, $f \in \mathscr{Y}'$ and $G \in \mathscr{X}'$ there exists a constant $C > 0$ which is independent of s, f and G such that*

$$\left|\int_U G(q(\boldsymbol{y}) - q^s(\boldsymbol{y}))\,\boldsymbol{\mu}(\mathrm{d}\boldsymbol{y})\right| \le C\|G\|_{\mathscr{X}'}\|f\|_{\mathscr{Y}'}s^{-2/p+1}. \tag{9}$$

Proof Assumption 1 implies bounded invertibility of $A(\boldsymbol{y})$ and $A^s(\boldsymbol{y})$ for any $\boldsymbol{y} \in U$, thus we can write (omitting the argument $\boldsymbol{y}$ for legibility) $A = A^s + A - A^s = A^s(I + (A^s)^{-1}(A - A^s))$. We aim to write the inverse of A given in this form as a Neumann series, which is justified for suitably large s by Lemma 1. Thus, we have

$$A^{-1} = \big(I + (A^s)^{-1}(A - A^s)\big)^{-1}(A^s)^{-1} = \sum_{k\ge 0}\big(-(A^s)^{-1}(A - A^s)\big)^k(A^s)^{-1}.$$

Fubini's theorem, together with linearity of G and of the integral then implies

$$\begin{aligned}
\int_U G(q(\boldsymbol{y}) - q^s(\boldsymbol{y}))\,\boldsymbol{\mu}(\mathrm{d}\boldsymbol{y}) &= \int_U G\big((A^{-1} - (A^s)^{-1})f\big)\,\boldsymbol{\mu}(\mathrm{d}\boldsymbol{y}) \\
&= \int_U G\Big(\sum_{k\ge 1}\Big(-\sum_{j>s} y_j(A^s)^{-1}A_j\Big)^k q^s\Big)\,\boldsymbol{\mu}(\mathrm{d}\boldsymbol{y}) \\
&= \sum_{k\ge 1}(-1)^k\int_U G\Big(\Big(\sum_{j>s} y_j(A^s)^{-1}A_j\Big)^k q^s\Big)\,\boldsymbol{\mu}(\mathrm{d}\boldsymbol{y}).
\end{aligned} \tag{10}$$

We assume now additionally Assumptions 2 and 3. Then, using a similar approach as in [15], we obtain for $k' \in \mathbb{N}$ and a constant $C_{f,G} > 0$ the bound

$$\Big|\sum_{k\geq k'}(-1)^k \int_U G\Big(\Big(\sum_{j>s} y_j (A^s)^{-1} A_j\Big)^k q^s\Big)\,\boldsymbol{\mu}(\mathrm{d}\boldsymbol{y})\Big| \tag{11}$$

$$\leq C\|G\|_{\mathscr{X}'}\|f\|_{\mathscr{Y}'} \sup_{\boldsymbol{y}\in U}\sum_{k\geq k'}\|(A^s)^{-1}(A-A^s)\|_{\mathscr{L}(\mathscr{X})}^k \leq C_{f,G} s^{k'(-1/p+1)}\,.$$

The above gives a bound for the remainder of the sum over k, starting at term k'; our goal now is to bound the terms up to k' by a better estimate. To this end, we use linearity of G and the integral, as well as the identity $\big(\sum_{j>s} y_j (A^s)^{-1}A_j\big)^k = \sum_{\boldsymbol{\eta}\in\{j>s\}^k}\prod_{i=1}^k (y_{\eta_i}(A^s)^{-1}A_{\eta_i})$, which respects the generally non-commutative nature of the operators. Denoting by $U_s = [-1/2, 1/2]^s$ the truncated parameter domain and setting $\boldsymbol{\mu}_s(\boldsymbol{y}_s) := \prod_{j=1}^s \mu_j(y_j)$ we obtain

$$\int_U G\Big(\Big(\sum_{j>s} y_j (A^s)^{-1}A_j\Big)^k q^s\Big)\,\boldsymbol{\mu}(\mathrm{d}\boldsymbol{y})$$

$$= \sum_{\boldsymbol{\eta}\in\{j>s\}^k}\int_U G\Big(\Big(\prod_{i=1}^k y_{\eta_i}\Big)\prod_{i=1}^k((A^s)^{-1}A_{\eta_i})q^s\Big)\,\boldsymbol{\mu}(\mathrm{d}\boldsymbol{y})$$

$$= \sum_{\boldsymbol{\eta}\in\{j>s\}^k}\int_U\Big(\prod_{i=1}^k y_{\eta_i}\Big)\,\boldsymbol{\mu}(\mathrm{d}\boldsymbol{y})\int_{U_s} G\Big(\prod_{i=1}^k((A^s)^{-1}A_{\eta_i})q^s\Big)\,\boldsymbol{\mu}_s(\mathrm{d}\boldsymbol{y}_s)\,.$$

It is important to note that the functional G in the last statement is applied to an expression that depends only on the first s dimensions (through A^s and q^s), allowing the integral with respect to y_j for $j > s$ to be separated out. See also the proof of [11, Proposition 5.1] for a similar argument. For any $\boldsymbol{\eta} \in \{j > s\}^k$ it holds that

$$\int_U G\Big(\Big(\prod_{i=1}^k y_{\eta_i}\Big)\prod_{i=1}^k((A^s)^{-1}A_{\eta_i})q^s\Big)\,\boldsymbol{\mu}(\mathrm{d}\boldsymbol{y})$$

$$= \int_U\int_{U_s}\Big(\prod_{i=1}^k y_{\eta_i}\Big)G\Big(\prod_{i=1}^k((A^s)^{-1}A_{\eta_i})q^s\Big)\,\boldsymbol{\mu}_s(\mathrm{d}\boldsymbol{y}_s)\,\boldsymbol{\mu}(\mathrm{d}\boldsymbol{y})$$

$$= \int_U\Big(\prod_{i=1}^k y_{\eta_i}\Big)\int_{U_s} G\Big(\prod_{i=1}^k((A^s)^{-1}A_{\eta_i})q^s\Big)\,\boldsymbol{\mu}_s(\mathrm{d}\boldsymbol{y}_s)\,\boldsymbol{\mu}(\mathrm{d}\boldsymbol{y})$$

$$= \int_U\Big(\prod_{i=1}^k y_{\eta_i}\Big)\,\boldsymbol{\mu}(\mathrm{d}\boldsymbol{y})\int_{U_s} G\Big(\prod_{i=1}^k((A^s)^{-1}A_{\eta_i})q^s\Big)\,\boldsymbol{\mu}_s(\mathrm{d}\boldsymbol{y}_s)\,.$$

We will now introduce various definitions required below: let $\boldsymbol{\nu}(\boldsymbol{\eta}) := (\#\{i = 1, \dots, k : \eta_i = j\})_{j\geq 1}$, i.e. $\nu_j(\boldsymbol{\eta}) \in \{0, \dots, k\}$ for $j \geq 1$, define the support of a multiindex $\boldsymbol{\nu} \in \mathscr{F}$ by $\operatorname{supp}(\boldsymbol{\nu}) := \{j \in \mathbb{N} : \nu_j \neq 0\}$ and let $\#S$ denote the cardinality of a set S. Note that $\min \operatorname{supp} \boldsymbol{\nu}(\boldsymbol{\eta}) > s$ as well as $\# \operatorname{supp}(\boldsymbol{\nu}(\boldsymbol{\eta})) \leq k$ for $\boldsymbol{\eta} \in \{j > s\}^k$ as in the sums above. Thus, for every such $\boldsymbol{\eta}$ we can write $\prod_{i=1}^k y_{\eta_i} = \prod_{j>s} y_j^{\nu_j(\boldsymbol{\eta})} = \boldsymbol{y}^{\boldsymbol{\nu}(\boldsymbol{\eta})}$, where the product is over a finite set since $\boldsymbol{\nu}(\boldsymbol{\eta})$ is finitely supported.

Assumption 4 now directly implies that all terms where $\nu_j = 1$ for at least one $j > s$ are zero. For $k = 1$, $\nu_j \in \{0, 1\}$ for all $j > s$, thus all terms contain at least one exponent equal to 1 and are zero. We consider in the following $k \geq 2$, and aim to rewrite the sum over $\boldsymbol{\eta}$ as a sum over the set $\mathscr{F}_{k,s} := \{\boldsymbol{\nu} \in \mathscr{F} : |\boldsymbol{\nu}| = k, \min \operatorname{supp}(\boldsymbol{\nu}) > s, \nu_j \neq 1\, \forall j\}$. For all $\boldsymbol{\nu} \in \mathscr{F}_{k,s}$ it holds that $\# \operatorname{supp}(\boldsymbol{\nu}) \leq k/2$, since the smallest nonzero element is 2, and we have the condition $|\boldsymbol{\nu}| = k$. We define $c_{\boldsymbol{\nu}} := \left| \int_U \boldsymbol{y}^{\boldsymbol{\nu}} \boldsymbol{\mu}(\mathrm{d}\boldsymbol{y}) \right|$, which, since $\# \operatorname{supp}(\boldsymbol{\nu}) \leq k/2$ and recalling the definition of C_k from Assumption 4 fulfills $c_{\boldsymbol{\nu}} \leq \prod_{j\in \operatorname{supp}(\boldsymbol{\nu})} C_{\nu_j} \leq \left(\max_{j>s} C_{\nu_j}\right)^{k/2}$. Defining $C_{f,G} := \|G\|_{\mathscr{X}'} \|f\|_{\mathscr{Y}'}$ and writing $\boldsymbol{\mu}_{\{1:s\}}(\boldsymbol{y}) = \prod_{j=1}^s \mu_j(y_j)$, we have

$$
\begin{aligned}
&\left| \sum_{\boldsymbol{\eta}\in\{j>s\}^k} \int_U \Big(\prod_{j>s} y_j^{\nu_j(\boldsymbol{\eta})}\Big) \boldsymbol{\mu}(\mathrm{d}\boldsymbol{y}) \int_{U_s} G\Big(\prod_{i=1}^k ((A^s)^{-1} A_{\eta_i}) q^s\Big) \boldsymbol{\mu}_{\{1:s\}}(\mathrm{d}\boldsymbol{y}) \right| \\
&\qquad \leq C_{f,G} \sum_{\boldsymbol{\eta}\in\{j>s\}^k} c_{\boldsymbol{\nu}(\boldsymbol{\eta})} \int_{U_s} \prod_{i=1}^k \left\| ((A^s)^{-1} A_{\eta_i}) \right\|_{\mathscr{L}(\mathscr{X})} \boldsymbol{\mu}_{\{1:s\}}(\mathrm{d}\boldsymbol{y}) \\
&\qquad = C_{f,G} \sum_{\boldsymbol{\nu}\in\mathscr{F}_{k,s}} \binom{k}{\boldsymbol{\nu}} c_{\boldsymbol{\nu}} \int_{U_s} \prod_{j>s} \left\| ((A^s)^{-1} A_j) \right\|_{\mathscr{L}(\mathscr{X})}^{\nu_j} \boldsymbol{\mu}_{\{1:s\}}(\mathrm{d}\boldsymbol{y})\,.
\end{aligned}
$$

We bound $\sup_{\boldsymbol{y}\in U} \|(A^s)^{-1} A_j\|_{\mathscr{L}(\mathscr{X})} \leq (1 - \kappa/2)^{-1} b_j =: \widetilde{b}_j$ similar to (8) and $\binom{k}{\boldsymbol{\nu}} c_{\boldsymbol{\nu}} \leq c_k := (\max_{j>s} C_{\nu_j})^{k/2} k!$. Let $\widetilde{\mathscr{F}}_{k,s} := \{\boldsymbol{\nu} \in \mathscr{F} : |\boldsymbol{\nu}|_\infty \leq k, \boldsymbol{\nu} \neq \boldsymbol{0}, \min \operatorname{supp}(\boldsymbol{\nu}) > s, \nu_j \neq 1\, \forall j\}$, where we observe that $\widetilde{\mathscr{F}}_{k,s} \supset \mathscr{F}_{k,s}$, which yields with $\widetilde{b}_j \geq 0$ for all $j \geq 1$

$$
C_{f,G} \sum_{\boldsymbol{\nu}\in\mathscr{F}_{k,s}} \binom{k}{\boldsymbol{\nu}} c_{\boldsymbol{\nu}} \int_{U_s} \prod_{j>s} \left\| ((A^s)^{-1} A_j) \right\|_{\mathscr{L}(\mathscr{X})}^{\nu_j} \boldsymbol{\mu}_{\{1:s\}}(\mathrm{d}\boldsymbol{y}) \leq c_k C_{f,G} \sum_{\boldsymbol{\nu}\in\widetilde{\mathscr{F}}_{k,s}} \prod_{j>s} \widetilde{b}_j^{\nu_j}\,.
$$

We now rewrite the sum over $\widetilde{\mathscr{F}}_{k,s}$ as the product of a sum, since we notice that every element of the set $\widetilde{\mathscr{F}}_{k,s}$ (resulting in a term $\widetilde{\boldsymbol{b}}^{\boldsymbol{\nu}}$) corresponds to one term of the product $\prod_{j>s}(1 + \sum_{\ell=2}^k \widetilde{b}_j^{\ell})$, with the exception of the additional term 1 (corresponding to $\boldsymbol{\nu} = \boldsymbol{0}$, which is excluded in $\widetilde{\mathscr{F}}_{k,s}$), that we subtract. Defining $\widehat{b}_j^2 := \widetilde{b}_j^2 (1 - \widetilde{b}_j)^{-1}$, it holds that $(\widehat{b}_j)_{j\geq 1} \in \ell^p(\mathbb{N})$ and $\widehat{b}_j < 1$ for all $j > s$ for suitably large s. The first term in parenthesis below can thus be rewritten using basic properties of the geometric series,

$$\sum_{\nu \in \widetilde{\mathscr{F}}_{k,s}} \widetilde{b}^{\nu} = \prod_{j>s}\Big(1+\sum_{\ell=2}^{k}\widetilde{b}_j^{\ell}\Big) - 1 = \prod_{j>s}\Big(1+\widetilde{b}_j^2\frac{1-\widetilde{b}_j^{k-1}}{1-\widetilde{b}_j}\Big) - 1 \le \prod_{j>s}\big(1+\widehat{b}_j^2\big) - 1$$

$$= \exp\Big(\sum_{j>s}\log\big(1+\widehat{b}_j^2\big)\Big) - 1 \le C\sum_{j>s}\widehat{b}_j^2 \le \frac{C}{1-s^{-2/p+1}}s^{-2/p+1}. \quad (12)$$

We recall the Neumann series (10), for which the $k = 1$ term is zero, and split it into a sum over $k = 2, \dots, k' - 1 < \infty$, where each term is bounded from above by (12) times the constants $c_k C_{f,G} < \infty$, and a remainder with $k \ge k'$ for which we use (11). For each $p < 1$, the choice $k' = k'(p) = \lceil (2-p)/(1-p) \rceil < \infty$ ensures that the remainder converges at least as rapidly as the estimate $s^{-2/p+1}$. Collecting terms then yields the statement. □

For fast decay of the sequence $\boldsymbol{b}$ from Assumption 2, i.e. for small values of p, the convergence rate of the dimension truncation error of the integral from Theorem 1 can be quite high. Balancing the required finite element and quadrature approximation errors with the possibly very small values of the dimension truncation error would result in a large number of samples, and consequently a large amount of work, even for moderate s. A standard Monte Carlo method converging like $N^{-1/2}$ in the number of samples N is thus not feasible, as the number of required samples is much too large. A QMC method converging at rate 1 is better, but for small p is still not accurate enough to allow these computations to be executed in a reasonable amount of time.

Other approaches may converge more quickly, for example adaptive Smolyak or sparse grid-type quadrature methods [12, 17]. However, these are inherently serial, and the computational cost due to the involved internal bookkeeping overhead also increases rapidly if high accuracies are required. Thus, the only method known to the author to perform well enough (in terms of convergence rate and amenability to parallel implementation) for such measurements to be performed for a large range of values of $0 < p < 1$ and in high enough dimension to yield conclusive results is the higher-order QMC method of [7, 13] based on interlaced polynomial lattice rules, which we now briefly describe.

4 Interlaced Polynomial Lattice Rules

For the presentation of interlaced polynomial lattice (IPL) rules, we require some definitions and notation. A polynomial lattice rule (without interlacing for the moment) is an equal-weight quadrature rule with $N = b^m$ points for some prime number b and positive integer m, and is given by a *generating vector* whose components are polynomials of degree less than m over the finite field $\mathbb{Z}_b$. Let $\mathbb{Z}_b[x]$ denote the set of all polynomials over $\mathbb{Z}_b$, i.e. polynomials of the form $\sum_{k=0}^{m-1} \xi_k x^k$ with $\xi_k \in \mathbb{Z}_b$. Then, the generating vector is denoted by $\boldsymbol{q} \in (\mathbb{Z}_b[x])^s$ with $\boldsymbol{q} = (q_j(x))_{j=1}^{s}$. We associate with each integer $n = 0, \dots, b^m - 1$ a polynomial $n(x) = \sum_{k=0}^{m-1} \xi_k x^k$, where ξ_k are the digits of n in base b, that is $n = \xi_0 + \xi_1 b + \xi_2 b^2 + \cdots + \xi_{m-1} b^{m-1}$.

To obtain points in $[0, 1]^s$ from the generating vector $\boldsymbol{q}$, we require the mapping $v_m : \mathbb{Z}_b(x^{-1}) \to [0, 1)$ which is given for any integer w by

$$v_m\left(\sum_{k=w}^{\infty} \xi_k x^{-k}\right) = \sum_{k=\max(1,w)}^{m} \xi_k b^{-k}.$$

For an irreducible polynomial $P \in \mathbb{Z}_b[x]$ of degree equal to m, the jth component of the nth point of the point set $\mathscr{P}_N = \{\boldsymbol{y}^{(0)}, \ldots, \boldsymbol{y}^{(N-1)}\}$ is given by

$$(\boldsymbol{y}^{(n)})_j = v_m\left(\frac{n(x)q_j(x)}{P(x)}\right), \quad n = 0, \ldots, N-1, \quad j = 1, \ldots, s.$$

To obtain orders of convergence higher than one, we require an additional interlacing step. To this end, we denote the digit interlacing function of $\alpha \in \mathbb{N}$ points as $D_\alpha : [0, 1)^\alpha \to [0, 1)$,

$$D_\alpha(x_1, \ldots, x_\alpha) = \sum_{a=1}^{\infty} \sum_{j=1}^{\alpha} \xi_{j,a} b^{-j-(a-1)\alpha},$$

where $\xi_{j,a}$ is the ath digit in the expansion of the jth point $x_j \in [0, 1)$ in base b^{-1}, $x_j = \xi_{j,1} b^{-1} + \xi_{j,2} b^{-2} + \ldots$. For vectors in αs dimensions, digit interlacing is defined block-wise and denoted by $\mathscr{D}_\alpha : [0, 1)^{\alpha s} \to [0, 1)^s$ with

$$\mathscr{D}_\alpha(x_1, \ldots, x_{\alpha s}) = \left(D_\alpha(x_1, \ldots, x_\alpha), D_\alpha(x_{\alpha+1}, \ldots, x_{2\alpha}), \ldots, D_\alpha(x_{(s-1)\alpha+1}, \ldots, x_{s\alpha})\right).$$

For a generating vector $\boldsymbol{q} \in (\mathbb{Z}_b[x])^{\alpha s}$ containing α components for each of the s dimensions, the interlaced polynomial lattice point set is $\mathscr{D}_\alpha(\widetilde{\mathscr{P}}_N) \subset [0, 1)^s$, where $\widetilde{\mathscr{P}}_N \subset [0, 1)^{\alpha s}$ denotes the (classical) polynomial lattice point set in αs dimensions with generating vector $\boldsymbol{q}$. For more details on this method, see e.g. [7, 10, 13]. The following proposition states the higher order rates that are obtainable under suitable sparsity assumptions of the form stated in Proposition 2.

Proposition 4 ([7, Theorem 3.1]) *For $m \geq 1$ and a prime number b, denote by $N = b^m$ the number of QMC points. Let $s \geq 1$ and $\boldsymbol{\beta} = (\beta_j)_{j\geq 1}$ be a sequence of positive numbers, and let $\boldsymbol{\beta}_s = (\beta_j)_{1\leq j\leq s}$ denote the first s terms of $\boldsymbol{\beta}$. Assume that $\boldsymbol{\beta} \in \ell^p(\mathbb{N})$ for some $p \in (0, 1)$.*

If there exists a $c > 0$ such that for $\alpha := \lfloor 1/p \rfloor + 1$ a function F satisfies

$$\forall \boldsymbol{\nu} \in \{0, 1, \ldots, \alpha\}^s, \forall s \in \mathbb{N} : \quad |(\partial_{\boldsymbol{y}}^{\boldsymbol{\nu}} F)(\boldsymbol{y})| \leq c\, |\boldsymbol{\nu}|!\, \boldsymbol{\beta}_s^{\boldsymbol{\nu}}, \tag{13}$$

then an interlaced polynomial lattice (IPL) rule of order α with N points can be constructed in $\mathcal{O}(\alpha\, s\, N \log N + \alpha^2 s^2 N)$ operations, such that for the quadrature error it holds that

$$|I_s(F) - \mathscr{Q}_{N,s}(F)| \leq C_{\alpha,\beta,b,p,F}\, N^{-1/p}, \tag{14}$$

where the constant $C_{\alpha,\beta,b,p,F} < \infty$ *is independent of* s *and* N.

5 Experiments

We consider three different examples that fit into the affine-parametric framework, and measure the dimension truncation error for each one, both for the pointwise case and for the integral case. For the latter, we verify in all three cases that the bound from Theorem 1 corresponds to the measured rates, for various values of the fluctuation summability exponent p. The results thus give concrete evidence supporting the sharpness of both estimates. Below, we specify the fluctuation decay rate $\zeta > 1$, which implies p-summability for any $p > 1/\zeta$. Using the limiting value $p = 1/\zeta$, we expect the dimension truncation convergence rate $s^{-\zeta+1}$ for the pointwise case and rate $s^{-2\zeta+1}$ for the integral case, cf. Proposition 3 and Theorem 1, respectively. For the computations of the pointwise error, we use the parameter value $\boldsymbol{y} = e^{-2}\mathbf{1} = e^{-2}(1, 1, \ldots) \in U$.

5.1 Example 1: Test Integrand

The first example is designed to serve as a simplified test case that does not require finite element discretization, since the operator equation simplifies to an algebraic equation, see also [10, Eq. 18]. We seek to approximate the integral over $U = [-1/2, 1/2]^{\mathbb{N}}$ of the "solution" $q(\boldsymbol{y})$ which is given in this case by

$$q(\boldsymbol{y}) = \Big(1 + \sum_{j\geq 1} y_j c_j\Big)^{1}, \quad c_j = \sigma j^{-\zeta}, \quad \sigma = 0.1\,. \tag{15}$$

We thus circumvent the finite element discretization error, but must still ensure that the QMC error is small enough. For a rough estimate of the number of QMC points $N = 2^m$, consider decay rate ζ and truncation dimension $s = 2^\beta$ for some $\beta > 1$. Then, by Theorem 1 the dimension truncation error is of order $s^{-2\zeta+1} = 2^{(-2\zeta+1)\beta}$. By Proposition 4, the QMC error converges like $N^{-\zeta} = 2^{-m\zeta}$. Assuming the constants in the error estimates to be equal, we thus require $2^{-m\zeta} < 2^{(-2\zeta+1)\beta}$, implying $m > (2 - 1/\zeta)\beta$. For $s = 1024$ we have $\beta = 10$ and obtain with $\zeta = 2$ the condition $m > 15$. Below, we use $m = 18$, which suffices for the considered values of ζ up to 3 and yields clear measurements of the integral dimension truncation error.

For this example, since no finite element solver is needed and efficiency is not such an issue, the implementation was conducted in Python with the higher-order QMC rules applied with the `pyQMC` library, see [8]. The required generating vectors

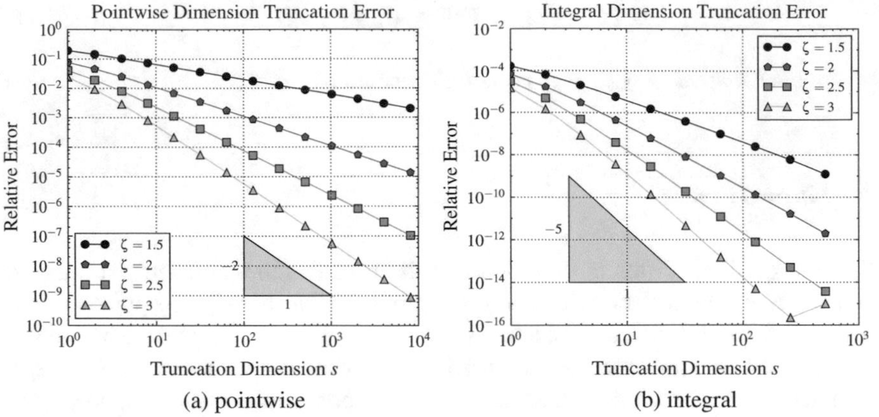

Fig. 1 Dimension truncation error for the test integrand (15). **a** pointwise dimension truncation error $|q(\boldsymbol{y}) - q^s(\boldsymbol{y})|$ for $\boldsymbol{y} = e^{-2}\mathbf{1}$ with $q(\boldsymbol{y})$ approximated by a reference value in $s = 2^{15}$ dimensions, with expected rate $-\zeta + 1$. **b** integral dimension truncation error $|\int_U q(\boldsymbol{y}) - q^s(\boldsymbol{y})\,\mathrm{d}\boldsymbol{y}|$ with reference dimension $s = 1024$ and expected rate $-2\zeta + 1$. Higher-order QMC based on IPL rules was used with $N = 2^{18}$ points.The expected rates are clearly observed in both cases

were obtained by fast CBC construction [10] and are available at [8] under the heading "Standard SPOD Weights". For the computations below, we used the parameters $\alpha = 2$, $C = 0.1$, and $\theta = 0.2$. The reference approximations in $s = 1024$ dimensions with an IPL rule based on $N = 2^{18}$ points are given for $\zeta = 1.5, 2, 2.5, 3$ by 1.0010038828766668, 1.0009035434306022, 1.0008655151823671, and 1.0008491 110838873, respectively.

As shown in the results in Fig. 1, the expected convergence rates can be clearly observed in both cases. We note that this example additionally allows straightforward computation of the integrals in (10) for arbitrary k, allowing verification of the individual estimates in the proof.

5.2 Example 2: Diffusion Equation in One Dimension

We formulate here a model diffusion equation in spatial dimension $d = 1, 2$ for use in this and the next example. Denoting by $D \subset \mathbb{R}^d$ a bounded domain, for any $\boldsymbol{y} \in U$ we seek $q(\cdot, \boldsymbol{y}) \in \mathscr{X} = \mathscr{Y} = H_0^1(D)$ such that

$$-\nabla \cdot \big(u(x, \boldsymbol{y})\nabla q(x, \boldsymbol{y})\big) = f(x) \text{ in } D, \quad q(x, \boldsymbol{y}) = 0 \text{ on } \partial D\,, \tag{16}$$

where $u(x, \boldsymbol{y}) \in \mathbb{R}$ denotes for each $\boldsymbol{y}$ a spatially varying diffusion coefficient. It is well-known that the following assumption implies that there exists a unique solution to (16) for any sequence $\boldsymbol{y} \in U$.

Assumption 5 *(Uniform ellipticity)* There exist constants $u^-, u^+ > 0$ such that for all $\mathbf{y} \in U$ and for almost every $x \in D$ it holds that $0 < u^- \leq u(x; \mathbf{y}) \leq u^+$.

An affine-parametric partial differential equation is obtained for example by the following choice of coefficient parametrization,

$$u(x; \mathbf{y}) = u_0 + \sum_{j \geq 1} y_j \psi_j(x), \quad x \in D, \ \mathbf{y} \in U. \tag{17}$$

In the one-dimensional case we use the parametric basis functions $\psi_{2j}(x) = (2j)^{-\zeta} \sin(j\pi x)$ and $\psi_{2j-1}(x) = (2j-1)^{-\zeta} \cos(j\pi x)$.

For simplicity of implementation, we compute the convergence of the pointwise dimension truncation error by applying a linear goal functional $G \in \mathcal{X}'$ and observing with Proposition 3 that

$$\left|G(q(\mathbf{y})) - G(q^s(\mathbf{y}))\right| \leq \|G\|_{\mathcal{X}'} \|(q - q^s)(\mathbf{y})\|_{\mathcal{X}} \leq C \|G\|_{\mathcal{X}'} \|f\|_{\mathcal{Y}'} s^{-1/p+1}.$$

We choose as goal functional integration over the spatial domain D, $G(q(\mathbf{y})) = \int_D q(x, \mathbf{y}) \, dx$ and set $f(x) = 10x$. Finite element discretization with standard piecewise linear finite elements on an equidistant mesh of $D = [0, 1]$ with meshwidth h is used. Since no exact solution is available, we resort to using a reference solution with truncation dimension chosen to be twice the number of dimensions in the most precise measurement, see the caption of Fig. 2 for details. The finite element meshwidth was the same for all computations and chosen to be $h = 2^{-18}$. Note that this is not sufficient to completely remove the finite element error; in the plots below,

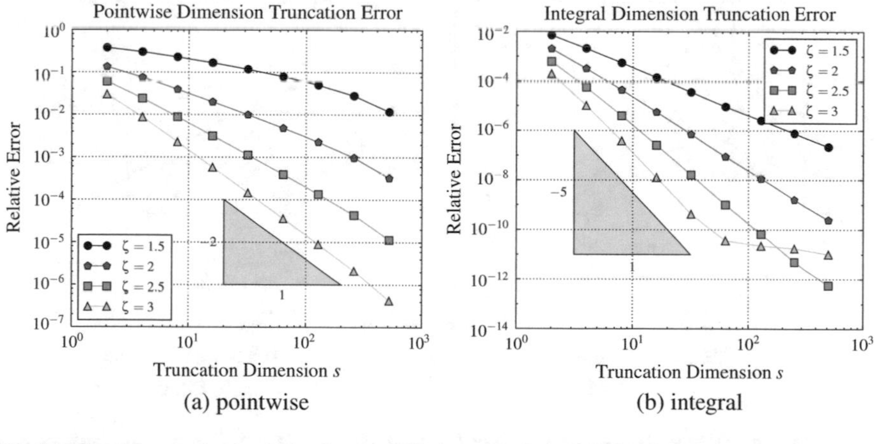

Fig. 2 Dimension truncation error for the diffusion equation with $d = 1$. **a** pointwise dimension truncation error $|q_h(\mathbf{y}) - q_h^s(\mathbf{y})|$ for $\mathbf{y} = e^{-2}\mathbf{1}$ with $q_h(\mathbf{y})$ approximated by a reference value in $s = 2^{15}$ dimensions, with expected rate $-\zeta + 1$. **b** integral dimension truncation error $|\int_U q_h(\mathbf{y}) - q_h^s(\mathbf{y}) \, d\mathbf{y}|$ with reference dimension $s = 1024$ and expected rate $-2\zeta + 1$. The expected rates are clearly observed in both cases

we consider convergence of $\int_U G(q_h^s(\boldsymbol{y}))\,\mathrm{d}\boldsymbol{y}$ to $\int_U G(q_h(\boldsymbol{y}))\,\mathrm{d}\boldsymbol{y}$, for fixed h. In this case, we can use the same generating vectors as for Example 1 but with $m = 20$.

5.3 Example 3: Diffusion Equation in Two Dimensions

We consider here again (16) as in Example 2, but in two spatial dimensions, requiring a different choice of fluctuation basis, see also related experiments in [4, 5]. The parametrization is given in terms of the eigenfunctions of the Dirichlet Laplacian on $D = (0, 1)^2$, where we choose the fluctuations $\psi_j(x) = (k_{1,j}^2 + k_{2,j}^2)^{-\zeta} \sin(\pi k_{1,j} x_1) \sin(\pi k_{2,j} x_2)$ by reordering the tuples $(k_{1,j}, k_{2,j}) \in \mathbb{N}^2$ such that $(\|\psi_j(x)\|_{L^\infty(D)})_{j\geq 1}$ is non-increasing. Choosing the deterministic right-hand side $f(x) = 100x_1$ results in an affine-parametric PDE satisfying Assumption 2 with $p > 1/\zeta$. The pointwise dimension truncation error is again computed by considering a goal functional as detailed in Example 2 above, where this time we choose as goal functional the integral over $\widetilde{D} = (1/2, 1)^2 \subset D$, i.e. $G(q(\boldsymbol{y})) = \int_{\widetilde{U}} q(x, \boldsymbol{y})\,\mathrm{d}x$.

For spatial discretization, we use a tensor product mesh with nodes obtained from the Cartesian product of equidistant nodes on (0, 1), with standard piecewise bilinear finite element basis functions. The one-dimensional meshwidth is 2^{-6}, resulting in $\mathcal{O}(2^{12})$ degrees of freedom. In this example and the previous one, a C++ implementation was used for efficiency reasons. The IPL rules were applied using the gMLQMC library [9] and the evaluation was conducted on the Piz Daint HPC system of CSCS with up to 1440 parallel processes. The generating vectors used here are available at [8] under the heading "SPOD Weights for 2d Diffusion Equation" (Fig. 3).

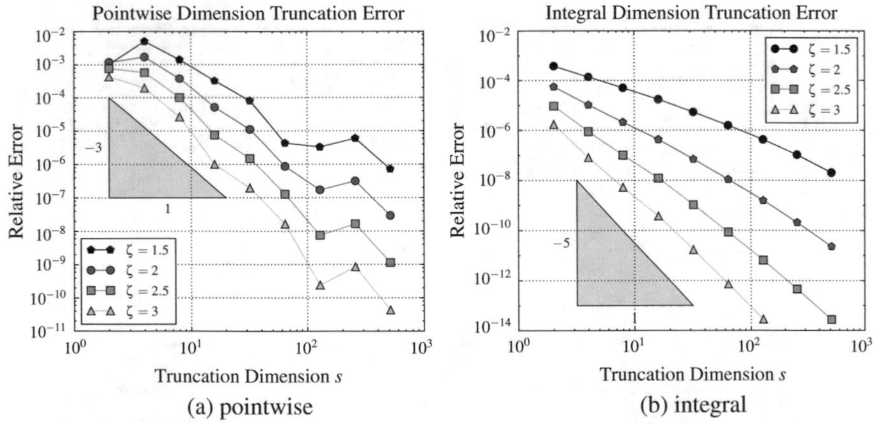

Fig. 3 Dimension truncation error for the diffusion equation with $d = 2$. **a** pointwise dimension truncation error $|q_h(\boldsymbol{y}) - q_h^s(\boldsymbol{y})|$ for $\boldsymbol{y} = e^{-2}\mathbf{1}$ with $q_h(\boldsymbol{y})$ approximated by a reference value in $s = 2^{15}$ dimensions, with expected rate $-\zeta + 1$. **b** integral dimension truncation error $|\int_U q_h(\boldsymbol{y}) - q_h^s(\boldsymbol{y})\,\mathrm{d}\boldsymbol{y}|$ with reference dimension $s = 1024$ and expected rate $-2\zeta + 1$. The expected rate is clearly observed in b, while in a the rate seems to be one order better than expected

6 Conclusions

We consider the error committed by truncating countably affine-parametric operator equations to a finite number of terms, and prove the convergence rate $s^{-2/p+1}$ where s is the truncation dimension and $p < 1$ the summability of the sequence of fluctuation operator norms, improving on the rate $s^{2(-1/p+1)}$ for the case of integration over a sequence of parameters. Numerical experiments verify this rate for a test integrand and two PDE examples, in one and two spatial dimensions, using up to $s = 1024$ parametric dimensions. Measurements of the pointwise dimension truncation error are also given, confirming the established theory.

Acknowledgements This work is supported by the Swiss National Science Foundation (SNSF) under project number SNF149819 and by the Swiss National Supercomputing Centre (CSCS) under Project ID d41.

References

1. Babuška, I., Tempone, R., Zouraris, G.E.: Galerkin finite element approximations of stochastic elliptic partial differential equations. SIAM J. Numer. Anal. **42**(2), 800–825 (2004). https://doi.org/10.1137/S0036142902418680
2. Barth, A., Schwab, Ch., Zollinger, N.: Multi-level Monte Carlo Finite Element method for elliptic PDEs with stochastic coefficients. Numer. Math. **119**(1), 123–161 (2011). https://doi.org/10.1007/s00211-011-0377-0
3. Dick, J., Le Gia, Q.T., Schwab, Ch.: Higher order Quasi-Monte Carlo integration for holomorphic, parametric operator equations. SIAM/ASA J. Uncertain. Quantif. **4**(1), 48–79 (2016). https://doi.org/10.1137/140985913
4. Dick, J., Gantner, R.N., Le Gia, Q.T., Schwab, Ch.: Higher order quasi-Monte Carlo integration for Bayesian estimation. Technical Report 2016-13, Seminar for Applied Mathematics, ETH Zürich, Switzerland (2016). https://www.sam.math.ethz.ch/sam_reports/reports_final/reports2016/2016-13.pdf
5. Dick, J., Gantner, R.N., Le Gia, Q.T., Schwab, Ch.: Multilevel higher-order quasi-Monte Carlo Bayesian estimation. Math. Model. Methods Appl. Sci. **27**(05), 953–995 (2017). https://doi.org/10.1142/S021820251750021X
6. Dick, J., Kuo, F.Y., Le Gia, Q.T., Schwab, Ch.: Multilevel higher order QMC Petrov-Galerkin discretization for affine parametric operator equations. SIAM J. Numer. Anal. **54**(4), 2541–2568 (2016). https://doi.org/10.1137/16M1078690
7. Dick, J., Kuo, F.Y., Le Gia, Q.T., Nuyens, D., Schwab, Ch.: Higher order QMC Petrov-Galerkin discretization for affine parametric operator equations with random field inputs. SIAM J. Numer. Anal. **52**(6), 2676–2702 (2014). https://doi.org/10.1137/130943984
8. Gantner, R.N.: Tools for higher-order Quasi-Monte Carlo. http://www.sam.math.ethz.ch/HOQMC/
9. Gantner, R.N.: A generic C++ library for multilevel Quasi-Monte Carlo. In: Proceedings of the Platform for Advanced Scientific Computing Conference, PASC '16, pp. 11:1–11:12. ACM, New York, NY, USA (2016). https://doi.org/10.1145/2929908.2929915
10. Gantner, R.N., Schwab, Ch.: Computational Higher Order Quasi-Monte Carlo Integration. In: Cools, R. Nuyens, D. (eds.) Monte Carlo and Quasi-Monte Carlo Methods: MCQMC, Leuven, Belgium, April 2014, pp. 271–288. Springer International Publishing, Cham (2016). https://doi.org/10.1007/978-3-319-33507-0_12

11. Gantner, R.N., Herrmann, L., Schwab, Ch.: Quasi-Monte Carlo integration for affine-parametric, elliptic PDEs: local supports and product weights. SIAM J. Numer. Anal. **56**(1), 111–135 (2018). https://doi.org/10.1137/16M1082597
12. Gerstner, T., Griebel, M.: Numerical integration using sparse grids. Numer. Algorithms **18**(3–4), 209–232 (1998). https://doi.org/10.1023/A:1019129717644
13. Goda, T., Dick, J.: Construction of interlaced scrambled polynomial lattice rules of arbitrary high order. Found. Comput. Math. **15**(5), 1245–1278 (2015). https://doi.org/10.1007/s10208-014-9226-8
14. Graham, I.G., Kuo, F.Y., Nichols, J.A., Scheichl, R., Schwab, Ch., Sloan, I.H.: Quasi-Monte Carlo finite element methods for elliptic PDEs with lognormal random coefficients. Numer. Math. **131**(2), 329–368 (2015). https://doi.org/10.1007/s00211-014-0689-y
15. Kuo, F.Y., Schwab, Ch., Sloan, I.H.: Quasi-Monte Carlo finite element methods for a class of elliptic partial differential equations with random coefficients. SIAM J. Numer. Anal. **50**(6), 3351–3374 (2012). https://doi.org/10.1137/110845537
16. Kuo, F.Y., Schwab, Ch., Sloan, I.H.: Multi-level quasi-Monte Carlo finite element methods for a class of elliptic PDEs with random coefficients. Found. Comput. Math. **15**(2), 411–449 (2015). https://doi.org/10.1007/s10208-014-9237-5
17. Schillings, C., Schwab, Ch.: Sparse, adaptive Smolyak quadratures for Bayesian inverse problems. Inverse Probl. **29**(6), 065011, 28 (2013). https://doi.org/10.1088/0266-5611/29/6/065011
18. Schwab, Ch.: QMC Galerkin discretization of parametric operator equations. In: Monte Carlo and quasi-Monte Carlo methods 2012. Springer Proceedings in Mathematics and Statistics, vol. 65, pp. 613–629. Springer, Heidelberg (2013). https://doi.org/10.1007/978-3-642-41095-6-32

Combining Sparse Grids, Multilevel MC and QMC for Elliptic PDEs with Random Coefficients

Michael B. Giles, Frances Y. Kuo and Ian H. Sloan

Abstract Building on previous research which generalized multilevel Monte Carlo methods using either sparse grids or Quasi-Monte Carlo methods, this paper considers the combination of all these ideas applied to elliptic PDEs with finite-dimensional uncertainty in the coefficients. It shows the potential for the computational cost to achieve an $O(\varepsilon)$ r.m.s. accuracy to be $O(\varepsilon^{-r})$ with $r < 2$, independently of the spatial dimension of the PDE.

Keywords Sparse grids · Multilevel · Quasi-Monte Carlo · Elliptic PDEs

1 Introduction

There has been considerable research in recent years into the estimation of the expected value of output functionals $P(u)$ arising from the solution of elliptic PDEs of the form

$$-\nabla \cdot \Big(a(\boldsymbol{x}, \boldsymbol{y}) \nabla u(\boldsymbol{x}, \boldsymbol{y}) \Big) = f(\boldsymbol{x}), \tag{1}$$

in the unit hypercube $[0, 1]^d$, with homogeneous Dirichlet boundary conditions. Here $\boldsymbol{x}$ represents the d-dimensional spatial coordinates and the gradients are with respect

M. B. Giles (✉)
Mathematical Institute, University of Oxford, Woodstock Road, Oxford OX2 6GG, UK
e-mail: mike.giles@maths.ox.ac.uk

F. Y. Kuo · I. H. Sloan
School of Mathematics and Statistics, University of New South Wales,
Sydney, NSW 2052, Australia
e-mail: f.kuo@unsw.edu.au

I. H. Sloan
e-mail: i.sloan@unsw.edu.au

A. B. Owen and P. W. Glynn (eds.), *Monte Carlo and Quasi-Monte Carlo Methods*, Springer Proceedings in Mathematics & Statistics 241,
https://doi.org/10.1007/978-3-319-91436-7_14

to these, while $\boldsymbol{y}$ represents the uncertainty. In this paper we will consider the simplest possible setting in which we have finite s-dimensional uncertainty where

$$a(\boldsymbol{x}, \boldsymbol{y}) = a_0(\boldsymbol{x}) + \sum_{j=1}^{s} y_j\, a_j(\boldsymbol{x}),$$

with the y_j independently and uniformly distributed on the interval $[-\frac{1}{2}, \frac{1}{2}]$, with $0 < a_{\min} \le a(\boldsymbol{x}, \boldsymbol{y}) \le a_{\max} < \infty$ for all $\boldsymbol{x}$ and $\boldsymbol{y}$. This is the so-called "uniform case".

In this paper we consider several grid-based sampling methods, in all of which the PDE (1) is solved approximately by full or sparse grid-based methods with respect to $\boldsymbol{x}$, for selected values of $\boldsymbol{y}$. We will consider both multilevel and multi-index methods [10, 15], and compare Monte Carlo (MC) and Quasi-Monte Carlo (QMC) methods for computing expected values with respect to $\boldsymbol{y}$. We pay attention to the dependence of the computational cost on the spatial dimension d, and we assume throughout this paper that the stochastic dimension s is fixed, though possibly large, and we do not track the dependence of the cost on s.

As a general approach in a wide range of stochastic applications, the multilevel Monte Carlo (MLMC) approach [10] computes solutions with different levels of accuracy, using the coarser solutions as a control variate for finer solutions. If the spatial dimension d is not too large, this can lead to an r.m.s. accuracy of ε being achieved at a computational cost which is $O(\varepsilon^{-2})$, which is much better than when using the standard MC method.

The earliest multilevel research on this problem was on the use of the MLMC method for both this "uniform case" [1, 17] and the harder "lognormal case" [5, 6, 18, 25] in which $a(\boldsymbol{x}, \boldsymbol{y})$ has a log-normal distribution with a specified spatial covariance so that $\log a(\boldsymbol{x}, \boldsymbol{y})$ has a Karhunen-Loève expansion of the form $\log a(\boldsymbol{x}, \boldsymbol{y}) = \kappa_0(\boldsymbol{x}) + \sum_{j=1}^{\infty} y_j \sqrt{\lambda_j}\, \kappa_j(\boldsymbol{x})$, where the y_j are independent with a standard normal distribution, and λ_j and $\kappa_j(\boldsymbol{x})$ are the non-decreasing eigenvalues and orthonormal eigenfunctions of integral operator involving the covariance kernel. For simplicity we will restrict our discussions to the uniform case here in this paper, but our results can be easily adapted for the lognormal case.

Subsequent research [7, 13, 19, 21, 22] combined the multilevel approach with the use of QMC points, to form multilevel Quasi-Monte Carlo (MLQMC). In the best cases, this can further reduce the computational cost to $O(\varepsilon^{-r})$ for $r < 2$.

The efficiency of both MLMC and MLQMC suffers when d is large, and the reason for this is easily understood. Suppose the numerical discretisation of the PDE has order of accuracy p, so that the error in the output functional is $O(h^p)$, where h is the grid spacing in each coordinate direction. To achieve an $O(\varepsilon)$ accuracy requires $h = O(\varepsilon^{1/p})$, but if this is the grid spacing in each direction then the total number of grid points is $O(\varepsilon^{-d/p})$. Hence, the computational cost of performing just one calculation on the finest level of resolution is $O(\varepsilon^{-d/p})$, and this then gives a lower bound on the cost of the MLMC and MLQMC methods.

This curse of dimensionality is well understood, and in the case of deterministic PDEs (i.e., without the uncertainty $\boldsymbol{y}$) it has been addressed through the development of sparse grid methods [4]. One variant of this, the sparse combination technique, was the inspiration for the development of the multi-index Monte Carlo (MIMC) method [15]. The latter is a generalization of MLMC which in the context of multi-dimensional PDEs uses a variety of regular grids, with differing resolutions in each spatial direction.

In this paper we have two main contributions:

- we present alternative ways of combining MLMC with sparse grids, and discuss their relationship to the MIMC method;
- we extend these approaches by considering the use of randomised QMC points, and derive the resulting computational cost if certain conditions are met.

The paper begins by reviewing sparse grid, MLMC/MIMC and randomised QMC methods [4, 8, 11]. Next we consider the combination of MLMC with sparse grids, before adding randomised QMC to the combination. In doing so, we present meta-theorems on the resulting computational cost, based on key assumptions about the asymptotic behaviour of certain quantities.

2 Sparse Grid Methods

There are two main classes of sparse grid methods for deterministic PDEs: sparse finite elements and the sparse combination technique [4].

2.1 Sparse Finite Element Method

The sparse finite element method for elliptic PDEs uses a standard Galerkin finite element formulation but with a sparse finite element basis. One advantage of this approach is that most of the usual finite element numerical analysis remains valid; the accuracy of the method can be bounded by using bounds on the accuracy in interpolating the exact solution using the sparse finite element basis functions. The main disadvantage of the approach compared to the sparse combination technique (see the next subsection) is the difficulty of its implementation.

Following the very clear description of the method in [3], suppose that we are interested in approximating the solution of an elliptic PDE in d-dimensions. For a non-negative multi-index $\boldsymbol{\ell} = (\ell_1, \ell_2, \ldots, \ell_d)$, let $\mathscr{V}_{\boldsymbol{\ell}}$ be the finite element space spanned by the usual d-linear hat functions on a grid with spacing $2^{-\ell_j}$ in dimension j for each $j = 1, \ldots, d$. The difference space $\mathscr{W}_{\boldsymbol{\ell}}$ is defined by

$$\mathscr{W}_{\boldsymbol{\ell}} = \mathscr{V}_{\boldsymbol{\ell}} \ominus \left(\bigoplus_{j=1}^{d} \mathscr{V}_{\boldsymbol{\ell}-\boldsymbol{e}_j} \right)$$

where $\boldsymbol{e}_j$ is the unit vector in direction j. Thus, $\mathscr{W}_{\boldsymbol{\ell}}$ has the minimal set of additional basis elements such that

$$\mathscr{V}_{\boldsymbol{\ell}} = \mathscr{W}_{\boldsymbol{\ell}} \oplus \left(\bigoplus_{j=1}^{d} \mathscr{V}_{\boldsymbol{\ell}-\boldsymbol{e}_j} \right).$$

A sparse finite element space is then defined by $\bigoplus_{\boldsymbol{\ell}\in\mathscr{L}} \mathscr{W}_{\boldsymbol{\ell}}$, for some index set $\mathscr{L}$. A simple and near-optimal choice for a given level of accuracy is the set $\mathscr{L} = \{\boldsymbol{\ell} : \|\boldsymbol{\ell}\|_1 \le L\}$ for some integer L; this is discussed in [3] (that paper also presents a slightly better choice). Having defined the finite element space used for both test and trial functions, the rest of the formulation is the standard Galerkin finite element method. In the following, the space $\mathscr{H}_1$ is the standard Sobolev space with mixed first derivatives in $\boldsymbol{x}$.

Theorem 1 (Sparse finite element method) *For fixed* $\mathbf{y}$*, if the PDE* (1) *is solved using the sparse finite element method with the index set specified by* $\|\boldsymbol{\ell}\|_1 \le L$*, then the computational cost is* $O(L^{d-1}\, 2^L)$*. Moreover, the* $\mathscr{H}_1$ *solution accuracy is* $O(2^{-L})$ *if the solution u has sufficient mixed regularity, and the accuracy of simple output functionals P (such as smoothly weighted averages of the solution) is* $O(2^{-2L})$*. Hence, the cost to achieve a functional accuracy of* ε *is* $O(\varepsilon^{-1/2} |\log \varepsilon|^{d-1})$*.*

Proof The cost and $\mathscr{H}_1$ solution accuracy are proved in [3, 14]. The super-convergence for output functionals is an immediate consequence of adjoint-based error analysis [12]. □

2.2 Sparse Combination Method

The sparse combination method combines the results of separate calculations on simple tensor product grids with different resolutions in each coordinate direction [14]. For a given output functional P and multi-index $\boldsymbol{\ell} = (\ell_1, \ldots, \ell_d)$, let $P_{\boldsymbol{\ell}}$ denote the approximate output functional obtained on a grid with spacing $2^{-\ell_j}$ in direction j for each $j = 1, \ldots, d$. For convenience, we define $P_{\boldsymbol{\ell}} := 0$ if any of the indices in $\boldsymbol{\ell}$ is negative.

The backward difference in the jth dimension is defined as $\Delta_j P_{\boldsymbol{\ell}} := P_{\boldsymbol{\ell}} - P_{\boldsymbol{\ell}-\boldsymbol{e}_j}$, and we define the d-dimensional mixed first difference as

$$\boldsymbol{\Delta} P_{\boldsymbol{\ell}} := \left(\prod_{j=1}^{d} \Delta_j \right) P_{\boldsymbol{\ell}}.$$

For an arbitrary multi-index $\boldsymbol{\ell}'$, it can be shown that

$$P_{\boldsymbol{\ell}'} = \sum_{\mathbf{0} \le \boldsymbol{\ell} \le \boldsymbol{\ell}'} \boldsymbol{\Delta} P_{\boldsymbol{\ell}}, \tag{2}$$

where the multi-index inequality $\boldsymbol{\ell} \leq \boldsymbol{\ell}'$ is applied element-wise (i.e. $\ell_j \leq \ell'_j, \forall j$). Taking the limit as $\boldsymbol{\ell}' \to \infty$ (i.e. $\ell'_j \to \infty, \forall j$) gives

$$P = \sum_{\boldsymbol{\ell} \geq 0} \boldsymbol{\Delta} P_{\boldsymbol{\ell}}. \tag{3}$$

The sparse combination method truncates the summation to a finite index set, with a simple and near-optimal choice again being $\|\boldsymbol{\ell}\|_1 \leq \ell$. This gives the approximation

$$P_\ell := \sum_{\|\boldsymbol{\ell}\|_1 \leq \ell} \boldsymbol{\Delta} P_{\boldsymbol{\ell}}, \tag{4}$$

where we are slightly abusing notation by distinguishing between the original $P_{\boldsymbol{\ell}}$ with a multi-index subscript (in bold type with a tilde underneath), and the new P_ℓ on the left-hand side of this equation with a scalar subscript (which is not in bold).

If we now define

$$S_\ell := \sum_{\|\boldsymbol{\ell}\|_1 = \ell} P_{\boldsymbol{\ell}} \tag{5}$$

and the backward difference $\Delta S_\ell := S_\ell - S_{\ell-1}$, then it can be shown [24] that

$$P_\ell = \Delta^{d-1} S_\ell = \sum_{k=0}^{d-1} (-1)^k \binom{d-1}{k} S_{\ell-k}.$$

Hence, the computation of P_ℓ requires $O(\ell^{d-1})$ separate computations, each on a grid with $O(2^\ell)$ grid points. This leads to the following theorem.

Theorem 2 (Sparse combination method) *For fixed* $\mathbf{y}$*, if the PDE* (1) *is solved using the sparse combination method with the index set specified by* $\|\boldsymbol{\ell}\|_1 \leq L$*, then the computational cost is* $O(L^{d-1} 2^L)$*. Moreover, if the underlying PDE approximation has second order accuracy and the solution u has sufficient mixed regularity, then* $\boldsymbol{\Delta} P_{\boldsymbol{\ell}}$ *has magnitude* $O(2^{-2\|\boldsymbol{\ell}\|_1})$ *so the error in* P_L *is* $O(L^{d-1} 2^{-2L})$*. Hence, the cost to achieve a functional accuracy of* ε *is* $O(\varepsilon^{-1/2} |\log \varepsilon|^{3(d-1)/2})$.

Proof For the results on the cost and accuracy see [24]. The cost result is an immediate consequence. □

3 MLMC and MIMC

3.1 MLMC

The multilevel Monte Carlo (MLMC) idea is very simple. As explained in a recent review article [11], given a sequence P_ℓ, $\ell = 0, 1, \ldots$ of approximations of an output functional P, with increasing accuracy and cost as ℓ increases, and defining $P_{-1} := 0$, we have the simple identity

$$\mathbb{E}[P] = \sum_{\ell=0}^{\infty} \mathbb{E}[\Delta P_\ell], \quad \Delta P_\ell := P_\ell - P_{\ell-1}.$$

The summation can be truncated to

$$\mathbb{E}[P] \approx \mathbb{E}[P_L] = \sum_{\ell=0}^{L} \mathbb{E}[\Delta P_\ell], \tag{6}$$

with L chosen to be sufficiently large to ensure that the weak error $\mathbb{E}[P - P_L]$ is acceptably small. Each of the expectations on the r.h.s. of (6) can be estimated independently using N_ℓ independent samples so that the MLMC estimator is

$$Y = \sum_{\ell=0}^{L} Y_\ell, \quad Y_\ell = \frac{1}{N_\ell} \sum_{i=1}^{N_\ell} \Delta P_\ell^{(i)}. \tag{7}$$

The computational savings comes from the fact that on the finer levels ΔP_ℓ is smaller and has a smaller variance, and therefore fewer samples N_ℓ are required to accurately estimate its expected value.

The optimal value for N_ℓ on level $\ell = 0, 1, \ldots, L$ can be estimated by approximately minimising the cost for a given overall variance. This results in the following theorem which is a slight generalization of the original in [10].

Theorem 3 (MLMC) *Let P denote an output functional, and let P_ℓ denote the corresponding level ℓ numerical approximation. Suppose there exist independent estimators Y_ℓ of $\mathbb{E}[\Delta P_\ell]$ based on N_ℓ Monte Carlo samples and positive constants $\alpha, \beta, \gamma, c_1, c_2, c_3$, with $\alpha \geq \frac{1}{2}\min(\beta, \gamma)$, such that*

(i) $\left|\mathbb{E}[P_\ell - P]\right| \longrightarrow 0$ *as* $\ell \longrightarrow \infty$,

(ii) $\left|\mathbb{E}[\Delta P_\ell]\right| \leq c_1 2^{-\alpha \ell}$,

(iii) $\mathbb{E}[Y_\ell] = \mathbb{E}[\Delta P_\ell]$,

(iv) $\mathbb{V}[Y_\ell] \leq c_2 N_\ell^{-1} 2^{-\beta \ell}$,

(v) $\mathrm{cost}(Y_\ell) \leq c_3 N_\ell 2^{\gamma \ell}$.

Then there exists a positive constant c_4 such that for any $\varepsilon < e^{-1}$ there are values L and N_ℓ for which the MLMC estimator (7) *achieves the mean-square-error bound $\mathbb{E}[(Y - \mathbb{E}[P])^2] < \varepsilon^2$ with the computational cost bound*

$$\mathrm{cost}(Y) \leq \begin{cases} c_4 \varepsilon^{-2}, & \beta > \gamma, \\ c_4 \varepsilon^{-2} |\log \varepsilon|^2, & \beta = \gamma, \\ c_4 \varepsilon^{-2-(\gamma-\beta)/\alpha}, & \beta < \gamma. \end{cases}$$

The proof of this theorem uses a constrained optimisation approach to optimise the number of samples N_ℓ on each level. This treats the N_ℓ as real variables, and then the optimal value is rounded up to the nearest integer. This rounding up improves the variance slightly, so that we still achieve our target mean-square-error accuracy, but it also increases the cost by at most one sample per level. This additional cost is dominated by the cost of one sample on the finest level, which is $O(\varepsilon^{-\gamma/\alpha})$ since the weak convergence condition requires that the finest level satisfies $2^{-\alpha L} = O(\varepsilon)$. The condition in the theorem that $\alpha \geq \frac{1}{2}\min(\beta, \gamma)$ ensures that this additional cost is negligible compared to the main cost.

When applied to our model elliptic PDE, if one uses a tensor product grid with spacing $2^{-\ell}$ in each direction, then if the numerical discretisation has second order accuracy it gives $\alpha=2$ and $\beta=4$, while with an ideal multigrid solver the cost is at best proportional to the number of grid points which is $2^{d\ell}$ so $\gamma=d$. Hence, the cost is $O(\varepsilon^{-r})$ where $r = \max(2, d/2)$, except for $d=4$ for which $\beta=\gamma$ and hence there is an additional $|\log \varepsilon|^2$ factor. It is the dependence on d which will be addressed by incorporating sparse grid methods.

3.2 *MIMC*

The multi-index Monte Carlo (MIMC) method [15] is inspired by the sparse combination technique. Starting from (3), if each of the $\Delta P_{\boldsymbol{\ell}}$ is now a random variable due to the random coefficients in the PDE, we can take expectations of each side and truncate the sum to give

$$\mathbb{E}[P] \approx \mathbb{E}[P_L] = \sum_{\|\boldsymbol{\ell}\|_1 \leq L} \mathbb{E}[\boldsymbol{\Delta} P_{\boldsymbol{\ell}}]. \tag{8}$$

This is now very similar to the telescoping sum (6) in MLMC, with the difference that the levels are now labelled by multi-indices, so allowing different discretizations in different directions. We can independently estimate each of the expectations on the r.h.s. of (8) using a number of independent samples $N_{\boldsymbol{\ell}}$ so that the MIMC estimator is

$$Y = \sum_{\|\boldsymbol{\ell}\|_1 \leq L} Y_{\boldsymbol{\ell}}, \quad Y_{\boldsymbol{\ell}} = \frac{1}{N_{\boldsymbol{\ell}}} \sum_{i=1}^{N_{\boldsymbol{\ell}}} \Delta P_{\boldsymbol{\ell}}^{(i)}. \tag{9}$$

The numbers $N_{\boldsymbol{\ell}}$ are optimised to minimise the cost of achieving a certain desired variance or mean-square-error.

The original paper [15] considers much more general circumstances: the different indices in $\boldsymbol{\ell}$ are not limited to the spatial discretizations in $\boldsymbol{x}$ but can also involve quantities such as the number of particles in a system, or the number of terms in a Karhunen–Loève expansion (arising from dimension truncation in the stochastic variables $\boldsymbol{y}$). Here in the isotropic PDE case, in which the behaviour in each space dimension is similar, this leads to the following theorem.

Theorem 4 (MIMC) *Let P denote an output functional, and for each multi-index $\boldsymbol{\ell}$ let $P_{\boldsymbol{\ell}}$ denote the approximate output functional indexed by $\boldsymbol{\ell}$. Suppose for each multi-index $\boldsymbol{\ell}$ there exist independent estimators $Y_{\boldsymbol{\ell}}$ of $\mathbb{E}[\boldsymbol{\Delta} P_{\boldsymbol{\ell}}]$ based on $N_{\boldsymbol{\ell}}$ Monte Carlo samples and positive constants $\alpha, \beta, \gamma, c_1, c_2, c_3$, with $\alpha \geq \frac{1}{2}\beta$, such that*

$$
\begin{aligned}
&(i) && \left|\mathbb{E}[P_{\boldsymbol{\ell}} - P]\right| \longrightarrow 0 \text{ as } \boldsymbol{\ell} \longrightarrow \infty\ (\ell_j \to \infty\,, \forall j),\\
&(ii) && \left|\mathbb{E}[\boldsymbol{\Delta} P_{\boldsymbol{\ell}}]\right| \leq c_1\, 2^{-\alpha \|\boldsymbol{\ell}\|_1},\\
&(iii) && \mathbb{E}[Y_{\boldsymbol{\ell}}] = \mathbb{E}[\boldsymbol{\Delta} P_{\boldsymbol{\ell}}],\\
&(iv) && \mathbb{V}[Y_{\boldsymbol{\ell}}] \leq c_2\, N_{\boldsymbol{\ell}}^{-1}\, 2^{-\beta \|\boldsymbol{\ell}\|_1},\\
&(v) && \mathrm{cost}(Y_{\boldsymbol{\ell}}) \leq c_3\, N_{\boldsymbol{\ell}}\, 2^{\gamma \|\boldsymbol{\ell}\|_1}.
\end{aligned}
$$

Then there exists a positive constant c_4 such that for any $\varepsilon < e^{-1}$ there are values L and $N_{\boldsymbol{\ell}}$ for which the MIMC estimator (9) *achieves the mean-square-error bound $\mathbb{E}[(Y - \mathbb{E}[P])^2] < \varepsilon^2$ with the computational cost bound*

$$
\mathrm{cost}(Y) \leq \begin{cases} c_4\, \varepsilon^{-2}\,, & \beta > \gamma,\\ c_4\, \varepsilon^{-2}\, |\log \varepsilon|^{e_1}, & \beta = \gamma,\\ c_4\, \varepsilon^{-2-(\gamma-\beta)/\alpha}\, |\log \varepsilon|^{e_2}, & \beta < \gamma, \end{cases}
$$

where

$$
\begin{aligned}
&e_1 = 2d, && e_2 = (d-1)\,(2+(\gamma-\beta)/\alpha), && \text{if } \alpha > \tfrac{1}{2}\beta,\\
&e_1 = \max(2d, 3(d-1)), && e_2 = (d-1)\,(1+\gamma/\alpha), && \text{if } \alpha = \tfrac{1}{2}\beta.
\end{aligned}
$$

Proof This is a particular case of the more general analysis in [15, Theorem 2.2]. □

In the case of MIMC, there are $O(L^{d-1})$ multi-indices on the finest level on which $\|\boldsymbol{\ell}\|_1 = L$. Hence the finest level is determined by the constraint $L^{d-1} 2^{-\alpha L} = O(\varepsilon)$, and the associated cost is $O(\varepsilon^{-\gamma/\alpha} |\log \varepsilon|^{(d-1)(1+\gamma/\alpha)})$. Given the assumption that $\alpha \geq \frac{1}{2}\beta$, this is not asymptotically bigger than the main cost except when $\alpha = \frac{1}{2}\beta$, in which case it is responsible for the e_2 and the $3(d-1)$ component in the maximum in e_1.

When applied to our model elliptic PDE, if one uses a tensor product grid with spacing $2^{-\ell_j}$ in the jth direction, and a numerical discretisation with second order accuracy, then we are likely to get $\alpha = 2$ and $\beta = 4$ if the solution has sufficient mixed regularity [24]. (Note that this is a much stronger statement than the $\alpha = 2$, $\beta = 4$ in the previous section; taking the case with $d = 3$ as an example, with grid spacing h_1, h_2, h_3 in the three dimensions, Sect. 3.1 requires only that $\Delta P_\ell = O(h^2)$ when all three spacings are equal to h, whereas in this section we require the product form $\boldsymbol{\Delta} P_{\boldsymbol{\ell}} = O(h_1^2\, h_2^2\, h_3^2)$ which is much smaller when $h_1, h_2, h_3 \ll 1$.) With an ideal multigrid solver, the cost is proportional to $2^{\|\boldsymbol{\ell}\|_1}$, so $\gamma = 1$. Since $\beta > \gamma$, the cost would then be $O(\varepsilon^{-2})$, regardless of the value of d.

4 Randomised QMC and MLQMC

4.1 Randomised QMC Sampling

A randomized QMC method with N deterministic points and R randomization steps approximates an s-dimensional integral over the unit cube $[-\frac{1}{2}, \frac{1}{2}]^s$ as follows

$$I := \int_{[-\frac{1}{2},\frac{1}{2}]^s} g(\boldsymbol{y})\,\mathrm{d}\boldsymbol{y} \quad \approx \quad \overline{Q} := \frac{1}{R}\sum_{k=1}^{R} Q_k, \qquad Q_k = \frac{1}{N}\sum_{i=1}^{N} g(\boldsymbol{y}^{(i,k)}).$$

For the purpose of this paper it suffices that we introduce briefly just a simple family of randomized QMC methods – randomly shifted lattice rules. We have

$$\boldsymbol{y}^{(i,k)} = \left\{\frac{i\boldsymbol{z}}{N} + \boldsymbol{\Delta}^{(k)}\right\} - \tfrac{1}{2},$$

where $\boldsymbol{z} \in \mathbb{N}^s$ is known as the generating vector; $\boldsymbol{\Delta}^{(1)}, \ldots, \boldsymbol{\Delta}^{(R)} \in (0,1)^s$ are R independent random shifts; the braces indicate that we take the fractional part of each component in the vector; and finally we subtract $\frac{1}{2}$ from each component of the vector to bring it into $[-\frac{1}{2}, \frac{1}{2}]^s$.

Randomly shifted lattice rules provide unbiased estimators of the integral. Indeed, it is easy to verify that $\mathbb{E}_{\boldsymbol{\Delta}}[\overline{Q}] = \mathbb{E}_{\boldsymbol{\Delta}}[Q_k] = I$, where we introduced the subscript $\boldsymbol{\Delta}$ to indicate that the expectation is taken with respect to the random shifts. In some appropriate function space setting for the integrand function g, it is known (see e.g., [8]) that good generating vectors $\boldsymbol{z}$ can be constructed so that the variance or mean-square-error satisfies $\mathbb{V}_{\boldsymbol{\Delta}}[\overline{Q}] = \mathbb{E}_{\boldsymbol{\Delta}}[(\overline{Q} - I)^2] \le C_\delta\, R^{-1}\, N^{-2(1-\delta)}$, for some $\delta \in (0, 1/2]$ with C_δ independent of the dimension s. In practical computations, we can estimate the variance by $\mathbb{V}_{\boldsymbol{\Delta}}[\overline{Q}] \approx \sum_{k=1}^{R}(Q_k - \overline{Q})^2/[R(R-1)]$. Typically we take a large value of N to benefit from the higher QMC convergence rate and use only a relatively small R (e.g., 20–50) for the purpose of estimating the variance.

There are other randomization strategies for QMC methods. For example, we can combine any digital net such as Sobol′ sequences or interlaced polynomial lattice rules with digital shift or Owen scrambling, to get an unbiased estimator with variance close to $O(N^{-2})$ or $O(N^{-3})$. We can also apply randomization to a higher order digital net to achieve $O(N^{-p})$ for $p > 2$ in an appropriate function space setting for smooth integrands. For detailed reviews of these results see see e.g., [8].

4.2 MLQMC

As a generalization of (7), the multilevel Quasi-Monte Carlo (MLQMC) estimator is

$$Y = \sum_{\ell=0}^{L} Y_\ell, \quad Y_\ell = \frac{1}{R_\ell} \sum_{k=1}^{R_\ell} \left(\frac{1}{N_\ell} \sum_{i=1}^{N_\ell} \Delta P_\ell^{(i,k)} \right). \tag{10}$$

Later in Theorem 5 we will state the corresponding generalization of Theorem 3.

The use of QMC instead of MC in a multilevel method was first considered in [13] where numerical experiments were carried out for a number of option pricing problems and showed convincingly that MLQMC improves upon MLMC. A meta-theorem similar to the MLMC theorem was proved in [9]. A slightly sharper version of the theorem, eliminating some $\log(\varepsilon)$ factors, will be stated and proved later in Sect. 6.

MLQMC methods have been combined with finite element discretizations for the PDE problems in [7, 21, 22]. The paper [22] studied the uniform case for the same elliptic PDE of this paper with randomly shifted lattice rules (which yield up to order 2 convergence in the variance); the paper [7] studied the uniform case for general operator equations with deterministic higher order digital nets; the paper [21] studied the lognormal case with randomly shifted lattice rules. A key analysis which is common among these papers is the required mixed regularity estimate of the solution involving both $\boldsymbol{x}$ and $\boldsymbol{y}$, see [20] for a survey of the required analysis in a unified framework.

5 Combining Sparse Grids and MLMC

After this survey of the three component technologies, sparse grid methods, MLMC and MIMC, and randomised QMC samples, the first novel observation in this paper is very simple: MIMC is not the only way in which MLMC can be combined with sparse grid methods.

An alternative is to use the standard MLMC approach, but with samples which are computed using sparse grid methods. The advantage of this is that it can be used with either sparse finite elements or the sparse combination technique.

5.1 MLMC with Sparse Finite Element Samples

In Theorem 3, if P_ℓ is computed using sparse finite elements as described in Sect. 2.1 based on grids with index set $\|\boldsymbol{\ell}\|_1 \le \ell$, and if the accuracy and cost are as given in Theorem 1, then we obtain $\alpha = 2-\delta$, $\beta = 4-\delta$, and $\gamma = 1+\delta$ for any $0 < \delta \ll 1$. Here δ arises due to the effect of some additional powers of ℓ. So $\beta > \gamma$ and therefore the computational cost is $O(\varepsilon^{-2})$.

Recall that with the full tensor product grid we had $\alpha = 2$, $\beta = 4$, and $\gamma = d$. Hence the improvement here is in the removal of the dependence of the cost parameter γ on d.

5.2 MLMC with Sparse Combination Samples

The aim in this section is to show that the MIMC algorithm is very similar to MLMC using sparse combination samples.

Suppose we have an MIMC application which satisfies the conditions of Theorem 4. For the MLMC version, we use (4) to define the P_ℓ in Theorem 3. Since

$$\mathbb{E}[\Delta P_\ell] = \sum_{\|\boldsymbol{\ell}\|_1=\ell} \mathbb{E}[\boldsymbol{\Delta} P_{\boldsymbol{\ell}}], \tag{11}$$

the two algorithms have exactly the same expected value if the finest level for each is given by $\|\boldsymbol{\ell}\|_1 = L$ for the same value of L. The difference between the two algorithms is that MIMC independently estimates each of the expectations on the r.h.s. of (11), using a separate estimator $Y_{\boldsymbol{\ell}}$ for each $\mathbb{E}[\boldsymbol{\Delta} P_{\boldsymbol{\ell}}]$ with independent samples of $\boldsymbol{y}$, whereas MLMC with sparse combination samples estimates the expectation on the l.h.s., using the combination

$$Y_\ell = \sum_{\|\boldsymbol{\ell}\|_1=\ell} Y_{\boldsymbol{\ell}},$$

with the $Y_{\boldsymbol{\ell}}$ all based on the same set of N_ℓ random samples $\boldsymbol{y}$.

There are no more than $(\ell+1)^{d-1}$ terms in the summation in (11), so if the cost of $Y_{\boldsymbol{\ell}}$ for MIMC is $O(N_{\boldsymbol{\ell}} 2^{\gamma\ell})$ when $\|\boldsymbol{\ell}\|_1 = \ell$, then the cost of the sparse combination estimator Y_ℓ for MLMC is $O(N_\ell \ell^{d-1} 2^{\gamma\ell}) = o(N_\ell 2^{(\gamma+\delta)\ell})$, for any $0 < \delta \ll 1$.

Likewise,

$$|\,\mathbb{E}[Y_\ell]\,| \leq \sum_{\|\boldsymbol{\ell}\|_1=\ell} |\,\mathbb{E}[Y_{\boldsymbol{\ell}}]\,|,$$

so if $|\,\mathbb{E}[Y_{\boldsymbol{\ell}}]\,| = O(2^{-\alpha\ell})$ when $\|\boldsymbol{\ell}\|_1 = \ell$, then $|\,\mathbb{E}[Y_\ell]\,| = o(2^{-(\alpha-\delta)\ell})$ for any $0 < \delta \ll 1$.

Furthermore, Jensen's inequality gives

$$\begin{aligned}
\mathbb{V}[Y_\ell] &= \mathbb{E}\left[(Y_\ell - \mathbb{E}[Y_\ell])^2\right] = \mathbb{E}\left[\left(\sum_{\|\boldsymbol{\ell}\|_1=\ell} (Y_{\boldsymbol{\ell}} - \mathbb{E}[Y_{\boldsymbol{\ell}}])\right)^2\right] \\
&\leq (\ell+1)^{d-1} \sum_{\|\boldsymbol{\ell}\|_1=\ell} \mathbb{E}\left[(Y_{\boldsymbol{\ell}} - \mathbb{E}[Y_{\boldsymbol{\ell}}])^2\right] \\
&= (\ell+1)^{d-1} \sum_{\|\boldsymbol{\ell}\|_1=\ell} \mathbb{V}[Y_{\boldsymbol{\ell}}],
\end{aligned}$$

so if $\mathbb{V}[Y_{\boldsymbol{\ell}}] = O(N_{\boldsymbol{\ell}}^{-1} 2^{-\beta\ell})$, then $\mathbb{V}[Y_\ell] = o(N_\ell^{-1}\, 2^{-(\beta-\delta)\ell})$, for any $0 < \delta \ll 1$.

This shows that the α, β, γ values for the MLMC algorithm using the sparse combination samples are almost equal to the α, β, γ for the MIMC method, which leads to the following lemma.

Lemma 1 *If a numerical method satisfies the conditions for the MIMC Theorem 4, then the corresponding MLMC estimator with sparse combination samples will have a cost which is* $O(\varepsilon^{-2})$, *if* $\beta > \gamma$, *and* $o(\varepsilon^{-2-(\gamma-\beta)/\alpha)-\delta})$, $\forall\, 0 < \delta \ll 1$, *if* $\beta \le \gamma$.

As with MLMC with sparse finite element samples, the key thing here is that the level ℓ MLMC samples use a set of grids in which the number of grid points is $O(2^{\|\boldsymbol{\ell}\|_1}) = O(2^{\ell})$. That is why the γ values for MIMC and MLMC are virtually identical.

If there is substantial cancellation in the summation, it is possible that $\mathbb{V}[Y_\ell]$ could be very much smaller than the $\mathbb{V}[Y_{\boldsymbol{\ell}}]$ for each of the $\boldsymbol{\ell}$ for which $\|\boldsymbol{\ell}\|_1 = \ell$. However, we conjecture that this is very unlikely, and therefore we are not suggesting that the MLMC with sparse combination samples is likely to be better than MIMC. The point of this section is to show that it cannot be significantly worse. In addition, this idea of combining MLMC with sparse grid samples works for sparse finite elements for which there seems to be no natural MIMC extension.

5.3 Nested MLMC

Another alternative to MIMC is nested MLMC. To illustrate this in 2D, suppose we start by using a single level index ℓ_1 to construct a standard MLMC decomposition

$$\mathbb{E}[P] \approx \mathbb{E}[P_{L_1}] = \sum_{\ell_1=0}^{L_1} \mathbb{E}[\Delta P_{\ell_1}].$$

Now, for each particular index ℓ_1 we can take $\mathbb{E}[\Delta P_{\ell_1}]$ and perform a secondary MLMC expansion with respect to a second index ℓ_2 to give

$$\mathbb{E}[\Delta P_{\ell_1}] \approx \sum_{\ell_2=0}^{L_2} \mathbb{E}[Q_{\ell_1,\ell_2} - Q_{\ell_1,\ell_2-1}],$$

with $Q_{\ell_1,-1} := 0$. If we allow L_2 to possibly depend on the value of ℓ_1, this results in an approximation which is very similar to the MIMC method,

$$\mathbb{E}[P] \approx \sum_{\boldsymbol{\ell}\in\mathcal{L}} \mathbb{E}\left[Q_{\ell_1,\ell_2} - Q_{\ell_1,\ell_2-1}\right],$$

with the summation over some finite set of indices $\mathcal{L}$. In contrast to the MIMC method, here $Q_{\ell_1,\ell_2} - Q_{\ell_1,\ell_2-1}$ is not necessarily expressible in the cross-difference form $\boldsymbol{\Delta} P_{\boldsymbol{\ell}}$ used in MIMC. Thus, this method is a generalization of MIMC.

This approach is currently being used in two new research projects. In one project, the second expansion is with respect to the precision of floating point computations; i.e. half, single or double precision. This follows ideas presented in Sect. 10.2 of

[11] and also in [2]. In the other project [16], the second expansion uses Rhee and Glynn's randomised multilevel Monte Carlo method [23] to provide an unbiased inner estimate in a financial nested expectation application.

6 MLQMC and MIQMC

The next natural step is to replace the Monte Carlo sampling with randomised QMC sampling to estimate $\mathbb{E}[\Delta P_\ell]$ or $\mathbb{E}[\boldsymbol{\Delta} P_{\boldsymbol{\ell}}]$.

6.1 MLQMC (Continued from Sect. 4.2)

In the best circumstances, using N_ℓ QMC deterministic points with $R_\ell = R$ randomisation steps to estimate $\mathbb{E}[\Delta P_\ell]$ gives a variance (with respect to the randomisation in the QMC points) which is $O(R^{-1} N_\ell^{-p} 2^{-\beta\ell})$, with $p > 1$. This leads to the following theorem which generalizes Theorem 3.

Theorem 5 (MLQMC) *Let P denote an output functional, and let P_ℓ denote the corresponding level ℓ numerical approximation. Suppose there exist independent estimators Y_ℓ of $\mathbb{E}[\Delta P_\ell]$ based on N_ℓ deterministic QMC points and $R_\ell = R$ randomization steps, and positive constants $\alpha, \beta, \gamma, c_1, c_2, c_3, p$, with $p > 1$ and $\alpha \geq \frac{1}{2}\beta$, such that*

$$\begin{aligned}
&(i) && \left|\mathbb{E}[P_\ell - P]\right| \longrightarrow 0 \text{ as } \ell \longrightarrow \infty,\\
&(ii) && \left|\mathbb{E}[\Delta P_\ell]\right| \leq c_1\, 2^{-\alpha\,\ell},\\
&(iii) && \mathbb{E}_{\boldsymbol{\Delta}}[Y_\ell] = \mathbb{E}[\Delta P_\ell],\\
&(iv) && \mathbb{V}_{\boldsymbol{\Delta}}[Y_\ell] \leq c_2\, R^{-1}\, N_\ell^{-p}\, 2^{-\beta\,\ell},\\
&(v) && \mathrm{cost}(Y_\ell) \leq c_3\, R\, N_\ell\, 2^{\gamma\,\ell}.
\end{aligned}$$

Then there exists a positive constant c_4 such that for any $\varepsilon < e^{-1}$ there are values L and N_ℓ for which the MLQMC estimator (10) *achieves the mean-square-error bound $\mathbb{E}_{\boldsymbol{\Delta}}[(Y - \mathbb{E}[P])^2] < \varepsilon^2$ with the computational cost bound*

$$\mathrm{cost}(Y) \leq \begin{cases} c_4\, \varepsilon^{-2/p}, & \beta > p\gamma,\\ c_4\, \varepsilon^{-2/p} |\log \varepsilon|^{(p+1)/p}, & \beta = p\gamma,\\ c_4\, \varepsilon^{-2/p-(p\gamma-\beta)/(p\alpha)}, & \beta < p\gamma. \end{cases}$$

Proof We omit the proof here because the theorem can be interpreted as a special case of Theorem 6 below for which we will provide an outline of the proof. □

6.2 MIQMC

As a generalization of (9), the MIQMC estimator is

$$Y = \sum_{\|\boldsymbol{\ell}\|_1 \le L} Y_{\boldsymbol{\ell}}, \quad Y_{\boldsymbol{\ell}} = \frac{1}{R_{\boldsymbol{\ell}}} \sum_{k=1}^{R_{\boldsymbol{\ell}}} \left(\frac{1}{N_{\boldsymbol{\ell}}} \sum_{i=1}^{N_{\boldsymbol{\ell}}} \Delta P_{\boldsymbol{\ell}}^{(i,k)} \right), \tag{12}$$

where $Y_{\boldsymbol{\ell}}$ is an estimator for $\mathbb{E}[\boldsymbol{\Delta} P_{\boldsymbol{\ell}}]$ based on $N_{\boldsymbol{\ell}}$ deterministic QMC points and $R_{\boldsymbol{\ell}}$ randomization steps.

Suppose that $Y_{\boldsymbol{\ell}}$ has variance and cost given by $\mathbb{V}_{\boldsymbol{\Delta}}[Y_{\boldsymbol{\ell}}] = N_{\boldsymbol{\ell}}^{-p} v_{\boldsymbol{\ell}}$ and $\mathrm{cost}(Y_{\boldsymbol{\ell}}) = N_{\boldsymbol{\ell}}\, c_{\boldsymbol{\ell}}$. The variance and total cost of the combined estimator Y are

$$\mathbb{V}_{\boldsymbol{\Delta}}[Y] = \sum_{\|\boldsymbol{\ell}\|_1 \le L} N_{\boldsymbol{\ell}}^{-p} v_{\boldsymbol{\ell}}, \quad \mathrm{cost}(Y) = \sum_{\|\boldsymbol{\ell}\|_1 \le L} N_{\boldsymbol{\ell}}\, c_{\boldsymbol{\ell}}.$$

Treating the $N_{\boldsymbol{\ell}}$ as real numbers, the cost can be minimised for a given total variance by introducing a Lagrange multiplier and minimising $\mathrm{cost}(Y) + \lambda \mathbb{V}_{\boldsymbol{\Delta}}[Y]$, which gives

$$N_{\boldsymbol{\ell}} = \left(\frac{\lambda\, p\, v_{\boldsymbol{\ell}}}{c_{\boldsymbol{\ell}}} \right)^{1/(p+1)}.$$

Requiring $\mathbb{V}_{\boldsymbol{\Delta}}[Y] = \frac{1}{2}\varepsilon^2$ to achieve a target accuracy determines the value of λ and then the total cost is

$$\mathrm{cost}(Y) = (2\,\varepsilon^{-2})^{1/p} \left(\sum_{\|\boldsymbol{\ell}\|_1 \le L} \left(c_{\boldsymbol{\ell}}^{p} v_{\boldsymbol{\ell}} \right)^{1/(p+1)} \right)^{(p+1)/p}.$$

This outline analysis shows that the behaviour of the product $c_{\boldsymbol{\ell}}^p v_{\boldsymbol{\ell}}$ as $\boldsymbol{\ell} \to \infty$ is critical. If $c_{\boldsymbol{\ell}} = O(2^{\gamma \ell})$ and $v_{\boldsymbol{\ell}} = O(2^{-\beta \ell})$ where $\ell = \|\boldsymbol{\ell}\|_1$, then $c_{\boldsymbol{\ell}}^p v_{\boldsymbol{\ell}} = O(2^{(p\gamma - \beta)\ell})$.

If $\beta > p\gamma$, then the total cost is dominated by the contributions from the coarsest levels, and we get a total cost which is $O(\varepsilon^{-2/p})$.

If $\beta = p\gamma$, then all levels contribute to the total cost, and it is $O(L^{d(p+1)/p} \varepsilon^{-2/p})$.

If $\beta < p\gamma$, then the total cost is dominated by the contributions from the finest levels, and we get a total cost which is $O(L^{(d-1)(p+1)/p}\, \varepsilon^{-2/p}\, 2^{(p\gamma-\beta)L/p})$.

To complete this analysis, we need to know the value of L which is determined by the requirement that the square of the bias is no more than $\frac{1}{2}\varepsilon^2$. This can be satisfied by ensuring that

$$\mathrm{bias}(Y) := \sum_{\|\boldsymbol{\ell}\|_1 > L} |\mathbb{E}[\boldsymbol{\Delta} P_{\boldsymbol{\ell}}]| \le \varepsilon/\sqrt{2}.$$

If $|\mathbb{E}[\boldsymbol{\Delta} P_{\boldsymbol{\ell}}]| = O(2^{-\alpha \|\boldsymbol{\ell}\|_1})$, then the contributions to $\mathrm{bias}(Y)$ come predominantly from the coarsest levels in the summation (i.e. $\|\boldsymbol{\ell}\|_1 = L + 1$), and hence $\mathrm{bias}(Y) =$

$O(L^{d-1}2^{-\alpha L})$. The bias constraint then gives $L^{d-1}2^{-\alpha L} = O(\varepsilon)$ and so $L = O(|\log \varepsilon|)$.

As discussed after the MLMC and MIMC theorems, the values for $N_{\boldsymbol{\ell}}$ need to be rounded up to the nearest integers, incurring an additional cost which is $O(\varepsilon^{-\gamma/\alpha}|\log \varepsilon|^{(d-1)(1+\gamma/\alpha)})$. If $\alpha > \frac{1}{2}\beta$ it is always negligible compared to the main cost, but it can become the dominant cost when $\alpha = \frac{1}{2}\beta$ and $\beta \le p\gamma$. This corresponds to the generalization of Cases C and D in Theorem 2.2 in the MIMC analysis in [15].

This outline analysis leads to the following theorem in which we make various assumptions and then draw conclusions about the resulting cost.

Theorem 6 (MIQMC) *Let P denote an output functional, and for each multi-index $\boldsymbol{\ell}$ let $P_{\boldsymbol{\ell}}$ denote the approximate output functional indexed by $\boldsymbol{\ell}$. Suppose for each multi-index $\boldsymbol{\ell}$ there exist independent estimators $Y_{\boldsymbol{\ell}}$ of $\mathbb{E}[\boldsymbol{\Delta} P_{\boldsymbol{\ell}}]$ based on $N_{\boldsymbol{\ell}}$ deterministic QMC samples and $R_{\boldsymbol{\ell}} = R$ randomization steps, and positive constants $\alpha, \beta, \gamma, c_1, c_2, c_3, p$, with $p > 1$ and $\alpha \ge \frac{1}{2}\beta$, such that*

(i) $\left|\mathbb{E}[P_{\boldsymbol{\ell}} - P]\right| \longrightarrow 0$ *as* $\boldsymbol{\ell} \longrightarrow \infty$ $(\ell_j \to \infty, \forall j)$,

(ii) $\left|\mathbb{E}[\boldsymbol{\Delta} P_{\boldsymbol{\ell}}]\right| \le c_1\, 2^{-\alpha \|\boldsymbol{\ell}\|_1}$

(iii) $\mathbb{E}_{\Delta}[Y_{\boldsymbol{\ell}}] = \mathbb{E}[\boldsymbol{\Delta} P_{\boldsymbol{\ell}}]$

(iv) $\mathbb{V}_{\Delta}[Y_{\boldsymbol{\ell}}] \le c_2\, R^{-1} N_{\boldsymbol{\ell}}^{-p}\, 2^{-\beta \|\boldsymbol{\ell}\|_1}$

(v) $\mathrm{cost}(Y_{\boldsymbol{\ell}}) \le c_3\, R\, N_{\boldsymbol{\ell}}\, 2^{\gamma \|\boldsymbol{\ell}\|_1}$.

Then there exists a positive constant c_4 such that for any $\varepsilon < e^{-1}$ there are values L and $N_{\boldsymbol{\ell}}$ for which the MIQMC estimator (12) *achieves the mean-square-error bound $\mathbb{E}_{\boldsymbol{\Delta}}[(Y - \mathbb{E}[P])^2] < \varepsilon^2$ with the computational cost bound*

$$\mathrm{cost}(Y) \le \begin{cases} c_4\, \varepsilon^{-2/p}, & \beta > p\gamma, \\ c_4\, \varepsilon^{-2/p}\, |\log \varepsilon|^{e_1}, & \beta = p\gamma, \\ c_4\, \varepsilon^{-2/p - (p\gamma - \beta)/p\alpha}\, |\log \varepsilon|^{e_2}, & \beta < p\gamma, \end{cases}$$

where

$$e_1 = d(p+1)/p, \quad e_2 = (d-1)((p+1)/p + (p\gamma - \beta)/p\alpha), \qquad \text{if } \alpha > \tfrac{1}{2}\beta,$$
$$e_1 = \max(d(p+1)/p,\ (d-1)(1+\gamma/\alpha)), \quad e_2 = (d-1)(1+\gamma/\alpha), \quad \text{if } \alpha = \tfrac{1}{2}\beta.$$

Proof The detailed proof follows the same lines as [15, Theorem 2.2]. □

The key observation here is that the dimension d does not appear in the exponent for ε in the cost bounds, so it is a significant improvement over the MLQMC result in which the cost is of the form ε^{-r} with $r = \max(2/p, d/2)$, which limits the multilevel benefits even for $d=3$ if $p > 4/3$.

It is interesting to compare the cost given by this theorem with that given by the MIMC Theorem 4. If $\beta > p\gamma$, then the use of QMC improves the cost from $O(\varepsilon^{-2})$ to $O(\varepsilon^{-2/p})$. This is because the dominant costs in this case are on the coarsest levels where many points have to be sampled, and therefore QMC will provide substantial benefits. On the other hand, if $\beta < \gamma$ then both approaches give a cost of approximately $O(\varepsilon^{-\gamma/\alpha})$ because in this case the dominant costs are on the finest levels, and on the finest levels the optimal number of QMC points is $O(1)$, which is why the additional cost of rounding up to the nearest integer often dominates the main cost. Hence the use of QMC points is almost irrelevant in this case. Fortunately, we expect that the favourable case $\beta > p\gamma$ is likely to be the more common one. It is clearly the case in our very simple elliptic model with $\beta = 4$ and $\gamma = 1$.

7 Concluding Remarks

In this paper we began by summarizing the meta-theorems for MLMC and MIMC in a common framework for elliptic PDEs with random coefficients, where we applied full or sparse grid methods with respect to the spatial variables $\boldsymbol{x}$ and used MC sampling for computing expected values with respect to the stochastic variables $\boldsymbol{y}$.

Following this, our novel contributions were

- showing that, in this context, MIMC is almost equivalent to the use of MLMC with sparse combination samples;
- introducing the idea of (a) MLMC with sparse finite element or sparse combination samples, and (b) nested MLMC, as other alternatives to MIMC;
- deriving the corresponding meta-theorems for MLQMC and MIQMC in this context, concluding that the computational cost to achieve $O(\varepsilon)$ r.m.s. accuracy can be reduced to $O(\varepsilon^{-r})$ with $r < 2$ independent of the spatial dimension d.

Natural extensions include allowing the different indices in $\boldsymbol{\ell}$ to cover also different levels of dimension truncation in the stochastic variables $\boldsymbol{y}$, as well as providing verifications of the precise parameters α, β, γ and p for specific PDE applications.

Acknowledgements The authors acknowledge the support of the Australian Research Council under the projects FT130100655 and DP150101770.

References

1. Barth, A., Schwab, Ch., Zollinger, N.: Multi-level Monte Carlo finite element method for elliptic PDEs with stochastic coefficients. Numerische Mathematik **119**(1), 123–161 (2011)
2. Brugger, C., de Schryver, C., Wehn, N., Omland, S., Hefter, M., Ritter, K., Kostiuk, A., Korn, R.: Mixed precision multilevel Monte Carlo on hybrid computing systems. In: Proceedings of the Conference CIFEr. IEEE (2014)
3. Bungartz, H.-J., Griebel, M.: A note on the complexity of solving Poisson's equation for spaces of bounded mixed derivatives. J. Complex. **15**(2), 167–199 (1999)
4. Bungartz, H.-J., Griebel, M.: Sparse grids. Acta Numerica **13**, 1–123 (2004)

5. Charrier, J., Scheichl, R., Teckentrup, A.: Finite element error analysis of elliptic PDEs with random coefficients and its application to multilevel Monte Carlo methods. SIAM J. Numer. Anal. **51**(1), 322–352 (2013)
6. Cliffe, K.A., Giles, M.B., Scheichl, R., Teckentrup, A.: Multilevel Monte Carlo methods and applications to elliptic PDEs with random coefficients. Comput. Vis. Sci. **14**(1), 3–15 (2011)
7. Dick, J., Kuo, F.Y., Le Gia, Q.T., Schwab, Ch.: Multi-level higher order QMC Petrov-Galerkin discretization for affine parametric operator equations. SIAM J. Numer. Anal. **54**(4), 2541–2568 (2016)
8. Dick, J., Kuo, F.Y., Sloan, I.H.: High-dimensional integration: the quasi-Monte Carlo way. Acta Numerica **22**, 133–288 (2013)
9. Gerstner, T., Noll, M.: Randomized multilevel quasi-Monte Carlo path simulation. In: Recent Developments in Computational Finance, pp. 349–372. World Scientific (2013)
10. Giles, M.B.: Multilevel Monte Carlo path simulation. Op. Res. **56**(3), 607–617 (2008)
11. Giles, M.B.: Multilevel Monte Carlo methods. Acta Numerica **24**, 259–328 (2015)
12. Giles, M.B., Süli, E.: Adjoint methods for PDEs: a posteriori error analysis and postprocessing by duality. Acta Numerica **11**, 145–236 (2002)
13. Giles, M.B., Waterhouse, B.J.: Multilevel quasi-Monte Carlo path simulation. Advanced Financial Modelling. Radon Series on Computational and Applied Mathematics, pp. 165–181. De Gruyter, Berlin (2009)
14. Griebel, M., Schneider, M., Zenger, C.: A combination technique for the solution of sparse grid problems. In: Iterative Methods in Linear Algebra, IMACS (1992)
15. Haji-Ali, A.-L., Nobile, F., Tempone, R.: Multi index Monte Carlo: when sparsity meets sampling. Numerische Mathematik **132**, 767–806 (2016)
16. Haji-Ali, A.-L., Giles, M.B.: MLMC for Value-at-Risk. Presentation at MCM (2017). http://people.maths.ox.ac.uk/hajiali/assets/files/hajiali-mcm2017-var.pdf
17. Harbrecht, H., Peters, M., Siebenmorgen, M.: On multilevel quadrature for elliptic stochastic partial differential equations. Sparse Grids and Applications. Lecture Notes in Computational Science and Engineering, vol. 88, pp. 161–179. Springer, Berlin (2013)
18. Harbrecht, H., Peters, M., Siebenmorgen, M.: Multilevel accelerated quadrature for PDEs with log-Normally distributed diffusion coefficient. SIAM/ASA J. Uncertain. Quantif. **4**(1), 520–551 (2016)
19. Harbrecht, H., Peters, M., Siebenmorgen, M.: On the quasi-Monte Carlo method with Halton points for elliptic PDEs with log-normal diffusion. Math. Comput. **86**, 771–797 (2017)
20. Kuo, F.Y., Nuyens, D.: Application of quasi-Monte Carlo methods to elliptic PDEs with random diffusion coefficients - a survey of analysis and implementation. Found. Comput. Math. **16**(6), 1631–1696 (2016)
21. Kuo, F.Y., Scheichl, R., Schwab, Ch., Sloan, I.H., Ullmann, E.: Multilevel quasi-Monte Carlo methods for lognormal diffusion problems. Math. Comput. **86**, 2827–2860 (2017)
22. Kuo, F.Y., Schwab, Ch., Sloan, I.H.: Multi-level quasi-Monte Carlo finite element methods for a class of elliptic partial differential equations with random coefficients. Found. Comput. Math. **15**(2), 411–449 (2015)
23. Rhee, C.-H., Glynn, P.W.: Unbiased estimation with square root convergence for SDE models. Op. Res. **63**(5), 1026–1043 (2015)
24. Reisinger, C.: Analysis of linear difference schemes in the sparse grid combination technique. IMA J. Numer. Anal. **33**(2), 544–581 (2013)
25. Teckentrup, A., Scheichl, R., Giles, M.B., Ullmann, E.: Further analysis of multilevel Monte Carlo methods for elliptic PDEs with random coefficients. Numerische Mathematik **125**(3), 569–600 (2013)

A Method to Compute an Appropriate Sample Size of a Two-Level Test for the NIST Test Suite

Hiroshi Haramoto and Makoto Matsumoto

Abstract Statistical testing of pseudorandom number generators (PRNGs) is indispensable for their evaluation. A common difficulty among statistical tests is how we consider the resulting probability values (p-values). When a suspicious p-value, such as 10^{-3}, is observed, it is unclear whether it is due to a defect of the PRNG or merely by chance. In order to avoid such a difficulty, testing the uniformity of p-values provided by a certain statistical test is widely used. This procedure is called a two-level test. The sample size at the second level requires a careful choice because too large sample leads to the erroneous rejection, but this choice is usually done through experiments. In this paper, we propose a criterion of an appropriate sample size when we use the Frequency test, the Binary Matrix Rank test and the Runs test at the first level in the NIST test suite. This criterion is based on χ^2-discrepancy, which measures the differences between the expected distribution of p-values and the exact distribution of those. For example, when we use the Frequency test with the sample size 10^6 as the first level test, an upper bound on the sample size at the second level derived by our criterion is 125,000.

Keywords Pseudorandom number generators · Statistical testing · NIST SP800-22 · Two-level tests · Chi-square discrepancy

1 Introduction

Pseudorandom number generators (PRNGs) are computer programs whose purpose is to produce sequences of numbers that seem to behave as if they were generated randomly from a specified probability distribution. We here consider the case that

H. Haramoto (✉)
Ehime University, Address 3 Bunkyo-cho, Matsuyama, Ehime, Japan
e-mail: haramoto@ehime-u.ac.jp

M. Matsumoto
Hiroshima University, 1-3-1 Kagamiyama, Higashi-Hiroshima, Hiroshima, Japan
e-mail: m-mat@math.sci.hiroshima-u.ac.jp

A. B. Owen and P. W. Glynn (eds.), *Monte Carlo and Quasi-Monte Carlo Methods*, Springer Proceedings in Mathematics & Statistics 241,
https://doi.org/10.1007/978-3-319-91436-7_15

the outputs of the PRNG imitate independent random variables from the uniform distribution over the interval $[0, 1)$ or over the integers in $\{0, 1, 2, \ldots, N\}$.

Since PRNGs have a deterministic and periodic output, it is clear that they do not produce independent random variables in the mathematical sense, and that they cannot pass all possible statistical tests of uniformity and independence. But some of them have huge period lengths and turn out to behave quite well in statistical tests that can be applied in reasonable time. On the other hand, some PRNGs, which are known to be defective, fail very simple tests [3].

Many statistical tests for PRNGs are proposed. Widely used examples are: the test suite of the National Institute of Standards and Technology (NIST) [1], and TestU01 by L'Ecuyer and Simard [4]. The usual way to test PRNGs is to generate an n-bit sequence and analyze it with a statistical test (one-level test) and report a probability value, called p-value. NIST and TestU01 suggest to check if the p-values are uniformly distributed in the $[0, 1]$ interval (two-level test).

These test suites have two-level tests, which may give a definitive p-value even if the one-level tests report moderate p-values. However, one may suffer from accumulated approximation error in computing p-values. We often compute p-values by using approximation formula: for example, the p-value of χ^2-test is computed by using an approximation. Therefore, some computing error exists in every p-value. Thus, if the p-values of the first level tests has 1% error in the same direction, and if the two-level test uses a large number of these p-values, then it may detect the systematic computing error, which may lead to a false rejection [10].

The aim of this paper is to give an appropriate sample size of the two-level tests which use the Frequency test, the Binary Matrix Rank test, and the Runs test as the one-level test implemented in the NIST test suite. The key of our criterion is χ^2-discrepancy which measures a discrepancy between two probability distributions.

The rest of this paper is organized as follows. Section 2 gives a brief explanation on statistical testing for PRNGs, especially two-level tests. Section 3 reviews the well-known χ^2-test for goodness-of-fit, and introduces χ^2-discrepancy. Section 4 shows the exact distribution of p-values of the Frequency test, the Binary Matrix Rank test, and the Runs test, and experimental results of the two-level tests which uses the above three tests as the one-level test.

2 Statistical Testing for PRNGs

Let $(a_1, \ldots, a_n)$ be a sequence that each a_i takes zero or one generated by a PRNG, and let Z_n be a test function of n variables from $\{0, 1\}^n$ to $\mathbb{R}$. A statistical test (one-level test) of $(a_1, \ldots, a_n)$ by Z_n is a function

$$T_{Z_n} : \{0, 1\}^n \to [0, 1], \qquad (a_1, \ldots, a_n) \mapsto \Pr(Z_n(X_1, \ldots, X_n) > Z_n(a_1, \ldots, a_n))$$

where $X_1, \ldots, X_n$ are random variables with identical, independent, uniform distribution on $\{0, 1\}$. The probability $\Pr(Z_n(X_1, \ldots, X_n) > Z_n(a_1, \ldots, a_n))$ is called the p-value of the test. If the p-value is extremely small (e.g., less than 10^{-10}), then the PRNG fails the test. On the other hand, if the p-value is suspicious but does not clearly indicate rejection (e.g. 10^{-3}), it is difficult to judge. When we apply several tests to a PRNG, p-values smaller than 0.01 or larger than 0.99 are often observed (since such values appear with probability 0.02). Therefore, users of test suites for PRNGs are often troubled by the interpretation of suspicious p-values.

In order to avoid such difficulty, a two-level test is often used. In a two-level test, we fix a test function Z_n. At the first level, we apply the test T_{Z_n} to the PRNG to be tested consecutively N times, then we obtain N p-values. At the second level, we test these N values under the null hypothesis of the uniform i.i.d. in the $[0, 1]$ interval by some statistical test such as a χ^2-test, Kolmogorov-Smirnov test, Anderson-Darling test, etc. The resulting p-value is the result of the two-level test. A merit of the two-level test is that it tends to give a clearer result, by accumulating the possibly existing deviation N times. Even if the one-level tests report moderate p-values, the two-level test may give a definitive p-values such as 10^{-10}.

On the other hand, one major problem of two-level tests is that in the most cases, the exact distribution of the one-level test is often not available, but only an approximation of it is. Hence, two-level tests may detect the lack-of-fit of that approximation, leading to rejection even if the PRNG is good. Moreover, in terms of the power of the test, a one-level test with the sample size nN is typically powerful than a two-level test with the same sample size. Therefore, for a given total computational budget nN, it is usually better to take N as small as possible, see [5, 6] for more details. Therefore, we are responsible for choosing an appropriate sample at the second level with the given one-level test and its sample size.

The NIST test suite is composed of 15 statistical tests at the first level. Furthermore, it adopts a χ^2 goodness-of-fit test at the second level: divide the interval $[0, 1]$ into 10 subintervals $I_1 := [0.0, 0.1)$, $I_2 := [0.1, 0.2)$, ..., $I_{10} := [0.9, 1.0]$. Determine the number Y_j of p-values in I_j, where $j = 1, 2, \ldots, 10$, then compute the χ^2-value χ^2 by

$$\chi^2 = \sum_{j=1}^{10} \frac{(Y_j - N/10)^2}{N/10}.$$

Let $p_1, p_2, \ldots, p_{10}$ be the probabilities that the p-values at the first level fall in each of the sub-intervals $I_1, I_2, \ldots, I_{10}$. Under the null hypothesis that

$$\mathscr{H}_0 : p_1 = p_2 = \cdots = p_{10} = 0.1$$

is true, the above χ^2 approximately has a χ^2 distribution with 9 degrees of freedom for large N. Let X be a random variable which conforms to a χ^2 distribution with 9 degrees of freedom. If the probability $\Pr(X > \chi^2) < 0.0001$, the null hypothesis $\mathscr{H}_0$ (and hence the tested PRNG) is rejected. NIST recommends that the sample size at the first level n is 10^6 and the sample size at the second level N is 10^3.

For the remainder of the paper, we write n for the sample size at the first level, and N for the sample size at the second level.

3 χ^2-discrepancy

In this section, we introduce χ^2-discrepancy that measures a discrepancy between two probability distributions. See [9] for details.

Consider a set of events $\{1, 2, \ldots, \nu + 1\}$. Let $\{p_k \mid k = 1, \ldots, \nu + 1\}$ be a probability distribution on $\{1, 2, \ldots, \nu + 1\}$, i.e.,

$$0 \leq p_k \leq 1 \text{ for } k = 1, 2, \ldots, \nu + 1 \quad \text{and} \quad \sum_{k=1}^{\nu+1} p_k = 1.$$

Suppose that we make a null hypothesis $\mathscr{H}_0$ that one trial of a probabilistic event conforms to the distribution $\{p_k\}$ and the different trials are independent identically distributed. To test the null hypothesis $\mathscr{H}_0$, we perform N trials and count the number Y_k of occurrences of each event $k \in \{1, 2, \ldots, \nu + 1\}$. It is well known that the χ^2-value of this experiment defined as

$$\chi^2 := \sum_{k=1}^{\nu+1} \frac{(Y_k - Np_k)^2}{Np_k}$$

approximately conforms to the χ^2-distribution with ν degrees of freedom under the null hypothesis $\mathscr{H}_0$, if Np_k is large enough for each k. The p-value corresponding to the observed χ^2-value χ^2_{obs} is defined by $\Pr(X > \chi^2_{obs})$, where X is a random variable with χ^2-distribution with ν degrees of freedom.

Suppose that the null hypothesis $\mathscr{H}_0$ is not correct, and the exact distribution is $\{q_k \mid k = 1, 2, \ldots, \nu + 1\}$. To measure the amount of discrepancy between two distributions, we define:

Definition 1 The χ^2-discrepancy δ between the two distributions $\{q_k\}$ and $\{p_k\}$ is defined by

$$\delta := \sum_{k=1}^{\nu+1} \frac{(q_k - p_k)^2}{p_k}.$$

Theorem 1 ([9]) *The absolute value of the difference between the expectation of* χ^2 *under the nonnull hypothesis* $\mathscr{H}_0$ *and* $\nu + N\delta$ *is bounded by*

$$|E(\chi^2) - (\nu + N\delta)| \leq \nu \max_{k=1,\ldots,\nu+1} \left| 1 - \frac{q_k}{p_k} \right|.$$

This theorem supports the fact that χ^2-value is shifted by $N\delta$ in average. Therefore, the p-value at the second level tends to be 0 if $\delta \neq 0$ and the sample size N is extremely large.

Definition 2 Let $m := \max_{k=1,\ldots,\nu+1} \left| 1 - \frac{q_k}{p_k} \right|$. We call the sample size N which satisfies

$$\Pr(X > \nu + N\delta - \nu m) = 0.0001 \quad (\text{resp.} \Pr(X > \nu + N\delta + \nu m) = 0.25,)$$

where $X \sim \chi^2_\nu$, the risky (resp. safe) sample size.

The risky (resp. safe) sample size can be computed by

$$(\chi^2_\nu(0.9999) - \nu + m\nu)/\delta \quad \left(\text{resp.}(\chi^2_\nu(0.75) - \nu - m\nu)/\delta\right)$$

where $\chi^2_\nu(\alpha)$ represents the (100α)th percentile.

At the risky (resp. safe) sample size N, the expectation of the χ^2-value corresponds to the p-value $p \sim 0.0001$ (resp. $p \sim 0.25$). As stated in Sect. 1, the NIST test suite rejects the PRNG when the p-value at the second level is smaller than 0.0001. Therefore, the two-level test with the risky sample size will tend to reject even if it the PRNG is good. Conversely, if we use the safe sample size, this erroneous rejection is unlikely to occur on average. For these reasons, we recommend the safe sample size as an upper bound on the sample size at the second level.

4 Computing of Exact Distributions and Experimental Results

In this section, we review the Frequency test, the Binary Matrix Rank test, and the Runs test. Recalling the purpose of these tests, we are able to compute the exact distribution of p-values $\{q_i\}$ of each test.

In order to verify the validity of our criterion, we apply two-level tests to two PRNGs: Mersenne Twister (MT) [8] and a PRNG from SHA1 implemented in the NIST test suite, which are well-known reliable generators.

4.1 Frequency test

The Frequency test examines the proportion of ones within an n-bit sequence generated by a PRNG to be tested. Let X_i be the ith bit in the sequence. The central limit theorem ensures that under the null hypothesis that X_i are independently identically distributed Bernoulli random variable with the probability of ones is 1/2 is true, the random variable

$$Z := \frac{\frac{X_1 + \cdots + X_n}{n} - \frac{1}{2}}{\frac{1}{2\sqrt{n}}} = \frac{2(X_1 + \cdots + X_n) - n}{\sqrt{n}}$$

is asymptotically normal with mean 0 and variance 1. Therefore for a realization z of Z, the p-value can be approximated by

$$\Pr(|z| < W) = 2\left(1 - \frac{1}{\sqrt{2\pi}} \int_{-\infty}^{|z|} e^{-\frac{t^2}{2}} dt\right),$$

where the random variable W conforms to $N(0, 1)$.

Let q_i $(1 \le i \le 10)$ be the probabilities that the approximated p-value falls in the interval I_i. Then q_i is the probability that

$$2\left(1 - \frac{1}{\sqrt{2\pi}} \int_{-\infty}^{|Z|} e^{-\frac{t^2}{2}} dt\right) \in I_i$$

for $i = 1, 2, \ldots, 10$. The exact distribution of the p-values $\{q_i\}$ can explicitly be computed by

$$q_i = \sum_{0 \le j \le n, \Pr(\frac{|2j-n|}{\sqrt{n}} < W) \in I_i} \binom{n}{j} / 2^n.$$

For example, at sample size $n = 10^6$, the exact distribution of p-values is $q_1 = 0.099969$, $q_2 = 0.100223$, $q_3 = 0.099542$, $q_4 = 0.100612$, $q_5 = 0.099327$, $q_6 = 0.099907$, $q_7 = 0.100654$, $q_8 = 0.100030$, $q_9 = 0.100255$, $q_{10} = 0.099948$. These quantities are slightly different from the expected value 0.1. The χ^2-discrepancy δ is 1.86×10^{-5}, hence the risky and safe sample size are 1,329,497 and 124,913 respectively. In the same way, the risky and safe sample sizes for $n = 10^4$ are 8874 and 592, and those for $n = 10^5$ are 48,986 and 3933.

We empirically test MT and SHA1 by the two level test with five different initial values randomly. The resulting p-values of the two-level test are shown in Tables 1, 2, and 3.

We also show the results of the two-level test to a Linear Congruential Generator (LCG), based on the recurrence

$$x_{n+1} = 33{,}952{,}834{,}046{,}453\, x_n \pmod{2^{48}}$$

and take the upper 32 bits. This generator is not an excellent generator; it is a toy-model to explain how our criterion works.

Table 1 Results on the two-level test for MT (above) and SHA1 (below) with $n = 10^4$

N	1st	2nd	3rd	4th	5th
Risky	1.5e-03	6.1e-04	4.6e-03	7.7e-06	6.8e-04
Safe	1.7e-01	5.8e-01	4.5e-02	8.8e-01	3.7e-01

N	1st	2nd	3rd	4th	5th
Risky	9.9e-07	6.8e-05	7.9e-07	3.5e-02	1.2e-05
Safe	8.2e-01	7.7e-01	9.3e-01	3.5e-01	4.5e-01

Table 2 Results on the two-level test for MT (above) and SHA1 (below) with $n = 10^5$

N	1st	2nd	3rd	4th	5th
Risky	7.8e-05	2.1e-06	1.8e-05	6.8e-04	1.8e-05
Safe	7.0e-01	5.2e-01	7.9e-01	3.0e-01	2.6e-01

N	1st	2nd	3rd	4th	5th
Risky	3.8e-05	3.4e-05	4.7e-02	1.5e-03	2.4e-02
Safe	9.4e-01	4.2e-01	8.8e-01	1.5e-01	1.6e-02

Table 3 Results on the two-level test for MT (above) and SHA1 (below) with $n = 10^6$

N	1st	2nd	3rd	4th	5th
Risky	7.7e-06	3.5e-03	6.3e-05	4.0e-03	1.4e-03
Safe	2.4e-01	6.3e-01	2.3e-03	1.9e-01	3.3e-02

N	1st	2nd	3rd	4th	5th
Risky	5.1e-06	6.4e-06	7.0e-04	3.5e-05	7.8e-07
Safe	5.4e-01	9.9e-01	8.3e-01	7.8e-01	5.4e-01

Table 4 Results on the two-level test for LCG with $n = 10^6$

N	1st	2nd	3rd	4th	5th
risky (1,329,497)	<2.2e-16	<2.2e-16	<2.2e-16	<2.2e-16	<2.2e-16
safe (124,913)	6.2e-11	1.0e-04	2.7e-02	4.9e-05	3.2e-05
NIST's recommendation (1000)	3.6e-01	9.4e-01	5.8e-02	5.5e-01	8.5e-01

Table 4 shows the results corresponding empirical two-level test for risky, safe and NIST's recommendation sample size 1000.

These results show that the LCG tends to provide smaller p-values than those of expected values, it indicates the defectiveness of the LCG, which is not revealed with $N = 1000$.

4.2 *Binary Matrix Rank test*

The Binary Matrix Rank test is to check for linear dependence among fixed-length subsequences of the original sequence to be tested.

We divide the n-bit sequence into k contiguous non-overlapping subsequences of m^2 bits. With these, build k of $m \times m$ matrices, and for each matrix, compute its rank r of it over the two element field. The probability that the rank R of an $m \times m$

random binary matrix is r is given by

$$\Pr(R=r)=\begin{cases}\dfrac{1}{2^{m^2}} & (r=0)\\ 2^{r(2m-r)-m^2}\displaystyle\prod_{i=0}^{r-1}\frac{(1-2^{i-m})^2}{1-2^{i-r}} & (1\le r\le m)\end{cases}$$

(see [7]). Then it compares this empirical distribution with the theoretical distribution of the rank of a random matrix, via a χ^2 test, after merging classes if needed.

The NIST test suite takes the size of matrix $m = 32$. Let F_m be the number of matrices with rank m, F_{m-1} the number of matrices with rank $m-1$, and F_{m-2} the number of matrices remaining (i.e., $F_{m-2} = k - F_m - F_{m-1}$). To apply a χ^2-test, we compute the statistic

$$X=\frac{(F_m-k\Pr(R=m))^2}{k\Pr(R=m)}+\frac{(F_{m-1}-k\Pr(R=m-1))^2}{k\Pr(R=m-1)}+\frac{(F_{m-2}-k\Pr(R\le m-2))^2}{k\Pr(R\le m-2)}.$$

Under the null hypothesis that the sequence is a sample of i.i.d. $U\{0,1\}$ random variable, X has an approximate χ^2-distribution with 2 degrees of freedom. Then the reported p-value is $\exp\left(-\frac{x}{2}\right)$ for a realization x of X.

The exact distribution of p-values $\{q_i\}$ can be derived by

$$q_i=\sum\frac{n!}{s!t!u!}\Pr(R=m)^s\Pr(R=m-1)^t\Pr(R\le m-2)^u,$$

for each $i = 1, 2, \ldots, 10$, here the sum is taken over all non-negative integers s, t, u such that $s+t+u=k$ and $\exp(-x/2)\in I_i$ where x is a realization of X with $F_m = s$, $F_{m-1} = t$, and $F_{m-2} = u$.

When $k = 976$, this test consumes $m^2 \times k = 32^2 \times 976 = 999{,}424$ bits, which is the nearest sample size for the NIST's recommendation 10^6, at a time; the NIST test suite admits the first 999,424 bits and discards the remaining 576 bits when we specify the sample size 10^6. In this case, the exact distribution of p-values is $q_1 = 0.099271$, $q_2 = 0.100959$, $q_3 = 0.100922$, $q_4 = 0.099342$, $q_5 = 0.101089$, $q_6 = 0.096403$, $q_7 = 0.100313$, $q_8 = 0.101222$, $q_9 = 0.100023$, $q_{10} = 0.100451$. Then $\delta = 1.865737 \times 10^{-4}$, the risky and safe sample size is 134,230 and 11,068 respectively.

Table 5 shows the results of the two-level test using the Binary Matrix Rank test at the first level with the sample size $n = 999{,}424$. Note that MT always fails the Binary Matrix Rank test when the matrix size large enough, this is in fact a limitation of all $\mathbb{F}_2$-linear generators; in this case, $m = 32$ is small.

Tables 6 and 7 show another type of results. We take the number of matrices k from 151 to 300. Each number means the number of rejection when we apply the two-level test with the corresponding risky sample size to MT and SHA1 five times.

Table 5 Results on the two-level test for MT (above) and SHA1 (below) with $n = 999{,}424$

N	1st	2nd	3rd	4th	5th
Risky	2.1e-04	1.4e-02	1.4e-09	1.3e-06	1.7e-03
Safe	4.2e-01	5.7e-01	2.3e-01	7.6e-01	5.3e-01

N	1st	2nd	3rd	4th	5th
Risky	2.1e-04	2.4e-04	9.8e-05	8.3e-06	5.3e-05
Safe	7.9e-01	8.3e-01	9.0e-01	5.2e-01	7.4e-01

Table 6 The number of rejections of the two-level test for MT when $k = 150$–300

3 5 3 1 2 1 2 2 4 3 2 0 3 1 2 2 2 4 2 2 2 4 4 3 3 3 3 4 3 5 4 2 4 2 1 5 1 3 2 2 2 2 1 3 2 2 3 2 1 4
4 3 1 2 2 2 4 2 2 4 4 2 4 1 3 4 4 1 2 3 1 3 4 2 3 3 3 1 3 4 1 3 3 3 3 2 4 2 3 3 3 2 3 3 1 1 2 3 2 3
2 2 4 4 3 2 2 3 1 1 3 3 1 2 4 3 2 3 3 1 4 3 1 4 2 2 3 2 4 2 2 2 4 2 3 1 3 2 4 4 4 4 3 3 2 2 4 4 3 1

Table 7 The number of rejections of the two-level test for SHA1 when $k = 150$–300

0 3 1 4 2 0 5 1 4 3 5 4 2 3 4 4 3 1 4 2 2 4 1 4 4 4 3 3 3 0 4 2 3 1 4 4 2 1 5 5 1 2 2 1 1 4 3 3 2 3
4 4 3 2 4 1 2 3 2 4 3 3 3 4 2 2 3 3 2 3 3 3 2 2 3 5 2 1 4 2 1 3 3 2 5 4 1 1 4 3 2 4 2 3 3 5 2 2 3 4
3 3 1 3 3 1 3 2 2 3 2 2 2 2 3 2 5 3 3 2 1 2 0 2 3 3 2 1 1 2 0 1 1 3 3 2 0 3 4 2 2 2 3 3 3 4 3 4 3 3

The average number of rejection is 2.62 for MT, and 2.61 for SHA1 respectively. We mention that no rejections occurred when the safe sample size is taken for each k.

4.3 *Runs test*

The purpose of the Runs test is to determine whether the number of runs of ones and zeros of various lengths is as expected for a random sequence.

For $r = 2, 3, \ldots n$, the probability distribution of R, the total number of runs of $n = n_1 + n_2$, n_1 of 1 and n_2 of 0, in a random sample is

$$\Pr(R = r) = 2\binom{n_1 - 1}{r/2 - 1}\binom{n_2 - 1}{r/2 - 1}\Big/\binom{n}{n_1}$$

if r is even, and

$$\Pr(R = r) = \left\{\binom{n_1 - 1}{(r-1)/2}\binom{n_2 - 1}{(r-3)/2} + \binom{n_1 - 1}{(r-3)/2}\binom{n_2 - 1}{(r-1)/2}\right\}\Big/\binom{n}{n_1}$$

if r is odd [2].

Table 8 Results on the two-level test for MT (above) and SHA1 (below) with $n = 10^3$

N	1st	2nd	3rd	4th	5th
Risky	3.9e-02	3.5e-03	5.0e-01	5.3e-01	3.5e-02
Safe	5.3e-01	4.8e-01	9.9e-01	3.8e-01	3.6e-01

N	1st	2nd	3rd	4th	5th
Risky	7.0e-07	1.9e-04	8.8e-02	2.8e-04	6.6e-03
Safe	5.3e-01	2.8e-01	5.9e-01	1.1e-01	4.7e-01

Table 9 Results on the two-level test for MT (above) and SHA1 (below) with $n = 10^4$

N	1st	2nd	3rd	4th	5th
Risky	2.8e-03	1.6e-05	4.6e-09	3.7e-04	1.1e-06
Safe	3.2e-01	2.2e-02	4.0e-01	4.4e-03	2.1e-01

N	1st	2nd	3rd	4th	5th
Risky	1.5e-04	1.0e-04	1.4e-07	9.8e-05	1.2e-02
Safe	7.1e-01	6.7e-01	1.1e-01	9.5e-01	1.9e-01

Table 10 Results on the two-level test for MT (above) and SHA1 (below) with $n = 10^5$

N	1st	2nd	3rd	4th	5th
Risky	1.1e-07	3.9e-05	1.8e-03	4.4e-05	7.5e-03
Safe	5.8e-02	9.2e-01	3.1e-01	7.3e-01	4.4e-01

N	1st	2nd	3rd	4th	5th
Risky	1.1e-04	2.9e-04	7.7e-06	1.1e-04	4.8e-03
Safe	4.0e-01	9.9e-01	2.9e-01	1.0e-02	4.6e-02

Let $\lambda = n_1/n$. For large n, the random variable

Table 11 Results on the two-level test for MT (above) and SHA1 (below) with $n = 10^6$

N	1st	2nd	3rd	4th	5th
Risky	1.4e-03	1.3e-04	2.2e-04	1.1e-06	8.0e-05
Safe	9.2e-01	8.6e-01	2.6e-01	3.1e-01	3.8e-01

N	1st	2nd	3rd	4th	5th
Risky	1.4e-04	5.3e-03	1.7e-04	4.0e-02	1.3e-03
Safe	1.5e-01	4.3e-01	6.3e-01	6.7e-01	1.1e-01

$$Z = \frac{R - 2n\lambda(1-\lambda)}{2\sqrt{n}\lambda(1-\lambda)}$$

is approximated by the standard normal distribution. Then we compute the p-value by

$$\Pr(|z| < W) = 2\left(1 - \frac{1}{\sqrt{2\pi}} \int_{-\infty}^{|z|} e^{-\frac{t^2}{2}} dt\right),$$

where $W \sim N(0, 1)$ and z is a realization of Z. Therefore the exact distribution of p-values $\{q_i\}$ is

$$q_i = \sum_{0 \le n_1 \le n, 1 \le r \le n, \Pr(|z|<W) \in I_i} N(r, n_1, n)/2^n$$

for $i = 1, 2, \ldots, 10$, where $N(r, n_1, n)$ is the numerator of the previous probability (i.e. the number of different sequences with r runs.)

It is difficult to calculate the χ^2-discrepancy δ for $n = 10^5$ because δ takes $O(n^2)$ computational cost. Pareschi, Rovatti, and Setti showed an approximation formula of the cumulative distribution function of the p-values of the Runs test with $O(n)$ computational cost [11]. By using it, the risky and safe sample size are approximately 781,442 and 72,527. In the same way, the risky and safe size for $n = 10^6$ are 13,229,762 and 1,267,025 respectively. The corresponding results are shown in Tables 8, 9, 10 and 11.

5 Conclusion

Although the exact distribution of the p-values of the Frequency test, the Binary Matrix Rank test and the Runs test can be computed, the NIST test suite does not use it but adopt a χ^2 goodness-of-fit test at the second level. It forces us to choose appropriate sample size, then we propose a criterion of an appropriate upper bound on the sample size of the two-level test. Experimental results support its reliability and statistical power.

On the other hand, we can hardly compute the exact distribution of p-values for the others tests in the NIST test suite, because these tests seem to require brute-force computation. Hence effective methods to compute or to approximate the exact distribution of p-values are needed.

The test statistic of each of the three tests we treat has a discrete distribution, then the p-value has a discrete distribution, it is not uniform over [0, 1]. If one wants to develop a test which uses the discrete distribution directly, it should be better to consider the left and right p-values separately, see, e.g., [4], Sect. 3.

In general, a Kolmogorov-Smirnov test is more appropriate than a χ^2 test [12]. Further work is to extend our proposal to a Kolmogorov-Smirnov test to increase the power of statistical tests for PRNGs, especially TestU01.

Acknowledgements We are thankful to Editor Professor Art Owen and the referees, who informed of numerous improvements on the manuscript. This research has been supported in part by JSPS Grant-In-Aid #26310211, #15K13460, #16K13750, #17K14234, #18K03213, and JST CREST "Theory of hyper uniformity and its development in randomness appeared in sciences."

References

1. Bassham III, L.E., Rukhin, A.L., Soto, J., Nechvatal, J.R., Smid, M.E., Barker, E.B., Leigh, S.D., Levenson, M., Vangel, M., Banks, D.L., Heckert, N.A., Dray, J.F., Vo, S.: Sp 800-22 rev. 1a. a statistical test suite for random and pseudorandom number generators for cryptographic applications. Technical report, National Institute of Standards & Technology, Gaithersburg, MD, United States (2010)
2. Gibbons, J.D., Chakraborti, S.: Nonparametric Statistical Inference, 5th edn. Chapman and Hall/CRC, Boca Raton (2010)
3. L'Ecuyer, P.: Software for uniform random number generation: distinguishing the good and the bad. In: Proceedings of the 2001 Winter Simulation Conference, vol. 95–105. IEEE Press (2001). http://dl.acm.org/citation.cfm?id=564124.564139
4. L'Ecuyer, P., Simard, R.: TestU01: a C library for empirical testing of random number generators. ACM Trans. Math. Softw. **33**(4), Art. 22, 40 (2007)
5. L'Ecuyer, P., Cordeau, J.F., Simard, R.: Close-point spatial tests and their application to random number generators. Op. Res. **48**(2), 308–317 (2000). https://doi.org/10.1287/opre.48.2.308.12385
6. L'Ecuyer, P., Simard, R., Wegenkittl, S.: Sparse serial tests of uniformity for random number generators. SIAM J. Sci. Comput. **24**(2), 652–668 (2002)
7. Marsaglia, G., Tsay, L.H.: Matrices and the structure of random number sequences. Linear Algebra Appl. **67**, 147–156 (1985). https://doi.org/10.1016/0024-3795(85)90192-2
8. Matsumoto, M., Nishimura, T.: Mersenne twister: a 623-dimensionally equidistributed uniform pseudo-random number generator. ACM Trans. Model. Comput. Simul. **8**(1), 3–30 (1998)
9. Matsumoto, M., Nishimura, T.: A nonempirical test on the weight of pseudorandom number generators. In: Monte Carlo and quasi-Monte Carlo methods 2000, pp. 381–395 (2000). Springer, Berlin (2002)
10. Pareschi, F., Rovatti, R., Setti, G.: Second-level NIST randomness tests for improving test reliability. In: 2007 IEEE International Symposium on Circuits and Systems, pp. 1437–1440 (2007). https://doi.org/10.1109/ISCAS.2007.378572
11. Pareschi, F., Rovatti, R., Setti, G.: On statistical tests for randomness included in the NIST SP800-22 test suite and based on the binomial distribution. IEEE Trans. Inf. Forensics Secur. **7**(2), 491–505 (2012). https://doi.org/10.1109/TIFS.2012.2185227
12. Simard, R., L'Ecuyer, P.: Computing the two-sided Kolmogorov-Smirnov distribution. J. Stat. Softw. **39**(11), 1–18 (2011). https://doi.org/10.18637/jss.v039.i11. https://www.jstatsoft.org/v039/i11

Lower Complexity Bounds for Parametric Stochastic Itô Integration

Stefan Heinrich

Abstract We study the complexity of pathwise approximation of parameter dependent stochastic Itô integration for C^r functions, with $r \in \mathbb{R}$, $r > 0$. Both definite and indefinite integration are considered. This complements previous results (Daun and Heinrich (J Complex 40:100–122, 2017, [2]) for classes of functions with dominating mixed smoothness. Upper bounds are obtained by embedding of function classes and applying some generalizations of these previous results. The emphasis of the present paper lies on lower bounds. While in Daun and Heinrich (J Complex 40:100–122, 2017), [2] only nonadaptive deterministic algorithms were considered, we prove here lower bounds for adaptive deterministic and randomized algorithms, both for the classes considered here as for those from Daun and Heinrich (J Complex 40:100–122, 2017), [2].

Keywords Stochastic integration · Complexity · Parametric problems · Deterministic and stochastic algorithms

1 Introduction

The complexity of stochastic integration of real-valued non-parametric functions was investigated in [8, 12, 13, 15]. In [2] the complexity of definite and indefinite stochastic Itô integration of parameter dependent random functions was studied. Classes of functions with smoothness of dominating mixed type $C^{r,\rho}$ with integer degree of differentiability r were considered there. A multilevel Euler–Maruyama scheme was developed and analyzed to obtain the upper bounds. Moreover, matching lower bounds were shown in the deterministic nonadaptive setting. The present paper extends and complements these results in a number of respects.

S. Heinrich (✉)
Department of Computer Science, University of Kaiserslautern,
67653 Kaiserslautern, Germany
e-mail: heinrich@informatik.uni-kl.de

A. B. Owen and P. W. Glynn (eds.), *Monte Carlo and Quasi-Monte Carlo Methods*, Springer Proceedings in Mathematics & Statistics 241,
https://doi.org/10.1007/978-3-319-91436-7_16

First of all, we study standard isotropic C^r-smoothness (definitions are given in the text below). This allows to compare the results with previous ones for (non-stochastic) parametric integration obtained in [3–5] and also in [1]. However, we consider real-valued r, thus differentiable functions whose derivatives of order $\lfloor r \rfloor$ satisfy suitable Hölder conditions. We discuss the extension of the results of [2] to fractional indices of smoothness. Then we derive upper bounds for C^r classes by studying their embedding into suitable $C^{r_1,\rho}$ classes and applying the algorithm and its analysis from [2].

The main results of the present paper concern lower bounds. First an abstract setting of algorithms and nth minimal errors is introduced, which extends respective approaches for deterministic problems. Then we prove lower bounds for adaptive algorithms both in the deterministic and randomized setting matching the upper bounds derived before (up to logarithmic factors, in general). We present a new technique, which involves exponential inequalities. It is also shown that the bounds obtained in [2] for nonadaptive deterministic algorithms hold true for adaptive deterministic and randomized algorithms, as well.

The structure of the paper is as follows: Sect. 2 contains notation and some preliminaries, including the needed function classes. In Sect. 3 we recall the multilevel Euler–Maruyama algorithm from [2] and derive error estimates. Section 4 is devoted to lower bounds.

2 Preliminaries

Let $\mathbb{N} = \{1, 2, \dots\}$ and $\mathbb{N}_0 = \{0, 1, 2, \dots\}$. Let X, Y be Banach spaces. The unit ball of X is denoted by B_X, the dual space by X^*, the σ-algebra of Borel subsets of X by $\mathscr{B}(X)$, and the space of bounded linear operators from Y to X by $\mathscr{L}(Y, X)$. Let $d \in \mathbb{N}$. The space of real-valued continuous functions on a compact set $Q \subset \mathbb{R}^d$ is denoted by $C(Q)$ and is equipped with the supremum norm. Furthermore, if Q is the closure of an open bounded set and $k \in \mathbb{N}$, $C^k(Q)$ denotes the space of all functions which are k-times continuously differentiable in the interior of Q and which together with their derivatives up to order k possess continuous extensions to all of Q. This space is equipped with the norm $\|f\|_{C^k(Q)} = \sup_{|\alpha|\le k,\, s\in Q} |D^\alpha f(s)|$ with $\alpha = (\alpha_1, \dots, \alpha_d) \in \mathbb{N}_0^d$ and $|\alpha| = |\alpha_1| + \dots + |\alpha_d|$. If $k = 0$, we put $C^0(Q) = C(Q)$. For $r \in \mathbb{R}$, $r > 0$, $r \notin \mathbb{N}$ put $k = \lfloor r \rfloor$, $\sigma = r - k$, and let $C^r(Q)$ be the space of all $f \in C^k(Q)$ satisfying $\|f\|_{C^r(Q)} < \infty$, where

$$\|f\|_{C^r(Q)} := \max\left(\|f\|_{C^k(Q)}, \max_{|\alpha|\le k} \sup_{s_1\ne s_2\in Q} |s_1 - s_2|^{-\sigma} \left|D^\alpha f(s_1) - D^\alpha f(s_2)\right| \right),$$

and $|\cdot|$ denotes the Euclidean norm on $\mathbb{R}^d$. For $1 \le p < \infty$ and $(M, \mathscr{M}, \mu)$ an arbitrary measure space, $L_p(M, \mathscr{M}, \mu, X)$, or shortly $L_p(M, X)$, is the space of Bochner p-integrable functions, equipped with the usual norm.

Throughout the paper the same symbol $c, c_1, c_2, \dots$ may denote different constants, even in a sequence of relations. Moreover, for nonnegative reals $(a_n)_{n\in\mathbb{N}}$ and $(b_n)_{n\in\mathbb{N}}$ we write $a_n \preceq b_n$ if there are constants $c > 0$ and $n_0 \in \mathbb{N}$ such that for all $n \ge n_0$, $a_n \le cb_n$. Furthermore, $a_n \asymp b_n$ means that $a_n \preceq b_n$ and $b_n \preceq a_n$. Finally, $a_n \preceq_{\log} b_n$ iff there are constants $c > 0$, $n_0 \in \mathbb{N}$, and $\theta \in \mathbb{R}$ such that for all $n \ge n_0$ $a_n \le cb_n(\log(n+1))^\theta$.

Throughout the rest of the paper we let $Q = [0, 1]^d$. Let $k \in \mathbb{N}$, $m \in \mathbb{N}$, and let $\Gamma_m^{k,d} = \left\{\frac{i}{km} : 0 \le i \le km\right\}^d$. Let P^k be Lagrange interpolation of degree k with respect to the uniform mesh of size $1/k$ on $[0, 1]$, let $P^{k,d} \in \mathscr{L}(C(Q), C(Q))$ be its d-fold tensor product, and let $P_m^{k,d} \in \mathscr{L}(C(Q), C(Q))$ be its composition with respect to the partition of Q into subcubes of sidelength $1/m$. Let $r \in \mathbb{R}$, $r > 0$, and set $k = \lceil r \rceil$. It is well-known that there are constants $c_0, c_1 > 0$ such that for all $m \in \mathbb{N}$

$$\|P_m^{k,d}\|_{\mathscr{L}(C(Q),C(Q))} \le c_0, \quad \|J - P_m^{k,d} J\|_{\mathscr{L}(C^r(Q),C(Q))} \le c_1 m^{-r}, \tag{1}$$

where $J : C^r(Q) \to C(Q)$ is the embedding.

Let $(\Omega, \Sigma, \mathbb{P})$ be a probability space, $(\Sigma_t)_{0\le t\le 1}$, $\Sigma_t \subseteq \Sigma$ a filtration, let $(W(t))_{0\le t\le 1}$, $W(t) = W(t, \omega)$ $(\omega \in \Omega)$ be a Wiener process on $(\Omega, \Sigma, \mathbb{P})$ adapted to (Σ_t) and such that for $0 \le t_1 \le t_2 \le 1$ the increments $W(t_2) - W(t_1)$ are independent of Σ_{t_1}. We assume w.l.o.g. that all trajectories of the Wiener process are continuous.

Next we introduce the class of random functions which we will study here. Let $r \in \mathbb{R}$, $r > 0$, $d \in \mathbb{N}$, $2 \le q < \infty$ and let $\mathscr{F}_q^r = \mathscr{F}_q^r(Q \times [0, 1] \times \Omega; \kappa)$ denote the set of all functions $f : Q \times [0, 1] \times \Omega \to \mathbb{R}$ such that for each $s \in Q$, $f(s, t, \omega)$ is progressively measurable, in other words, for each $\tau \in [0, 1]$ the restriction $f(s, \cdot, \cdot)|_{[0,\tau]\times\Omega}$ is $\mathscr{B}([0, \tau]) \times \Sigma_\tau$ measurable,

$$f(\cdot, \cdot, \omega) \in C^r(Q \times [0, 1]) \quad (\omega \in \Omega), \tag{2}$$

$$\left(\mathbb{E}\, \|f(\cdot, \cdot, \omega)\|^q_{C^r(Q\times[0,1])}\right)^{1/q} \le \kappa. \tag{3}$$

We need to recall the definition of related classes in [2]. Let $r_1 \in \mathbb{R}$, $r_1 > 0$, $0 \le \rho \le 1$ and let $F^{r_1,\rho} = F^{r_1,\rho}(Q \times [0, 1] \times \Omega; \kappa)$ denote the set of all functions $f : Q \times [0, 1] \times \Omega \to \mathbb{R}$ such that for each $s \in Q$, $f(s, t, \omega)$ is progressively measurable and

$$f(\cdot, t, \omega) \in C^{r_1}(Q) \quad ((t, \omega) \in [0, 1] \times \Omega), \tag{4}$$

$$\left(\mathbb{E}\, \|f(\cdot, 0, \omega)\|^2_{C^{r_1}(Q)}\right)^{1/2} \le \kappa, \tag{5}$$

$$\left(\mathbb{E}\, \|f(\cdot, t_1, \omega) - f(\cdot, t_2, \omega)\|^2_{C^{r_1}(Q)}\right)^{1/2} \le \kappa|t_1 - t_2|^\rho \quad (t_1, t_2 \in [0, 1]). \tag{6}$$

Let $F^{r_1,\rho}(Q \times [0, 1] \times \Omega) = \cup_{\kappa>0} F^{r_1,\rho}(Q \times [0, 1] \times \Omega; \kappa)$ be the respective linear space. Moreover, for $2 < q < \infty$ let $F_q^{r_1,\rho} = F_q^{r_1,\rho}(Q \times [0, 1] \times \Omega; \kappa)$ be the subset of those $f \in F^{r_1,\rho}(Q \times [0, 1] \times \Omega; \kappa)$ which fulfill

$$\left(\mathbb{E}\max_{t\in M}\|f(\cdot\,,t,\omega)\|^q_{C^{r_1}(Q)}\right)^{1/q}\le\kappa\quad(M\subset[0,1],\,|M|<\infty).\tag{7}$$

Let us consider the relation between the two types of function classes.

Lemma 1 *Let $r, r_1 > 0$, $0 \le \rho \le 1$, $r \ge r_1 + \rho$, $2 < q < \infty$. Then there are constants $c_1, c_2 > 0$ such that*

$$\mathscr{F}_2^r(Q\times[0,1]\times\Omega;\kappa)\subseteq F^{r_1,\rho}(Q\times[0,1]\times\Omega;c_1\kappa)\tag{8}$$
$$\mathscr{F}_q^r(Q\times[0,1]\times\Omega;\kappa)\subseteq F_q^{r_1,\rho}(Q\times[0,1]\times\Omega;c_2\kappa).\tag{9}$$

Proof Let $2 \le q < \infty$ and $f \in \mathscr{F}_q^r(Q\times[0,1]\times\Omega;\kappa)$. Clearly, (4) follows from (2), while (3) implies (5) and (7). It remains to show that (6) holds. We can assume $r = r_1 + \rho$. Let $r = k + \sigma$, $r_1 = k_1 + \sigma_1$ $(k, k_1 \in \mathbb{N}_0,\ 0 \le \sigma, \sigma_1 < 1)$. Fix $\omega \in \Omega$ and set $\kappa(\omega) = \|f(\cdot\,,\,\cdot\,,\omega)\|_{C^r(Q\times[0,1])}$. Let $s_1, s_2 \in Q$, $s_1 \ne s_2$, $t_1, t_2 \in [0,1]$, $t_1 \ne t_2$, $\alpha \in \mathbb{N}_0^d$, $|\alpha| \le k_1$, and put

$$\begin{aligned}\Delta_1(t_1,t_2,\omega)&:=\|f(\cdot\,,t_1,\omega)-f(\cdot\,,t_2,\omega)\|_{C^{k_1}(Q)}\\ \Delta_2^\alpha(s_1,s_2,t_1,t_2,\omega)&:=|D_s^\alpha f(s_1,t_1,\omega)-D_s^\alpha f(s_1,t_2,\omega)\\&\qquad -D_s^\alpha f(s_2,t_1,\omega)+D_s^\alpha f(s_2,t_2,\omega)|.\end{aligned}$$

Using the definition of $\kappa(\omega)$ above, it is readily checked that

$$\Delta_1(t_1,t_2,\omega)\le\kappa(\omega)|t_1-t_2|^\rho.\tag{10}$$

To estimate Δ_2, we first we assume $\sigma_1+\rho=\sigma$, thus $k_1=k$. Then, taking into account $|t_1-t_2|\le 1$,

$$\begin{aligned}&\Delta_2^\alpha(s_1,s_2,t_1,t_2,\omega)\\&\quad\le 2\kappa(\omega)\min(|s_1-s_2|,|t_1-t_2|)^\sigma\le 2\kappa(\omega)|s_1-s_2|^{\sigma_1}|t_1-t_2|^\rho.\end{aligned}\tag{11}$$

Now we assume $\sigma_1+\rho=1+\sigma$, hence $k_1=k-1$. Let e_i be the ith unit vector in $\mathbb{R}^d$. If $|t_1-t_2|\ge|s_1-s_2|$, then, denoting $s_1=(s_{1,i})_{i=1}^d$, $s_2=(s_{2,i})_{i=1}^d$,

$$\begin{aligned}&\Delta_2^\alpha(s_1,s_2,t_1,t_2,\omega)\\&=\Bigg\|\int_0^1\sum_{i=1}^d\Big(D_s^{\alpha+e_i}f(s_2+\theta(s_1-s_2),t_1,\omega)\\&\qquad -D_s^{\alpha+e_i}f(s_2+\theta(s_1-s_2),t_2,\omega)\Big)(s_{1,i}-s_{2,i})d\theta\Bigg\|_{C^{k_1}(Q)}\\&\le\sqrt{d}\kappa(\omega)|s_1-s_2||t_1-t_2|^\sigma\le\sqrt{d}\kappa(\omega)|s_1-s_2|^{\sigma_1}|t_1-t_2|^\rho.\end{aligned}\tag{12}$$

Similarly, if $|t_1-t_2|<|s_1-s_2|$, then

$$\begin{aligned}
&\Delta_2^\alpha(s_1, s_2, t_1, t_2, \omega) \\
&= \left\| \int_0^1 \Big(D_{s,t}^{\alpha,1} f(s_1, t_2 + \theta(t_1 - t_2), \omega) \right. \\
&\qquad \left. - D_{s,t}^{\alpha,1} f(s_2, t_2 + \theta(t_1 - t_2), \omega) \Big)(t_1 - t_2) d\theta \right\|_{C^{k_1}(Q)} \\
&\le \kappa(\omega)|t_1 - t_2||s_1 - s_2|^\sigma \le \kappa(\omega)|s_1 - s_2|^{\sigma_1}|t_1 - t_2|^\rho. \qquad (13)
\end{aligned}$$

It follows from (10)–(13) and (3) that

$$\begin{aligned}
&\left(\mathbb{E}\, \|f(\cdot\,, t_1, \omega) - f(\cdot\,, t_2, \omega)\|_{C^{r_1}(Q)}^2\right)^{1/2} \\
&= \left(\mathbb{E} \max \left(\Delta_1(t_1, t_2, \omega), \max_{|\alpha| \le k_1} \sup_{s_1 \ne s_2 \in Q} \frac{\Delta_2^\alpha(s_1, s_2, t_1, t_2, \omega)}{|s_1 - s_2|^{\sigma_1}} \right)^2 \right)^{1/2} \\
&\le \max(\sqrt{d}, 2) \left(\mathbb{E}\, \kappa(\omega)^2\right)^{1/2} |t_1 - t_2|^\rho \le \max(\sqrt{d}, 2)\kappa|t_1 - t_2|^\rho,
\end{aligned}$$

which shows (6).

Now we consider parametric indefinite stochastic integration $\int_0^t f(s, \tau) dW(\tau)$ ($s \in Q, t \in [0, 1]$). This is a stochastic process indexed by $Q \times [0, 1]$. It was shown in [2] (for $r \in \mathbb{N}$, but the argument is the same for real $r > 0$) that we can find a continuous version in the sense that there is a mapping

$$\hat{\mathscr{S}} : F^{r,0}(Q \times [0, 1] \times \Omega) \to L_2(\Omega, C(Q \times [0, 1]))$$

such that for $s \in Q, t \in [0, 1]$

$$(\hat{\mathscr{S}}(f))(s, t) = \int_0^t f(s, \tau) dW(\tau). \qquad (14)$$

It follows from the linearity of the stochastic integral (and a standard density and continuity argument) that the operator $\hat{\mathscr{S}}$ is linear. For our purposes we need a (pathwise) mapping

$$\mathscr{S} : F^{r,0}(Q \times [0, 1] \times \Omega) \times \Omega \to C(Q \times [0, 1]) \qquad (15)$$

such that $\mathscr{S}(f, \cdot) = \hat{\mathscr{S}}(f)$, with equality meant in $L_2(\Omega, C(Q \times [0, 1]))$, and $\mathscr{S}$ is linear in f. Let $(f_i)_{i \in I}$, I a suitable index set, be a Hamel basis of $F^{r,0}(Q \times [0, 1] \times \Omega)$ (i.e., each element can be written uniquely as linear combination of a finite number of basis vectors). For each $i \in I$ let $g_i = g_i(\omega)$ be a representative of the equivalence class $\hat{\mathscr{S}}(f_i) \in L_2(\Omega, C(Q \times [0, 1]))$. Then we set $\mathscr{S}(f_i, \omega) = g_i(\omega)$ for $i \in I$ and $\omega \in \Omega$ and extend the so-defined mapping by linearity to all of $F^{r,0}(Q \times [0, 1] \times \Omega)$. It follows from the linearity of $\hat{\mathscr{S}}$ that $\mathscr{S}$ is as required.

For parametric definite stochastic integration $\int_0^1 f(s,\tau)dW(\tau)$ $(s \in Q)$ we define $\mathscr{S}_1 : F^{r,0}(Q \times [0,1] \times \Omega) \times \Omega \to C(Q)$ by setting

$$(\mathscr{S}_1(f,\omega))(s) = (\mathscr{S}(f,\omega))(s,1) \quad (s \in Q, \omega \in \Omega). \tag{16}$$

It follows that $\mathscr{S}_1(f,\,\cdot\,) \in L_2(\Omega, C(Q))$ and

$$(\mathscr{S}_1(f,\,\cdot\,))(s) = \int_0^1 f(s,\tau)dW(\tau) \quad (s \in Q), \tag{17}$$

with equality (17) meant in $L_2(\Omega)$. Due to Lemma 1, the operators $\mathscr{S}$ and $\mathscr{S}_1$ are also defined on the respective sets $\mathscr{F}_q^r(Q \times [0,1] \times \Omega; \kappa)$.

3 An Algorithm for Parametric Stochastic Integrals

First we recall the algorithm from [2]. Let $n \in \mathbb{N}$, $t_k = k/n$ $(k = 0, \ldots, n)$, and define $A_n(f,\omega) \in C([0,1])$ for any function $f : [0,1 \times \Omega \to \mathbb{R}$ and $\omega \in \Omega$ by

$$A_n(f,\omega) = P_n^{1,1}\left(\sum_{j=0}^{k-1} f(t_j,\omega)(W(t_{j+1},\omega) - W(t_j,\omega))\right)_{k=0}^{n}.$$

This is the piecewise linear interpolation of the Euler–Maruyama scheme. Furthermore, we set $A_{n,1}(f,\omega) := (A_n(f,\omega))(1)$. Next we pass to the multilevel scheme of [2]. Put $k = \lceil r \rceil$, fix $l_1 \in \mathbb{N}_0$, $n_0, \ldots, n_{l_1} \in \mathbb{N}$, let $f : Q \times [0,1] \times \Omega \to \mathbb{R}$ be any function and $\omega \in \Omega$. For the indefinite problem we define

$$\mathscr{A}(f,\omega) = \sum_{l=0}^{l_1} \left(P_{2^l}^{k,d} - P_{2^{l-1}}^{k,d}\right) \left(A_{n_l}(f_s,\omega)\right)_{s \in \Gamma_{2^l}^{k,d}},$$

where f_s is given by $f_s(t,\omega) := f(s,t,\omega)$ $(t \in [0,1], \omega \in \Omega)$ and $P_{2^{-1}} := 0$. In the definite case we define $\mathscr{A}_1(f,\omega)$ analogously, using the $A_{n_l,1}$. Let $\mathrm{card}(\mathscr{A})$ denote the number of evaluations of f and W used in algorithm $\mathscr{A}$ (see Section 4 for a general definition). We have

$$\mathrm{card}(\mathscr{A}) = \mathrm{card}(\mathscr{A}_1) \le c \sum_{l=0}^{l_1} n_l 2^{dl}.$$

On the basis of the considerations above and the results of [2] we can now derive error estimates for algorithms $\mathscr{A}_1$ and $\mathscr{A}$ on the classes $\mathscr{F}_q^r(Q \times [0,1] \times \Omega; \kappa)$.

Theorem 1 *Let* $r \in \mathbb{R}$, $r > 0$, $d \in \mathbb{N}$, $2 < q < \infty$, $\kappa > 0$. *Then there are constants* $c_{1-4} > 0$ *such that for each* $n \in \mathbb{N}$ *with* $n \geq 2$ *there is a choice of* $l_1 \in \mathbb{N}_0$ *and* $n_0, \dots, n_{l_1} \in \mathbb{N}_0$ *such that* $\operatorname{card}(\mathscr{A}_1) \leq c_1 n$ *and*

$$\sup_{f \in \mathscr{F}_2^r(Q\times[0,1]\times\Omega;\kappa)} \left(\mathbb{E}\,\|\mathscr{S}_1(f,\omega) - \mathscr{A}_1(f,\omega)\|_{C(Q)}^2\right)^{1/2}$$
$$\leq c_2 \begin{cases} n^{-\frac{r}{d+1}} (\log n)^{\frac{r}{d+1}+\frac{3}{2}} & \text{if } \frac{r}{d+1} \leq 1, \\ n^{-1} & \text{if } \frac{r}{d+1} > 1. \end{cases} \tag{18}$$

Moreover, for each $n \in \mathbb{N}$ *with* $n \geq 2$ *there are* $l_1 \in \mathbb{N}_0$ *and* $n_0, \dots, n_{l_1} \in \mathbb{N}_0$ *such that* $\operatorname{card}(\mathscr{A}) \leq c_3 n$ *and*

$$\sup_{f \in \mathscr{F}_q^r(Q\times[0,1]\times\Omega;\kappa)} \left(\mathbb{E}\,\|\mathscr{S}(f,\omega) - \mathscr{A}(f,\omega)\|_{C(Q\times[0,1])}^2\right)^{1/2}$$
$$\leq c_4 \begin{cases} n^{-\frac{r}{d+1}} (\log n)^{\frac{r}{d+1}+\frac{3}{2}} & \text{if } \frac{r}{d+1} \leq 1/2, \\ n^{-\frac{1}{2}} (\log n)^{\frac{1}{2}} & \text{if } \frac{r}{d+1} > \frac{1}{2}. \end{cases} \tag{19}$$

Proof We shall use Theorem 5.3 of [2], which was shown there for $r_1 \in \mathbb{N}$. It is easily seen that it also holds for real $r_1 > 0$. Indeed, Lemma 5.1 and Proposition 5.2 of [2] are readily extended to non-integer $r_1 > 0$, using (1) of the present paper. The proof of Theorem 5.3 in [2] relies only on Proposition 5.2 and does not use the assumption of r_1 being integer. We define

$$r_1 = \frac{dr}{d+1}, \quad \rho = \min\left(\frac{r}{d+1}, 1\right), \tag{20}$$

hence $r \geq r_1 + \rho$ and the conclusions (8)–(9) of Lemma 1 hold. Denote the left-hand side of (18) and (19) by E_1 and E, respectively.

First we derive (18). If $\frac{r}{d+1} \leq 1$, then $\frac{r_1}{d} = \rho$, thus (8) together with the second relation of (78) in [2] yields

$$E_1 \leq cn^{-\frac{r_1}{d}} (\log n)^{\frac{r_1}{d}+\frac{3}{2}} = cn^{-\frac{r}{d+1}} (\log n)^{\frac{r}{d+1}+\frac{3}{2}},$$

which is the first relation of (18). If $\frac{r}{d+1} > 1$, we have by (20) $\frac{r_1}{d} > \rho = 1$, so the third relation of (78) in [2] gives $E_1 \leq cn^{-1}$ and thus the second part of (18).

Next we prove (19). If $\frac{r}{d+1} \leq \frac{1}{2}$, we conclude from (20) $\frac{r_1}{d} = \rho = \min(\rho, 1/2)$, so (9) together with the second relation of (79) in [2] implies

$$E \leq cn^{-\frac{r_1}{d}} (\log n)^{\frac{r_1}{d}+\frac{3}{2}} = cn^{-\frac{r}{d+1}} (\log n)^{\frac{r}{d+1}+\frac{3}{2}},$$

thus the first relation of (19). Finally, if $\frac{r}{d+1} > \frac{1}{2}$, then by (20), $\rho > 1/2$, hence $\frac{r_1}{d} > \min(\rho, 1/2)$, and the third relation of (79) in [2] implies $E \leq cn^{-\frac{1}{2}} (\log n)^{\frac{1}{2}}$, showing the second relation of (19) and completing the proof. □

4 Lower Bounds and Complexity

In this section we extend the approach of [6, 7] to stochastic problems. An abstract stochastic numerical problem is described by a tuple $\mathscr{P} = (F, (\Omega, \Sigma, \mathbb{P}), G, S, K, \Lambda)$. The set F is an arbitrary non-empty set, $(\Omega, \Sigma, \mathbb{P})$ a probability space, G is a Banach space and $S : F \times \Omega \to G$ an arbitrary mapping, the solution operator, which maps the input $(f, \omega) \in F$ to the exact solution $S(f, \omega)$. We assume that for each $f \in F$ the mapping $\omega \to S(f, \omega)$ is Σ-to-Borel-measurable and $\mathbb{P}$-almost surely separably valued, the latter meaning that for each $f \in F$ there is a separable subspace G_f of G such that $\mathbb{P}\{\omega : S(f, \omega) \in G_f\} = 1$. Furthermore, K is a nonempty set and Λ a set of mappings from $F \times \Omega$ to K, the set of information functionals.

A deterministic algorithm for $\mathscr{P}$ is a tuple $A = ((L_i)_{i=1}^\infty, (\tau_i)_{i=0}^\infty, (\varphi_i)_{i=0}^\infty)$ such that $L_1 \in \Lambda$, $\tau_0 \in \{0, 1\}$, $\varphi_0 \in G$, and for $i \in \mathbb{N}$

$$L_{i+1} : K^i \to \Lambda, \quad \tau_i : K^i \to \{0, 1\}, \quad \varphi_i : K^i \to G$$

are arbitrary mappings, where K^i is the i-fold Cartesian product of K. Given $(f, \omega) \in F \times \Omega$, we associate with it a sequence $(a_i)_{i=1}^\infty$ defined as follows:

$$a_1 = L_1(f, \omega), \quad a_i = \big(L_i(a_1, \dots, a_{i-1})\big)(f, \omega) \quad (i \ge 2). \tag{21}$$

Define $\operatorname{card}(A, f, \omega)$, the cardinality of A at input (f, ω), to be 0 if $\tau_0 = 1$. If $\tau_0 = 0$, let $\operatorname{card}(A, f, \omega)$ be the first integer $n \in \mathbb{N}$ with $\tau_n(a_1, \dots, a_n) = 1$, if there is such an n. If no such $n \in \mathbb{N}$ exists, set $\operatorname{card}(A, f, \omega) = \infty$. We define the output $A(f, \omega)$ of algorithm A at input (f, ω) as

$$A(f, \omega) = \begin{cases} \varphi_0 & \text{if } \operatorname{card}(A, f, \omega) = 0 \\ \varphi_n(a_1, \dots, a_n) & \text{if } \operatorname{card}(A, f, \omega) = n < \infty \\ \varphi_0 & \text{if } \operatorname{card}(A, f, \omega) = \infty. \end{cases}$$

Informally, the algorithm starts with evaluating an information functional $L_1 \in \Lambda$ at input (f, ω), that is, $L_1(f, \omega) := a_1$. Depending on this value, another functional $L_2(a_1) \in \Lambda$ is chosen and $\big(L_2(a_1)\big)(f, \omega)$ is evaluated, etc., until stopping, ruled by the τ_i. Finally, a mapping φ_n is applied, representing the computations performed on the information, leading to the approximation $\varphi_n(a_1, \dots, a_n)$ of $S(f, \omega)$ in G.

Given $n \in \mathbb{N}_0$, we define $\mathscr{A}_n^{\det}(\mathscr{P})$ as the set of those deterministic algorithms A for $\mathscr{P}$ with the following properties: For each $f \in F$ the mapping $\omega \to \operatorname{card}(A, f, \omega)$ is Σ-measurable, $\mathbb{E}\operatorname{card}(A, f, \omega) \le n$, and the mapping $\omega \to A(f, \omega) \in G$ is Σ-to-Borel-measurable and $\mathbb{P}$-almost surely separably valued. The cardinality of $A \in \mathscr{A}_n^{\det}(\mathscr{P})$ is defined as

$$\operatorname{card}(A) = \sup_{f \in F} \mathbb{E}\operatorname{card}(A, f, \omega),$$

the error of A in approximating S as

$$e(S, A, F \times \Omega, G) = \sup_{f \in F} \mathbb{E}\, \|S(f, \omega) - A(f, \omega)\|_G$$

and the deterministic nth minimal error of S is defined for $n \in \mathbb{N}_0$ as

$$e_n^{\det}(S, F \times \Omega, G) = \inf_{A \in \mathscr{A}_n^{\det}(\mathscr{P})} e(S, A, F \times \Omega, G).$$

It follows that no deterministic algorithm that uses (on the average with respect to $\mathbb{P}$) at most n information functionals can have a smaller error than $e_n^{\det}(S, F \times \Omega, G)$.

A randomized algorithm for $\mathscr{P}$ is a tuple $A = ((\Omega_1, \Sigma_1, \mathbb{P}_1), (A_{\omega_1})_{\omega_1 \in \Omega_1})$, where $(\Omega_1, \Sigma_1, \mathbb{P}_1)$ is another probability space and for each $\omega_1 \in \Omega_1$, A_{ω_1} is a deterministic algorithm for $\mathscr{P}$. Let $(\Omega_1 \times \Omega, \Sigma_1 \times \Sigma, \mathbb{P}_1 \times \mathbb{P})$ be the product probability space. For $n \in \mathbb{N}_0$ we define $\mathscr{A}_n^{\mathrm{ran}}(\mathscr{P})$ as the class of those randomized algorithms A for $\mathscr{P}$ which possess the following properties: For each $f \in F$ the mapping $(\omega_1, \omega) \to \mathrm{card}(A_{\omega_1}, f, \omega)$ is $\Sigma_1 \times \Sigma$-measurable, $\mathbb{E}_{\mathbb{P}_1 \times \mathbb{P}} \mathrm{card}(A_{\omega_1}, f, \omega) \le n$, and the mapping $(\omega_1, \omega) \to A_{\omega_1}(f, \omega)$ is $\Sigma_1 \times \Sigma$-to-Borel-measurable and $\mathbb{P}_1 \times \mathbb{P}$-almost surely separably valued. We define the cardinality of $A \in \mathscr{A}_n^{\mathrm{ran}}(\mathscr{P})$ as

$$\mathrm{card}(A) = \sup_{f \in F} \mathbb{E}_{\mathbb{P}_1 \times \mathbb{P}} \mathrm{card}(A_{\omega_1}, f, \omega),$$

the error as

$$e(S, A, F \times \Omega, G) = \sup_{f \in F} \mathbb{E}_{\mathbb{P}_1 \times \mathbb{P}} \|S(f, \omega) - A_{\omega_1}(f, \omega)\|_G$$

and the randomized nth minimal error of S as

$$e_n^{\mathrm{ran}}(S, F \times \Omega, G) = \inf_{A \in \mathscr{A}_n^{\mathrm{ran}}(\mathscr{P})} e(S, A, F \times \Omega, G).$$

Similarly to the above, this means that no randomized algorithm that uses (on the average with respect to $\mathbb{P}_1 \times \mathbb{P}$) at most n information functionals can have a smaller error than $e_n^{\mathrm{ran}}(S, F \times \Omega, G)$. Deterministic algorithms can be viewed as a special case of randomized ones, namely by considering trivial one-point probability spaces $\Omega_1 = \{\omega_1\}$. Hence,

$$e_n^{\mathrm{ran}}(S, F \times \Omega, G) \le e_n^{\det}(S, F \times \Omega, G). \tag{22}$$

Now we study the complexity of definite and indefinite stochastic integration. Let $r, r_1 > 0$, $0 \le \rho \le 1$, $2 < q < \infty$. We set $K = \mathbb{R}$ and

$$\Lambda = \Lambda_1 \cup \Lambda_2, \quad \Lambda_1 = \{\delta_{st} : s \in Q, t \in [0, 1]\}, \quad \Lambda_2 = \{\delta_t : t \in [0, 1]\}, \tag{23}$$

where $\delta_{st}(f,\omega)=f(s,t,\omega)$ and $\delta_t(f,\omega)=W(t,\omega)$ $(f\in F,\ \omega\in\Omega)$. For definite integration we choose $F=\mathscr{F}_2^r(Q\times[0,1]\times\Omega;\kappa)$ or $F=F^{r_1,\rho}(Q\times[0,1]\times\Omega;\kappa)$, $G=C(Q)$, $S=\mathscr{S}_1$. For the indefinite problem we set $F=\mathscr{F}_q^r(Q\times[0,1]\times\Omega;\kappa)$ or $F=F_q^{r_1,\rho}(Q\times[0,1]\times\Omega;\kappa)$, $G=C(Q\times[0,1])$, $S=\mathscr{S}$.

Theorem 2 *Let $r,r_1\in\mathbb{R}$, $r,r_1>0$, $0\le\rho\le 1$, $d\in\mathbb{N}$, $\kappa>0$, and $2<q<\infty$. Then*

$$e_n^{\rm ran}(\mathscr{S}_1,\mathscr{F}_2^r\times\Omega,C(Q))\succeq\max\left(n^{-\frac{r}{d+1}},n^{-1}\right) \tag{24}$$

$$e_n^{\rm ran}(\mathscr{S}_1,F^{r_1,\rho}\times\Omega,C(Q))\succeq\max\left(n^{-\frac{r_1}{d}},n^{-\rho}\right). \tag{25}$$

$$e_n^{\rm ran}(\mathscr{S},\mathscr{F}_q^r\times\Omega,C(Q\times[0,1]))\succeq\max\left(n^{-\frac{r}{d+1}},n^{-\frac{1}{2}}(\log n)^{\frac{1}{2}}\right) \tag{26}$$

$$e_n^{\rm ran}(\mathscr{S},F_q^{r_1,\rho}\times\Omega,C(Q\times[0,1]))\succeq\max\left(n^{-\frac{r_1}{d}},n^{-\frac{1}{2}}(\log n)^{\frac{1}{2}},n^{-\rho}\right). \tag{27}$$

Theorem 1 above and Theorem 5.3 of [2] show that, up to logarithmic factors, these bounds match the upper bounds.

Corollary 1 *Relations (24)–(27) also hold with $\succeq$ replaced by $\preceq_{\log}$.*

Moreover, by (22) and since the algorithms $\mathscr{A}$ and $\mathscr{A}_1$ are deterministic, the conclusions of Theorem 2 and Corollary 1 hold for $e_n^{\rm det}$ in place of $e_n^{\rm ran}$, as well.

To prove Theorem 2 we need a number of auxiliary results. For this we return to the general setting. The first observation concerns the case that F consists of a single element, in other words, S is essentially independent of F and $\mathscr{P}$ is a pure average case problem. Then the above inequality (22) has a certain converse. This is a version of the well-known principle that, in general, for pure average case problems randomized algorithms do not bring essential gains.

Lemma 2 *If $F=F_0=\{f_0\}$, then*

$$e_n^{\rm ran}(S,F_0\times\Omega,G)\ge\frac{1}{2}e_{2n}^{\rm det}(S,F_0\times\Omega,G). \tag{28}$$

Proof Let $\delta>0$ and $A\in\mathscr{A}_n^{\rm ran}(\mathscr{P})$ with

$$e(S,A,F_0\times\Omega,G)\le e_n^{\rm ran}(S,F_0\times\Omega,G)+\delta.$$

This means

$$\mathbb{E}_{\mathbb{P}_1\times\mathbb{P}}\|S(f_0,\omega)-A_{\omega_1}(f_0,\omega)\|_G\le e_n^{\rm ran}(S,F_0\times\Omega,G)+\delta,$$
$$\mathbb{E}_{\mathbb{P}_1\times\mathbb{P}}\,{\rm card}(A_{\omega_1},f_0,\omega)\le n.$$

Consequently, setting

$$\Omega_{1,1}=\{\omega_1:\mathbb{E}_{\mathbb{P}}\|S(f_0,\omega)-A_{\omega_1}(f_0,\omega)\|_G\le 2e_n^{\rm ran}(S,F_0\times\Omega,G)+2\delta\},$$
$$\Omega_{1,2}=\{\omega_1:\mathbb{E}_{\mathbb{P}}\,{\rm card}(A_{\omega_1},f_0,\omega)\le 2n\},$$

we conclude that $\mathbb{P}_1(\Omega_{1,1}) > 1/2$ and $\mathbb{P}_1(\Omega_{1,2}) > 1/2$. It follows that for $\omega_1 \in \Omega_{1,1} \cap \Omega_{1,2} \neq \emptyset$ we have $A_{\omega_1} \in \mathscr{A}_{2n}^{\det}(\mathscr{P})$ and

$$e(S, A_{\omega_1}, F_0 \times \Omega, G) \leq 2e_n^{\text{ran}}(S, F_0 \times \Omega, G) + 2\delta,$$

which implies (28). □

Next we explore the connection between the original stochastic problem and the deterministic problem we obtain by fixing the random input. For this purpose, we assume that we are given a decomposition of the set Λ

$$\Lambda = \Lambda_F \cup \Lambda_\Omega, \quad \Lambda_F \neq \emptyset, \quad \Lambda_F \cap \Lambda_\Omega = \emptyset$$

such that for all $\lambda \in \Lambda_\Omega$ we have $\lambda(f, \omega) = \lambda(g, \omega)$ $(f, g \in F, \omega \in \Omega)$, that is, all $\lambda \in \Lambda_\Omega$ depend only on $\omega \in \Omega$ (the $\lambda \in \Lambda_F$ may depend on both f and ω). For $\lambda \in \Lambda_\Omega$ we use both the notation $\lambda(f, \omega)$ as well as $\lambda(\omega)$. Note that there is always the trivial splitting $\Lambda_F = \Lambda$, $\Lambda_\Omega = \emptyset$. An example of a nontrivial splitting is (23) above. Fix $\omega \in \Omega$. We define the restricted problem $\mathscr{P}_\omega = (F, G, S_\omega, K, \Lambda_{F,\omega})$ by setting

$$S_\omega : F \to G, \quad S_\omega(f) = S(f, \omega), \quad \Lambda_{F,\omega} = \{\lambda(\,\cdot\,, \omega) : \lambda \in \Lambda_F\}.$$

To a given a deterministic algorithm A for $\mathscr{P}$ and $\omega \in \Omega$ we want to associate a restricted algorithm A_ω for the respective problem $\mathscr{P}_\omega$ in a rigorous way.

Lemma 3 *Let A be a deterministic algorithm for $\mathscr{P}$ and let $\omega \in \Omega$. Then there is a deterministic algorithm A_ω for $\mathscr{P}_\omega$ such that for all $f \in F$*

$$\text{card}(A_\omega, f) = \text{card}(A, f, \omega), \quad A_\omega(f) = A(f, \omega). \tag{29}$$

Proof Let $\mu_0 \in \Lambda_F$ be any element, let $A = ((L_i)_{i=1}^\infty, (\tau_i)_{i=0}^\infty, (\varphi_i)_{i=0}^\infty)$, and fix $\omega \in \Omega$. We define $A_\omega = ((L_{i,\omega})_{i=1}^\infty, (\tau_{i,\omega})_{i=0}^\infty, (\varphi_{i,\omega})_{i=0}^\infty)$ and a sequence $(\xi_i)_{i=1}^\infty$ of functions $\xi_i : K^i \to K^i$ by induction. Put

$$\tau_{0,\omega} = \tau_0, \quad \varphi_{0,\omega} = \varphi_0, \quad L_{1,\omega} = \begin{cases} L_1(\,\cdot\,, \omega) \text{ if } & L_1 \in \Lambda_F \\ \mu_0(\,\cdot\,, \omega) \text{ if } & L_1 \in \Lambda_\Omega \end{cases} \tag{30}$$

and define for $z_1 \in K$

$$\xi_1(z_1) = \begin{cases} z_1 & \text{if } \ L_1 \in \Lambda_F \\ L_1(\omega) & \text{if } \ L_1 \in \Lambda_\Omega. \end{cases} \tag{31}$$

Now let $i \geq 1$ and assume that $(L_{j,\omega})_{j \leq i}$, $(\tau_{j,\omega})_{j<i}$, $(\varphi_{j,\omega})_{j<i}$, and $(\xi_j)_{j \leq i}$ have been defined. Let $z_1, \ldots, z_i, z_{i+1} \in K$ and set

$$\lambda_{i+1} = L_{i+1}(\xi_i(z_1, \ldots, z_i)) \tag{32}$$

$$L_{i+1,\omega}(z_1, \ldots, z_i) = \begin{cases} \lambda_{i+1}(\,\cdot\,, \omega) \text{ if } & \lambda_{i+1} \in \Lambda_F \\ \mu_0(\,\cdot\,, \omega) \quad \text{if } & \lambda_{i+1} \in \Lambda_\Omega \end{cases} \tag{33}$$

$$\tau_{i,\omega}(z_1, \ldots, z_i) = \tau_i(\xi_i(z_1, \ldots, z_i)) \tag{34}$$

$$\varphi_{i,\omega}(z_1, \ldots, z_i) = \varphi_i(\xi_i(z_1, \ldots, z_i)) \tag{35}$$

$$\xi_{i+1}(z_1, \ldots, z_i, z_{i+1}) = \begin{cases} (\xi_i(z_1, \ldots, z_i), z_{i+1}) & \text{if } \lambda_{i+1} \in \Lambda_F \\ (\xi_i(z_1, \ldots, z_i), \lambda_{i+1}(\omega)) & \text{if } \lambda_{i+1} \in \Lambda_\Omega. \end{cases} \tag{36}$$

Now let $f \in F$, let $(a_i)_{i=1}^{\infty}$ be the sequence given by (21) and define, respectively

$$a_{1,\omega} = L_{1,\omega}(f), \quad a_{i,\omega} = \big(L_{i,\omega}(a_{1,\omega}, \ldots, a_{i-1,\omega})\big)(f) \quad (i \geq 2). \tag{37}$$

We show by induction that for all $i \in \mathbb{N}$.

$$\xi_i(a_{1,\omega}, \ldots, a_{i,\omega}) = (a_1, \ldots, a_i). \tag{38}$$

For $i = 1$ this follows directly from (21), (30), (31), and (37). Now let $i \in \mathbb{N}$ and assume that (38) holds. Let

$$\lambda_{i+1} = L_{i+1}(\xi_i(a_{1,\omega}, \ldots, a_{i,\omega})) = L_{i+1}(a_1, \ldots, a_i). \tag{39}$$

First assume $\lambda_{i+1} \in \Lambda_F$. Then by (37), (33), (39) and (21)

$$\begin{aligned} a_{i+1,\omega} &= \big(L_{i+1,\omega}(a_{1,\omega}, \ldots, a_{i,\omega})\big)(f) = \lambda_{i+1}(f, \omega) \\ &= \big(L_{i+1}(a_1, \ldots, a_i)\big)(f, \omega) = a_{i+1}. \end{aligned}$$

With (36) this gives

$$\xi_{i+1}(a_{1,\omega}, \ldots, a_{i,\omega}, a_{i+1,\omega}) = (\xi_i(a_{1,\omega}, \ldots, a_{i,\omega}), a_{i+1,\omega}) = (a_1, \ldots, a_i, a_{i+1}).$$

In the case $\lambda_{i+1} \in \Lambda_\Omega$ we have, using (39) and (21)

$$\lambda_{i+1}(\omega) = \lambda_{i+1}(f, \omega) = \big(L_{i+1}(a_1, \ldots, a_i)\big)(f, \omega) = a_{i+1}.$$

By (36),

$$\xi_{i+1}(a_{1,\omega}, \ldots, a_{i,\omega}, a_{i+1,\omega}) = (\xi_i(a_{1,\omega}, \ldots, a_{i,\omega}), \lambda_{i+1}(\omega)) = (a_1, \ldots, a_i, a_{i+1}).$$

This proves (38). From (38) we conclude that for all $i \in \mathbb{N}_0$

$$\tau_{i,\omega}(a_{1,\omega}, \ldots, a_{i,\omega}) = \tau_i(a_1, \ldots, a_i), \quad \varphi_{i,\omega}(a_{1,\omega}, \ldots, a_{i,\omega}) = \varphi_i(a_1, \ldots, a_i),$$

which implies (29). □

Next we derive a lower bound for the randomized nth minimal errors. Analogous to the classical one it uses the average setting with respect to a probability measure on F. However, due to the additional stochastic component, it is somewhat more involved. For the notation of the average case setting we refer to [6, 7], $\int^*$ denotes the upper integral.

Lemma 4 *Let ν be a probability measure on F supported by a finite set. Then for all $n \in \mathbb{N}_0$,*

$$e_n^{\mathrm{ran}}(S, F \times \Omega, G) \geq \frac{1}{3} \inf_{D \in \Sigma, \mathbb{P}(D) \geq 1/4} \int_D^* e_{2n}^{\mathrm{avg}}(S_\omega, \nu, G) d\mathbb{P}(\omega). \tag{40}$$

Proof Let $A \in \mathscr{A}_n^{\mathrm{ran}}(\mathscr{P})$, $A = ((\Omega_1, \Sigma_1, \mathbb{P}_1), (A_{\omega_1})_{\omega_1 \in \Omega_1})$. Then

$$\begin{aligned} n &\geq \sup_{f \in F} \int_{\Omega_1 \times \Omega} \mathrm{card}(A_{\omega_1}, f, \omega) d\mathbb{P}_1(\omega_1) d\mathbb{P}(\omega) \\ &\geq \int_{\Omega_1 \times \Omega} \int_F \mathrm{card}(A_{\omega_1}, f, \omega) d\nu(f) d\mathbb{P}_1(\omega_1) d\mathbb{P}(\omega). \end{aligned} \tag{41}$$

Let

$$B = \left\{ (\omega_1, \omega) \in \Omega_1 \times \Omega : \int_F \mathrm{card}(A_{\omega_1}, f, \omega) d\nu(f) \leq 2n \right\} \tag{42}$$

and for $\omega_1 \in \Omega_1$, $B_{\omega_1} = \{\omega : (\omega_1, \omega) \in B\}$. Since ν is of finite support, it follows that $B \in \Sigma_1 \times \Sigma$ and $B_{\omega_1} \in \Sigma$. We also set $B' = \{\omega_1 : \mathbb{P}(B_{\omega_1}) \geq 1/4\}$, then $B' \in \Sigma_1$. Moreover, (41) and (42) yield $(\mathbb{P}_1 \times \mathbb{P})(B) \geq 1/2$, hence $\frac{1}{2} \leq \mathbb{P}_1(B') + \frac{1}{4}(1 - \mathbb{P}_1(B'))$, which implies $\mathbb{P}_1(B') \geq 1/3$.

Now we estimate the error of $A = (A_{\omega_1})_{\omega_1 \in \Omega_1}$ from below. For each $\omega_1 \in \Omega_1$ and $\omega \in \Omega$, let $A_{\omega_1,\omega}$ be the respective algorithm for S_ω resulting from A_{ω_1} according to Lemma 3.

$$\begin{aligned} e(S, A, F \times \Omega, G) &= \sup_{f \in F} \int_{\Omega_1 \times \Omega} \|S(f, \omega) - A_{\omega_1}(f, \omega)\|_G d\mathbb{P}_1(\omega_1) d\mathbb{P}(\omega) \\ &\geq \int_{B'} \int_{B_{\omega_1}} \int_F \|S_\omega(f) - A_{\omega_1,\omega}(f)\|_G d\nu(f) d\mathbb{P}(\omega) d\mathbb{P}_1(\omega_1) \\ &\geq \frac{1}{3} \inf_{D \in \Sigma, \mathbb{P}(D) \geq 1/4} \int_D \int_F \|S_\omega(f) - A_{\omega_1,\omega}(f)\|_G d\nu(f) d\mathbb{P}(\omega) \\ &\geq \frac{1}{3} \inf_{D \in \Sigma, \mathbb{P}(D) \geq 1/4} \int_D^* e_{2n}^{\mathrm{avg}}(S_\omega, \nu, G) d\mathbb{P}(\omega). \end{aligned}$$

□

Let $(\gamma_j)_{j=1}^\infty$ be a sequence of independent standard Gaussian random variables. For $m \in \mathbb{N}$ we set $\mathscr{J}_m = \{1, 2, \dots, m\}$.

Lemma 5 *There is a constant $c > 0$ such that for all $m \in \mathbb{N}$*

$$\mathbb{P}\left\{\omega \in \Omega : \min_{\mathscr{J} \subseteq \mathscr{J}_m, |\mathscr{J}| \geq m/2} \left(\sum_{j \in \mathscr{J}} \gamma_j(\omega)^2\right)^{1/2} \geq c m^{1/2}\right\} \geq 7/8.$$

Proof Let $k = \lceil m/2 \rceil$ and let $c_0 > 0$ be a constant to be fixed later on. Then

$$\begin{aligned}
&\mathbb{P}\left\{\min_{\mathscr{J} \subseteq \mathscr{J}_m, |\mathscr{J}| \geq m/2} \sum_{j \in \mathscr{J}} \gamma_j(\omega)^2 \geq c_0^2 m\right\} \\
&= \mathbb{P}\left\{\min_{\mathscr{J} \subseteq \mathscr{J}_m, |\mathscr{J}| = k} \sum_{j \in \mathscr{J}} \gamma_j(\omega)^2 \geq c_0^2 m\right\} \\
&\geq 1 - \sum_{\mathscr{J} \subseteq \mathscr{J}_m, |\mathscr{J}| = k} \mathbb{P}\left\{\sum_{j \in \mathscr{J}} \gamma_j(\omega)^2 < c_0^2 m\right\} \\
&\geq 1 - 2^{2k}\, \mathbb{P}\left\{\sum_{j=1}^{k} \gamma_j(\omega)^2 < 2c_0^2 k\right\}.
\end{aligned} \tag{43}$$

Furthermore, let B_2^k denote the unit ball of $\mathbb{R}^k$, endowed with the Euclidean norm $|\cdot|$. There is a constant $c_1 > 0$ such that for all $k \in \mathbb{N}$

$$\operatorname{Vol}\left(B_2^k\right) \leq c_1^k k^{-k/2}, \tag{44}$$

see, e.g., [11], relation 1.18 on p. 11. Consequently,

$$\begin{aligned}
\mathbb{P}\left\{\sum_{j=1}^{k} \gamma_j(\omega)^2 < 2c_0^2 k\right\} &= (2\pi)^{-k/2} \int_{|x| \leq c_0 (2k)^{1/2}} e^{-|x|^2/2} dx \\
\leq \operatorname{Vol}\left(c_0 (2k)^{1/2} B_2^k\right) &= c_0^k (2k)^{k/2} \operatorname{Vol}\left(B_2^k\right) \leq 2^{k/2} c_0^k c_1^k.
\end{aligned} \tag{45}$$

Joining (43) and (45) and setting $c_0 = 2^{-11/2} c_1^{-1}$, we arrive at

$$\mathbb{P}\left\{\min_{\mathscr{J} \subseteq \mathscr{J}_m, |\mathscr{J}| \geq m/2} \sum_{j \in \mathscr{J}} \gamma_j(\omega)^2 \geq c_0^2 m\right\} \geq 1 - 2^{-3k} \geq 7/8.$$

Proof of Theorem 2 Two parts of the lower bound estimates easily reduce to known results. Let $\theta_0 \in C(Q)^*$ be defined by $\theta_0(f) = f(0)$ $(f \in C(Q))$. Firstly, we let $f_1(s, t, \omega) = \kappa t$ and put $F_1 := \{f_1\}$. Then $F_1 \subseteq \mathscr{F}_2^r(Q \times [0, 1] \times \Omega; \kappa)$. We have

$$(\theta_0 \circ \mathscr{S}_1)(f_1, \omega) = \kappa \left(\int_0^1 t dW(t)\right)(\omega) \quad (\mathbb{P}\text{-almost surely}).$$

Consequently,

$$e_n^{\text{ran}}(\mathscr{S}_1, \mathscr{F}_2^r \times \Omega, C(Q)) \geq e_n^{\text{ran}}(\theta_0 \circ \mathscr{S}_1, F_1 \times \Omega, \mathbb{R}) \geq cn^{-1}, \tag{46}$$

where the last relation follows from Theorem 1 in [15] (who considered deterministic algorithms) and Lemma 2 above. Secondly, we set $f_2(s,t,\omega) \equiv \kappa$ and $F_2 := \{f_2\}$. We have

$$F_2 \subseteq \mathscr{F}_q^r(Q \times [0,1] \times \Omega; \kappa), \quad F_2 \subseteq F_q^{r_1,\rho}(Q \times [0,1] \times \Omega; \kappa),$$

and

$$\big((\theta_0 \circ \mathscr{S})(f_2, \omega)\big)(t) = \kappa W(t, \omega).$$

Therefore we get from [9, 14], using Lemma 2 again,

$$\begin{aligned} & e_n^{\text{ran}}(\mathscr{S}, \mathscr{F}_q^r(Q \times [0,1] \times \Omega; \kappa) \times \Omega, C(Q \times [0,1])) \\ & \geq e_n^{\text{ran}}(\theta_0 \circ \mathscr{S}, F_2 \times \Omega, C([0,1])) \geq cn^{-1/2}(\log n)^{1/2}, \end{aligned} \tag{47}$$

and similarly

$$e_n^{\text{ran}}(\mathscr{S}, F_q^{r_1,\rho}(Q \times [0,1] \times \Omega; \kappa), C(Q \times [0,1])) \geq cn^{-1/2}(\log n)^{1/2}. \tag{48}$$

Now we consider a third subclass. For the purpose of constructing a suitable measure ν in order to apply Lemma 4 we let φ_0 be a C^∞ function on $\mathbb{R}^d$ with support in Q and $\|\varphi_0\|_{C(Q)} = 1$ and let φ_1 be a C^∞ function on $\mathbb{R}$ with support in $[0,1]$ and $\|\varphi_1\|_{L_2(\mathbb{R})} = 1$. Let $m_0, m_1 \in \mathbb{N}$ and let Q_i $(i = 1, \ldots, m_0^d)$ be the subdivision of Q into m_0^d cubes of disjoint interior of side-length m_0^{-1}. Let s_i be the point in Q_i with minimal coordinates. Put $t_j = j/m_1$ and define for $s \in Q, t \in [0,1], i = 1, \ldots, m_0^d$, $j = 1, \ldots, m_1$

$$\varphi_{0,i}(s) = \varphi_0(m_0(s - s_i)), \quad \varphi_{1,j}(t) = \varphi_1(m_1(t - t_j)), \quad \psi_{ij}(s,t) = \varphi_{0,i}(s)\varphi_{1,j}(t).$$

Denote $\mathscr{K}_{m_0 m_1} = \{1, \ldots, m_0^d\} \times \{1, \ldots, m_1\}$ and

$$\Psi_{m_0 m_1} = \left\{ \sum_{(i,j) \in \mathscr{K}_{m_0 m_1}} \delta_{ij}\psi_{ij} : \delta_{ij} \in \{-1, 0, 1\} \right\}.$$

The stochastic integral $m_1^{1/2} \int_0^1 \varphi_{1,j}(t)dW(t)$ is an element of $L_2(\Omega)$, hence an equivalence class of functions. Let the function $\gamma_j = \gamma_j(\omega)$ be any representative of it. Since $\|\varphi_{1,j}\|_{L_2([0,1])} = m_1^{-1/2}$, the $(\gamma_j)_{j=1}^{m_1}$ are independent standard Gaussian random variables. By (17) and the linearity of the stochastic integral, for $(i,j) \in \mathscr{K}_{m_0 m_1}$ and each $s \in Q$ we have $\mathbb{P}$-almost surely $(\mathscr{S}_1(\psi_{ij}, \omega))(s) = m_1^{-1/2}\varphi_{0,i}(s)\gamma_j(\omega)$. Using continuity and a density argument yields that there is an $\Omega_0 \in \Sigma$ with $\mathbb{P}(\Omega_0) = 1$ such that for all $\omega \in \Omega_0$ and $(i,j) \in \mathscr{K}_{m_0 m_1}$ we have $\mathscr{S}_1(\psi_{ij}, \omega) = m_1^{-1/2}\varphi_{0,i}\gamma_j(\omega)$. We conclude, using the linearity of $\mathscr{S}_1$ that for all $\delta_{ij} \in \{-1, 0, 1\}$ and $\omega \in \Omega_0$

$$\mathscr{S}_1\Bigg(\sum_{(i,j)\in\mathscr{K}_{m_0m_1}} \delta_{ij}\psi_{ij}, \omega\Bigg) = m_1^{-1/2}\sum_{i=1}^{m_0^d}\varphi_{0,i}\sum_{j=1}^{m_1}\delta_{ij}\gamma_j(\omega). \tag{49}$$

Let $\{\epsilon_{ij} : (i, j) \in \mathscr{K}_{m_0m_1}\}$ be independent Bernoulli random variables with $\mathbb{P}_2\{\epsilon_{ij} = -1\} = \mathbb{P}_2\{\epsilon_{ij} = +1\} = 1/2$ on a probability space $(\Omega_2, \Sigma_2, \mathbb{P}_2)$ and define the measure ν to be the distribution of the $\Psi_{m_0m_1}$ valued random variable $\sum_{(i,j)\in\mathscr{K}_{m_0m_1}} \epsilon_{ij}\psi_{ij}$. Let $n \in \mathbb{N}$ be such that

$$m_0^d m_1 \ge 8n. \tag{50}$$

Now we are ready to apply Lemma 4, which yields

$$e_n^{\mathrm{ran}}(\mathscr{S}_1, \Psi_{m_0m_1} \times \Omega, C(Q)) \ge \frac{1}{3}\inf_{D\in\Sigma,\mathbb{P}(D)\ge 1/4}\int_D^* e_{2n}^{\mathrm{avg}}(\mathscr{S}_{1,\omega}, \nu, C(Q))d\mathbb{P}(\omega). \tag{51}$$

Lemma 6 of [6] together with (49) and (50) allow to estimate the right-hand side in terms of the γ_j. For $\omega \in \Omega_0$ we have

$$\begin{aligned}
&e_{2n}^{\mathrm{avg}}(\mathscr{S}_{1,\omega}, \nu, C(Q))\\
&\ge \frac{1}{2}\min_{\mathscr{K}\subseteq\mathscr{K}_{m_0m_1},|\mathscr{K}|\ge m_0^d m_1-4n} \mathbb{E}_{\mathbb{P}_2}\Bigg\|\mathscr{S}_1\Bigg(\sum_{(i,j)\in\mathscr{K}}\epsilon_{ij}\psi_{ij},\omega\Bigg)\Bigg\|_{C(Q)}\\
&\ge \frac{1}{2}\min_{\mathscr{K}\subseteq\mathscr{K}_{m_0m_1},|\mathscr{K}|\ge m_0^d m_1-4n} \mathbb{E}_{\mathbb{P}_2}\Bigg\|m_1^{-1/2}\sum_{i=1}^{m_0^d}\varphi_{0,i}\sum_{j:(i,j)\in\mathscr{K}}\epsilon_{ij}\gamma_j(\omega)\Bigg\|_{C(Q)}\\
&\ge \frac{1}{2}m_1^{-1/2}\min_{\mathscr{K}\subseteq\mathscr{K}_{m_0m_1},|\mathscr{K}|\ge m_0^d m_1-4n} \mathbb{E}_{\mathbb{P}_2}\max_{1\le i\le m_0^d}\Bigg|\sum_{j:(i,j)\in\mathscr{K}}\epsilon_{ij}\gamma_j(\omega)\Bigg|\\
&\ge \frac{1}{2}m_1^{-1/2}\min_{\mathscr{K}\subseteq\mathscr{K}_{m_0m_1},|\mathscr{K}|\ge m_0^d m_1-4n}\max_{1\le i\le m_0^d} \mathbb{E}_{\mathbb{P}_2}\Bigg|\sum_{j:(i,j)\in\mathscr{K}}\epsilon_{ij}\gamma_j(\omega)\Bigg|.
\end{aligned} \tag{52}$$

By Khintchine's inequality, see [10], Theorem 2.b.3,

$$\mathbb{E}_{\mathbb{P}_2}\Bigg|\sum_{j:(i,j)\in\mathscr{K}}\epsilon_{ij}\gamma_j(\omega)\Bigg| \ge c\Bigg(\sum_{j:(i,j)\in\mathscr{K}}\gamma_j(\omega)^2\Bigg)^{1/2}.$$

So we obtain from (52)

$$\begin{aligned}
&e_{2n}^{\mathrm{avg}}(\mathscr{S}_{1,\omega}, \nu, C(Q))\\
&\ge c m_1^{-1/2}\min_{\mathscr{K}\subseteq\mathscr{K}_{m_0m_1},|\mathscr{K}|\ge m_0^d m_1-4n}\max_{1\le i\le m_0^d}\Bigg(\sum_{j:(i,j)\in\mathscr{K}}\gamma_j(\omega)^2\Bigg)^{1/2}.
\end{aligned}$$

For each $\mathscr{K} \subseteq \mathscr{K}_{m_0 m_1}$ with $|\mathscr{K}| \ge m_0^d m_1 - 4n$ we have by (50) $|\mathscr{K}| \ge |\mathscr{K}_{m_0 m_1}|/2$, hence there is an i with $1 \le i \le m_0^d$ such that $|\{j : (i, j) \in \mathscr{K}\}| \ge m_1/2$. With $\mathscr{J}_{m_1} = \{1, 2, \dots, m_1\}$ it follows that

$$e_{2n}^{\mathrm{avg}}(\mathscr{S}_{1,\omega}, \nu, C(Q)) \ge c m_1^{-1/2} \min_{\mathscr{J} \subseteq \mathscr{J}_{m_1}, |\mathscr{J}| \ge m_1/2} \left(\sum_{j \in \mathscr{J}} \gamma_j(\omega)^2 \right)^{1/2},$$

and therefore, by (51) and Lemma 5, for n satisfying (50),

$$\begin{aligned} &e_n^{\mathrm{ran}}(\mathscr{S}_1, \Psi_{m_0 m_1} \times \Omega, C(Q)) \\ &\ge c m_1^{-1/2} \inf_{D \in \Sigma, \mathbb{P}(D) \ge 1/4} \int_D \min_{\mathscr{J} \subseteq \mathscr{J}_{m_1}, |\mathscr{J}| \ge m_1/2} \left(\sum_{j \in \mathscr{J}} \gamma_j(\omega)^2 \right)^{1/2} d\mathbb{P}(\omega) \ge c. \quad (53) \end{aligned}$$

Let $2 \le q < \infty$ and observe that there is a constant $c_0 > 0$ such that for $m_0, m_1 \in \mathbb{N}$

$$c_0 (\max(m_0, m_1))^{-r} \Psi_{m_0 m_1} \subseteq \mathscr{F}_q^r(Q \times [0, 1] \times \Omega; \kappa). \quad (54)$$

For $n \in \mathbb{N}$ put $m_0 = m_1 = \left\lceil 4n^{\frac{1}{d+1}} \right\rceil$, hence (50) is satisfied, and therefore (16), (53), and (54) imply

$$\begin{aligned} &e_n^{\mathrm{ran}}(\mathscr{S}, \mathscr{F}_q^r(Q \times [0, 1] \times \Omega; \kappa) \times \Omega, C(Q \times [0, 1])) \\ &\ge e_n^{\mathrm{ran}}(\mathscr{S}_1, \mathscr{F}_q^r(Q \times [0, 1] \times \Omega; \kappa) \times \Omega, C(Q)) \\ &\ge c_0^{-1} m_0^{-r} e_n^{\mathrm{ran}}(\mathscr{S}_1, \Psi_{m_0 m_1} \times \Omega, C(Q)) \ge c n^{-\frac{r}{d+1}}. \quad (55) \end{aligned}$$

Combining (46)–(47) and (55) proves the lower bounds (24) and (26).

Next let $2 < q < \infty$ and note that there is a constant $c_1 > 0$ such that for all $m_0, m_1 \in \mathbb{N}$

$$c_1 m_0^{-r_1} m_1^{-\rho} \Psi_{m_0 m_1} \subseteq F_q^{r_1,\rho}(Q \times [0, 1] \times \Omega; \kappa) \subseteq F^{r_1,\rho}(Q \times [0, 1] \times \Omega; \kappa). \quad (56)$$

Let $n \in \mathbb{N}$. First we put $m_0 = \left\lceil 8n^{\frac{1}{d}} \right\rceil$, $m_1 = 1$. Again (50) is satisfied, thus (16), (53), and (56) yield

$$\begin{aligned} &e_n^{\mathrm{ran}}(\mathscr{S}, F_q^{r_1,\rho} \times \Omega, C(Q \times [0, 1])) \ge e_n^{\mathrm{ran}}(\mathscr{S}_1, F_q^{r_1,\rho} \times \Omega, C(Q)) \\ &\ge c_1^{-1} m_0^{-r} e_n^{\mathrm{ran}}(\mathscr{S}_1, \Psi_{m_0 m_1} \times \Omega, C(Q)) \ge c n^{-\frac{r}{d}}. \quad (57) \end{aligned}$$

Now we set $m_0 = 1$, $m_1 = 8n$. Clearly, (50) holds and, using again (16), (53), and (56), we conclude

$$e_n^{\text{ran}}(\mathscr{S}, F_q^{r_1,\rho} \times \Omega, C(Q \times [0,1])) \ge e_n^{\text{ran}}(\mathscr{S}_1, F_q^{r_1,\rho} \times \Omega, C(Q))$$
$$\ge c_1^{-1} m_1^{-\rho} e_n^{\text{ran}}(\mathscr{S}_1, \Psi_{m_0 m_1} \times \Omega, C(Q)) \ge c n^{-\rho}. \quad (58)$$

Now the lower bounds (25) and (27) follow from (48) and (57)–(58), which completes the proof. □

Acknowledgements I want to thank Paweł Przybyłowicz for valuable comments.

References

1. Daun, Th., Heinrich, S.: Complexity of Banach space valued and parametric integration. In: Dick, J., Kuo, F.Y., Peters, G.W., Sloan, I.H. (eds.) Monte Carlo and Quasi-Monte Carlo Methods 2012, pp. 297–316. Springer (2013)
2. Daun, Th, Heinrich, S.: Complexity of Banach space valued and parametric stochastic Itô integration. J. Complex. **40**, 100–122 (2017). https://doi.org/10.1016/j.jco.2017.01.004
3. Heinrich, S.: The multilevel method of dependent tests. In: Balakrishnan, N., Melas, V.B., Ermakov, S.M. (eds.) Advances in Stochastic Simulation Methods, pp. 47–62. Birkhäuser (2000)
4. Heinrich, S.: Multilevel Monte Carlo methods. In: Margenov, S., Waśniewski, J., Yalamov, P. (eds.) Large-Scale Scientific Computing, pp. 58–67. Springer (2001)
5. Heinrich, S., Sindambiwe, E.: Monte Carlo complexity of parametric integration. J. Complex. **15**, 317–341 (1999)
6. Heinrich, S.: Monte Carlo approximation of weakly singular integral operators. J. Complex. **22**, 192–219 (2006)
7. Heinrich, S.: The randomized information complexity of elliptic PDE. J. Complex. **22**, 220–249 (2006)
8. Hertling, P.: Nonlinear Lebesgue and Itô Integration Problems of High Complexity. J. Complex. **17**, 366–387 (2001)
9. Hofmann, N., Müller-Gronbach, T., Ritter, K.: Step size control for the uniform approximation of systems of stochastic differential equations with additive noise. Ann. Appl. Prob. **10**(2), 616–633 (2000)
10. Lindenstrauss, J., Tzafriri, L.: Classical Banach Spaces, 1: Sequence Spaces. Springer (1977)
11. Pisier, G.: The Volume of Convex Bodies and Banach Space Geometry. Cambridge University Press, Cambridge (1989)
12. Przybyłowicz, P.: Linear information for approximation of the Itô integrals. Numer. Algorithms **52**(4), 677–699 (2009)
13. Przybyłowicz, P.: Adaptive Itô-Taylor algorithm can optimally approximate the Itô integrals of singular functions. J. Comput. Appl. Math. **235**(1), 203–217 (2010)
14. Ritter, K.: Approximation and optimization on the Wiener space. J. Complex. **6**(4):337–364 (1990)
15. Wasilkowski, G.W., Woźniakowski, H.: On the complexity of stochastic integration. Math. Comput. **70**(234), 685–698 (2001)

QMC Algorithms with Product Weights for Lognormal-Parametric, Elliptic PDEs

Lukas Herrmann and Christoph Schwab

Abstract We survey recent convergence rate bounds for single-level and multilevel QMC Finite Element (FE for short) algorithms for the numerical approximation of linear, second order elliptic PDEs in divergence form in a bounded, polygonal domain D. The diffusion coefficient a is assumed to be an isotropic, log-Gaussian random field (GRF for short) in D. The representation of the GRF $Z = \log a$ is assumed affine-parametric with i.i.d. standard normal random variables, and with *locally supported* functions ψ_j characterizing the spatial variation of the GRF Z. The goal of computation is the evaluation of expectations (i.e., of so-called "ensemble averages") of (linear functionals of) the random solution. The QMC rules employed are randomly shifted lattice rules proposed in Nichols, Kuo (J Complex 30:444–468, 2014, [19]) as used and analyzed previously in a similar setting (albeit for globally in D supported spatial representation functions ψ_j as arise in Karhunen-Loève expansions) in Graham et al. (Numer Math 131:329–368, 2015, [9]), Kuo et al. (Math Comput 86:2827–2860, 2017, [14]). The multilevel QMC-FE approximation Q_L^* analyzed here for locally supported ψ_j was proposed first in Kuo, Schwab, Sloan (Found Comput Math 15:411–449, 2015, [17]) for affine-parametric operator equations. As shown in Gantner, Herrmann, Schwab (SIAM J Numer Anal 56:111–135, 2018, [7]), Gantner, Herrmann, Schwab (Contemporary computational mathematics - a celebration of the 80th birthday of Ian Sloan. Springer, Cham, 2018, [6]), Herrmann, Schwab (QMC integration for lognormal-parametric, elliptic PDEs: local supports and product weights. Technical Report 2016-39, Seminar for Applied Mathematics, ETH Zürich, Switzerland, 2016, [10]), Herrmann, Schwab (Multilevel quasi-Monte Carlo integration with product weights for elliptic PDEs with lognormal coefficients. Technical Report 2017-19, Seminar for Applied Mathematics, ETH Zürich, Switzerland, 2017, [11]) localized supports of the ψ_j (which appear in multiresolution representations of GRFs Z of Lévy–Ciesielski type in D) allow for the use of product weights, originally proposed in construction of QMC rules in Sloan, Woźniakowski

L. Herrmann (✉) · C. Schwab
Seminar for Applied Mathematics, ETH Zürich, Rämistrasse 101, 8092 Zurich, Switzerland
e-mail: lukas.herrmann@sam.math.ethz.ch

C. Schwab
e-mail: christoph.schwab@sam.math.ethz.ch

A. B. Owen and P. W. Glynn (eds.), *Monte Carlo and Quasi-Monte Carlo Methods*, Springer Proceedings in Mathematics & Statistics 241,
https://doi.org/10.1007/978-3-319-91436-7_17

(J Complex 14:1–33, 1998, [23]) (cf. the survey (Dick, Kuo, Sloan in Acta Numer 22:133–288, 2013, [4]) and references there). The present results from Herrmann, Schwab (Multilevel quasi-Monte Carlo integration with product weights for elliptic PDEs with lognormal coefficients. Technical Report 2017-19, Seminar for Applied Mathematics, ETH Zürich, Switzerland, 2017, [11]) on convergence rates for the multilevel QMC FE algorithm allow for general polygonal domains D and for GRFs Z whose realizations take values in weighted spaces containing $W^{1,\infty}(D)$. Localized support assumptions on ψ_j are shown to allow QMC rule generation by the fast, FFT based CBC constructions in Nuyens, Cools (J Complex 22:4–28, 2006, [21]), Nuyens, Cools (Math Comput 75:903–920, 2006, [20]) which scale linearly in the integration dimension which, for multiresolution representations of GRFs, is proportional to the number of degrees of freedom used in the FE discretization in the physical domain D. We show numerical experiments based on public domain QMC rule generating software in Gantner (A generic c++ library for multilevel quasi-Monte Carlo. In: Proceedings of the Platform for Advanced Scientific Computing Conference, PASC '16, ACM, New York, USA, pp 11:1–11:12 2016, [5]), Kuo, Nuyens (Found Comput Math 16:1631–1696, 2016, [13]).

Keywords Quasi-Monte Carlo methods · Multilevel quasi-Monte Carlo · Uncertainty quantification · Error estimates · High-dimensional quadrature · Elliptic partial differential equations with lognormal input

1 Introduction

The numerical solution of partial differential equations (PDEs for short) with random input data is a core task in the field of computational uncertainty quantification. Particular models of randomness in the PDEs' input parameters entail particular requirements for efficient computational uncertainty quantification algorithms. A basic case arises when there are only a finite number of random variables whose densities have bounded support and which parametrize the uncertain input in the forward PDE model: computation of statistical moments of responses and also Bayesian inversion then amounts to numerical integration over a bounded domain of finite dimension s. Statistical independence and scaling implies numerical integration over the unit cube $[0, 1]^s$ against a product probability measure. In the context of PDEs, so-called *distributed random inputs* such as spatially heterogeneous diffusion coefficients, uncertain physical domains, etc. imply, via *uncertainty parametrizations* (such as Fourier- , B-spline or wavelet expansions) in physical domains D, a countably-infinite number of random parameters (being, for example, Fourier- or wavelet coefficients). This, in turn, renders the problem of estimation of response statistics of solutions of a problem of infinite-dimensional numerical integration. Assuming again statistical independence of the system of (countably many) random input parameters results in the problem of numerical integration against a product measure. In case of Gaussian random field (GRF for short) inputs under consideration in this note, in addition the domain of integration is the countable product of real lines $\mathbb{R}^{\mathbb{N}}$, endowed with a Gaussian measure (GM for short); see, e.g., [3] for details on GMs on $\mathbb{R}^{\mathbb{N}}$.

Here, as in [9, 14] and the references there, we analyze QMC quadratures in the FE solution of linear, second order elliptic PDEs in a bounded, polygonal domain D, with isotropic, log-Gaussian diffusion coefficient $a = \exp(Z)$, where Z is a GRF in D. As in [9, 14], we confine the analysis to first order, randomly shifted lattice rules proposed originally in [19], and to continuous, piecewise linear "Courant" FE methods in D. We adopt the setting of our analysis [10] of the single-level QMC-FE algorithm: consider

$$-\nabla \cdot (a\nabla u) = f \text{ in } D, \quad u = 0 \text{ on } \partial D \tag{1}$$

where D is a bounded interval in space dimension $d = 1$ or a bounded polygon with J straight sides and J corners $c_j, i = 1, \ldots, J$, in space dimension $d = 2$. We endow $\Omega := \mathbb{R}^{\mathbb{N}}$ with the Gaussian product measure and the corresponding product sigma algebra, cf. [3]

$$\mu(\mathrm{d}\boldsymbol{y}) := \bigotimes_{j\geq 1} \frac{1}{\sqrt{2\pi}} e^{-\frac{y_j^2}{2}} \mathrm{d}y_j, \quad \boldsymbol{y} = (y_j)_{j\geq 1} \in \Omega.$$

The random input is modelled on $(\Omega, \bigotimes_{j\geq 1} \mathscr{B}(\mathbb{R}), \mu)$ which is a probability space (cf. for example [3, Example 2.3.5]). The GRF $Z = \log(a) : \Omega \to L^\infty(D)$ is assumed to be affine-parametric:

$$Z := \sum_{j\geq 1} y_j \psi_j. \tag{2}$$

In order to render the random coefficient $a = \exp(Z)$ in (1) meaningful, we imposed in [10] on the $(\psi_j)_{j\geq 1}$ in (2) the summability condition

$$\left\| \sum_{j\geq 1} \frac{|\psi_j|}{b_j} \right\|_{L^\infty(D)} < \infty \tag{A1}$$

such that $(b_j)_{j\geq 1} \in \ell^{p_0}(\mathbb{N})$ for some $p_0 \in (0, \infty)$, and the positive sequence $(b_j)_{j\geq 1}$ encodes decay of $(\psi_j)_{j\geq 1}$. We observe that (A1) is weaker than the summability conditions imposed in [9, 14] in the case that the ψ_j have local supports, as observed in [2] in the context on N-term polynomial chaos approximation rate analysis of the random field solution u of (1). The assumption of local supports in (A1) allows for the use of product weights, cf. [6, 7, 10, 11]. The QMC points result from generating vectors that are constructed with the component-by-component (CBC for short) construction. The CBC construction for product weights, cf. [20, 21], has computational cost which scales linearly with respect to the dimension s of the domain of integration. Reproducing kernel Hilbert spaces (RKHS for short) with product weights were introduced in [23]. For general surveys on QMC we refer to [4, 15] and the references there. A finite dimension s of integration results from the truncation of the expansion of the GRF Z which, if e.g. $(\psi_j)_{j\geq 1}$ is a multiresolution analysis, couples with the FE discretization. The main theoretical results on single-level and multilevel QMC with product weights that we survey in this note are proven in our manuscripts [10, 11], which are in review at the time of writing the present paper. These are backed here with numerical experiments whose implementation uses public domain software [13].

2 Spatial Approximation

The spatial approximation of the PDE (1) by the FE method is based on its (primal) variational formulation in D, while considering the coefficient sequence $\boldsymbol{y}$ in the random input as "a parameter". Let $V := H_0^1(D)$ with dual V^*. Find $u : \Omega \to V$ such that

$$\int_D a\nabla u \cdot \nabla v \mathrm{d}x = f(v), \quad v \in V. \tag{3}$$

The Assumption (A1) and that for some $p_0 \in (0, \infty)$, $(b_j)_{j\geq 1} \in \ell^{p_0}(\mathbb{N})$ implies that $Z \in L^q(\Omega; L^\infty(D))$ for every $q \in [1, \infty)$, cf. [10, Theorem 2]. This implies that μ-a.s. $0 < \operatorname{ess\,inf}_{x\in D}\{a(x)\} \leq \|a\|_{L^\infty(D)} < \infty$. For the ensuing presentation, we define the random variables

$$a_{\min} := \operatorname{ess\,inf}_{x\in D}\{a(x)\} \quad \text{and} \quad a_{\max} := \|a\|_{L^\infty(D)}\ .$$

Hence, the random bilinear form $(w, v) \mapsto \int_D a\nabla w \cdot \nabla v \mathrm{d}x$ on $V \times V$ is continuous and coercive with coercivity constant $a_{\min}$ and continuity constant $a_{\max}$. By the Lax–Milgram lemma, the solution u exists and solves (3) uniquely. Also due to [10, Corollary 6 and Eq. (16)], we obtain the estimate for every $q \in [1, \infty)$,

$$\|u\|_{L^q(\Omega;V)} \leq \|1/a_{\min}\|_{L^q(\Omega)}\|f\|_{V^*} < \infty,$$

where the strong measurability of u follows, since u depends continuously on a (by the second Strang lemma). To obtain a finite dimensional integration domain, we consider dimension truncation. For every $s \in \mathbb{N}$, let $a^s := \exp(Z^s) = \exp(\sum_{j=1}^s y_j\psi_j)$ denote the truncated lognormal field and define the random variables

$$a^s_{\min} := \operatorname{ess\,inf}_{x\in D}\{a^s(x)\} \quad \text{and} \quad a^s_{\max} := \|a^s\|_{L^\infty(D)}.$$

Let $u^s : \Omega \to V$ be the solution with respect to the coefficient a^s, i.e.,

$$\int_D a^s\nabla u^s \cdot \nabla v \mathrm{d}x = f(v), \quad v \in V.$$

Assuming that $(b_j)_{j\geq 1} \in \ell^{p_0}(\mathbb{N})$ for some $p_0 \in (0, \infty)$ by [10, Proposition 7], for every $\varepsilon > 0$, there exists a constant $C_\varepsilon > 0$ such that for every $G(\cdot) \in V^*$

$$|\mathbb{E}(G(u)) - \mathbb{E}(G(u^s))| \leq C_\varepsilon \|G(\cdot)\|_{V^*} \|f\|_{V^*} \sup_{j>s} \{b_j^{1-\varepsilon}\}. \tag{4}$$

Approximations with Finite Elements in a polygon $D \subset \mathbb{R}^2$ with respect to uniformly refined triangulations may result in suboptimal convergence rates. We therefore consider certain weighted Sobolev spaces, cf. [1]. For a J-tuple $\boldsymbol{\beta} = (\beta_1, \dots, \beta_J)$ of weight exponents, we define the *corner weight function* $\Phi_{\boldsymbol{\beta}}(x) := \prod_{i=1}^{J} |c_i - x|^{\beta_i}$, $x \in D$, where $\beta_i \in [0, 1)$, $i = 1, \dots, J$, and $\{c_1, \dots, c_J\} \subset \partial D$ are the corners of D. Here and in the following, the Euclidean norm in $\mathbb{R}^2$ is denoted by $|\cdot|$. We define the function spaces $L^2_{\boldsymbol{\beta}}(D)$ and $H^2_{\boldsymbol{\beta}}(D)$ as closures of $C^\infty(\overline{D})$ with respect to the norms $\|v\|_{L^2_{\boldsymbol{\beta}}(D)} := \|v\Phi_{\boldsymbol{\beta}}\|_{L^2(D)}$ and $\|v\|^2_{H^2_{\boldsymbol{\beta}}(D)} := \|v\|^2_{H^1(D)} + \sum_{|\alpha|=2} \||\partial_x^\alpha v|\Phi_{\boldsymbol{\beta}}\|^2_{L^2(D)}$.

Lemma 1 *There is $C > 0$ such that for every $f \in L^2_{\boldsymbol{\beta}}(D)$ holds $\|f\|_{V^*} \leq C\|f\|_{L^2_{\boldsymbol{\beta}}(D)}$.*

Proof The statement of the lemma is equivalent to the continuity of the embedding $L^2_{\boldsymbol{\beta}}(D) \subset V^*$. By duality, this is equivalent to the continuity of the embedding $V \subset (L^2_{\boldsymbol{\beta}}(D))^*$. We therefore identify $L^2(D)$ with its dual $(L^2(D))^*$, and obtain for an arbitrary $w \in (L^2_{\boldsymbol{\beta}}(D))^*$ with the Cauchy–Schwarz inequality

$$\begin{aligned}\|w\|_{(L^2_{\boldsymbol{\beta}}(D))^*} = \sup_{v \in L^2_{\boldsymbol{\beta}}(D), \|v\|_{L^2_{\boldsymbol{\beta}}(D)}=1} w(v) &= \sup_{v \in L^2_{\boldsymbol{\beta}}(D), \|v\|_{L^2_{\boldsymbol{\beta}}(D)}=1} \int wv \mathrm{d}x \\ &\leq \|w/\Phi_{\boldsymbol{\beta}}\|_{L^2(D)} = \|w\|_{L^2_{-\boldsymbol{\beta}}(D)} .\end{aligned}$$

By the Hardy inequality (see, e.g., [22, Theorem 21.3] with the choices $p = q = 2$, $\alpha = -p$, $\beta = 0$, $\kappa = 1$), there exists a constant $C' > 0$ such that for every $\tilde{w} \in V$, with $\mathrm{dist}_{\partial D}(x)$ denoting for $x \in D$ the regularized distance of x to the (Lipschitz) boundary ∂D, as defined e.g. in [24, Chap. 6.2], $\|\tilde{w}/\mathrm{dist}_{\partial D}\|_{L^2(D)} \leq C'\|\tilde{w}\|_V$, we conclude that the embedding $V \subset (L^2_{\boldsymbol{\beta}}(D))^*$ is continuous. This implies the assertion of this lemma. □

In the weighted spaces $H^2_{\boldsymbol{\beta}}(D)$ there holds a full regularity shift for the Dirichlet Laplacean, cf. [1, Theorem 3.2]: there exists a constant $C > 0$ such that for every $w \in V$ with $\Delta w \in L^2_{\boldsymbol{\beta}}(D)$,

$$\|w\|_{H^2_{\boldsymbol{\beta}}(D)} \leq C\|\Delta w\|_{L^2_{\boldsymbol{\beta}}(D)}, \tag{5}$$

provided that the weight exponent J-tuple $\boldsymbol{\beta}$ satisfies $0 \leq \beta_i$ and $1 - \pi/\omega_i < \beta_i < 1$, $i = 1, \dots, J$. The interior angle of the corner c_i is denoted by ω_i, $i = 1, \dots, J$. Since in [1] the Poisson boundary value problem with a zero order term is considered, i.e.,

$-\Delta w + w = f$, we also used the estimate that for constants $C_1, C_2, C_3 > 0$ independent of $w \in V \cap H^2_\beta(D)$, $\|w\|_{L^2_\beta(D)} \le C_1 \|w\|_{L^2(D)} \le C_2 \|w\|_V = C_2 \|\Delta w\|_{V^*} \le C_3 \|\Delta w\|_{L^2_\beta(D)}$, which is a consequence of Lemma 1. Also in FE spaces $V_\ell := \{v \in V : v|_K \in \mathbb{P}^1(K), K \in \mathscr{T}_\ell\}$ there is an approximation property, cf. [1, Lemmas 4.1 and 4.5], where $\mathbb{P}^1(K)$ are the affine functions on K and $\{\mathscr{T}_\ell\}_{\ell \ge 0}$ are sequences of regular, simplicial triangulations with proper mesh refinement near the corners c_i of D. Specifically, there exists a constant C such that for every $w \in H^2_\beta(D)$ there is $w_\ell \in V_\ell$ satisfying

$$\|w - w_\ell\|_V \le C M_\ell^{-1/d} \|w\|_{H^2_\beta(D)}, \tag{6}$$

where $M_\ell := \dim(V_\ell)$. Let $u^{s,\mathscr{T}_\ell} : \Omega \to V_\ell$ be the FE solution, i.e.,

$$\int_D a^s \nabla u^{s,\mathscr{T}_\ell} \cdot \nabla v \mathrm{d}x = f(v), \quad \forall v \in V_\ell. \tag{7}$$

Let $W^{1,\infty}_\beta(D)$ denote the Banach space of measurable functions $v : D \to \mathbb{R}$ that have finite $W^{1,\infty}_\beta(D)$-norm, where $\|v\|_{W^{1,\infty}_\beta(D)} := \max\{\|v\|_{L^\infty(D)}, \||\nabla v|\Phi_\beta\|_{L^\infty(D)}\}$. We introduce the following mixed sparsity assumption on the function system $(\psi_j)_{j\ge1}$. Let $(\bar{b}_j)_{j\ge1}$ be a positive sequence such that

$$\left\| \sum_{j\ge1} \frac{\max\{|\nabla \psi_j|\Phi_\beta, |\psi_j|\}}{\bar{b}_j} \right\|_{L^\infty(D)} < \infty. \tag{A2}$$

The Assumption (A2) (which is stronger than (A1)) is essential in obtaining improved error vs. work bounds for the multilevel QMC approximation Q^*_L as compared to the bounds for the single-level QMC approximation in [10, Theorem 17]. The following proposition is obtained as [10, Theorem 2], we omit the details of its proof here.

Proposition 1 *Let the assumption in (A2) be satisfied for some sequence* $(\bar{b}_j)_{j\ge1}$ *such that* $(\bar{b}_j)_{j\ge1} \in \ell^{p_0}(\mathbb{N})$ *for some* $p_0 \in (0, \infty)$. *For every* $\varepsilon > 0$ *and* $q \in [1, \infty)$ *there exists a constant* $C > 0$ *such that for every* $s \in \mathbb{N}$,

$$\|Z - Z^s\|_{L^q(\Omega:W^{1,\infty}_\beta(D))} \le C \sup_{j>s}\{\bar{b}_j^{1-\varepsilon}\}.$$

We obtain with [10, Corollary 6], that the identity $(\nabla a)\Phi_\beta = (a \nabla Z)\Phi_\beta$ holds in $L^\infty(D)^d$, μ-a.s. With the Cauchy–Schwarz inequality, it implies that for every $q \in [1, \infty)$ there exists a constant $C > 0$ such that for every $s \in \mathbb{N}$,

$$\|a\|_{L^q(\Omega; W^{1,\infty}_\beta(D))} < \infty \quad \text{and} \quad \|a^s\|_{L^q(\Omega; W^{1,\infty}_\beta(D))} \le C < \infty.$$

We observe that μ-a.s holds, that for every subset $\widetilde{D} \subset\subset D$, $|\nabla a| \in L^\infty(\widetilde{D})$ and also that for every $q \in [1, \infty)$, $|\nabla a| \in L^q(\Omega; L^\infty(\widetilde{D}))$. We assume that $f, G(\cdot) \in L^2_\beta(D)$.

Then, by the divergence theorem and product rule

$$\int_D f v \mathrm{d}x = \int_D a \nabla u \cdot \nabla v \mathrm{d}x = -\int_D [a \Delta u + \nabla a \cdot \nabla u] v \mathrm{d}x, \quad \forall v \in C_0^\infty(D).$$

Formally testing the corresponding pointwise identity (which holds for pointwise a.e. $x \in D$) with $-\Delta u \Phi_\beta^2/a$, we obtain the following estimate, valid μ-a.s.

$$\|\Delta u\|_{L^2_\beta(D)} \leq \frac{\|f\|_{L^2_\beta(D)}}{a_{\min}} + \|Z\|_{W^{1,\infty}_\beta(D)} \|u\|_V \leq C \frac{\|f\|_{L^2_\beta(D)}}{a_{\min}} (1 + \|Z\|_{W^{1,\infty}_\beta(D)}). \quad (8)$$

Note that we may test with $-\Delta u \Phi_\beta^2/a$, since it can be approximated by elements of $C_0^\infty(D)$ in $L^2(D)$. Here we used Lemma 1, i.e., $\|f\|_{V^*} \leq C\|f\|_{L^2_\beta(D)}$ with a constant $C > 0$ depending only on the domain D, which is independent of f. By an Aubin–Nitsche argument, by (4), (5), (6), Proposition 1, and (8), for every $\varepsilon > 0$ there exists a constant $C > 0$ such that for every $s \in \mathbb{N}$, $\ell \in \mathbb{N}_0$

$$|\mathbb{E}(G(u)) - \mathbb{E}(G(u^{s,\mathcal{T}_\ell}))| \leq C \left(\sup_{j>s}\{b_j^{1-\varepsilon}\} + M_\ell^{-2/d} \right) \|G(\cdot)\|_{L^2_\beta(D)} \|f\|_{L^2_\beta(D)}. \quad (9)$$

Remark 1 The regularity shift in (5) and the estimate in (8) can be interpolated between the interpolation couple $L^2_\beta(D) \subset V^*$ as well as the approximation property in (6). If $f \in (V^*, L^2_\beta(D))_{t,\infty}$ and if $G(\cdot) \in (V^*, L^2_\beta(D))_{t',\infty}$ for some $t, t' \in [0, 1]$, then the estimate (9) holds with the term $M_\ell^{-2/d}$ that bounds the error contribution from the FE discretization replaced by $M_\ell^{-(t+t')/d}$. Here and throughout what follows, interpolation spaces shall be understood with respect to the real method of interpolation; we refer to [25, Chap. 1] and the references there for definitions and basic properties of interpolation spaces.

3 Single-Level QMC

Dimension independent convergence rates of QMC with randomly shifted lattice rules can be shown by estimating the *worst-case error* of a particular weighted Sobolev space of type $\mathcal{W}_\gamma$ and the norm in this Sobolev space of the integrand. We generally seek to approximate s-dimensional integrals with respect to the multivariate normal distribution

$$I_s(F) := \int_{\mathbb{R}^s} F(\mathbf{y}) \prod_{j=1}^{s} \phi(y_j) \mathrm{d}\mathbf{y},$$

where the univariate, standard normal density is denoted by $\phi(\cdot)$.

For every $s \in \mathbb{N}$ and product weights $\boldsymbol{\gamma} = (\gamma_{\mathfrak{u}})_{\mathfrak{u} \subset \mathbb{N}}$, we introduce the weighted Sobolev spaces $\mathcal{W}_\gamma(\mathbb{R}^s)$, which are given by the norm

$$\|F\|_{\mathscr{W}_{\boldsymbol{\gamma}}(\mathbb{R}^s)}$$
$$:= \left(\sum_{\mathfrak{u} \subset \{1:s\}} \gamma_{\mathfrak{u}}^{-1} \int_{\mathbb{R}^{|\mathfrak{u}|}} \left| \int_{\mathbb{R}^{s-|\mathfrak{u}|}} \partial_{\boldsymbol{y}}^{\mathfrak{u}} F(\boldsymbol{y}) \prod_{j \in \{1:s\} \setminus \mathfrak{u}} \phi(y_j) \mathrm{d}\boldsymbol{y}_{\{1:s\} \setminus \mathfrak{u}} \right|^2 \prod_{j \in \mathfrak{u}} w_j^2(y_j) \mathrm{d}\boldsymbol{y}_{\mathfrak{u}} \right)^{1/2} . \tag{10}$$

The considered weights $\boldsymbol{\gamma}$ are of product type, i.e., for some positive sequence $(\gamma_j)_{j \geq 1}$

$$\gamma_{\mathfrak{u}} = \prod_{j \in \mathfrak{u}} \gamma_j, \quad \mathfrak{u} \subset \mathbb{N}, |\mathfrak{u}| < \infty.$$

The weight functions in (10) are either unnormalized Gaussians or exponentially decaying, i.e.,

$$w_{\mathrm{g},j}^2(y) := e^{-\frac{y^2}{2\alpha_{\mathrm{g}}}}, \quad y \in \mathbb{R}, j \geq 1, \quad \text{and} \quad w_{\mathrm{exp},j}^2(y) := e^{-\alpha_{\mathrm{exp}}|y|}, \quad y \in \mathbb{R}, j \geq 1,$$

where $\alpha_{\mathrm{g}} > 1$ and $\alpha_{\mathrm{exp}} > 0$. The QMC quadrature in $s \in \mathbb{N}$ dimensions with N points is denoted by $Q_{s,N}(\cdot)$. Using randomly shifted lattice rules, there exist QMC points such that for every $F \in \mathscr{W}_{\boldsymbol{\gamma}}(\mathbb{R}^s)$ the mean squared error integrated over all random shifts $\boldsymbol{\Delta}$ (w.r.t. to the uniform measure, cf. [19]) satisfies

$$\sqrt{\mathbb{E}^{\boldsymbol{\Delta}}(|I_s(F) - Q_{s,N}(F)|^2)} \leq C_{\boldsymbol{\gamma}}(\varphi(N))^{-1/(2\lambda)} \|F\|_{\mathscr{W}_{\boldsymbol{\gamma}}(\mathbb{R}^s)}, \tag{11}$$

where the constant $C_{\boldsymbol{\gamma}}$ is finite if $(\gamma_j)_{j \geq 1} \in \ell^\lambda(\mathbb{N})$ and then uniformly bounded in the dimension s (and in particular independent of F) for $\lambda \in (1/(2r), 1]$. This follows by [19, Theorem 8], [16, Lemma 6.3], and [18, Examples 4 and 5], where

$$r = \begin{cases} 1 - 1/(2\alpha_{\mathrm{g}}) & \text{for Gaussian weight functions,} \\ 1 - \delta & \text{for exponential weight functions and any } \delta \in (0, 1/2). \end{cases}$$

The Euler totien function is denoted by $\varphi(\cdot)$. In the following, the solution u^s and the coefficient a^s are viewed as mappings from $\mathbb{R}^s$ to V and $L^\infty(D)$, respectively. In the analysis of bounds of the $\mathscr{W}_{\boldsymbol{\gamma}}(\mathbb{R}^s)$-norm of the specific integrand $F(\boldsymbol{y}) = G(u^s(\boldsymbol{y}))$, $\boldsymbol{y} \in \mathbb{R}^s$, global bounds of the function system $(\psi_j)_{j \geq 1}$ have been used in [9] with POD weights. The theory in [2] is able to derive parametric regularity estimates taking into account possible locality of the supports of ψ_j. Specifically, [2, Theorem 4.1] states that if for a positive sequence $(\rho_j)_{j \geq 1}$

$$\left\| \sum_{j \geq 1} \rho_j |\psi_j| \right\|_{L^\infty(D)} < \log(2), \tag{12}$$

then there exists a constant C that is independent of s such that for every $\boldsymbol{y} \in \mathbb{R}^s$,

$$\sum_{\mathfrak{u}\subset\{1:s\}} \|\partial_y^{\mathfrak{u}} u^s(\mathbf{y})\|_{a^s(\mathbf{y})}^2 \prod_{j\in\mathfrak{u}} \rho_j^2 \le C\|u^s(\mathbf{y})\|_{a^s(\mathbf{y})}^2, \tag{13}$$

where $\|\cdot\|_{a(\mathbf{y})}$ denotes the parametric energy norm. In [10], this estimate is used to prove dimension independent convergence rates of randomly shifted lattice rules with product weights. The product weights will be defined using the sequence $(b_j)_{j\ge1}$, where the smallness assumption in (12) does not affect the QMC weights nor the function system $(\psi_j)_{j\ge1}$. Some of the sparsity of the sequence $(b_j)_{j\ge1}$ is used to control the weight functions in the norm (10).

Theorem 1 ([10, Theorems 11 and 13]) *For $p' \in (0,1]$, consider the weight sequence*

$$\gamma_j := b_j^{2p'}, \quad j \ge 1.$$

Let the Assumption (A1) be satisfied and let the conditions below hold, respectively:

1. *Gaussian weight functions: $(b_j)_{j\ge1} \in \ell^p(\mathbb{N})$ for some $p \in (2/3, 2)$ with $\chi = 1/(2p) + 1/4 - \delta$. The weight sequence $(\gamma_j)_{j\ge1}$ is applied with $p' = p/4 + 1/2 - \delta p$ for $\delta \in (0, 3/4 - 1/(2p))$.*
2. *Exponential weight functions: $(b_j)_{j\ge1} \in \ell^p(\mathbb{N})$ for some $p \in (2/3, 1]$ with $\chi = 1/p - 1/2$. The weight sequence $(\gamma_j)_{j\ge1}$ is applied with $p' = 1 - p/2$.*

Then, there exists a constant C independent of N and s such that

$$\sqrt{\mathbb{E}^{\Delta}(|I_s(G(u^s)) - Q_{s,N}(G(u^s))|^2)} \le C(\varphi(N))^{-\chi}.$$

4 Multilevel QMC

The multilevel QMC quadrature is for a maximum level $L \in \mathbb{N}_0$ defined by a telescoping sum expansion

$$Q_L^*(G(u^L) := \sum_{\ell=0}^{L} Q_{s_\ell,N_\ell}(G(u^\ell) - G(u^{\ell-1})), \tag{14}$$

where $G(u^{-1}) := 0$ and $u^\ell := u^{s_\ell,\mathcal{T}_\ell}$, $\ell \ge 0$. It requires choices of dimensions $(s_\ell)_{\ell=0,\dots,L}$ and numbers of QMC points $(N_\ell)_{\ell=0,\dots,L}$. The random shifts between the different levels in (14) are assumed to be independent. This implies with (11)

$$\mathbb{E}^{\Delta}(|I_{s_L}(G(u^L)) - Q_L^*(G(u^L))|^2) \le C_\gamma^2 \sum_{\ell=0}^{L} (\varphi(N_\ell))^{-1/\lambda} \|G(u^\ell) - G(u^{\ell-1})\|_{\mathcal{W}_\gamma(\mathbb{R}^{s_\ell})}^2.$$

According to this error estimate, it is crucial to find suitable bounds of the $\mathcal{W}_\gamma(\mathbb{R}^{s_\ell})$-norm of the difference $G(u^\ell) - G(u^{\ell-1})$ in order that the multilevel QMC quadrature benefits from the coupling between the levels $\ell = 1, \dots, L$.

4.1 Error Estimate

Parametric regularity estimates of the type of (13) can be shown for dimensionally truncated and FE differences between two consecutive levels.

Proposition 2 *Let a positive sequence* $(\rho_j)_{j\geq 1}$ *satisfy* (12) *and for some* $\eta > 0$

$$K_\eta := \left\| \sum_{j\geq 1} \rho_j^{1+\eta} |\psi_j| \right\|_{L^\infty(D)} < \infty.$$

Then, there exists a constant $C > 0$ *such that for every* $s' < s \in \mathbb{N}_0$ *and every* $\boldsymbol{y} \in \mathbb{R}^s$,

$$\sum_{\mathfrak{u}\subset\{1:s\}} \|\partial_{\boldsymbol{y}}^{\mathfrak{u}}(u^s(\boldsymbol{y}) - u^{s'}(\boldsymbol{y}))\|_{a^s(\boldsymbol{y})}^2 \prod_{j\in\mathfrak{u}} \rho_j^2$$
$$\leq C \left(\left\| \frac{a^s(\boldsymbol{y}) - a^{s'}(\boldsymbol{y})}{a^s(\boldsymbol{y})} \right\|_{L^\infty(D)}^2 \|u^{s'}(\boldsymbol{y})\|_{a^s(\boldsymbol{y})}^2 + \sup_{j>s'}\{\rho_j^{-2\eta}\} \|u^{s'}(\boldsymbol{y})\|_{a^s(\boldsymbol{y})}^2 \right).$$

Proposition 3 *Let* $G(\cdot) \in L^2_{\beta}(D)$ *and let a positive sequence* $(\rho_j)_{j\geq 1}$ *satisfy*

$$\left\| \sum_{j\geq 1} \rho_j \max\{|\nabla\psi_j|\Phi_{\beta}, |\psi_j|\} \right\|_{L^\infty(D)} < \sup\left\{c > 0 : ce^c \leq 1\right\} \frac{1}{\sqrt{2}}.$$

Then, there exists a constant $C > 0$ *such that for every* $s \in \mathbb{N}_0$, $\ell \in \mathbb{N}_0$, *and for every* $\boldsymbol{y} \in \mathbb{R}^s$,

$$\sum_{\mathfrak{u}\subset\{1:s\}} |\partial_{\boldsymbol{y}}^{\mathfrak{u}}(G(u^s(\boldsymbol{y})) - G(u^{s,\mathcal{T}_\ell}(\boldsymbol{y})))|^2 \prod_{j\in\mathfrak{u}} \rho_j^2$$
$$\leq C \left(\frac{\|a^s(\boldsymbol{y})\|_{L^\infty(D)}^2}{(a_{\min}^s(\boldsymbol{y}))^4} (1 + \|Z^s(\boldsymbol{y})\|_{W_{\beta}^{1,\infty}(D)}^2) \right)^2 M_\ell^{-4/d} \|G(\cdot)\|_{L_{\beta}^2(D)}^2 \|f\|_{L_{\beta}^2(D)}^2.$$

Propositions 2 and 3 are versions restricted to first order mixed derivatives $\partial_{\boldsymbol{y}}^{\mathfrak{u}}$ of [11, Theorems 4.3 and 4.8]. The parametric regularity estimates in Propositions 2 and 3 are used to show the following multilevel QMC error estimate analogously to the proof of [10, Theorems 11, and 13].

Theorem 2 ([11, Theorem 5.1]) *For* $p' \in (0, 1]$, $\theta \in (0, 1)$, *consider the weight sequence*

$$\gamma_j := (b_j^{1-\theta} \vee \bar{b}_j)^{2\bar{p}'}, \quad j \geq 1.$$

Consider sequences $(s_\ell)_{\ell=0,\ldots,L}$ *and* $(N_\ell)_{\ell=0,\ldots,L}$, $L \in \mathbb{N}_0$, *under the conditions:*

1. *Gaussian weight functions:* $(b_j^{1-\theta} \vee \bar{b}_j)_{j\geq 1} \in \ell^{\bar{p}}(\mathbb{N})$ *for some* $\bar{p} \in (2/3, 2)$ *with* $\bar{\chi} = 1/(2\bar{p}) + 1/4 - \bar{\delta}$. *The weight sequence in* $(\gamma_j)_{j\geq 1}$ *is applied with* $\bar{p}' = \bar{p}/4 + 1/2 - \bar{\delta}\bar{p}$ *for* $\bar{\delta} \in (0, 3/4 - 1/(2\bar{p}))$.
2. *Exponential weight functions:* $(b_j^{1-\theta} \vee \bar{b}_j)_{j\geq 1} \in \ell^{\bar{p}}(\mathbb{N})$ *for some* $\bar{p} \in (2/3, 1]$ *with* $\bar{\chi} = 1/\bar{p} - 1/2$. *The weight sequence in* $(\gamma_j)_{j\geq 1}$ *is applied with* $\bar{p}' = 1 - \bar{p}/2$.

Then, for any $\varepsilon \in (0, 1)$, *there exists a constant* $C > 0$ *that is in particular independent of* $(s_\ell)_{\ell=0,\dots,L}$, $(N_\ell)_{\ell=0,\dots,L}$ *and* $L \in \mathbb{N}_0$ *such that, with* $\xi_{\ell,\ell-1} := 0$ *if* $s_\ell = s_{\ell-1}$ *and* $\xi_{\ell,\ell-1} := 1$ *otherwise,*

$$\sqrt{\mathbb{E}^{\Delta}(|\mathbb{E}(G(u)) - Q_L^*(G(u^L))|^2)}$$
$$\leq C\Bigg(\sup_{j>s_L}\{b_j^{2(1-\varepsilon)}\} + M_L^{-4/d} + \sum_{\ell=0}^{L}(\varphi(N_\ell))^{-2\bar{\chi}}\left(\xi_{\ell,\ell-1} \sup_{j>s_{\ell-1}}\{b_j^{2\theta}\} + M_{\ell-1}^{-4/d}\right)\Bigg)^{1/2}.$$

Remark 2 For $f \in (V^*, L^2_{\beta}(D))_{t,\infty}$ and $G(\cdot) \in (V^*, L^2_{\beta}(D))_{t',\infty}$, with some $t, t' \in [0, 1]$, the error estimate in Theorem 2 holds with the term $M_\ell^{4/d}$ that bounds the FE discretization error replaced by $M_\ell^{-2\tau/d}$, $\ell = 0, \dots, L$, where $\tau = t + t'$.

4.2 Error Versus Work

We discuss in some detail the use of Multiresolution Analyses (MRAs for short) to model the GRF $\log(a) = Z$, analogous to the Lévy–Ciesielski representation of the Wiener process. To this end, we assume that $(\psi_\lambda)_{\lambda\in\nabla}$ constitute a MRA which is generated by a finite number of sufficiently smooth mother wavelets, i.e.,

$$\psi_\lambda(x) = \psi_{(|\lambda|,k))}(x) := \psi(2^{|\lambda|}x - k), \quad k \in \nabla_{|\lambda|}, x \in D.$$

We use the usual notation, where in the index $\lambda = (|\lambda|, k)$ refers to the level $|\lambda| \in \mathbb{N}_0$ and the translation $k \in \nabla_{|\lambda|}$. The index set ∇_ℓ has cardinality $|\nabla_\ell| = \mathcal{O}(2^{d\ell}), \ell \in \mathbb{N}_0$. We assume that the overlap on a fixed level $\ell \in \mathbb{N}_0$ is uniformly bounded, i.e., there exists K such that for every $\ell \in \mathbb{N}_0$ and every $x \in D$,

$$|\{\lambda \in \nabla : |\lambda| = \ell, \psi_\lambda(x) \neq 0\}| \leq K.$$

Additionally, we introduce the scaling that for some $\widehat{\alpha}, \sigma > 0$,

$$\|\psi_\lambda\|_{L^\infty(D)} \leq \sigma 2^{-\widehat{\alpha}|\lambda|}, \quad \lambda \in \nabla,$$

and assume that there exists a constant $C > 0$ such that the mother wavelet satisfies $\||\nabla\psi|\|_{L^\infty(D)} \leq C\|\psi\|_{L^\infty(D)}$. For this MRA the Assumption (A1) is satisfied with the sequence

$$b_{j(\lambda)} = b_\lambda := 2^{-\widehat{\beta}|\lambda|}, \quad \lambda \in \nabla,$$

for $\widehat{\alpha} > \widehat{\beta} > 0$, where $j : \mathbb{N} \to \nabla$ is a suitable enumeration. In this setting the work to compute one sample of the stiffness matrix is $\mathscr{O}(M_\ell \log(s_\ell))$, where s_ℓ denotes the truncation level of the coefficient. We assume that the work to solve the linear system resulting from the FE discretization satisfies that for some $\eta \geq 0$

$$\text{work}_{\text{PDEsolve}} = \mathscr{O}(M_\ell^{1+\eta}). \tag{A3}$$

Therefore, the overall work of the multilevel QMC quadrature satisfies for $L \in \mathbb{N}_0$,

$$\text{work} = \mathscr{O}\left(\sum_{\ell=0}^{L} N_\ell (M_\ell \log(s_\ell) + M_\ell^{1+\eta})\right).$$

The cost of the CBC construction is here excluded from the work model for simplicity. We refer to [11, Sect. 6] for a discussion of a work model that considers the cost of the CBC construction. For $\widehat{\alpha} > \widehat{\beta} > 1$, the MRA $(\psi_\lambda)_{\lambda\in\nabla}$ and the sequence

$$\bar{b}_j := b_j^{(\widehat{\beta}-1)/\widehat{\beta}}, \quad j \in \mathbb{N},$$

satisfies the Assumption (A2). We assume in this section

$$f \in (V^*, L^2_\beta(D))_{t,\infty} \quad \text{and} \quad G(\cdot) \in (V^*, L^2_\beta(D))_{t',\infty}, \quad t, t' \in [0, 1], \tag{A4}$$

and set $\tau := t + t'$. Also, assume that $M_\ell = \mathscr{O}(2^{d\ell})$, $\ell \in \mathbb{N}_0$. We suppose that $(s_\ell)_{\ell=0,\dots,L}$, θ, and $(M_\ell)_{\ell\geq 0}$ are given such that the truncation error in the multilevel QMC error estimate in Theorem 2 is controlled by the FE discretization error on levels $\ell = 0, \dots, L$. Specifically, we suppose that s_ℓ depends algebraically on M_ℓ; two concrete strategies on how to choose s_ℓ are discussed in [11, Sect. 6]. Analogous to the analysis in [11, Sect. 6] (see also [6, 14, 17]), explicit expressions for the QMC sample numbers $(N_\ell)_{\ell=0,\dots,L}$ are found by optimizing work versus the (estimated) error:

$$N_\ell = \begin{cases} \left\lceil N_0 M_\ell^{-(2\tau/d+1+\eta)/(1+2\bar{\chi})} \right\rceil & \text{if } \eta > 0, \\ \left\lceil N_0 \left(M_\ell^{-1-2\tau/d} \log(s_\ell)^{-1}\right)^{1/(1+2\bar{\chi})} \right\rceil & \text{if } \eta = 0, \end{cases} \quad \ell = 1, \dots, L. \tag{15}$$

and

$$N_0 = \begin{cases} \lceil 2^{\tau L/\bar{\chi}} \rceil & \text{if } 1+\eta < \tau/(d\bar{\chi}), \\ \lceil 2^{\tau L/\bar{\chi}} L^{1/(2\bar{\chi})} \rceil & \text{if } 1+\eta = \tau/(d\bar{\chi}), \eta > 0, \\ \lceil 2^{\tau L/\bar{\chi}} L^{(1+4\bar{\chi})/(\bar{\chi}(2+4\bar{\chi}))} \rceil & \text{if } d = \tau/\bar{\chi}, \eta = 0, \\ \lceil 2^{(2\tau+d(1+\eta))L/(1+2\bar{\chi})} & \text{if } 1+\eta > \tau/(d\bar{\chi}), \eta > 0, \\ \lceil 2^{(d+2\tau)L/(1+2\bar{\chi})} L^{1/(1+2\bar{\chi})} \rceil & \text{if } d > \tau/\bar{\chi}, \eta = 0. \end{cases} \tag{16}$$

Theorem 3 ([11, Theorem 6.2]) *Let the Assumptions (A4) and (A3) be satisfied for $\eta \geq 0$. The sample numbers for $Q_L^*(\cdot)$ are given by* (15) *and* (16)*, $L \in \mathbb{N}_0$.*

1. *Gaussian weight functions: for $\bar{p} \in (\max\{2/3, d/(\widehat{\beta}-1)\}, 2)$, $\bar{\chi} = 1/(2\bar{p}) + 1/4 - \bar{\delta}$ for $\bar{\delta} > 0$ sufficiently small assuming $d/(\widehat{\beta}-1) < 2$.*
2. *Exponential weight functions: for $\bar{p} \in (\max\{2/3, d/(\widehat{\beta}-1)\}, 1]$, $\bar{\chi} = 1/\bar{p} - 1/2$ assuming $d/(\widehat{\beta}-1) < 1$.*

For an error threshold $\varepsilon > 0$, we obtain

$$\sqrt{\mathbb{E}^{\boldsymbol{\Delta}}(|\mathbb{E}(G(u)) - Q_L^*(G(u^L))|^2)} = \mathcal{O}(\varepsilon),$$

which is achieved with

$$\text{work} = \begin{cases} \mathcal{O}(\varepsilon^{-1/\bar{\chi}}) & \textit{if } 1+\eta < \tau/(d\bar{\chi}), \\ \mathcal{O}(\varepsilon^{-1/\bar{\chi}} \log(\varepsilon^{-1})^{(1+2\bar{\chi})/(2\bar{\chi})}) & \textit{if } 1+\eta = \tau/(d\bar{\chi}), \eta > 0, \\ \mathcal{O}(\varepsilon^{-1/\bar{\chi}} \log(\varepsilon^{-1})^{(1+4\bar{\chi})/(2\bar{\chi})}) & \textit{if } d = \tau/\bar{\chi}, \eta = 0, \\ \mathcal{O}(\varepsilon^{-d/\tau(1+\eta)}) & \textit{if } 1+\eta > \tau/(d\bar{\chi}), \eta > 0, \\ \mathcal{O}(\varepsilon^{-d/\tau} \log(\varepsilon^{-1})) & \textit{if } d > \tau/\bar{\chi}, \eta = 0. \end{cases}$$

5 Numerical Experiments

Consider (1) in space dimension $d = 1$ with $D = (0, 1)$, i.e.,

$$-\partial_x(a\,\partial_x u) = f \text{ in } D, \quad u(0) = u(1) = 0. \tag{17}$$

The coefficient a is given by $a = \exp(Z)$, where $Z = \sum_{j\geq 1} y_j \psi_j$. We consider two possible cases for the MRA $(\psi_j)_{j\geq 1}$: the Haar system and a family of biorthogonal, continuous, piecewise linear spline wavelets.

5.1 Single-Level QMC

We suppose that the GRF Z is represented by the Haar system $(\psi_{j(\ell,k)})_{j\geq 1}$, i.e., it is generated by the mother wavelet $\psi(x) = \mathbf{1}_{[0,1/2)}(x) - \mathbf{1}_{[1/2,1)}(x)$, $x \in \mathbb{R}$. Haar wavelets are for $\widehat{\alpha} > 0$ and $\sigma > 0$ given by

$$\psi_{\ell,k}(x) := \sigma 2^{-\widehat{\alpha}\ell} \psi(2^\ell x - k), \quad \ell \geq 0, k = 0, \ldots, 2^\ell - 1.$$

In our computations, we consider truncated fields Z^s with $s = 2^{L+1} - 1$, $L \geq 0$. In this way, the expansion of Z^s consists of full partial sums over activated levels. Realizations of the coefficient a^s are piecewise constant on D. For a constant right hand side $f \equiv$ constant, the solution u^s of (17) takes values in the piecewise quadratic functions on D. Hence, for such a^s, the corresponding FE solution of (17) also solves (17) if $\mathscr{P}^2$ Lagrange FE is applied. Therefore, *in this example we are able to study the QMC error in the absence of spatial discretization errors*.

This Haar system $(\psi_j)_{j\geq 1}$ and the sequence $(b_j)_{j\geq 1}$ given by

$$b_{j(\ell,k)} := c 2^{-\widehat{\beta}\ell}, \quad \ell \geq 0, k = 0, \ldots, 2^\ell - 1,$$

satisfy the Assumption (A1) for every $\widehat{\beta}$ such that $\widehat{\alpha} > \widehat{\beta} > 0$ and $c > 0$. The enumeration $j : \mathbb{N} \to \nabla$ is given by $j(\ell, k) = 2^\ell + k$, $\ell \geq 0$, $k = 0, \ldots, 2^\ell - 1$. Since $b_j \sim j^{-\widehat{\beta}}$, $j \geq 1$, $(b_j)_{j\geq 1} \in \ell^p(\mathbb{N})$ for every $p > 1/\widehat{\beta}$. For $p > 1/\widehat{\beta}$ and exponential weight functions, we will use the product weights $\boldsymbol{\gamma} = (\gamma_{\mathrm{u}})_{\mathrm{u}\subset\mathbb{N}}$ given by

$$\gamma_{\mathrm{u}} = \prod_{j\in\mathrm{u}} b_j^{2-p}, \quad \mathrm{u} \subset \mathbb{N}, |\mathrm{u}| < \infty.$$

For the computation of the QMC generating vectors, we use the Python code `QMC4PDE`, cf. [13], which is also able to compute generating vectors for product weights with exponential weight functions, where we take $c = 0.1$ as the scaling of the sequence $(b_j)_{j\geq 1}$. It has been observed in [8] (there for interlaced polynomial lattice rules) that using QMC weight sequences that are scaled by a constant smaller than one, may result in better suited generating vectors. In our experiments, for dimensions of the order $\mathscr{O}(10^3)$ the value $c = 0.1$ resulted in better suited generating vectors than $c = 1$. Any scaling of the sequence $(b_j)_{j\geq 1}$ is justified by our theory. However, a smaller value of c may lead to larger dimension-independent constants in the presented error estimates. We observe that the theoretical bounds for α_{exp} may be overly conservative and the resulting generating vectors may be ill-suited for practical QMC quadrature; we refer to the discussion in [8]. Therefore, smaller values of α_{exp} are considered, i.e., the code `QMC4PDE` uses parameter-dependent values α_j (in the notation of [13]) according to [13, p. 1672], where we have set $a_3 = 1$.

We present results for a right hand side $f \equiv 15$ and $G(\cdot)$ is the function evaluation at $\bar{x} = 0.7$, which is not a FE node for all discretization levels. Convergence of the QMC approximation using randomly shifted lattice rules with $N = 2^m$ points, $m = 1, \ldots, 18$, is presented in Fig. 1a, b. The results with $m = 19$ averaged over R_0 random shifts is used as the reference value $\bar{Q}$. The mean squared error over $R \geq 2$ random shifts is approximated by the unbiased estimator $\sum_{j=1}^{R}(Q_j - \bar{Q})^2/(R - 1) \approx \mathbb{E}^{\boldsymbol{\Delta}}(|\mathbb{E}(G(u^s)) - Q_{s,N}(G(u^s))|^2)$, where Q_j, $j = 1, \ldots, R$, are the results of

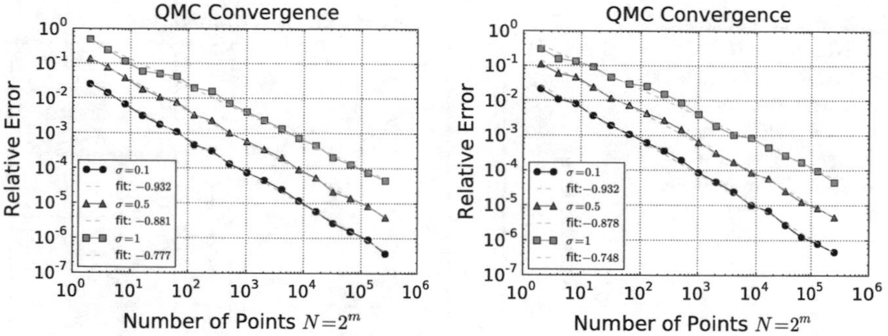

(a) Parameter choices $\widehat{\alpha} = 1.61, \widehat{\beta} = 1.51$ (b) Parameter choices $\widehat{\alpha} = 1.36, \widehat{\beta} = 1.26$

Fig. 1 Here, $s_L = 8191$, $R_0 = R = 20$. The convergence rates expected from our error analysis in these examples are $1 - \delta$ and $0.76 - \delta$, respectively

$Q_{s,N}(G(u^s))$ for R i.i.d. random shifts. For all data points, the truncations level $L = 12$ is used. This results in $s_L = 2^{13} - 1 = 8191$ dimensions of integration and FEM meshwidth $h = 2^{-13}$. In Fig. 1a and 1b, we observe that the convergence rate is depending on the variance of $\log(a) = Z$, which is equal to $\sigma^2/(1 - 2^{-2\widehat{\alpha}})$. Also the convergence rate is in both cases little different and not larger than 0.95. A dependence of the convergence rate on the variance has also been observed in numerical experiments with randomly shifted lattice rules using POD weights in [9, Tables 1 and 2].

5.2 *Multilevel QMC*

The multilevel QMC convergence analysis requires higher spatial regularity of the solution, which may not hold if the coefficient is expanded in the Haar system. We consider here continuous, piecewise linear spline wavelets $(\psi_j)_{j\geq 1}$, e.g. [12, Chap. 12], and assume that Z is expanded in this MRA. We suppose the decay for $\widehat{\alpha} > 1$

$$\|\psi_{j(\ell,k)}\|_{L^\infty(D)} = \sigma 2^{-\widehat{\alpha}\ell}, \quad \ell \geq 0, k = 1, \ldots, 2^\ell.$$

These $(\psi_j)_{j\geq 1}$ and the sequences

$$b_{j(\ell,k)} := c 2^{-\widehat{\beta}\ell}, \quad \ell \geq 0, k = 1, \ldots, 2^\ell, \quad \text{and} \quad \bar{b}_j := b_j^{(\widehat{\beta}-1)/\widehat{\beta}}, \quad j \geq 1,$$

satisfy the assumption in (A1) and in (A2), if $\widehat{\beta}$ is such that $\widehat{\alpha} > \widehat{\beta} > 1$. We present numerical experiments for a right hand side $f \equiv 15$ and $G(\cdot)$ is the function evaluation at the point $\bar{x} = 0.7$. Note that $G(\cdot) \in H^{-1/2+\varepsilon}$ for every $\varepsilon > 0$, which implies a FE convergence rate of $\tau = 3/2 - \varepsilon$ for every $\varepsilon > 0$. We will use the limiting value

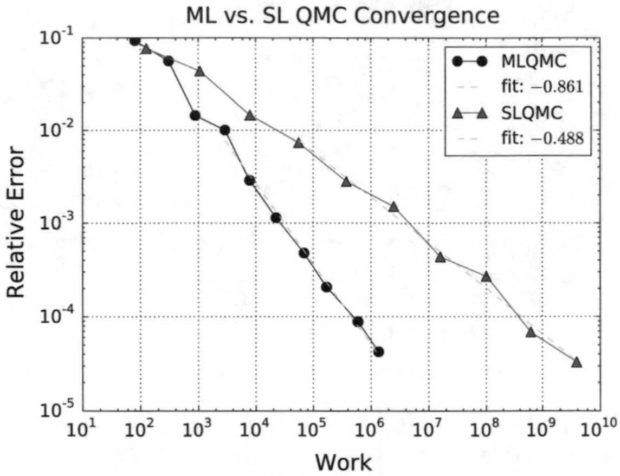

Fig. 2 Parameter choices $\widehat{\alpha} = 3.11$, $\widehat{\beta} = 3.01$, $\sigma = 0.1$, $R_0 = R = 20$. The convergence rate for multilevel QMC expected from our error vs. work analysis in this example is $0.9 - \delta$

$\tau = 3/2$ for the sample numbers $(N_\ell)_{\ell \geq 0}$. Let us assume that MRA and FE meshes are aligned. This refers to Strategy 2 in [11, Sect. 6] or [6, Sect. 6] and requires $\widehat{\beta} > \tau$. Hence, for $\theta = \tau/\widehat{\beta}$, the product weights are considered with respect to the sequence

$$(b_j^{1-\theta} \vee \bar{b}_j)_{j \geq 1} = (b_j^{1-\theta})_{j \geq 1}.$$

For simplicity, we will consider sample numbers $N_\ell = 2^{m_\ell}$, which upper bound the choices from (15) and (16), where

$$m_\ell = \max\left\{ \left\lceil \frac{\tau}{\bar{\chi}} L - \frac{1+2\tau}{1+2\bar{\chi}} \left(\ell + \log_2(\ell+1)\right) \right\rceil, 1 \right\}, \quad \ell = 0, \ldots, L.$$

Convergence of single-level and multilevel QMC is presented in Fig. 2 for $L = 2, \ldots, 11$. There multilevel and single-level QMC is applied to the same integration problem with respect to continuous, piecewise linear spline wavelets. Here, we use piecewise linear $\mathscr{P}^1$ FE. For the single-level QMC, the QMC sample numbers N_L are chosen to equilibrate the errors $N_L^{-\chi}$ and $h_L^{-\tau}$, cf. [10, Theorem 17], which leads to the choice $N_L = 2^{\lceil \tau L/\chi \rceil}$. The measured error vs. work convergence rates are displayed in Fig. 2 for comparison. As a reference solution, the approximation on the level $L = 12$ with a total of $s_L = 8191$ dimensions was used, respectively. For the single-level QMC, the same weight sequence may be applied. Since the generating vector is constructed by CBC iterating over the dimension, the generating vector for the highest dimension may be truncated and used for smaller dimensional randomly shifted lattice rules as well. The measured rates were obtained by a linear least squares fit on the last 7 data points. The total work (for one realization of the random shift per discretization level) is, for multilevel QMC, given by

$$W_L^{ML} = N_0 h_0^{-1} \log_2(s_0) + \sum_{\ell=1}^{L} N_\ell (h_\ell^{-1} \log_2(s_\ell) + h_{\ell-1}^{-1} \log_2(s_{\ell-1}))$$

and for single-level QMC by $W_L^{SL} = N_L h_L^{-1} \log_2(s_L)$. The convergence result in Theorem 3 is asymptotic and implies a convergence rate of $1 - \delta$ for multilevel QMC in Fig. 2. The error estimate in Theorem 2 and the chosen work model for multilevel QMC are used to monitor error vs. work in numerical experiments which are then fitted with least squares. For the range of L corresponding to the data points in Fig. 2, which are used in the computation of the measured convergence rate, this results in a "predicted" rate of $0.9 - \delta$ for arbitrary small $\delta > 0$. Predicted rates have been used in the literature e.g. [5, Table 1].

Acknowledgements This work was supported in part by the Swiss National Science Foundation (SNSF) under grant SNF 159940. The authors acknowledge the computational resources provided by the EULER cluster of ETH Zürich. The authors thank Robert N. Gantner for letting them use parts of his Python code.

References

1. Babuška, I., Kellogg, R.B., Pitkäranta, J.: Direct and inverse error estimates for finite elements with mesh refinements. Numer. Math. **33**(4), 447–471 (1979)
2. Bachmayr, M., Cohen, A., DeVore, R., Migliorati, G.: Sparse polynomial approximation of parametric elliptic PDEs. part II: lognormal coefficients. ESAIM. Math. Model. Numer. Anal. **51**(1), 341–363 (2017)
3. Bogachev, V.I.: Gaussian Measures, Mathematical Surveys and Monographs, vol. 62. American Mathematical Society, Providence (1998)
4. Dick, J., Kuo, F.Y., Sloan, I.H.: High-dimensional integration: the quasi-Monte Carlo way. Acta Numer. **22**, 133–288 (2013)
5. Gantner, R.N.: A generic c++ library for multilevel quasi-Monte Carlo. In: Proceedings of the Platform for Advanced Scientific Computing Conference, PASC '16, ACM, New York, NY, USA, pp. 11:1–11:12 (2016)
6. Gantner, R.N., Herrmann, L., Schwab, C.: Multilevel QMC with product weights for affine-parametric, elliptic PDEs. In: Dick, J., Kuo, F.Y., Woźniakowski, H. (eds.) Contemporary Computational Mathematics - a celebration of the 80th birthday of Ian Sloan. Springer, Cham (2018). https://doi.org/10.1007/978-3-319-72456-0_18
7. Gantner, R.N., Herrmann, L., Schwab, C.: Quasi-Monte Carlo integration for affine-parametric, elliptic PDEs: local supports and product weights. SIAM J. Numer. Anal. **56**(1), 111–135 (2018)
8. Gantner, R.N., Schwab, C.: Computational higher order quasi-Monte Carlo integration. Monte Carlo and Quasi-Monte Carlo Methods: MCQMC, vol. 163, pp. 271–288. Springer, Leuven, Belgium (2016)
9. Graham, I.G., Kuo, F.Y., Nichols, J.A., Scheichl, R., Schwab, C., Sloan, I.H.: Quasi-Monte Carlo finite element methods for elliptic PDEs with lognormal random coefficients. Numer. Math. **131**(2), 329–368 (2015)
10. Herrmann, L., Schwab, C.: QMC integration for lognormal-parametric, elliptic PDEs: local supports and product weights. Technical Report 2016-39, Seminar for Applied Mathematics, ETH Zürich, Switzerland (2016). https://www.sam.math.ethz.ch/sam_reports/reports_final/reports2016/2016-39_rev1.pdf

11. Herrmann, L., Schwab, C.: Multilevel quasi-Monte Carlo integration with product weights for elliptic PDEs with lognormal coefficients. Technical Report 2017-19, Seminar for Applied Mathematics, ETH Zürich, Switzerland (2017). https://www.sam.math.ethz.ch/sam_reports/reports_final/reports2017/2017-19.pdf
12. Hilber, N., Reichmann, O., Schwab, C., Winter, C.: Computational Methods for Quantitative Finance. Finite Element Methods for Derivative Pricing. Springer, Heidelberg (2013)
13. Kuo, F.Y., Nuyens, D.: Application of quasi-Monte Carlo Methods to elliptic PDEs with random diffusion coefficients: a survey of analysis and implementation. Found. Comput. Math. **16**(6), 1631–1696 (2016)
14. Kuo, F.Y., Scheichl, R., Schwab, Ch., Sloan, I.H., Ullmann, E.: Multilevel quasi-Monte Carlo methods for lognormal diffusion problems. Math. Comput. **86**(308), 2827–2860 (2017)
15. Kuo, F.Y., Schwab, Ch., Sloan, I.H.: Quasi-Monte Carlo methods for high-dimensional integration: the standard (weighted Hilbert space) setting and beyond. ANZIAM J. **53**(1), 1–37 (2011)
16. Kuo, F.Y., Schwab, Ch., Sloan, I.H.: Quasi-Monte Carlo finite element methods for a class of elliptic partial differential equations with random coefficients. SIAM J. Numer. Anal. **50**(6), 3351–3374 (2012)
17. Kuo, F.Y., Schwab, Ch., Sloan, I.H.: Multi-level quasi-Monte Carlo finite element methods for a class of elliptic PDEs with random coefficients. Found. Comput. Math. **15**(2), 411–449 (2015)
18. Kuo, F.Y., Sloan, I.H., Wasilkowski, G.W., Waterhouse, B.J.: Randomly shifted lattice rules with the optimal rate of convergence for unbounded integrands. J. Complex. **26**(2), 135–160 (2010)
19. Nichols, J.A., Kuo, F.Y.: Fast CBC construction of randomly shifted lattice rules achieving $\mathcal{O}(n^{-1+\delta})$ convergence for unbounded integrands over $\mathbb{R}^s$ in weighted spaces with POD weights. J. Complex. **30**(4), 444–468 (2014)
20. Nuyens, D., Cools, R.: Fast algorithms for component-by-component construction of rank-1 lattice rules in shift-invariant reproducing kernel Hilbert spaces. Math. Comput. **75**(254), 903–920 (2006). (electronic)
21. Nuyens, D., Cools, R.: Fast component-by-component construction of rank-1 lattice rules with a non-prime number of points. J. Complex. **22**(1), 4–28 (2006)
22. Opic, B., Kufner, A.: Hardy-Type Inequalities. Pitman Research Notes in Mathematics Series, vol. 219. Longman Scientific Technical, Harlow (1990)
23. Sloan, I.H., Woźniakowski, H.: When are quasi-Monte Carlo algorithms efficient for high-dimensional integrals? J. Complex. **14**(1), 1–33 (1998)
24. Stein, E.M.: Singular Integrals and Differentiability Properties of Functions. Princeton Mathematical Series. Princeton University Press, Princeton (1970). No. 30
25. Triebel, H.: Interpolation Theory, Function Spaces, Differential Operators, 2nd edn. Johann Ambrosius Barth, Heidelberg (1995)

QMC Designs and Determinantal Point Processes

Masatake Hirao

Abstract In this paper, we deal with two types of determinantal point processes (DPPs) for equal weight numerical integration (quasi-Monte Carlo) rules on the sphere, and discuss the behavior of the worst-case numerical integration error for functions from Sobolev space over the d-dimensional unit sphere $\mathbb{S}^d$. As by-products, we know the spherical ensemble, a well-studied DPP on $\mathbb{S}^2$, generates asymptotically on average QMC design sequences for Sobolev space over $\mathbb{S}^2$ with smoothness $1 < s < 2$. Moreover, compared to i.i.d. uniform random points, we also know harmonic ensembles on $\mathbb{S}^d$ for $d \geq 2$, which are DPPs defined by reproducing kernels for polynomial spaces over $\mathbb{S}^d$, generate on average faster convergent sequences of the square worst-case error for Sobolev space over $\mathbb{S}^d$ with smoothness $d/2 + 1/2 < s < d/2 + 1$.

Keywords QMC design · Determinantal point process · Spherical ensemble · Harmonic ensemble

1 Introduction

Let $\mathbb{R}^{d+1}$ be the $(d+1)$-dimensional Euclidean space with the usual inner product and norm; $\boldsymbol{x} \cdot \boldsymbol{y} = x_1 y_1 + \cdots + x_{d+1} y_{d+1}$ and $|\boldsymbol{x}| = \sqrt{\boldsymbol{x} \cdot \boldsymbol{x}}$ for $\boldsymbol{x} = (x_1, \ldots, x_{d+1})$, $\boldsymbol{y} = (y_1, \ldots, y_{d+1}) \in \mathbb{R}^{d+1}$, $d \geq 2$. Let $\mathbb{S}^d = \{\boldsymbol{x} \in \mathbb{R}^{d+1} \mid |\boldsymbol{x}| = 1\}$ be the d-dimensional unit sphere, and $X_N = \{\boldsymbol{x}_1, \ldots, \boldsymbol{x}_N\}$ be a finite subset of $\mathbb{S}^d$ with N points. We denote by $\mathscr{P}_t(\mathbb{S}^d)$ the vector space of all polynomials of degree at most t in $d+1$ variables restricted to $\mathbb{S}^d$.

We first give the definition of spherical design introduced in the fundamental paper Delsarte et al. [11]; see also, e.g., the recent survey by Bannai and Bannai [3].

M. Hirao (✉)
Department of Information and Science Technology, Aichi Prefectural University, 1522-3 Ibaragabasama, Nagakute, Aichi 480-1198, Japan
e-mail: hirao@ist.aichi-pu.ac.jp

A. B. Owen and P. W. Glynn (eds.), *Monte Carlo and Quasi-Monte Carlo Methods*, Springer Proceedings in Mathematics & Statistics 241,
https://doi.org/10.1007/978-3-319-91436-7_18

Definition 1 *(spherical design)* Let t be a natural number. Then X_N is a spherical t-design, if the following condition is satisfied:

$$\frac{1}{N}\sum_{i=1}^{N} f(\boldsymbol{x}_i) = \int_{\mathbb{S}^d} f(\boldsymbol{x})\, d\sigma_d(\boldsymbol{x})$$

for any polynomial $f \in \mathscr{P}_t(\mathbb{S}^d)$, where σ_d is the normalized surface measure on $\mathbb{S}^d$.

A fundamental question concerns the explicit construction of spherical designs. Seymour and Zaslavsky [18] show there always exists a spherical t-design X_N on $\mathbb{S}^d$ for sufficiently large points N. In fact, Bondarenko et al. [4] show the following conjecture of Korevaar and Meyers [14]: Given $d \geq 2$, there exists a spherical t-design on $\mathbb{S}^d$ with N points for every $N \geq c_d t^d$, where the constant c_d does only depend on d; see also [6] for the recent progress. However, their results give no information about how to explicitly construct such designs. Although there exist many articles on numerical search of spherical designs, e.g., Chen and Womersly [10], Chen et al. [9], it seems to be not easy to give constructions of such designs in general.

Brauchart et al. [7] discuss the asymptotic behavior of the bounds of the worst-case numerical integration error for functions from Sobolev space $\mathbb{H}^s(\mathbb{S}^d)$ with smoothness $s > d/2$, and introduce the new concept of QMC (quasi-Monte Carlo) design sequences for $\mathbb{H}^s(\mathbb{S}^d)$. (The definition of Sobolev space is introduced in Sect. 2.2.)

Definition 2 Let $s > d/2$. A sequence $\{X_N\}$ of N-point configurations on $\mathbb{S}^d$ with $N \to \infty$ is said to be a QMC design sequence for $\mathbb{H}^s(\mathbb{S}^d)$ if there exists $c(s,d) > 0$, independent of N, such that

$$\sup_{\substack{f \in \mathbb{H}^s(\mathbb{S}^d) \\ \|f\|_{\mathbb{H}^s} \leq 1}} \left| \frac{1}{N}\sum_{i=1}^{N} f(\boldsymbol{x}_i) - \int_{\mathbb{S}^d} f(\boldsymbol{x})\, d\sigma_d(\boldsymbol{x}) \right| \leq \frac{c(s,d)}{N^{s/d}}, \tag{1}$$

where $\|\cdot\|_{\mathbb{H}^s}$ is the $\mathbb{H}^s(\mathbb{S}^d)$-norm.

Brauchart et al. [7] show that i.i.d. random points on the sphere give a slower rate of convergence for the expected worst-case error as given in (1), and hence do not form QMC designs. However, if one compartmentalizes the random point selection process with respect to a partition of the sphere into N equal area regions with small diameter, then one does get an average worst-case error rate appropriate to QMC designs for $d/2 < s < d/2 + 1$ but not for $s > d/2 + 1$.

Thus in this paper, we also focus on the behavior of the worst-case numerical integration error for $\mathbb{H}^s(\mathbb{S}^d)$ with smoothness $d/2 < s < d/2 + 1$ by using determinantal point processes(DPPs). We show that two types of DPPs give a faster rate of convergence for the expected square worst-case error for Sobolev spaces over the sphere (Theorems 1 and 2). By using these results, we know the spherical ensemble, a well-studied DPP on $\mathbb{S}^2$, generates asymptotically on average QMC design sequences for Sobolev space $\mathbb{H}^s(\mathbb{S}^2)$ with $1 < s < 2$. Moreover, in comparison with

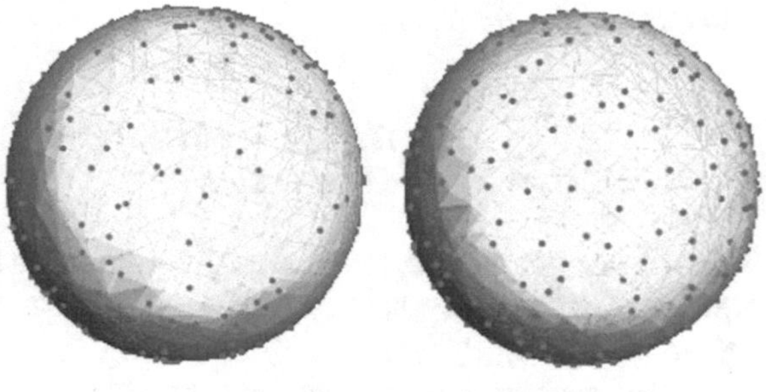

Fig. 1 Simulations of 300 random points on $\mathbb{S}^2$ (see Example 1)

i.i.d. uniform random points, we also know harmonic ensembles on $\mathbb{S}^d$ for $d \geq 2$, which are DPPs defined by reproducing kernels for polynomial spaces over $\mathbb{S}^d$, generate on average a faster rate of convergent sequence of the square worst-case error for $\mathbb{H}^s(\mathbb{S}^d)$ with $d/2 + 1/2 < s < d/2 + 1$.

Example 1 Here, we give two simulations of 300 random points on $\mathbb{S}^2$. The left display shows independently and uniformly distributed random points whereas the right display the points thrown on the sphere by a DPP. It is easy to see that the points of a determinantal point process are distributed more evenly (Fig. 1).

Outline of the paper.

The next section provides some preliminaries for QMC design sequences for Sobolev space $\mathbb{H}^s(\mathbb{S}^d)$ and two types of DPPs, i.e., spherical ensemble and harmonic ensembles on the sphere. Section 3 is the main section. We discuss the square worst-case error for Sobolev spaces over the sphere by using some sequences given by spherical and harmonic ensembles.

2 Preliminaries

2.1 Spherical Harmonics

Let $\mathrm{Harm}_\ell(\mathbb{S}^d)$ be the vector space of all the harmonic and homogeneous polynomials of exact degree ℓ in $d+1$ variables restricted to $\mathbb{S}^d$. Let $\{Y_{\ell,k} \mid k = 1, \ldots, Z(d,\ell)\}$ be a real orthonormal basis of $\mathrm{Harm}_\ell(\mathbb{S}^d)$, where

$$Z(d,\ell) = \dim(\mathrm{Harm}_\ell(\mathbb{S}^d)) = \binom{d+\ell}{\ell} - \binom{d+\ell-2}{\ell-2}.$$

The set $\{Y_{\ell,k} \mid k = 1, \ldots, Z(d,\ell), \ell = 0, 1, \ldots\}$ forms a complete orthonormal system for the Hilbert space $\mathbb{L}_2(\mathbb{S}^d)$ with the usual inner product and induced norm

$$\langle f, g\rangle_{\mathbb{L}_2(\mathbb{S}^d)} = \int_{\mathbb{S}^d} f(\boldsymbol{x})g(\boldsymbol{x})\, d\sigma_d(\boldsymbol{x}), \quad \|f\|_{\mathbb{L}_2(\mathbb{S}^d)} = \sqrt{\langle f, f\rangle_{\mathbb{L}_2(\mathbb{S}^d)}}.$$

The identity

$$\sum_{k=1}^{Z(d,\ell)} Y_{\ell,k}(\boldsymbol{x})Y_{\ell,k}(\boldsymbol{y}) = Z(d,\ell)P_\ell^{(d)}(\boldsymbol{x}\cdot\boldsymbol{y}), \quad \boldsymbol{x},\boldsymbol{y}\in\mathbb{S}^d,$$

is known as *addition theorem*, where $P_\ell^{(d)}$ is the normalized Gegenbauer polynomial, orthogonal on the interval $[-1,1]$ with respect to the weight function $(1-x^2)^{d/2-1}$, and normalized by $P_\ell^{(d)}(1)=1$.

For any positive integer L, we denote by

$$\mathscr{K}_L(\boldsymbol{x},\boldsymbol{y}) = R_L(\boldsymbol{x}\cdot\boldsymbol{y}), \quad R_L(x) = \sum_{\ell=0}^{L} Z(d,\ell)P_\ell^{(d)}(x), \tag{2}$$

the Lth reproducing kernel for the polynomial space $\mathscr{P}_L(\mathbb{S}^d)$. The polynomial R_L is a multiple of the Jacobi polynomial $P_L^{(d/2,d/2-1)}$ of degree L and parameters $d/2$ and $d/2-1$ with standard normalizations; see, e.g., [11, 12];

$$R_L(x) = \frac{2^L(d+2L-1)!!(d+L-1)!}{(d-1)!!(d+2L-1)!}P_L^{(d/2,d/2-1)}(x). \tag{3}$$

Thus, the polynomials $R_0, R_1, \ldots$ are also orthogonal on the interval $[-1,1]$ with respect to the weight function $(1-x)^{d/2}(1+x)^{d/2-1}$. Moreover, it holds that

$$R_L(1) = \sum_{\ell=0}^{L} Z(d,\ell) = \dim(\mathscr{P}_L(\mathbb{S}^d)) = \binom{d+L}{d} + \binom{d+L-1}{d}.$$

The following lemma is useful to show Theorem 2; see also, e.g., [19].

Lemma 1 ([12]) *Let L be a nonnegative integer. For any $\alpha,\beta\geq 0$, there exists a constant $C>0$ such that*

$$\left(\sin\frac{\theta}{2}\right)^{\alpha+1/2}\left(\cos\frac{\theta}{2}\right)^{\beta+1/2}|P_L^{(\alpha,\beta)}(\cos\theta)|$$
$$\leq \frac{C}{\sqrt{2}}(2L+\alpha+\beta+1)^{-1/4}\left(\frac{\Gamma(L+\alpha+1)\Gamma(L+\beta+1)}{\Gamma(L+1)\Gamma(L+\alpha+\beta+1)}\right)^{1/2}, \quad 0\leq\theta\leq\pi,$$

where Γ is the Gamma function.

2.2 Sobolev Space

The Sobolev space $\mathbb{H}^s(\mathbb{S}^d)$ with smoothness s is the vector space of all functions $f\in\mathbb{L}_2(\mathbb{S}^d)$, whose Laplace-Fourier coefficients

$$\widehat{f}_{\ell,k} = \langle f, Y_{\ell,k}\rangle_{\mathbb{L}_2(\mathbb{S}^d)} = \int_{\mathbb{S}^d} f(\boldsymbol{x}) Y_{\ell,k}(\boldsymbol{x})\, d\sigma_d(\boldsymbol{x})$$

satisfy

$$\sum_{\ell=0}^{\infty} \sum_{k=1}^{Z(d,\ell)} (1+\lambda_\ell)^s |\widehat{f}_{\ell,k}|^2 < \infty,$$

where $\lambda_\ell = \ell(\ell + d - 1)$. We note that $\mathbb{H}^0(\mathbb{S}^d) = \mathbb{L}_2(\mathbb{S}^d)$ and $\mathbb{H}^s(\mathbb{S}^d) \subset \mathbb{H}^{s'}(\mathbb{S}^d)$ for $s > s'$.

An inner product on $\mathbb{H}^s(\mathbb{S}^d)$ is defined by

$$\langle f, g\rangle_{\mathbb{H}^s} = \sum_{\ell=0}^{\infty} \sum_{k=1}^{Z(d,\ell)} \frac{1}{a_\ell^{(s)}} \widehat{f}_{\ell,k} \widehat{g}_{\ell,k},$$

and the corresponding norm in the Sobolev space is

$$\|f\|_{\mathbb{H}^s} = \left[\sum_{\ell=0}^{\infty} \sum_{k=1}^{Z(d,\ell)} \frac{1}{a_\ell^{(s)}} |\widehat{f}_{\ell,k}|^2 \right]^{1/2},$$

where $\{a_\ell^{(s)}\}_{\ell\geq 0}$ is a sequence of positive real numbers satisfying

$$a_\ell^{(s)} \asymp (1+\lambda_\ell)^{-s} \asymp (1+\ell)^{-2s}.$$

Here we write $a_n \asymp b_n$ to mean that there exist positive constants c_1 and c_2 independent of n such that $c_1 a_n \leq b_n \leq c_2 a_n$ for all n.

2.3 Worst-Case Errors and Reproducing Kernels in $\mathbb{H}^s(\mathbb{S}^d)$

For an integer $L > 0$, we let

$$V_{d-2s}(\mathbb{S}^d) = \int_{\mathbb{S}^d} \int_{\mathbb{S}^d} |\boldsymbol{x} - \boldsymbol{y}|^{2s-d}\, d\sigma_d(\boldsymbol{x}) d\sigma_d(\boldsymbol{y}) = 2^{2s-1} \frac{\Gamma((d+1)/2)\Gamma(s)}{\sqrt{\pi}\,\Gamma(d/2+s)},$$

$$\alpha_\ell^{(s)} = V_{d-2s}(\mathbb{S}^d) \frac{(-1)^{L+1}(d/2-s)_\ell}{(d/2+s)_\ell}, \quad \ell \geq 1,$$

where $(a)_\ell$ is the Pochhammer symbol defined by

$$(a)_0 = 1, \quad (a)_{\ell+1} = (a)_\ell(\ell + a) = \frac{\Gamma(\ell + a)}{\Gamma(a)}, \quad \ell = 0, 1, \ldots.$$

It is mentioned in [7] that $\alpha_\ell^{(s)}$ satisfies

$$\alpha_\ell^{(s)} \sim 2^{2s-1} \frac{\Gamma((d+1)/2)\Gamma(s)}{\sqrt{\pi}[(-1)^{L+1}\Gamma(d/2-s)]} \ell^{-2s} \quad \text{as } \ell \to \infty.$$

Here, the notation $a_n \sim b_n$ means that $\lim_{n\to\infty} a_n/b_n = 1$. Thus, when choosing the above sequence $\{\alpha_\ell^{(s)}\}_{\ell\geq 0}$, Sobolev space $\mathbb{H}^s(\mathbb{S}^d)$ is a reproducing kernel Hilbert space. In this case, Brauchart and Womersley [8] introduce the "generalized distance" kernel as follows (see also [7]): For $d/2 < s < d/2+1$,

$$\mathscr{K}_{\mathrm{gd}}^{(s)}(\boldsymbol{x}, \boldsymbol{y}) = 2V_{d-2s}(\mathbb{S}^2) - |\boldsymbol{x} - \boldsymbol{y}|^{2s-d}, \quad \boldsymbol{x}, \boldsymbol{y} \in \mathbb{S}^d. \tag{4}$$

Now, we define the *worst-case error in* $\mathbb{H}^s(\mathbb{S}^d)$ for a cubature sum $Q[X_N](f) = N^{-1}\sum_{x\in X_N} f(\boldsymbol{x})$ for an integral $I(f) = \int_{\mathbb{S}^d} f(\boldsymbol{x})\, d\sigma_d(\boldsymbol{x})$ as follows:

$$\mathrm{wce}(Q[X_N]; \mathbb{H}^s(\mathbb{S}^d)) = \sup_{\substack{f\in\mathbb{H}^s(\mathbb{S}^d)\\ \|f\|_{\mathbb{H}}\leq 1}} |Q[X_N](f) - I(f)|.$$

It satisfies a *Koksma–Hlawka type inequality*:

$$|Q[X_N](f) - I(f)| \leq \mathrm{wce}(Q[X_N]; \mathbb{H}^s(\mathbb{S}^d)) \|f\|_{\mathbb{H}^s}.$$

By using the kernel (4), we have the following worst-case errors with X_N as follows (see also [7]): For $d/2 < s < d/2+1$,

$$\mathrm{wce}(Q[X_N]; \mathbb{H}^s(\mathbb{S}^d)) = \left(V_{d-2s}(\mathbb{S}^d) - \frac{1}{N^2}\sum_{i\neq j} |\boldsymbol{x}_j - \boldsymbol{x}_i|^{2s-d} \right)^{1/2}. \tag{5}$$

Remark 1 For $s \geq d/2+1$, one can also choose $\{a_\ell^{(s)}\}_{\ell\geq 0}$ in such a way to have a very simple yet technically more involved closed formula for reproducing kernel and this general case will be left for future work.

2.4 Determinantal Point Processes on the Sphere

In this section we give a brief overview of determinantal point processes. Readers are referred to, e.g., Hough et al. [13], Krishnapur [15], Lavancier et al. [16] for further details.

Let S be a locally compact Hausdorff space with a countable basis and μ be a Radon measure on S. We mainly consider the cases $S = \mathbb{C}$ and $S = \mathbb{S}^d$ in this paper. We denote by $\mathscr{X}$ a random point process on S. $\mathscr{X}$ is said to be *simple* if there are no coincidence points almost surely. For any simple point process $\mathscr{X}$, $\mathscr{X}$ can be identified with a random discrete subset of S and $\mathscr{X}(D)$ is to specify the random variable of the number of points fallen in D; see, e.g., Møller and Waagepetersen [17] for measure theoretical details.

Let $\mathscr{K}(\boldsymbol{x}, \boldsymbol{y}) : S^2 \to \mathbb{C}$ be a measurable function.

Definition 3 A point process $\mathscr{X}$ on S is said to be a determinantal point process with kernel $\mathscr{K}$ if it is simple and its k-point correlation functions $\rho_k : S^k \to \mathbb{R}_{\geq 0}$ with respect to the measure μ satisfy

$$\rho_k(\boldsymbol{x}_1, \ldots, \boldsymbol{x}_k) = \det(\mathscr{K}(\boldsymbol{x}_i, \boldsymbol{x}_j))_{1 \leq i,j \leq k},$$

for every $k \geq 1$, that is, for any Borel function $h : S^k \to [0, \infty)$, we have

$$\mathbb{E}[\sum_{\boldsymbol{x}_1,\ldots,\boldsymbol{x}_k \in \mathscr{X}}^{\neq} h(\boldsymbol{x}_1, \ldots, \boldsymbol{x}_k)] = \int_S \cdots \int_S \rho_k(\boldsymbol{x}_1, \ldots, \boldsymbol{x}_k) h(\boldsymbol{x}_1, \ldots, \boldsymbol{x}_k) \, d\mu(\boldsymbol{x}_1) \cdots d\mu(\boldsymbol{x}_k).$$

Here $\sum_{\boldsymbol{x}_1,\ldots,\boldsymbol{x}_k \in \mathscr{X}}^{\neq}$ means a multi-sum over the n-tuples of $\mathscr{X}$ whose components are all pairwise distinct.

In the next subsections, we prepare two typical types of determinantal point processes.

2.4.1 Spherical and Harmonic Ensembles

We first consider the *spherical ensemble*. There are several studies on the spherical ensemble; see, e.g., [1, 13, 15]. Let A_N and B_N be independent $N \times N$ random matrices with independent and identically distributed standard complex Gaussian entries. Then, the set of eigenvalues $\{\lambda_1, \lambda_2, \ldots, \lambda_N\}$ of $A_N^{-1} B_N$ form a DPP on the complex plane with kernel

$$\mathscr{K}(z, w) = (1 + z\bar{w})^{N-1}$$

with respect to the measure $\frac{N}{\pi(1+|z|^2)^{N+1}} \, dm(z)$ and m denotes the Lebesgue measure on the complex plane $\mathbb{C}$. We note that $A_N^{-1} B_N$ has N eigenvalues almost surely; see also Krishnapur [15].

Moreover, let g be the stereographic projection of the sphere $\mathbb{S}^2$ from the North Pole onto the plane $\{(t_1, t_2, 0) \mid t_1, t_2 \in \mathbb{R}\}$. Then, the set $\{\boldsymbol{x}_i = g^{-1}(\lambda_i) \mid 1 \leq i \leq N\}$ form a DPP on the sphere $\mathbb{S}^2$. We call such a DPP a *spherical ensemble on* $\mathbb{S}^2$. For example, we know that 2-point correlation function on $\mathbb{S}^2$ is given as the follows:

$$\rho_2(\boldsymbol{x}, \boldsymbol{y}) = N^2 \left\{ 1 - \left(\frac{|\boldsymbol{x} - \boldsymbol{y}|^2}{4} \right)^{N-1} \right\}, \quad \boldsymbol{x}, \boldsymbol{y} \in \mathbb{S}^2;$$

see also Alishahi and Zamani [1] for further details.

Next we consider another types of DPPs associated by reproducing kernels for polynomial spaces over $\mathbb{S}^d$. We use the following two lemmas for the existence of this type of DPP.

Lemma 2 ([13]) *Suppose $\{\phi_k\}_{k=1}^n$ is an orthonormal set in $\mathbb{L}^2(S)$. Then there exists a determinantal process with kernel $\mathscr{K}(\boldsymbol{x}, \boldsymbol{y}) = \sum_{k=1}^n \phi_k(\boldsymbol{x})\overline{\phi_k(\boldsymbol{y})}$.*

Lemma 3 ([13]) *Suppose $\mathscr{X}$ is a determinantal point process on S, with kernel $\mathscr{K}(\boldsymbol{x}, \boldsymbol{y}) = \sum_{k=1}^n \phi_k(\boldsymbol{x})\overline{\phi_k(\boldsymbol{y})}$, where $\{\phi_k\}_{1\le k\le n}$ is a finite orthonormal set in $\mathbb{L}^2(S)$. Then the number of points in $\mathscr{X}$ is equal to n, almost surely.*

By using Lemmas 2 and 3, for the Lth reproducing kernel $\mathscr{K}_L(\boldsymbol{x}, \boldsymbol{y})$ for $\mathscr{P}_L(\mathbb{S}^d)$, there exist DPPs $\mathscr{X}_N$ on $\mathbb{S}^d$, associated with $\mathscr{K}_L(\boldsymbol{x}, \boldsymbol{y})$, with $N = \dim(\mathscr{P}_L(\mathbb{S}^d)) \asymp L^d$ points, almost surely. Such DPPs are known as *harmonic ensembles on* $\mathbb{S}^d$ by Beltrán et al. [5] and have been studied on, e.g., the relationships of Riesz energies.

In this case, by (2) we can calculate the 2-point correlation function as follows:

$$\rho_2(\boldsymbol{x}, \boldsymbol{y}) = \det \begin{bmatrix} R_L(1) & R_L(\boldsymbol{x}\cdot\boldsymbol{y}) \\ R_L(\boldsymbol{x}\cdot\boldsymbol{y}) & R_L(1) \end{bmatrix} = R_L(1)^2 - R_L(\boldsymbol{x}\cdot\boldsymbol{y})^2. \qquad (6)$$

We deal with such DPPs in the next section.

3 DPPs and Worst-Case Errors

In this section we discuss the rates of convergence for the expected square worst-case error for Sobolev spaces over the sphere by using two types of DPPs, spherical and harmonic ensembles. As I mentioned before, Brauchart et al. [7] show that i.i.d. uniform random points $\mathscr{X}_N$ on the sphere give the following rate of convergence for the expected square worst-case error: Given $s > d/2$,

$$\mathbb{E}\left[\{\mathrm{wce}(Q[\mathscr{X}_N]; \mathbb{H}^s(\mathbb{S}^d))\}^2\right] = \frac{c'(s, d)}{N},$$

for some explicit constant $c'(s, d) > 0$. Compared to the above i.i.d. case, we show that sequences generated by the two types of DPPs give a faster rate of convergence for the expected square worst-case errors.

3.1 Spherical Ensemble on $\mathbb{S}^2$

Theorem 1 *Let $N \ge 2$ be an integer. Let $\mathscr{X}_N$ be an N-point spherical ensemble on $\mathbb{S}^2$. For $1 < s < 2$, we have*

$$\mathbb{E}\left[\{\text{wce}(Q[\mathscr{X}_N]; \mathbb{H}^s(\mathbb{S}^2))\}^2\right] = 2^{2s-2}B(s, N),$$

where B is the Beta function.

Remark 2 For a given s, by using Stirling's formula, we have $B(s, N) \sim \Gamma(s)N^{-s}$ as $N \to \infty$. Thus, for sufficient large N, we obtain an average worst case error (1) appropriate to QMC design for any $d/2 < s < d/2 + 1$.

In order to prove this theorem, we use the following lemma. For any $\gamma \in \mathbb{R}$, the Riesz γ-energy of n-points $\boldsymbol{x}_1, \boldsymbol{x}_2, \ldots, \boldsymbol{x}_n$ on $\mathbb{S}^2$ is defined by

$$E_\gamma(\boldsymbol{x}_1, \ldots, \boldsymbol{x}_n) = \sum_{i \neq j} |\boldsymbol{x}_i - \boldsymbol{x}_j|^{-\gamma}. \tag{7}$$

Lemma 4 ([1]) *Let $\boldsymbol{x}_1, \ldots, \boldsymbol{x}_N$ form a spherical ensemble on $\mathbb{S}^2$. For any $\gamma \in (-2, 0)$, we have*

$$\mathbb{E}E_\gamma(\boldsymbol{x}_1, \ldots, \boldsymbol{x}_N) = \frac{2^{1-\gamma}}{2-\gamma}N^2 - \frac{2^{-\gamma}\Gamma(N)\Gamma(1-\gamma/2)}{\Gamma(N+1-\gamma/2)}N^2. \tag{8}$$

Proof of Theorem 1. By combining (5), (7) and (8), we obtain

$$\mathbb{E}[E_{2-2s}(\boldsymbol{x}_1, \ldots, \boldsymbol{x}_N)] = \mathbb{E}\left[\sum_{i \neq j} |\boldsymbol{x}_j - \boldsymbol{x}_i|^{2s-2}\right] = \frac{2^{2s-2}}{s}N^2 - 2^{2s-2}B(s, N)N^2.$$

Thus we have

$$\begin{aligned}
\mathbb{E}[\{\text{wce}(Q[\mathscr{X}_N]; \mathbb{H}^s(\mathbb{S}^2))\}^2] &= \frac{2^{2s-2}}{s} - \frac{1}{N^2}\mathbb{E}[\sum_{i \neq j} |\boldsymbol{x}_j - \boldsymbol{x}_i|^{2s-2}] \\
&= \frac{2^{2s-2}}{s} - \frac{2^{2s-2}}{s} + 2^{2s-2}B(s, N) \\
&= 2^{2s-2}B(s, N).
\end{aligned}$$

□

3.2 Harmonic Ensembles on $\mathbb{S}^d$

Theorem 2 *Let L be a positive integer. Let $\mathscr{X}_N$ be a harmonic ensemble on $\mathbb{S}^d$, associated with the Lth reproducing kernel $\mathscr{K}_L(\boldsymbol{x}, \boldsymbol{y})$, with $N = \dim(\mathscr{P}_L(\mathbb{S}^d))$ points. For $d/2 + 1/2 < s < d/2 + 1$, there exists $C(s, d) > 0$, independent of N, such that*

$$\mathbb{E}[\{\text{wce}(Q[\mathscr{X}_N]; \mathbb{H}^s(\mathbb{S}^d))\}^2] \leq \frac{C(s, d)}{N^{1+1/(2d)}}.$$

Proof By recalling (5) and (6), for $d/2 + 1/2 < s < d/2 + 1$, we have

$$\begin{aligned}
&\mathbb{E}[\{\mathrm{wce}(Q[\mathscr{X}_N]; \mathbb{H}^s(\mathbb{S}^d))\}^2] \\
&= V_{d-2s}(\mathbb{S}^d) - \frac{1}{N^2}\mathbb{E}[\sum_{i\neq j} |\boldsymbol{x}_j - \boldsymbol{x}_i|^{2s-d}] \\
&= V_{d-2s}(\mathbb{S}^d) - \frac{1}{N^2}\int_{\mathbb{S}^d\times\mathbb{S}^d} |\boldsymbol{x} - \boldsymbol{y}|^{2s-d}\rho_2(\boldsymbol{x}, \boldsymbol{y})\, d\sigma_d(\boldsymbol{x})d\sigma_d(\boldsymbol{y}) \\
&= V_{d-2s}(\mathbb{S}^d) - \int_{\mathbb{S}^d\times\mathbb{S}^d} |\boldsymbol{x} - \boldsymbol{y}|^{2s-d}\, d\sigma_d(\boldsymbol{x})d\sigma_d(\boldsymbol{y}) \\
&\quad + \frac{1}{N^2}\int_{\mathbb{S}^d\times\mathbb{S}^d} R_L(\boldsymbol{x}\cdot\boldsymbol{y})^2|\boldsymbol{x} - \boldsymbol{y}|^{2s-d}\, d\sigma_d(\boldsymbol{x})d\sigma_d(\boldsymbol{y}) \\
&= \frac{1}{N^2}\int_{\mathbb{S}^d\times\mathbb{S}^d} R_L(\boldsymbol{x}\cdot\boldsymbol{y})^2\,(2 - 2\boldsymbol{x}\cdot\boldsymbol{y})^{s-d/2}\, d\sigma_d(\boldsymbol{x})d\sigma_d(\boldsymbol{y}),
\end{aligned}$$

where the last equality follows from $|\boldsymbol{x} - \boldsymbol{y}|^2 = 2 - 2\boldsymbol{x}\cdot\boldsymbol{y}$ for all $\boldsymbol{x}, \boldsymbol{y} \in \mathbb{S}^d$. Letting $\boldsymbol{a} = (0, \ldots, 0, 1) \in \mathbb{S}^d$, by using Funk–Hecke formula, we have

$$\begin{aligned}
&\mathbb{E}[\{\mathrm{wce}(Q[\mathscr{X}_N]; \mathbb{H}^s(\mathbb{S}^d))\}^2] \\
&= \frac{1}{N^2}\int_{\mathbb{S}^d} R_L(\boldsymbol{x}\cdot\boldsymbol{a})^2\,(2 - 2\boldsymbol{x}\cdot\boldsymbol{a})^{s-d/2}\, d\sigma_d(\boldsymbol{x}) \\
&= \frac{|\mathbb{S}^{d-1}|}{|\mathbb{S}^d|N^2}\int_{-1}^{1} R_L(x)^2\,(2 - 2x)^{s-d/2}\,(1 - x^2)^{d/2-1}\, dx \\
&= \frac{|\mathbb{S}^{d-1}|2^{s-d/2}}{|\mathbb{S}^d|N^2}\int_{-1}^{1} R_L(x)^2(1 - x)^{s-1}(1 + x)^{d/2-1}\, dx.
\end{aligned}$$

Here by using Lemma 1 and (3), we have

$$\begin{aligned}
&\int_{-1}^{1} R_L(x)^2(1 - x)^{s-1}(1 + x)^{d/2-1}\, dx \\
&= 2^{s+d/2-1}\int_0^{\pi} R_L(\cos\theta)^2(\sin\frac{\theta}{2})^{2s-1}(\cos\frac{\theta}{2})^{d-1}\, d\theta \\
&= 2^{s+d/2-1+2L}\left\{\frac{(d + 2L - 1)!!}{(d - 1)!!}\frac{(d + L - 1)!}{(d + 2L - 1)!}\right\}^2 \\
&\quad \times \int_0^{\pi}\left\{P_L^{(d/2,d/2-1)}(\cos\theta)^2(\sin\frac{\theta}{2})^{d+1}(\cos\frac{\theta}{2})^{d-1}\right\}(\sin\frac{\theta}{2})^{2s-d-2}\, d\theta \\
&\leq C^2 2^{s+d/2-1+2L}\left\{\frac{(d + 2L - 1)!!}{(d - 1)!!}\frac{(d + L - 1)!}{(d + 2L - 1)!}\right\}^2 \\
&\quad \times \int_0^{\pi}\left\{\frac{1}{2\sqrt{2L + d}}\frac{\Gamma(L + d/2 + 1)\Gamma(L + d/2)}{L!(L + d - 1)!}\right\}(\sin\frac{\theta}{2})^{2s-d-2}\, d\theta
\end{aligned}$$

$$= \frac{C^2 2^{s+d/2+2L-2}}{\sqrt{2L+d}} \left\{ \frac{(d+2L-1)!!}{(d-1)!!} \frac{(d+L-1)!}{(d+2L-1)!} \right\}^2 \frac{\Gamma(L+d/2+1)\Gamma(L+d/2)}{L!(L+d-1)!}$$
$$\times \int_0^\pi (\sin \frac{\theta}{2})^{2s-d-2}\, d\theta.$$

Thus we have

$$\mathbb{E}[\{\mathrm{wce}(Q[\mathscr{X}_N]; \mathbb{H}^s(\mathbb{S}^d))\}^2]$$
$$\leq \frac{C^2 2^{2s-2} \int_0^\pi (\sin \frac{\theta}{2})^{2s-d-2}\, d\theta}{N^2 (d-1)!! \int_0^\pi \sin^{d-1}\theta\, d\theta}$$
$$\times \frac{2^{2L}}{\sqrt{2L+d}} \left\{ \frac{(d+2L-1)!!(d+L-1)!}{(d+2L-1)!} \right\}^2 \frac{\Gamma(L+d/2+1)\Gamma(L+d/2)}{L!(L+d-1)!}.$$

Now, in order to estimate the above, we divide into two parts (i) $d = 2m,\ m \geq 1$ case, and (ii) $d = 2m+1,\ m \geq 1$ case.
(i) By letting $d = 2m$, we have

$$\frac{2^{2L}}{\sqrt{2L+d}} \left\{ \frac{(d+2L-1)!!(d+L-1)!}{(d+2L-1)!} \right\}^2 \frac{\Gamma(L+d/2+1)\Gamma(L+d/2)}{L!(L+d-1)!}$$
$$= \frac{(2m+L-1)!(L+m)}{2^{2m-2} L! \sqrt{2L+2m}} \asymp L^{d-1/2} \asymp N^{(d-1/2)/d}.$$

(ii) By letting $d = 2m+1$, we have

$$\frac{2^{2L}}{\sqrt{2L+d}} \left\{ \frac{(d+2L-1)!!(d+L-1)!}{(d+2L-1)!} \right\}^2 \frac{\Gamma(L+d/2+1)\Gamma(L+d/2)}{L!(L+d-1)!}$$
$$= \frac{\pi(2m+L)!(2L+2m+1)}{2^{2m+1}\sqrt{2L+2m+1L!}} \asymp L^{d-1/2} \asymp N^{(d-1/2)/d}.$$

Therefore, by combining the above two cases and some calculations, we obtain

$$\mathbb{E}[\{\mathrm{wce}(Q[\mathscr{X}_N]; \mathbb{H}^s(\mathbb{S}^d))\}^2] \leq \frac{C(s,d)}{N^{1+1/(2d)}}.$$

□

Remark 3 Since $\int_0^\pi (\sin \frac{\theta}{2})^{2s-d-2}\, d\theta$ does not converge to a finite value for $d/2 < s \leq d/2 + 1/2$, we restricted to consider $d/2 + 1/2 < s < d/2 + 1$ in the above theorem.

Remark 4 Recalling the definition of harmonic ensembles, we know that there exist DPPs on other manifolds, e.g., the unit cube, the unit ball and so on. In fact, there exist a few articles on numerical computations by using DPPs. For example, Bardenet and

Hardy [2] discuss Monte Carlo methods by using DPPs on the cube. Thus we believe that our approach will be generalized so as to deal with other function spaces.

On the other hand, there exist many types of DPPs even when we only consider the sphere case. Moreover, the present paper only discuss an equal weight numerical integration rule on the sphere. Once again focusing on Bardenet and Hardy [2], they deal with DPPs defined by a specific measure on the cube, and give very efficient results by using a stochastic analog of Gaussian quadrature or a similar method of importance sampling. Can we find the best DPP for numerical integration the sphere and the best way to use DPPs effectively? This problem is left for future works.

Acknowledgements The author would like to thank Professor I.H. Sloan, Professor R.S. Womersley, Professor Hirofumi Osada and Professor Tomoyuki Shirai for fruitful discussions and valuable comments. The author is partially supported by Grant-in-Aid for Young Scientists (B) 16K17645 by the Japan Society for the Promotion of Science.

References

1. Alishahi, K., Zamani, M.S.: The spherical ensemble and uniform distribution of points on the sphere. Electron. J. Probab. **20**(23), 1–27 (2015)
2. Bardenet, R., Hardy, A.: Monte Carlo with Determinantal Point Processes. arXiv:1605.00361
3. Bannai, Ei: Bannai, Etsu: A survey on spherical designs and algebraic combinatorics on spheres. Eur. J. Comb. **30**(6), 1392–1425 (2009)
4. Bondarenko, A., Radchenko, D., Viazovska, M.: Optimal asymptotic bounds for spherical designs. Ann. Math. **178**(2), 443–452 (2013)
5. Beltrán, C., Marzo, J., Ortega-Cedrà, J.: Energy and discrepancy of rotationally invariant determinantal point processes in high dimensional spheres. J. Complex. **37**, 76–109 (2016)
6. Bondarenko, A., Radchenko, D., Viazovska, M.: Well-separated spherical designs. Constr. Approx. **41**(1), 93–112 (2015)
7. Brauchart, J.S., Saff, E.B., Sloan, I.H., Womersley, R.S.: QMC designs: optimal order quasi-Monte Carlo integration schemes on the sphere. Math. Comput. **83**(290), 2821–2851 (2014)
8. Brauchart, J.S., Womersley, R.S.: Numerical integration over the unit sphere, $\mathbb{L}_2$-discrepancy and sum of distances. In preparation
9. Chen, X., Frommer, A., Lang, B.: Computational existence proofs for spherical t-designs. Numer. Math. **117**(2), 289–305 (2011)
10. Chen, X., Womersley, R.S.: Existence of solutions to systems of underdetermined equations and spherical designs. SIAM J. Numer. Anal. **44**(6), 2326–2341 (2006)
11. Delsarte, P., Goethals, J.M., Seidel, J.J.: Spherical codes and designs. Geom. Dedicata **6**(3), 363–388 (1977)
12. Haagerup, U., Schlichtkrull, H.: Inequalities for Jacobi polynomials. Ramanujan J. **33**(2), 227–246 (2014)
13. Hough, J.B., Krishnapur, M., Peres, Y., Virág, B.: Zeros of Gaussian Analytic Functions and Determinantal Point Processes. American Mathematical Society, Providence (2009)
14. Korevaar, J., Meyers, J.L.H.: Spherical Faraday cage for the case of equal point charges and Chebyshev-type quadrature on the sphere. Integr. Transforms Spec. Funct. **1**(2), 105–117 (1993)
15. Krishnapur, M.: Zeros of random analytic functions. Ph.D. thesis. University of California, Berkeley (2006)
16. Lavancier, K., Møller, J., Rubak, E.: Determinantal point process models and statistical inference. J. R. Stat. Soc. Ser. B **77**(4), 853–877 (2015)

17. Møller, J.: Statistical Inference and Simulation for Spatial Point Processes. Chapman and Hall/CRC, Boca Raton (2004)
18. Seymour, P.D., Zaslavsky, T.: Averaging sets: a generalization of mean values and spherical designs. Adv. Math. **52**(3), 213–240 (1984)
19. Szogő, G.: Orthogonal polynomials. In: Colloquium publications/American mathematical society, vol. 23, Providence (1975)

Efficient Monte Carlo for Diffusion Processes Using Ornstein–Uhlenbeck Bridges

Adam W. Kolkiewicz

Abstract We present an extension of a recently proposed method of sampling Brownian paths conditionally on integrated bridges. By combining the Brownian bridge construction with conditioning on integrals, the method turns out to be a very effective way of capturing important dimensions in problems that involve integral functionals of Brownian motions. In this paper we show that by conditioning on integrated Ornstein–Uhlenbeck bridges, combined with a proper change of measure, we can eliminate variability due to integrals of the squared process. This result forms the theoretical basis for an efficient Monte Carlo method applicable to problems involving exponential functions of integrated diffusion processes. We illustrate the method by applying it to the problem of bond pricing under the exponential Vasicek model.

Keywords Sampling Brownian motion · Dimension reduction · Bond pricing

1 Introduction and Motivation

In this paper we propose a new method of sampling an underlying process with the objective of constructing efficient Monte Carlo simulation methods for estimating expectations of the form

$$\mathrm{E}[e^{\int_0^T g(t,U(t))dt}], \tag{1}$$

where g is a given function and $\{U(t), t \geq 0\}$ is either an Ornstein–Uhlenbeck (OU) process or an Ornstein–Uhlenbeck bridge. The former, which we shall denote by $\{X(t), t \geq 0\}$, solves the following stochastic differential equation

$$dX(t) = \theta(\mu - X(t))dt + \sigma dW(t), \quad X(0) = x, \tag{2}$$

A. W. Kolkiewicz (✉)
Department of Statistics and Actuarial Science, University of Waterloo,
Waterloo, ON, Canada
e-mail: wakolkie@uwaterloo.ca

A. B. Owen and P. W. Glynn (eds.), *Monte Carlo and Quasi-Monte Carlo Methods*, Springer Proceedings in Mathematics & Statistics 241,
https://doi.org/10.1007/978-3-319-91436-7_19

where $\sigma > 0$ and $\{W(t); t \in [0, T]\}$ is a standard Brownian motion. In many cases of interest, the process $\{U(t), t \geq 0\}$ in (1) will be just Brownian motion, but the main motivation for including OU-processes is the fact that they arise naturally in the context of the method proposed in this paper.

Recently [7] has introduced an efficient method of sampling Brownian paths for integrands that involve an integrated Brownian motion. The method, to which we shall refer as BBI and explain in greater detail in Sect. 2.1, combines the Brownian bridge construction with sampling bridges conditionally on their integrals. In a nutshell, it takes advantage of the observation that if g is a smooth function, then for small values of T most of the variability of

$$G\left(\int_0^T g(t, W_{x,y}^T(t))dt\right), \tag{3}$$

where G is a given function and $W_{x,y}^T$ is a Brownian bridge from x to y, is due to the integral $\int_0^T W_{x,y}^T(t)dt$. Therefore, efficient simulation methods can be constructed by applying, for example, a low-discrepancy sequence to integrate with respect to the distribution of $\int_0^T W_{x,y}^T(t)dt$, and then sampling paths of the process conditionally on a value of this integral. The method that we are proposing in this paper enhances efficiency of the BBI method by combining it with importance sampling, where instead of sampling Brownian bridges we rather sample OU-bridges. When G is an exponential function, this technique eliminates variability due to integrals of the squared process.

One area of applications where expectations of the form (1) arise frequently is modern finance, where popular models for short interest rates solve stochastic differential equations of the form

$$dr(t) = \mu(r(t))dt + \sigma(r(t))dW(t), \quad t \geq 0, \tag{4}$$

with $r(0)$ equal to the current interest rate r_0. The well-know examples include the Vasicek model, where $\mu(x) = k[\theta - x]$ and the diffusion term is constant, and the Cox-Ingersol-Ross model, which has the same drift term but $\sigma(x) = \sigma\sqrt{x}$. In the context of fixed income markets, the basic problem is the one of pricing bonds, which in the case of default-free bonds amounts to finding

$$P(t, T) := \mathrm{E}\left[e^{-\int_t^T r(s)ds}|\mathscr{F}_t\right], \quad t < T, \tag{5}$$

where $\{\mathscr{F}_t\}$ is the filtration generated by $\{r(t)\}$.

The two particular models mentioned above allow for analytical bond prices, but most others do not. The latter include the exponential Vasicek (EV) model, which assumes that the logarithm of the short interest rate follows an OU-process. For this model, the dynamic of the short rate $\{r(t)\}$ is described by the stochastic differential equation

$$dr(t) = r(t)[\eta - \alpha \ln r(t)]dt + \sigma r(t)dW(t), \quad r(0) = r_0, \tag{6}$$

solution of which admits the following explicit form

$$r(t) = \exp\left[\ln r_0 \cdot e^{-\alpha t} + \frac{\eta - \sigma^2/2}{a}(1 - e^{-\alpha t}) + \sigma \int_0^t e^{\alpha(s-t)} dW(s)\right]. \quad (7)$$

This can be rewritten as

$$r(t) = c(t)e^{\sigma Z(t)} \quad (8)$$

with

$$c(t) := \exp\left[\ln r_0 \cdot e^{-\alpha t} + \frac{\eta - \sigma^2/2}{\alpha}(1 - e^{-\alpha t})\right], \quad Z(t) := \int_0^t e^{-\alpha(t-s)} dW(s). \quad (9)$$

It can be verified that $\{Z(t)\}$ is an Ornstein–Uhlenbeck process with $\mu = 0$ and $\theta = \alpha$, and hence finding a bond price amounts to finding

$$\mathrm{E}[e^{-\int_0^T r(t)dt}] = \mathrm{E}[e^{-\int_0^T c(t)e^{\sigma Z(t)}dt}], \quad (10)$$

which is of the form (1).

Although the expectation in (5) involves an integral of the diffusion process (4), it is possible to modify the problem so that (5) reduces to (1) with U being a Brownian motion or a Brownian bridge. To see this, we first use the transformation $\beta(x) := \int^x 1/\sigma(z)dz$ to obtain a process with constant diffusion term. By the Itô's formula, the process $Y(t) := \beta(r(t))$ satisfies the following equation

$$dY(t) = \bar{\mu}(Y(t))dt + dW(t), \quad Y(0) = \beta(r(0)),$$

where

$$\bar{\mu}(y) = \frac{\mu(\alpha(y))}{\sigma(\alpha(y))} - \frac{1}{2}\sigma'(\alpha(y))$$

and α is the inverse function of β. From the Girsanov theorem we can find that the likelihood ratio of the measure Q generated by the process $\{Y(t)\}$ with respect to the one induced by the Brownian motion $\{Y(0) + W(t)\}$ is given by[1]

$$\frac{dQ}{dP}(\omega) := \exp\left[\int_0^T \bar{\mu}(\omega(u))d\omega(u) - \frac{1}{2}\int_0^T \bar{\mu}^2(\omega(u))du\right]. \quad (11)$$

By applying the Itô's formula to $D(W(t))$, with $D(x) := \int^x \bar{\mu}(z)dz$, the likelihood can be rewritten as

$$\frac{dQ}{dP}(\omega) := \exp\left[D(\omega(T)) - D(Y(0)) - \frac{1}{2}\int_0^T \bar{\sigma}(\omega(u))du\right] \quad (12)$$

[1]For regularity conditions, we refer to Beskos et al. [2].

with $\bar{\sigma} := \bar{\mu}^2 + \bar{\mu}'$. Thus, we arrive at the following representation of the expectation (5) with $t = 0$:

$$E[e^{-\int_0^T r(u)du}|\mathscr{F}(0)]$$
$$= \mathrm{E}\left[\exp\left[D(W(T)) - D(Y(0)) - \int_0^T (\alpha(W(u)) + \frac{1}{2}\bar{\sigma}(W(u)))du\right]\right], \quad (13)$$

which, by conditioning on $W(T)$, can be reduced to the form (1) with $\{U(t), t \geq 0\}$ being a Brownian bridge.

Expectations of the form (1) with $\{U(t), t \geq 0\}$ being a Brownian bridge also arise in the context of maximum likelihood estimation methods for diffusion processes sampled at discrete time intervals. For details, we refer to [2, 3].

The remainder of the paper is organized as follows. In Sect. 2.1 we outline the BBI method and motivate further our approach. In Sect. 2.2 we gather some facts about OU-bridges. Section 3 presents theoretical foundations of the proposed method, which is formulated in Sect. 4. In Sect. 5 we illustrate the method by applying it to the problem of bond pricing under the EV model.

2 Preliminaries

2.1 Dimension Reduction by Conditioning on Integrals.

Here we specialize the BBI method proposed in [7] to the case of expectations of the form (1). Let $W_{x,y}^{a,b} := \{W_{x,y}^{a,b}(t); t \in [a, b]\}$ denote a Brownian bridge from x to y on the interval $[a, b]$. To simplify the notation, in cases when it does not lead to any ambiguity, we will use $W_{x,y}^{b}$ when $a = 0$, or remove the superscript completely. Let $\underline{W}_{d_0}^{BB} := (W(t_1), \ldots, W(t_{d_0}))$, for $t_i \equiv t_i(d_0) := iT/d_0, i = 1, \ldots, d_0$, and $d_0 := d/2$, where d is an even integer. We also define a vector of integrals along path of a process as

$$\underline{A}_{d_0} := (A(0, t_1), A(t_1, t_2), \ldots, A(t_{d_0-1}, T))$$

with

$$(t, s) \equiv A(t, s)(\omega) := \int_t^s \omega(z)dz, \quad \text{for } t, s \in [0, T], t < s .$$

The BBI method of sampling Brownian paths is motivated by the following representation

$$\mathrm{E}\left[G\left(\int_0^T g(s, W(s))ds\right)\right] = \mathrm{E}\left[\mathrm{E}\left[G\left(\int_0^T g(s, W(s))ds\right) |\underline{W}_{d_0}^{BB}, \underline{A}_{d_0}\right]\right]. \quad (14)$$

The basic fact behind the method is that conditionally on the vector $\underline{W}_{d_0}^{BB}$ the integral $\int_0^T g(s, W(s))ds$ depends on Brownian motion only through local Brownian bridges. Therefore, for functions g that locally can be closely approximated by linear functions, conditioning jointly on $\underline{W}_{d_0}^{BB}$ and the integrals $\int_{t_{i-1}}^{t_i} W(z)dz$, $i = 1, \ldots, d_0$, eliminates most of the variability of $\int_0^T g(s, W(s))ds$. This observation suggests applying more efficient integration methods to the vector $V(d) := (\underline{W}_{d_0}^{BB}, \underline{A}_{d_0})$, like quasi-Monte Carlo, and then randomly sampling Brownian paths conditionally on $V(d)$.

Quasi-Monte Carlo (QMC) integration methods have deterministic error bound typically in the order $O(n^{-1}(\log n)^d)$, where n is the number of integration points and d is the dimension of the integration region. Hence, QMC may not be superior to Monte Carlo if d is large (the role the dimension plays in financial applications, and methods of identifying important coordinates, are discussed, among others, by Sloan and Wang [8–10]). However, as demonstrated in [7], the vector $V(d)$ with small values of d will often capture a large part of the overall variability of $G(\int_0^T g(s, W(s))ds)$. This finding explains high efficiency of the BBI method when compared with other methods of sampling Brownian paths, like the Brownian bridge construction.

When G is an exponential function, Eq. (14) reduces to

$$\begin{aligned} \mathrm{E}[e^{\int_0^T g(s,W(s))ds}] &= \mathrm{E}\left[\mathrm{E}\left[\prod_{i=1}^{d_0} e^{\int_{t_{i-1}}^{t_i} g(s,W(s))ds} | \underline{W}_{d_0}^{BB}, \underline{A}_{d_0}\right]\right] \\ &= \mathrm{E}\left[\prod_{i=1}^{d_0} \mathrm{E}\left[e^{\int_{t_{i-1}}^{t_i} g(s,W(s))ds} | W(t_{i-1}), W(t_i), A(t_{i-1}, t_i)\right]\right], \end{aligned} \tag{15}$$

where in (15) we have used conditional independence of the bridges $W_{W(t_{i-1}),W(t_i)}^{t_{i-1},t_i}$, $i = 1, \ldots, d_0$, given $\underline{W}_{d_0}^{BB}$. Under the assumption that locally g can be closely approximated by polynomials, the integrand for the ith interval $[t_{i-1}, t_i]$ will depend on Brownian motion mostly through the variables

$$I_{q,p}^{(i)} := \int_{t_{i-1}}^{t_i} s^q W_{x,y}^{t_{i-1},t_i}(s)^p ds \quad \text{for } q, p = 0, 1, 2, \ldots, \tag{16}$$

where x and y are fixed. In order to describe the impact each variable in (16) has on the overall variability of the integrand, it is convenient to represent (16) in terms of a standard Brownian bridge over the interval $[0, \Delta]$, with $\Delta := T/d_0$. Using the following well-known representation of Brownian bridge

$$W_{x,y}^{t_{i-1},t_i}(s) = x + \frac{y-x}{\Delta}(s - t_{i-1}) + W_0^{\Delta}(s - t_{i-1}), \quad s \in [t_{i-1}, t_i], \tag{17}$$

where $\{W_0^{\Delta}(s),\ s \in [0, \Delta]\}$ is a standard Brownian bridge, it is easy to verify that the variables $I_{q,p}^{(i)}$ can be represented as linear functions of

$$I_{q,p} := \int_0^{\Delta} s^q W_0^{\Delta}(s)^p ds \ \text{ for } q, p = 0, 1, 2, \ldots. \tag{18}$$

If we consider only terms $I_{q,p}$, $q + p \leq 2$, then the following lemma describes the rates at which the variances of $I_{q,p}$ converge to zero. The proof of this result is similar to the one presented in [7] for $I_{0,1}$, and hence is omitted.

Lemma 1 *As $\Delta \to 0$, we have*

$$Var[\int_0^{\Delta} W_0^{\Delta}(s)ds] = O(\Delta^3) \tag{19}$$

$$Var[\int_0^{\Delta} W_0^{\Delta}(s)^2 ds] = O(\Delta^4) \tag{20}$$

$$Var[\int_0^{\Delta} s W_0^{\Delta}(s)ds] = O(\Delta^5). \tag{21}$$

These results suggest that by conditioning on $I_{0,1}^{(i)}$ we can eliminate completely the main source of variability in the subinterval $[t_{i-1}, t_i]$, $i = 1, \ldots, d_0$. The same technique, however, does not work so well for the terms $I_{0,2}^{(i)}$, $i = 1, \ldots, d_0$, since conditioning on $I_{0,1}^{(i)}$, $i = 1, \ldots, d_0$, reduces their variability only by a constant [7]. Clearly, the variability due to the quadratic term $I_{0,2}$ can be reduced by increasing the dimension d of the conditioning vector $V(d)$, which leads to smaller values of Δ. However, in Sect. 3 we present a method that entirely eliminates variability due to this term. The method combines conditioning on integrals with changes of measure in each subinterval, and it relies on the existence of an efficient sampling method for Ornstein–Uhlenbeck bridges conditioned on integrals, which we describe in Sect. 2.2.

2.2 Ornstein–Uhlenbeck Bridge Conditioned on Its Integral.

It is well known that the solution to (2) admits the following explicit form

$$X(t) = xe^{-\theta t} + \mu(1 - e^{-\theta t}) + \sigma \int_0^t e^{\theta(u-t)} dW(u),\ t \geq 0, \tag{22}$$

which can be used to find

$$\mathrm{E}[X(t)] = xe^{-\theta t} + \mu(1 - e^{-\theta t}), \tag{23}$$

$$\mathrm{Cov}[X(s), X(t)] = \frac{\sigma^2}{2\theta} e^{-\theta(s+t)}(e^{2\theta s \wedge t} - 1). \tag{24}$$

When we condition $\{X(s),\ s < T\}$ on $X(T)$, we get an Ornstein–Uhlenbeck bridge. For $X_0 = x$ and $X(T) = y$, we shall denote it by $X^T_{x,y} \equiv \{X^T_{x,y}(t)\}$, while a similar bridge over a general interval $[t, s],\ t < s$, will be denoted by $X^{t,s}_{x,y}$. Since by (22) the process $\{X(t)\}$ is Gaussian, it can be shown that $X^T_{x,y}$ is also Gaussian with the following moments for $\theta \neq 0$:

$$\mathrm{E}[X^T_{x,y}(s)] = \mu + (x - \mu)e^{-\theta s} + \frac{e^{\theta(s-T)} - e^{-\theta(s+T)}}{1 - e^{-2\theta T}} \left[y - (x - \mu)e^{-\theta T} - \mu \right] \tag{25}$$

$$\mathrm{Cov}[X^T_{x,y}(s), X^T_{x,y}(t)] = \frac{\sigma^2}{2\theta} \left[e^{-\theta(s+t)}(e^{2\theta s\wedge t} - 1) - \frac{(e^{\theta t} - e^{-\theta t})(e^{\theta s} - e^{-\theta s})}{e^{2\theta T} - 1} \right], \tag{26}$$

where $s, t \in [0, T]$. When $\theta = 0$, then $\{X_t\}$ is a Brownian motion with zero drift, and (25)–(26) must be replaced with $x + (y - x)s/T$ and $\sigma^2(s \wedge t - st/T)$, respectively. In the remainder of this paper, we shall denote the mean $\mathrm{E}[X^T_{x,y}(s)]$ by $\mu_{x,y}(s)$ and the covariance $\mathrm{Cov}[X^T_{x,y}(s), X^T_{x,y}(t)]$ by $K_{x,y}(s, t)$.

It can be verified that the integral

$$I = I(x, y, T) := \int_0^T X^T_{x,y}(u)du$$

is normally distributed with the following moments when $\theta \neq 0$:

$$\mathrm{E}[I] = \mu T + (x - \mu)\frac{1 - e^{-\theta T}}{\theta} + \frac{e^{\theta T} + e^{-\theta T} - 2}{(e^{\theta T} - e^{-\theta T})\theta} \left[y - (x - \mu)e^{-\theta T} - \mu \right] \tag{27}$$

$$\mathrm{Var}[I] = \frac{\sigma^2}{2\theta^3} \left[2\theta T - e^{-2\theta T} + 4e^{-\theta T} - 3 - \frac{e^{2\theta T} - 4e^{\theta T} - 4e^{-\theta T} + e^{-2\theta T} + 6}{e^{2\theta T} - 1} \right]. \tag{28}$$

When $\theta = 0$, we have

$$\mathrm{E}[I] = \frac{x + y}{2} T \quad \text{and} \quad \mathrm{Var}[I] = \frac{\sigma^2}{12} T^3. \tag{29}$$

In the method we describe in Sect. 4, we need the law of $\{X^T_{x,y}(t)\}$ conditional on $I(x, y, T)$. Suppose that ν is a given function on $[0, T]$, and let

$$k(t) := \int_0^T K_{x,y}(t, s)\nu(s)ds \quad \text{and} \quad \bar{k}(t) := k(t) / \int_0^T k(s)\nu(s)ds,\ t \in [0, T]. \tag{30}$$

By the result presented in [7] for Gaussian processes, the conditional law of $\{X^T_{x,y}(t)\}$ given $\int_0^T X^T_{x,y}(s)\nu(s)ds = l, l \in \mathscr{R}$, coincides with the law of the following process:

$$Z^l_{x,y}(t) := X^T_{x,y}(t) - \bar{k}(t)\left[\int_0^T X^T_{x,y}(s)\nu(s)ds - l\right], \quad t \in [0, T]. \tag{31}$$

This characterization provides a practical way of generating paths of an Ornstein–Uhlenbeck bridge conditioned on $\int_0^T X^T_{x,y}(s)\nu(s)ds = l$, since it suffices to add to a path of the OU-bridge a function of t that depends on the bridge only through the integral $\int_0^T X^T_{x,y}(s)\nu(s)ds$.

It can be shown, through direct calculations, that $\bar{k}$ defined in (30) for $\nu \equiv 1$ is of the following form

$$\bar{k}_\theta(t) := \frac{1 + e^{\theta T} - e^{\theta(T-t)} - e^{\theta t}}{(1 + e^{\theta T})T + \frac{2}{\theta}(1 - e^{\theta T})}, \quad t \in [0, T]. \tag{32}$$

This function has the property that $\bar{k}_\theta(0) = \bar{k}_\theta(T) = 0$, and it converges to

$$\bar{k}_0(t) := 6\frac{t(T-t)}{T^3}, \quad t \in [0, T], \tag{33}$$

as $\theta \to 0$, which recovers the corresponding function $\bar{k}$ for a standard Brownian bridge over the interval $[0, T]$.

In the rest of the paper we will deal mostly with the particular case of an OU process corresponding to $\mu = 0$, $x = 0$, and $\sigma = 1$. We will refer to this case as a basic OU-process.[2] Similarly, a basic OU-process conditioned on its terminal value will be referred to as the basic OU-bridge.

3 Dimension Reduction by Change of Measure

Motivated by our discussion in Sect. 2.1, here we consider the problem of eliminating variability due to the quadratic term in the expectation of the form

$$\mathrm{E}[e^{\lambda_0 \int_t^s X^{t,s}_{x,y}(u)du - \lambda \int_t^s X^{t,s}_{x,y}(u)^2 du}], \tag{34}$$

where $\lambda_0 \in \mathscr{R}$ and $\lambda \in \bar{\mathscr{R}}^+ := \mathscr{R}^+ \cup \{0\}$. The method that we are proposing is based on a change of measure for Ornstein–Uhlenbeck bridges, which include Brownian bridges as particular cases. We first present results for an Ornstein–Uhlenbeck pro-

[2] The description "standard OU-process" is usually applied to the case when $\mu = 0$, $\theta = 1$, and $\sigma = \sqrt{2}$ (e.g., Baldeaux and Platen [1]).

cess, and then for an Ornstein–Uhlenbeck bridge. For ease of notation, we consider the processes on the interval $[0, T]$.

Let us denote by $Q(\theta)$ the probability measure induced on the space of continuous functions by a basic OU process with parameter θ. We use the superscript $Q(\theta)$ when an expectation is taken with respect to this measure, and no superscript when $\theta = 0$ (i.e., when the process is a standard Brownian motion).

Proposition 1 *For $\lambda \in \bar{\mathscr{R}}^+$ and an integrable functional G on the space of continuous functions, we have*

$$E^{Q(\theta)}[G(\omega)e^{-\lambda\int_0^T (a_0+\omega(u))^2 du}] \\ = e^{-\lambda a_0^2 T} E^{Q(\theta^*)}[G(\omega)e^{-\frac{1}{2}(\theta-\theta^*)(\omega^2(T)-T)-2\lambda a_0 \int_0^T \omega(u)du}], \quad (35)$$

where $\theta^* := -\sqrt{2\lambda + \theta^2}$.

Proof Let $\bar{\lambda} := \lambda + \theta^2/2$. We have

$$\begin{aligned}
& e^{\lambda a_0^2 T}\mathrm{E}^{Q(\theta)}[G(\omega)e^{-\lambda\int_0^T (a_0+\omega(u))^2 du}] \\
&= \mathrm{E}^{Q(\theta)}[G(\omega)e^{-2\lambda a_0 \int_0^T \omega(u)du-\lambda\int_0^T \omega(u)^2 du}] \\
&= \mathrm{E}[G(W)e^{-2\lambda a_0 \int_0^T W(u)du-\lambda\int_0^T W(u)^2 du-\theta\int_0^T W(u)dW(u)-\frac{\theta^2}{2}\int_0^T W(u)^2 du}] \\
&= \mathrm{E}[G(W)e^{-2\lambda a_0 \int_0^T W(u)du-\frac{\theta}{2}[W(T)^2-T]-\bar{\lambda}\int_0^T W(u)^2 du}] \\
&= \mathrm{E}[G(W)e^{-2\lambda a_0 \int_0^T W(u)du-\frac{\theta}{2}[W(T)^2-T]-\int_0^T \sqrt{2\bar{\lambda}}W(u)dW(u)+\int_0^T \sqrt{2\bar{\lambda}}W(u)dW(u)-\bar{\lambda}\int_0^T W(u)^2 du}] \\
&= \mathrm{E}^{Q(\theta^*)}[G(\omega)e^{-2\lambda a_0 \int_0^T \omega(u)du-\frac{1}{2}(\theta+\sqrt{2\lambda+\theta^2})(\omega^2(T)-T)}],
\end{aligned}$$

where in the third line we use the Girsanov theorem with the likelihood ratio (11) corresponding to the basic OU process. In the fourth line we use $\int_0^T W(u)dW(u) = (W(T)^2 - T)/2$, while in the fifth line we add and subtract the same term in the exponent. In the last line we use again the Girsanov theorem. □

According to this result, we can eliminate an integral of the squared OU-process by properly changing the parameter θ. Therefore, if we ignore for a moment the functional G, the integrand on the right-hand side of (35) depends on the path of the process only through two jointly normally distributed variables, $\omega(T)$ and $\int_0^T \omega(u)du$, and hence, for integration purposes, the dimension of the integrand has been reduced to just two. This reduction of the dimension will also lead to efficient simulation methods for non-constant G under the assumption that conditionally on $\omega(T)$ and $\int_0^T \omega(u)du$ its variability is relatively low.

We would like to note that effectiveness of any Monte Carlo method that takes advantage of the representation (35) can be further improved by using a biased OU bridge, where the distribution of the terminal value of the process $\omega(T)$ is modified by including the term $\exp[-\frac{1}{2}(\theta - \theta^*)\omega(T)^2]$ into the density of $\omega(T)$.

Corollary 1 *When we apply Proposition 1 to a standard Brownian motion we get for $\lambda_0 \in \mathscr{R}$ and $\lambda \in \bar{\mathscr{R}}^+$:*

$$E\left[G(W)e^{\lambda_0 \int_0^T W(u)du - \lambda \int_0^T W(u)^2 du}\right] = E^{Q(-\sqrt{2\lambda})}\left[G(\omega)e^{\lambda_0 \int_0^T \omega(u)du - \sqrt{\frac{\lambda}{2}}(\omega(T)^2 - T)}\right]. \tag{36}$$

We would like to have analogous results for an OU bridge. Let us denote the transition density of the OU process (2) with $\mu = 0$ and $\sigma = 1$ by p_θ. By (22)–(24) we have

$$p_\theta(y; x, T) = \frac{1}{\sqrt{2\pi\sigma_\theta^2}} e^{-\frac{1}{2\sigma_\theta^2}(y-\mu_\theta)^2}, \tag{37}$$

where

$$\mu_\theta = xe^{-\theta T} \quad \text{and} \quad \sigma_\theta^2 = \frac{1 - e^{-2\theta T}}{2\theta},$$

with $\sigma_0^2 := T$. In addition, let $Q(\theta, x, y)$ be the probability measure induced on the space of continuous functions by the basic OU bridge from x to y over $[0, T]$.

Proposition 2 *For any functional G on the space of continuous functions for which the expectations below exist, we have*

$$E^{Q(\theta,x,y)}[G(\omega)e^{-\lambda \int_0^T \omega(u)^2 du}] = C \cdot E^{Q(\theta^*,x,y)}[G(\omega)], \tag{38}$$

where $\lambda \in \mathscr{R}^+$, $\theta^* := -\sqrt{2\lambda + \theta^2}$, and

$$C = \frac{p_{\theta^*}(y; x, T)}{p_\theta(y; x, T)} e^{-\frac{1}{2}(\theta + \sqrt{2\lambda+\theta^2})(y^2 - x^2 - T)}. \tag{39}$$

The above result can be proven by combining the method we use in Proposition 1 with an explicit form of the likelihood ratio for conditioned diffusion processes obtained by Dacunha-Castelle and Florens-Zmirou [5].

Corollary 2 *By properly selecting G in (38), we get the following formula*

$$E^{Q(\theta,x,y)}[e^{\int_0^T g(\omega(u))du}] = C \cdot E^{Q(\theta^*,x,y)}[e^{\int_0^T g(\omega(u))du + \lambda \int_0^T \omega(u)^2 du}], \tag{40}$$

where C is defined in (39), and g and $\lambda \in \mathscr{R}^+$ are such that the expectations in (40) exist.

Corollary 3 *When we apply Proposition 2 to the Brownian bridge $\{W_{x,y}^T\}$, then, for $\lambda_0 \in \mathscr{R}$ and $\lambda \in \bar{\mathscr{R}}^+$, we get*

$$E[G(W_{x,y}^T)e^{\lambda_0 \int_0^T W_{x,y}^T(u)du - \lambda \int_0^T W_{x,y}^T(u)^2 du}] = C_0 \cdot E^{Q(\bar{\theta},x,y)}[G(\omega)e^{\lambda_0 \int_0^T \omega(u)du}], \tag{41}$$

where $\bar{\theta} := -\sqrt{2\lambda}$ *and*

$$C_0 \equiv C_0(x, y, T, \lambda) := \frac{p_{\bar{\theta}}(y; x, T)}{p_0(y; x, T)} e^{-\sqrt{\frac{\lambda}{2}}(y^2 - x^2 - T)}. \tag{42}$$

Analogously to (40), by properly selecting G, formula (41) becomes

$$E[e^{\int_0^T g(W_{x,y}^T(u))du}] = C_0 \cdot E^{Q(\bar{\theta},x,y)}[e^{\int_0^T g(\omega(u))du + \lambda \int_0^T \omega(u)^2 du}], \tag{43}$$

where g *and* $\lambda \in \bar{\mathscr{R}}^+$ *are such that the expectations in (43) exist.*

4 The Method

In the method we are proposing in this paper we also combine the Brownian bridge construction with conditioning on integrals, but in each subinterval we change the measure according to (40) to eliminate variability due to the quadratic component in the function g. We shall refer to this method as OUBIM (Ornstein–Uhlenbeck Bridges with Integrals and Measure change). For ease of explanation, we present the method assuming that $\{U(t)\}$ in (1) is a standard Brownian motion, but in the next section we extend the method to the case when $\{U(t)\}$ is a basic OU-process.

Using the notation of Sect. 2.1, the method is based on the following modification of (15)

$$\mathrm{E}\left[e^{\int_0^T g(u,W(u))du}\right] = \mathrm{E}\left[\prod_{i=1}^{d_0} \mathrm{E}\left[e^{\int_{t_{i-1}}^{t_i} g(u,W_{W_{i-1},W_i}(u))du} | W_{i-1}, W_i\right]\right] \tag{44}$$

$$= \mathrm{E}\left[\prod_{i=1}^{d_0} C_i \cdot \mathrm{E}^{Q(\bar{\theta},W_{i-1},W_i)}\left[e^{\int_{t_{i-1}}^{t_i} g(u,\omega(u))du + \lambda_i \int_{t_{i-1}}^{t_i} \omega(u)^2 du} | W_{i-1}, W_i, A^{m_i}(t_{i-1}, t_i)\right]\right], \tag{45}$$

where $W_i = W(t_i)$, $i = 1, \ldots, d_0$, and under the measure $Q(\bar{\theta}, W_{i-1}, W_i)$ the process $\{\omega(t)\}$ is the basic OU-bridge with the end values given by W_{i-1} and W_i, and the parameter $\bar{\theta}$ equal to $-\sqrt{2\lambda_i}$. In (45), the coefficient C_i is defined according to (42)

$$C_i = C_0(W_{i-1}, W_i, t_i - t_{i-1}, \lambda), \tag{46}$$

and the variable $A^{m_i}(t_{i-1}, t_i)$ is

$$A^{m_i}(t_{i-1}, t_i) = \int_{t_{i-1}}^{t_i} m_i(u)\omega(u)du, \tag{47}$$

where m_i is a non-stochastic function of time. The outer expectation in (45) is taken with respect to the join distribution of $W_1, \ldots, W_{d_0}$ and $A^{m_1}(t_0, t_1), \ldots, A^{m_{d_0}}(t_{d_0-1},$

t_{d_0}), which is Gaussian and can be determined in a similar way we use for $(\underline{W}_{d_0}^{BB}, \underline{A}_{d_0})$ in the BBI method.

For each i, $i = 1, \ldots, d_0$, the parameters λ_i and functions m_i are allowed to depend on W_{i-1} and W_i. In the next section we discuss some methods of selecting them.

4.1 Selection of λ_i and m_i

In (45) we use the following representation for the ith subinterval:

$$\begin{aligned} &\mathrm{E}[e^{\int_{t_{i-1}}^{t_i} g(u,W_{x_i,y_i}(u))du}] \\ &\quad = C_i \cdot \mathrm{E}^{Q(\bar{\theta},x_i,y_i)}\left[\mathrm{E}^{Q(\bar{\theta},x_i,y_i)}[e^{\int_{t_{i-1}}^{t_i} g(u,\omega(u))du+\lambda_i \int_{t_{i-1}}^{t_i} \omega(u)^2 du} | A^{m_i}(t_{i-1}, t_i)]\right] \end{aligned} \tag{48}$$

where $\bar{\theta} = -\sqrt{2\lambda_i}$, and x_i and y_i are fixed. To device a simulation method based on such a representation, we need to select λ_i and m_i, with the objective of eliminating variability of the integrand due to terms of the form $\int_{t_{i-1}}^{t_i} n(u)W_{x,y}(u)du$ and $\int_{t_{i-1}}^{t_i} W_{x,y}(u)^2 du$, where n is a deterministic function of time. Results presented in Sect. 3 show how to select λ_i when g is a quadratic function. Below we discuss a possible approach for more general functions g. For ease of notation we remove the subscript i from x_i, y_i, m_i, and λ_i.

A simple way of identifying a quadratic component in g is to use its Taylor series expansion in a neighborhood of a selected point $p_e = (t_e, z_e)$ from $[t_{i-1}, t_i] \times [x, y]$, where we assume without loss of generality that $x \le y$. Assuming that $g(u, z) : [t_{i-1}, t_i] \times \mathscr{R} \to \mathscr{R}$ is smooth enough to admit such an expansion, we have

$$\begin{aligned} \int_{t_{i-1}}^{t_i} g(u,\omega(u))du = a_0 &+ \int_{t_{i-1}}^{t_i} (g_z^{(1)}(p_e) - g_{zz}^{(2)}(p_e)z_e + g_{uz}^{(2)}(p_e)(u-u_e))\omega(u)du \\ &+ \frac{1}{2} g_{zz}^{(2)}(p_e) \int_{t_{i-1}}^{t_i} \omega(u)^2 du + \int_{t_{i-1}}^{t_i} \varepsilon(\omega(u))du, \end{aligned} \tag{49}$$

where a_0 includes all the terms that do not depend on $\{\omega(t)\}$, ε is the remainder term, and $g_{zz}^{(2)}$, $g_{uz}^{(2)}$ denote partial derivatives of g. By substituting (49) into (48) and ignoring the remainder term, it is easy to verify that we can eliminate integrals involving $\omega(u)$ and $\omega(u)^2$ by selecting λ and m equal respectively to

$$\lambda_E(x, y) := -\frac{1}{2} g_{zz}^{(2)}(p_e) \tag{50}$$

and

$$m_E(u) = \begin{cases} 0 & \text{if } g_z^{(1)}(p_e) - g_{zz}^{(2)}(p_e)z_e = 0 \text{ and } g_{uz}^{(2)}(p_e) = 0, \\ u - u_e & \text{if } g_z^{(1)}(p_e) - g_{zz}^{(2)}(p_e)z_e = 0 \text{ and } g_{uz}^{(2)}(p_e) \neq 0, \\ 1 + \frac{g_{uz}^{(2)}(p_e)(u-u_e)}{g_z^{(1)}(p_e) - g_{zz}^{(2)}(p_e)z_e} & \text{if } g_z^{(1)}(p_e) - g_{zz}^{(2)}(p_e)z_e \neq 0. \end{cases} \tag{51}$$

In (50) we assume that $g_{zz}^{(2)}(p_e) \leq 0$, otherwise we do not change the measure. When g does not depend on time, and $g_z^{(1)}(p_e) - g_{zz}^{(2)}(p_e)z_e \neq 0$, we have

$$m_E \equiv m_C := 1. \tag{52}$$

In the case of a constant m, as in (52), sampling the basic OU-bridge conditionally on its integral can be carried out by using the method described in Sect. 2.1 with $\bar{k}$ given by (32).

For other selections of m, finding the corresponding $\bar{k}$ defined in (30) is more difficult. This is still possible for the linear function defined in (51), but the resulting function $\bar{k}$ is significantly more involved that the one given in (32). We can expect, however, that in some situations the use of m_C will lead to a method that has similar efficiency as that of methods that utilize linear functions m. In the context of conditioning on Brownian bridges, this can be explained by two facts. Firstly, by (19)–(21), the contribution to the overall variance given by $\int_{t_{i-1}}^{t_i}(u - t_{i-1})W_0(u)du$ is of two order smaller than that of $\int_{t_{i-1}}^{t_i} W_0(u)du$, and hence, in presence of the latter term, conditioning on $\int_{t_{i-1}}^{t_i} W_0(u)du$ should always be considered first. In such cases, additional conditioning on $\int_{t_{i-1}}^{t_i}(u - t_{i-1})W_0(u)du$ increases the dimension of the conditioning vector, which potentially may lead to a deterioration of the efficiency of the integration method we apply to this vector.

Secondly, conditioning on $\int_{t_{i-1}}^{t_i} W_0(u)du$ not only eliminates variability due to this term, but it also reduces variance due to the term $\int_{t_{i-1}}^{t_i}(u - t_{i-1})W_0(u)du$. This is captured more formally by the following result. Let

$$J_1 = \int_{t_{i-1}}^{t_i} (u - t_{i-1})W_0(u)du \quad \text{and} \quad J_0 = \int_{t_{i-1}}^{t_i} W_0(u)du.$$

By using the fact that J_0 and J_1 are jointly normally distributed with known moments, one can quantify the reduction of variance when conditioning on J_0 as follows.

Lemma 2 *As $t_i - t_{i-1} \to 0$, we have*

$$E[Var[J_1|J_0]] = \frac{1}{16}Var[J_1] + \varepsilon_1, \tag{53}$$

where ε_1 converges to zero faster than $Var[J_1]$.

This result shows that although conditioning on $\int_{t_{i-1}}^{t_i} W_0(u)du$ does not eliminate completely variability due to $\int_{t_{i-1}}^{t_i}(u - t_{i-1})W_0(u)du$, it reduces its magnitude significantly.

The above arguments led us to use in our numerical study the values of λ and m given in (50) and (52), respectively. We should mention that it is possible to devise a more accurate than (49) expansion, which, in the context of a Brownian bridge, is based on the observation that values of the process at any time $s \in [t_{i-1}, t_i]$ are centered around the mean $x + \frac{y-x}{t_i - t_{i-1}}(s - t_{i-1})$. Due to the space constraint, and the fact that the selection given by (50) and (52) works well in the cases we have considered, we leave a detailed study of the problem of proper selection of m as a topic for future research.

4.2 The Algorithm

We denote the Brownian bridge construction that uses conditioning on $\underline{W}_d^{BB}$ as BBd, the construction that uses conditioning on $(\underline{W}_{d/2}^{BB}, \underline{A}_{d/2})$ as BBId, and the proposed method, described at the beginning of this section, as OUBIMd. In all cases, d refers to the dimension of the conditioning vector.

Since the BBId method is described in [7], below we only outline the OUBIMd algorithm, and for this we assume that the functional of interest is $\omega \to \mathscr{G}(\omega) := \exp\{\int_0^T g(u, \omega(u))du\}$ and $\{U(t)\}$ in (1) is a standard Brownian motion. In the next section we present a modification of the algorithm when $\{U(t)\}$ is a basic OU-process.

In order to approximate integrals along paths of the process, we divide T into L equally spaced subintervals, where we assume that L is a multiple of $d/2$. We represent the mesh points at which the integrand $\mathscr{G}$ is evaluated as

$$M := \{m\frac{T}{L} : m = 1, \ldots, L\} = \cup_{l=1}^{d_0} M_l, \;\; M_l := M \cap (t_{l-1}, t_l], t_l := l\frac{T}{d_0}, \; l = 1, \ldots, d_0.$$

By N_{LD} we denote the number of low-discrepancy points used to evaluate the expectation in (45) with respect to the distribution of the conditioning vector, and by N_{MC} the number of random paths used for each internal expectation in (45).

Algorithm for the OUBIMd method Set $Res = 0$ and $d_0 = d/2$.
For $i = 1, \ldots, N_{LD}$ **do**

Step 1. Obtain $u_i := (u_i(1), \ldots, u_i(d))$ from a d-dimensional LD sequence.
Step 2. Using the first d_0 coordinates of u_i, find $\Phi^{-1}((u_i(1)), \ldots, \Phi^{-1}(u_i(d_0))$ and then compute values of the process at $T/d_0, 2T/d_0, \ldots, T$ using the Brownian bridge construction. Denote these values as $w_1, \ldots, w_{d_0}$. Set $w_0 = 0$.
Step 3. For subinterval $l, l = 1, \ldots, d_0$, calculate λ_l from (50) and C_l from (46). Use the last d_0 coordinates of u_i to obtain

$$a_l = \mu_l + \sigma_l \Phi^{-1}(u_i(d_0 + l)), \;\; l = 1, \ldots, d_0,$$

where for each $\lambda_l > 0$ the values μ_l and σ_l are obtained from the formulae (27)–(28) with $x = 0$, $\mu = 0$, $T = t_l - t_{l-1}$ and θ equal to $\bar{\theta} = -\sqrt{2\lambda_l}$. When $\lambda_l = 0$, the corresponding values μ_l and σ_l are obtained from (29).
Step 4. **For** $j = 1, \ldots, N_{MC}$ **do**
• For each subinterval $[t_{l-1}, t_l]$, $l = 1, \ldots, d_0$, conditionally on $\omega(t_{l-1}) = w_{l-1}$, $\omega(t_l) = w_l$, and $\int_{t_{l-1}}^{t_l} \omega(u)du = a_l$, generate a random path ω at the mesh points M_l using representation (31), with $\bar{k}$ given by $\bar{k}_\theta$ in (32) modified to the current sub-interval.
• Take

$$Res \leftarrow Res + \prod_{k=1}^{d_0} C_k \cdot \exp\left\{\sum_{l=1}^{d_0} \sum_{u \in M_l} [g(u, \omega(u)) + \lambda_l \omega(u)^2] T/L\right\} /(N_{LD} N_{MC}).$$

end for

end for

In Step 3 of the algorithm, we change the measure only if $\lambda_l > 0$. Otherwise we sample the process using formulae in Sect. 2.2 corresponding to a Brownian bridge.

5 Numerical Example: EV Model

Motivated by the example given in Sect. 1, here we consider the problem of finding

$$\mathrm{E}[e^{-\int_0^T c(u)e^{\sigma Z(u)}du}], \tag{54}$$

where c is given in (9) and $\{Z(t)\}$ is a basic Ornstein–Uhlenbeck process with $\theta = \alpha$. For this we use three methods, which we refer to as OUBd, OUBId, and OUBIMd. The first two methods are the same as the Brownian bridge construction BBd and the BBId method described in Sect. 2.1, except that now we use the process $\{Z(t)\}$ instead of a Brownian motion. The OUBIMd method follows the algorithm described in Sect. 4, with the modification that $\{U(t)\}$ is now the basic OU-process $\{Z(t)\}$. Below we provide additional details about each method.

For the OUBd method, we sample first the terminal value $Z(T)$ using the normal distribution with

$$\mathrm{E}[Z_T] = 0 \quad \text{and} \quad \mathrm{Var}[Z(T)] = \frac{1}{2\alpha}[1 - e^{-2\alpha T}], \tag{55}$$

and then the remaining points using conditional normal distributions. In particular, $Z(t_l)$ is sampled conditionally on $Z(t_{l-1}) = x_l$ and $Z(T) = y$, $t_l > t_{l-1}$, by using normal distribution with the moments

$$\mathrm{E}[Z_{x_l,y}(t_l)] = x_l e^{-\alpha\Delta} + \frac{e^{\alpha(\Delta-(T-t_{l-1}))} - e^{-\alpha(\Delta+(T-t_{l-1}))}}{1-e^{-2\alpha(T-t_{l-1})}}[y - x_l e^{-\alpha(T-t_{l-1})}], \tag{56}$$

$$\mathrm{Var}[Z_{x_l,y}(t_l)] = \frac{1}{2\alpha}\left[(1-e^{-2\alpha\Delta}) - \frac{(e^{\alpha\Delta}-e^{-\alpha\Delta})^2}{e^{2\alpha(T-t_{l-1})}-1}\right], \tag{57}$$

where $\Delta = t_l - t_{l-1}$ and $\{Z_{x_l,y}\}$ is the process Z conditioned on its values at the end-points. In order to make the method comparable with OUBId and OUBIMd, we obtain $Z(t_1), \ldots, Z(t_d)$ using a d-dimensional point from a low-discrepancy sequence, and then the remaining points, for each of the N_{MC} paths, are sampled using a pseudo-random generator.

For the OUBId method, we sample $Z(t_1), \ldots, Z(t_{d_0})$ using the OUBd_0 construction described above. Then, for the lth subinterval $[t_{l-1}, t_l]$, $l = 1, \ldots, d_0$, we generate, using the $d_0 + l$th coordinate of a low-discrepancy point, a value a_l of the integral $\int_{t_{l-1}}^{t_l} Z_{x_l,y_l}(u)du$. For this, we use normal distribution with moments given by the formulae (27)–(28) with $x = 0$, $\mu = 0$, $\sigma = 1$, $T = t_l - t_{l-1}$, and θ equal to α. Then, N_{MC} paths are simulated randomly and conditionally given $\int_{t_{l-1}}^{t_l} Z_{x_l,y_l}(u)du = a_l$ by using (31) with $\bar{k}$ of the form (32) modified to the current sub-interval.

The method OUBIMd follows the algorithm described in Sect. 4 with the following modifications. In Step 2 we obtain $Z(t_1), \ldots, Z(t_{d_0})$ using the OUBd_0 construction described above. For each subinterval $[t_{l-1}, t_l]$, with $Z(t_{l-1}) = x_l$ and $Z(t_l) = y_l$, $l = 1, \ldots, d_0$, sampling of the process is based on the following representation derived from Proposition 2:

$$\mathrm{E}\left[e^{-\int_{t_{l-1}}^{t_l} c(u)e^{\sigma Z_{x_l,y_l}(u)}du}\right]$$
$$= C_l \cdot \mathrm{E}\left[\mathrm{E}\left[e^{\int_{t_{l-1}}^{t_l}[-c(u)e^{\sigma Z_{x_l,y_l}(u)} + \frac{1}{2}c(t_{l-1})e^{u^*}\sigma^2 Z_{x_l,y_l}(u)^2]du} \,\Big|\, \int_{t_{l-1}}^{t_l} Z_{x_l,y_l}(u)du\right]\right], \tag{58}$$

where u^* is selected from $[x_l, y_l]$, $\{Z_{x_l,y_l}(u)\}$ is a basic OU-bridge with θ equal to

$$\theta^* = -\sqrt{c(t_{l-1})e^{u^*}\sigma^2 + \alpha^2}, \tag{59}$$

and

$$C_l = \frac{p_{\theta^*}(y_l; x_l, \Delta)}{p_\alpha(y_l; x_l, \Delta)} e^{-\frac{1}{2}(\alpha-\theta^*)(y_l^2 - x_l^2 - \Delta)}. \tag{60}$$

Based on (58), Steps 3 and 4 of the algorithm are modified as follows:

Step 3. Use the last d_0 coordinates of u_i to obtain

$$a_l = \mu_l + \sigma_l \Phi^{-1}(u_i(d_0 + l)), \quad l = 1, \ldots, d_0,$$

where μ_l and σ_l are obtained from (27)–(28) with $x = 0$, $\mu = 0$, $\sigma = 1$, $T = t_l - t_{l-1}$, and θ equal to θ^* given in (59).

Table 1 Estimates of ratios of variances

Method of sampling	LD-sequence	Maturity T			
		1	4	7	10
OUB2	Faure	12	3	4	4
OUBI2	Faure	130	22	27	16
OUBIM2	Faure	946	75	69	54
OUB2	Halton	9	4	4	4
OUBI2	Halton	181	30	23	16
OUBIM2	Halton	768	98	59	48

Table 2 Estimates of ratios of variances when c is constant (Faure sequence)

Method of sampling	Maturity			
	1	4	7	10
OUB2	11	3	4	4
OUBI2	239	51	47	18
OUBIM2	2389	314	158	107

Step 4. For $j = 1, \ldots, N_{MC}$ do

- For each subinterval $[(t_{l-1}, t_l]$, $l = 1, \ldots, d_0$, conditionally on w_{l-1}, w_l, and $\int_{t_{l-1}}^{t_l} \omega(u)du - \bar{a}_l$, generate a random path ω at the mesh points M_l by using (31), with $\bar{k}$ given in (32) modified to the current sub-interval.
- Take

$$Res \leftarrow Res + \prod_{k=1}^{d_0} C_k \cdot \exp\left\{\sum_{l=1}^{d_0} \sum_{u \in M_l} [-c(u)e^{\sigma\omega(u)} + \frac{1}{2}c(t_{l-1})e^{u^*}\sigma^2\omega(u)^2]]T/L\right\} /(N_{LD}N_{MC}),$$

with $C_l, l = 1, \ldots, d_0$, obtained from (60).

In our implementation, we used parameters from Table 3.2 in [4] [3] which in our parametrization of the model are: $\alpha = 0.4512$, $\eta = -1.2321$, $\sigma = 0.6927$, and $r_0 = 1.0094$. To sample normal variates, we used the inverse transform method, where the inverse cumulative normal distribution was approximated using the algorithm presented in [6]. The parameters for the three methods were: $d = 2$, $L = 250 * T$, $N_{MC} = 20$, $N_{LD} = 1000$, and T varied from 1 to 10. In each subinterval, the point u^* in (58) was taken equal to $(x_l + y_l)/2$. Each experiment was repeated 50 times, and the ratios of crude Monte Carlo variances divided by the estimated variances of each of the three methods are reported in Table 1. The results show that by combining conditioning on integrals with the proposed change of measure we can significantly

[3]They were derived from the model calibration to the actual Euro ATM caps volatility curve.

improve efficiency of the OUBI2 method, which already is quite competitive when compared with some of the existing methods.

For comparison, in Table 2 we present results of a similar experiment but now we assume that c in (54) is constant. The results show that in this case the UOBIM2 method leads to even more pronounced reduction of variance.

Acknowledgements The author would like to thank an anonymous referee for constructive comments. He also acknowledges support from the Natural Sciences and Engineering Research Council of Canada.

References

1. Baldeaux, J., Platen, E.: Functionals of Multidimensional Diffusions with Applications to Finance. Bocconi and Springer Series, vol. 5. Springer, Berlin (2013)
2. Beskos, A., Papaspiliopoulos, O., Roberts, G.O., Fearnhead, P.: Exact and computationally efficient likelihood-based estimation for discretely observed diffusion processes (with discussion). J. R. Stat. Soc. B **68**(3), 333–382 (2006)
3. Beskos, A., Roberts, G.O.: Exact simulation of diffusions. Ann. Appl. Probab. **15**(4), 2422–2444 (2005)
4. Brigo, D., Mercurio, F.: Interest Rate Models - Theory and Practice, 2nd edn. Springer, Berlin (2006)
5. Dacunha-Castelle, D., Florens-Zmirou, D.: Estimation of the coefficients of a diffusion from discrete observations. Stoch. Int. J. Probab. Stoch. Process. **19**(4), 263–284 (1986)
6. Glasserman, P.: Monte Carlo Methods in Financial Engineering. Springer, Berlin (2004)
7. Kolkiewicz, A.W.: Efficient Monte Carlo simulation for integral functionals of Brownian motion. J. Complex. **30**, 255–278 (2014)
8. Sloan, I.H., Wang, X.: Quasi-Monte Carlo methods in financial engineering: an equivalence principle and dimension reduction. Oper. Res. **59**(1), 80–95 (2011)
9. Wang, X., Tan, K.S.: How do path generation methods affect the accuracy of quasi-Monte Carlo methods for problems in finance? J. Complex. **28**(2), 250–277 (2012)
10. Wang, X.: On the effects of dimension reduction techniques on some high-dimensional problems in finance. Oper. Res. **54**(6), 1063–1078 (2006)

Optimal Discrepancy Rate of Point Sets in Besov Spaces with Negative Smoothness

Ralph Kritzinger

Abstract We consider the local discrepancy of a symmetrized version of Hammersley type point sets in the unit square. As a measure for the irregularity of distribution we study the norm of the local discrepancy in Besov spaces with dominating mixed smoothness. It is known that for Hammersley type points this norm has the best possible rate provided that the smoothness parameter of the Besov space is nonnegative. While these point sets fail to achieve the same for negative smoothness, we will prove in this note that the symmetrized versions overcome this defect. We conclude with some consequences on discrepancy in further function spaces with dominating mixed smoothness and on numerical integration based on quasi-Monte Carlo rules.

Keywords Besov spaces · Discrepancy · Hammersley point set · Haar functions

1 Introduction

For a multiset $\mathscr{P}$ of $N \geq 1$ points in the unit square $[0, 1]^2$ we define the local discrepancy as

$$D_{\mathscr{P}}(\boldsymbol{t}) := \frac{1}{N} \sum_{z \in \mathscr{P}} \mathbf{1}_{[\mathbf{0},\boldsymbol{t})}(z) - t_1 t_2.$$

Here $\mathbf{1}_I$ denotes the indicator function of an interval $I \subseteq [0, 1)^2$. For $\boldsymbol{t} = (t_1, t_2) \in [0, 1]^2$ we set $[\mathbf{0}, \boldsymbol{t}) := [0, t_1) \times [0, t_2)$ with volume $t_1 t_2$. To obtain a global measure for the irregularity of a point distribution $\mathscr{P}$, one usually considers a norm of the local discrepancy in some function space. A popular choice are the L_p spaces for $p \in [1, \infty]$, which are defined as the collection of all functions f on $[0, 1)^2$ with finite $L_p([0, 1)^2)$ norm. For $p = \infty$ this norm is the supremum norm, i.e.

$$\left\| f | L_\infty([0, 1)^2) \right\| := \sup_{\boldsymbol{t} \in [0,1]^2} |f(\boldsymbol{t})|,$$

R. Kritzinger (✉)
Johannes Kepler University, Altenbergerstr. 69, Linz, Austria
e-mail: ralph.kritzinger@jku.at

A. B. Owen and P. W. Glynn (eds.), *Monte Carlo and Quasi-Monte Carlo Methods*, Springer Proceedings in Mathematics & Statistics 241,
https://doi.org/10.1007/978-3-319-91436-7_20

and for $p \in [1, \infty)$ these norms are given by

$$\|f|L_p([0,1)^2)\| := \left(\int_{[0,1)^2} |f(\boldsymbol{t})|^p \mathrm{d}\boldsymbol{t}\right)^{\frac{1}{p}}.$$

Throughout this note, for functions $f, g : \mathbb{N} \to \mathbb{R}^+$, we write $g(N) \lesssim f(N)$ and $g(N) \gtrsim f(N)$, if there exists a constant $C > 0$ independent of N such that $g(N) \leq Cf(N)$ or $g(N) \geq Cf(N)$ for all $N \in \mathbb{N}$, $N \geq 2$, respectively. We write $f(N) \asymp g(N)$ to express that $g(N) \lesssim f(N)$ and $g(N) \gtrsim f(N)$ holds simultaneously. It is a well-known fact that for every $p \in [1, \infty]$ and $N \in \mathbb{N}$ any N-element point set $\mathscr{P}$ in $[0, 1)^2$ satisfies

$$\|D_{\mathscr{P}}|L_p([0,1)^2)\| \gtrsim N^{-1}(\log N)^{\frac{1}{2}}. \tag{1}$$

This inequality was shown by Roth [12] for $p = 2$ (and therefore for $p \in (2, \infty]$ because of the monotonicity of the L_p norms) and Schmidt [13] for $p \in (1, 2)$. From the work of Halász [3] we know that it also holds for $p = 1$. In recent years several other norms of the local discrepancy have been studied. In this note we would like to investigate the discrepancy of certain point sets in Besov spaces $S^r_{p,q}B([0, 1)^2)$ with dominating mixed smoothness. The parameter p describes the integrability of functions belonging to this space, while r is related to the smoothness of these functions. The third parameter q is a regulation parameter. A definition of $S^r_{p,q}B([0, 1)^2)$ can be found in Sect. 2. We denote the Besov norm of a function f by $\|f|S^r_{p,q}B([0, 1)^2)\|$. The study of discrepancy in function spaces with dominating mixed smoothness was initiated by Triebel [14, 15], since it is directly connected to numerical integration (see Sect. 4). He could show lower and upper bounds on the Besov norm of the local discrepancy of point sets in the unit square. His results are valid for the parameter range $1 \leq p, q \leq \infty$, where $q < \infty$ if $p = 1$ and $q > 1$ if $p = \infty$, and for those $r \in \mathbb{R}$ such that $\frac{1}{p} - 1 < r < \frac{1}{p}$. For these choices of p, q and r he proved that for all $N \in \mathbb{N}$ the local discrepancy of any N-element point set $\mathscr{P}$ in $[0, 1)^2$ satisfies

$$\|D_{\mathscr{P}}|S^r_{p,q}B([0,1)^2)\| \gtrsim N^{r-1}(\log N)^{\frac{1}{q}}. \tag{2}$$

Concerning lower bounds of this kind, the natural question arises whether there exist point sets which match such a bound. Triebel was able to show that for any $N \geq 2$ there exists a point set $\mathscr{P}$ in $[0, 1)^2$ with N points such that

$$\|D_{\mathscr{P}}|S^r_{p,q}B([0,1)^2)\| \lesssim N^{r-1}(\log N)^{\left(\frac{1}{q}+1-r\right)}.$$

Hence, there remained a gap between the exponents of the lower and the upper bounds. This gap was closed by Hinrichs in [5] for the smoothness range $0 \leq r < \frac{1}{p}$, showing that the lower bound (2) is sharp in this case. He used Hammersley type point sets $\mathscr{R}_n$ as introduced below. It follows from his proof that these point sets can not be used to close the gap also for the parameter range $1/p - 1 < r < 0$. It remained an open problem to find a point set which closes this gap for negative smoothness. This problem was again mentioned in [6, Problem 3] (here also for higher dimensions) and [16, Remark 6.8]. It is the aim of this note to show that a solution is possible by applying some simple modifications to the point sets $\mathcal{R}_n$, which will lead to the main result of this note.

In [5] Hinrichs studied the class of Hammersley type point sets

$$\mathcal{R}_n := \left\{ \left(\frac{t_n}{2} + \frac{t_{n-1}}{2^2} + \cdots + \frac{t_1}{2^n}, \frac{s_1}{2} + \frac{s_2}{2^2} + \cdots + \frac{s_n}{2^n} \right) | t_1, \ldots, t_n \in \{0, 1\} \right\}$$

for $n \in \mathbb{N}$, where $s_i = t_i$ or $s_i = 1 - t_i$ depending on i. It is obvious that $\mathcal{R}_n$ has 2^n elements. We fix $\mathcal{R}_n$ and introduce three connected point sets by

$$\begin{aligned}\mathcal{R}_{n,1} &:= \{(x, 1-y) | (x, y) \in \mathcal{R}_n\},\\ \mathcal{R}_{n,2} &:= \{(1-x, y) | (x, y) \in \mathcal{R}_n\},\\ \mathcal{R}_{n,3} &:= \{(1-x, 1-y) | (x, y) \in \mathcal{R}_n\}.\end{aligned}$$

We set $\widetilde{\mathcal{R}}_n := \mathcal{R}_n \cup \mathcal{R}_{n,1} \cup \mathcal{R}_{n,2} \cup \mathcal{R}_{n,3}$ and call $\widetilde{\mathcal{R}}_n$ a symmetrized Hammersley type point set. In literature one often finds a symmetrization in the sense of Davenport [1], which would be $\mathcal{R}_n \cup \mathcal{R}_{n,1}$. However, for our purposes we need to work with the point sets $\widetilde{\mathcal{R}}_n$, which have $N = 2^{n+2}$ elements, where some points might coincide. With the point sets $\widetilde{\mathcal{R}}_n$ we have the following main result of this note.

Theorem 1 *Let* $1 \le p, q \le \infty$ *and* $r \in \mathbb{R}$ *such that* $1/p - 1 < r < 1/p$. *Then the point sets* $\widetilde{\mathcal{R}}_n$ *in* $[0, 1)^2$ *with* $N = 2^{n+2}$ *elements satisfy*

$$\|D_{\widetilde{\mathcal{R}}_n} | S^r_{p,q} B([0, 1)^2)\| \lesssim N^{r-1} (\log N)^{1/q}.$$

We would like to stress again that our result improves on [5, Theorem 1.1] in the sense that we extended the range for the smoothness parameter r to negative values.

The rest of this note is structured as follows. In Sect. 2 we introduce the Besov spaces $S^r_{p,q} B([0, 1)^2)$ of dominating mixed smoothness and explain how these function spaces can be characterized in terms of Haar functions. We will employ this characterization in Sect. 3 to proof Theorem 1. While Sects. 2 and 3 only cover the Besov spaces, we introduce further functions spaces with dominating mixed smoothness in Sect. 4, namely Triebel–Lizorkin spaces $S^r_{p,q} F([0, 1)^2)$ and Sobolev spaces $S^r_p H([0, 1)^2)$. It is possible to derive discrepancy results in these function spaces from Theorem 1 via embedding theorems. In the same section, we also explain the relation between discrepancy and numerical integration of functions in the mentioned spaces using quasi-Monte Carlo algorithms. We close with a discussion on a possible generalization of Theorem 1 to higher dimensions in the final Sect. 5.

2 Preliminaries

We give a definition of the Besov spaces with dominating mixed smoothness. Let therefore $\mathscr{S}(\mathbb{R}^2)$ denote the Schwartz space and $\mathscr{S}'(\mathbb{R}^2)$ the space of tempered distributions on $\mathbb{R}^2$. For $f \in \mathscr{S}'(\mathbb{R}^2)$ we denote by $\mathscr{F} f$ the Fourier transform of f and by $\mathscr{F}^{-1} f$ its inverse. Let $\phi_0 \in \mathscr{S}(\mathbb{R})$ satisfy $\phi_0(t) = 1$ for $|t| \leq 1$ and $\phi_0(t) = 0$ for $|t| > \frac{3}{2}$. Let

$$\phi_k(t) = \phi_0(2^{-k}t) - \phi_0(2^{-k+1}t),$$

where $t \in \mathbb{R}$, $k \in \mathbb{N}$, and $\phi_{\boldsymbol{k}}(\boldsymbol{t}) = \phi_{k_1}(t_1)\phi_{k_2}(t_2)$ for $\boldsymbol{k} = (k_1, k_2) \in \mathbb{N}_0^2$, $\boldsymbol{t} = (t_1, t_2) \in \mathbb{R}^2$. We note that $\sum_{\boldsymbol{k}\in\mathbb{N}_0^2} \phi_{\boldsymbol{k}}(\boldsymbol{t}) = 1$ for all $\boldsymbol{t} \in \mathbb{R}^2$. The functions $\mathscr{F}^{-1}(\phi_{\boldsymbol{k}}\mathscr{F} f)$ are entire analytic functions for any $f \in \mathscr{S}'(\mathbb{R}^2)$. Let $0 < p, q \leq \infty$ and $r \in \mathbb{R}$. The Besov space $S^r_{p,q}B(\mathbb{R}^2)$ with dominating mixed smoothness consists of all $f \in \mathscr{S}'(\mathbb{R}^2)$ with finite quasi-norm

$$\left\| f | S^r_{p,q}B(\mathbb{R}^2) \right\| = \left(\sum_{\boldsymbol{k}\in\mathbb{N}_0^2} 2^{r(k_1+k_2)q} \left\| \mathscr{F}^{-1}(\phi_{\boldsymbol{k}}\mathscr{F} f) | L_p(\mathbb{R}^2) \right\|^q \right)^{\frac{1}{q}},$$

with the usual modification if $q = \infty$. Let $\mathscr{D}([0, 1)^2)$ be the set of all complex-valued infinitely differentiable functions on $\mathbb{R}^2$ with compact support in the interior of $[0, 1)^2$ and let $\mathscr{D}'([0, 1)^2)$ be its dual space of all distributions in $[0, 1)^2$. The Besov space $S^r_{p,q}B([0, 1)^2)$ of dominating mixed smoothness on the domain $[0, 1)^2$ consists of all functions $f \in \mathscr{D}'([0, 1)^2)$ with finite quasi norm

$$\left\| f | S^r_{p,q}B([0, 1)^2) \right\| = \inf \left\{ \left\| g | S^r_{p,q}B(\mathbb{R}^2) \right\| : g \in S^r_{p,q}B(\mathbb{R}^2),\ g|_{[0,1)^2} = f \right\}.$$

These function spaces are independent of the choice of ϕ_0, as mentioned for instance in [14, Remark 1.39].

Actually, we will not make use of this technical definition. For our approach it is more convenient to employ a characterization of Besov spaces via Haar functions, which we define in the following.

A dyadic interval of length 2^{-j}, $j \in \mathbb{N}_0$, in $[0, 1)$ is an interval of the form

$$I = I_{j,m} := \left[\frac{m}{2^j}, \frac{m+1}{2^j} \right) \quad \text{for } m = 0, 1, \ldots, 2^j - 1.$$

We also define $I_{-1,0} = [0, 1)$. The left and right half of $I_{j,m}$ are the dyadic intervals $I_{j+1,2m}$ and $I_{j+1,2m+1}$, respectively. For $j \in \mathbb{N}_0$, the Haar function $h_{j,m}$ is the function on $[0, 1)$ which is $+1$ on the left half of $I_{j,m}$, -1 on the right half of $I_{j,m}$ and 0 outside of $I_{j,m}$. The L_∞-normalized Haar system consists of all Haar functions $h_{j,m}$ with $j \in \mathbb{N}_0$ and $m = 0, 1, \ldots, 2^j - 1$ together with the indicator function $h_{-1,0}$ of $[0, 1)$. Normalized in $L_2([0, 1))$ we obtain the orthonormal Haar basis of $L_2([0, 1))$.

Let $\mathbb{N}_{-1} = \mathbb{N}_0 \cup \{-1\}$ and define $\mathbb{D}_j = \{0, 1, \ldots, 2^j - 1\}$ for $j \in \mathbb{N}_0$ and $\mathbb{D}_{-1} = \{0\}$. For $\boldsymbol{j} = (j_1, j_2) \in \mathbb{N}_{-1}^2$ and $\boldsymbol{m} = (m_1, m_2) \in \mathbb{D}_{\boldsymbol{j}} := \mathbb{D}_{j_1} \times \mathbb{D}_{j_2}$, the Haar function $h_{\boldsymbol{j},\boldsymbol{m}}$ is given as the tensor product $h_{\boldsymbol{j},\boldsymbol{m}}(\boldsymbol{t}) = h_{j_1,m_1}(t_1)h_{j_2,m_2}(t_2)$ for $\boldsymbol{t} = (t_1, t_2) \in [0, 1)^2$. We speak of $I_{\boldsymbol{j},\boldsymbol{m}} = I_{j_1,m_1} \times I_{j_2,m_2}$ as dyadic boxes.

We have the following crucial result [14, Theorem 2.41].

Proposition 1 *Let $0 < p, q \leq \infty$, $1 < q \leq \infty$ if $p = \infty$, and $\frac{1}{p} - 1 < r < \min\left\{\frac{1}{p}, 1\right\}$. Let $f \in \mathscr{D}'([0, 1)^2)$. Then $f \in S^r_{p,q}B([0, 1)^2)$ if and only if it can be represented as*

$$f = \sum_{\boldsymbol{j}\in\mathbb{N}_{-1}^2} \sum_{\boldsymbol{m}\in\mathbb{D}_{\boldsymbol{j}}} \mu_{\boldsymbol{j},\boldsymbol{m}} 2^{\max\{0,j_1\}+\max\{0,j_2\}} h_{\boldsymbol{j},\boldsymbol{m}}$$

for some sequence $(\mu_{\boldsymbol{j},\boldsymbol{m}})$ satisfying

$$\left(\sum_{\boldsymbol{j}\in\mathbb{N}_{-1}^2} 2^{(j_1+j_2)\left(r-\frac{1}{p}+1\right)q} \left(\sum_{\boldsymbol{m}\in\mathbb{D}_{\boldsymbol{j}}} \left|\mu_{\boldsymbol{j},\boldsymbol{m}}\right|^p \right)^{\frac{q}{p}} \right)^{\frac{1}{q}} < \infty,$$

where the convergence is unconditional in $\mathscr{D}'([0, 1)^2)$ and in any $S^\rho_{p,q}B([0, 1)^2)$ with $\rho < r$. This representation of f is unique with the Haar coefficients

$$\mu_{\boldsymbol{j},\boldsymbol{m}} = \mu_{\boldsymbol{j},\boldsymbol{m}}(f) = \int_{[0,1)^2} f(\boldsymbol{t}) h_{\boldsymbol{j},\boldsymbol{m}}(\boldsymbol{t}) \mathrm{d}\boldsymbol{t}.$$

The expression on the left-hand-side of the above inequality provides an equivalent quasi-norm on $S^r_{p,q}B([0, 1)^2)$, i.e.

$$\left\| f | S^r_{p,q}B([0, 1)^2) \right\| \asymp \left(\sum_{\boldsymbol{j}\in\mathbb{N}_{-1}^2} 2^{(j_1+j_2)\left(r-\frac{1}{p}+1\right)q} \left(\sum_{\boldsymbol{m}\in\mathbb{D}_{\boldsymbol{j}}} \left|\mu_{\boldsymbol{j},\boldsymbol{m}}\right|^p \right)^{\frac{q}{p}} \right)^{\frac{1}{q}}.$$

We will follow the same approach as Hinrichs and first estimate the Haar coefficients of $D_{\widetilde{\mathscr{R}}_n}$ and then apply Proposition 1. This note is therefore similar in structure to [5] and uses several results from there.

3 Proof of Theorem 1

To begin with, we state several auxiliary results from [5, Lemmas 3.2–3.4, 3.6].

Lemma 1 *Let $f(\boldsymbol{t}) = t_1 t_2$ for $\boldsymbol{t} = (t_1, t_2) \in [0, 1)^2$. For $\boldsymbol{j} \in \mathbb{N}_{-1}^2$ and $\boldsymbol{m} \in \mathbb{D}_{\boldsymbol{j}}$ let $\mu_{\boldsymbol{j},\boldsymbol{m}}$ be the Haar coefficients of f. Then*

(i) *If* $\boldsymbol{j} = (j_1, j_2) \in \mathbb{N}_0^2$ *then* $\mu_{\boldsymbol{j},\boldsymbol{m}} = 2^{-2(j_1+j_2+2)}$.
(ii) *If* $\boldsymbol{j} = (-1, k)$ *or* $\boldsymbol{j} = (k, -1)$ *with* $k \in \mathbb{N}_0$ *then* $\mu_{\boldsymbol{j},\boldsymbol{m}} = -2^{-(2k+3)}$.

Lemma 2 *Fix* $\boldsymbol{z} = (z_1, z_2) \in [0, 1)^2$ *and let* $f(\boldsymbol{t}) = \mathbf{1}_{[0,\boldsymbol{t})}(\boldsymbol{z})$ *for* $\boldsymbol{t} = (t_1, t_2) \in [0, 1)^2$. *For* $\boldsymbol{j} \in \mathbb{N}_{-1}^2$ *and* $\boldsymbol{m} = (m_1, m_2) \in \mathbb{D}_{\boldsymbol{j}}$ *let* $\mu_{\boldsymbol{j},\boldsymbol{m}}$ *be the Haar coefficients of* f. *Then* $\mu_{\boldsymbol{j},\boldsymbol{m}} = 0$ *whenever* $\boldsymbol{z} \notin \mathring{I}_{\boldsymbol{j},\boldsymbol{m}}$, *where* $\mathring{I}_{\boldsymbol{j},\boldsymbol{m}}$ *denotes the interior of* $I_{\boldsymbol{j},\boldsymbol{m}}$. *If* $\boldsymbol{z} \in \mathring{I}_{\boldsymbol{j},\boldsymbol{m}}$ *then*

(i) *If* $\boldsymbol{j} = (j_1, j_2) \in \mathbb{N}_0^2$ *then*

$$\mu_{\boldsymbol{j},\boldsymbol{m}} = 2^{-(j_1+j_2+2)}(1 - |2m_1 + 1 - 2^{j_1+1}z_1|)(1 - |2m_2 + 1 - 2^{j_2+1}z_2|).$$

(ii) *If* $\boldsymbol{j} = (-1, k)$, $k \in \mathbb{N}_0$, *then* $\mu_{\boldsymbol{j},\boldsymbol{m}} = -2^{-(k+1)}(1 - z_1)(1 - |2m_2 + 1 - 2^{k+1} z_2|)$.
(iii) *If* $\boldsymbol{j} = (k, -1)$, $k \in \mathbb{N}_0$, *then* $\mu_{\boldsymbol{j},\boldsymbol{m}} = -2^{-(k+1)}(1 - z_2)(1 - |2m_1 + 1 - 2^{k+1} z_1|)$.

Lemma 3 *Let* $\mathscr{R}_n$ *be a Hammersley type point set with* 2^n *points. Let* $\boldsymbol{j} = (j_1, j_2) \in \mathbb{N}_0^2$ *and* $\boldsymbol{m} = (m_1, m_2) \in \mathbb{D}_{\boldsymbol{j}}$. *Then, if* $j_1 + j_2 < n$,

$$\sum_{\boldsymbol{z}\in\mathscr{R}_n\cap\mathring{I}_{\boldsymbol{j},\boldsymbol{m}}} (1 - |2m_1 + 1 - 2^{j_1+1}z_1|) = \sum_{\boldsymbol{z}\in\mathscr{R}_n\cap\mathring{I}_{\boldsymbol{j},\boldsymbol{m}}} (1 - |2m_2 + 1 - 2^{j_2+1}z_2|) = 2^{n-j_1-j_2-1}$$

and, if $j_1 + j_2 < n - 1$,

$$\sum_{\boldsymbol{z}\in\mathscr{R}_n\cap\mathring{I}_{\boldsymbol{j},\boldsymbol{m}}} (1 - |2m_1 + 1 - 2^{j_1+1}z_1|)(1 - |2m_2 + 1 - 2^{j_2+1}z_2|) = 2^{n-j_1-j_2-2} + 2^{j_1+j_2-n}.$$

Now we are ready to compute the Haar coefficients of $D_{\widetilde{\mathscr{R}}_n}$.

Proposition 2 *Let* $\widetilde{\mathscr{R}}_n$ *be a symmetrized Hammersley type point set with* $N = 2^{n+2}$ *elements and let* f *be the local discrepancy of* $\widetilde{\mathscr{R}}_n$ *and* $\mu_{\boldsymbol{j},\boldsymbol{m}}$ *the Haar coefficients of* f *for* $\boldsymbol{j} \in \mathbb{N}_{-1}^2$ *and* $\boldsymbol{m} = (m_1, m_2) \in \mathbb{D}_{\boldsymbol{j}}$.

Let $\boldsymbol{j} = (j_1, j_2) \in \mathbb{N}_0^2$. *Then*

(i) *if* $j_1 + j_2 < n - 1$ *and* $j_1, j_2 \geq 0$ *then* $|\mu_{\boldsymbol{j},\boldsymbol{m}}| = 2^{-2(n+1)}$.
(ii) *if* $j_1 + j_2 \geq n - 1$ *and* $0 \leq j_1, j_2 \leq n$ *then* $|\mu_{\boldsymbol{j},\boldsymbol{m}}| \leq 2^{-(n+j_1+j_2)}$ *and* $|\mu_{\boldsymbol{j},\boldsymbol{m}}| = 2^{-2(j_1+j_2+2)}$ *for all but at most* 2^{n+2} *coefficients* $\mu_{\boldsymbol{j},\boldsymbol{m}}$ *with* $\boldsymbol{m} \in \mathbb{D}_{\boldsymbol{j}}$.
(iii) *if* $j_1 \geq n$ *or* $j_2 \geq n$ *then* $|\mu_{\boldsymbol{j},\boldsymbol{m}}| = 2^{-2(j_1+j_2+2)}$.

Now let $\boldsymbol{j} = (-1, k)$ *or* $\boldsymbol{j} = (k, -1)$ *with* $k \in \mathbb{N}_0$. *Then*

(iv) *if* $k < n$ *then* $\mu_{\boldsymbol{j},\boldsymbol{m}} = 0$.
(v) *if* $k \geq n$ *then* $|\mu_{\boldsymbol{j},\boldsymbol{m}}| = 2^{-(2k+3)}$.

Finally,

(vi) $\mu_{(-1,-1),(0,0)} = 0$.

Proof The cases (iii) and (v) follow from the fact that no elements of $\widetilde{\mathscr{R}}_n$ are contained in the interior of a dyadic box $I_{(j_1,j_2),\boldsymbol{m}}$ if $j_1 \geq n$ or $j_2 \geq n$, together with Lemma 1. We consider the case (ii). For a fixed $\boldsymbol{j} = (j_1, j_2)$ the interiors of the dyadic boxes $I_{\boldsymbol{j},\boldsymbol{m}}$ for $\boldsymbol{m} \in \mathbb{D}_{\boldsymbol{j}}$ are mutually disjunct and at most 2^{n+2} of these boxes can contain points from $\widetilde{\mathscr{R}}_n$. We have $\mu_{\boldsymbol{j},\boldsymbol{m}} = 2^{-2(j_1+j_2+2)}$ if the corresponding box $I_{\boldsymbol{j},\boldsymbol{m}}$ is empty. The other boxes contain at most 8 points (because the volume of $I_{\boldsymbol{j},\boldsymbol{m}}$ is at most $2^{-(n-1)}$ due to the condition $j_1 + j_2 \geq n - 1$ and because of the net property of $\mathscr{R}_n$ and its connected point sets). Together with the first part of Lemma 2 and the triangle inequality this yields $|\mu_{\boldsymbol{j},\boldsymbol{m}}| \leq 8 \cdot 2^{-(n+2)} 2^{-(j_1+j_2+2)} + 2^{-2(j_1+j_2+2)} \leq 2^{-(n+j_1+j_2)}$.

The case (vi) can be seen as follows:

$$\begin{aligned}
\mu_{(-1,-1),(0,0)} &= \int_0^1 \int_0^1 D_{\widetilde{\mathscr{R}}_n}(t_1, t_2) \mathrm{d}t_1 \mathrm{d}t_2 = \frac{1}{N} \sum_{z \in \widetilde{\mathscr{R}}_n} \int_{z_1}^1 \int_{z_2}^1 1 \mathrm{d}t_1 \mathrm{d}t_2 - \int_0^1 \int_0^1 t_1 t_2 \mathrm{d}t_1 \mathrm{d}t_2 \\
&= \frac{1}{2^{n+2}} \sum_{z \in \widetilde{\mathscr{R}}_n} (1 - z_1)(1 - z_2) - \frac{1}{4} \\
&= \frac{1}{2^{n+2}} \sum_{(x,y) \in \mathscr{R}_n} [(1-x)(1-y) + (1-x)y + x(1-y) + xy] - \frac{1}{4} \\
&= \frac{1}{2^{n+2}} \sum_{(x,y) \in \mathscr{R}_n} 1 - \frac{1}{4} = \frac{1}{2^{n+2}} 2^n - \frac{1}{4} = 0.
\end{aligned}$$

To show the claim in (iv) for the case $\boldsymbol{j} = (k, -1)$ with $k \in \mathbb{N}_0$, $k < n$, we have to consider the expression

$$S := \sum_{z \in \widetilde{\mathscr{R}}_n \cap \mathring{I}_{(k,-1),(m_1,0)}} (1 - |2m_1 + 1 - 2^{k+1} z_1|)(1 - z_2)$$

for any $m_1 \in \{0, \ldots, 2^k - 1\}$. We can write

$$\begin{aligned}
S = &\sum_{(x,y) \in \mathscr{R}_n \cap \mathring{I}_{(k,-1),(m_1,0)}} (1 - |2m_1 + 1 - 2^{k+1} x|)(1 - y) \\
&+ \sum_{(x,1-y) \in \mathscr{R}_n \cap \mathring{I}_{(k,-1),(m_1,0)}} (1 - |2m_1 + 1 - 2^{k+1} x|) y \\
&+ \sum_{(1-x,y) \in \mathscr{R}_n \cap \mathring{I}_{(k,-1),(m_1,0)}} (1 - |2m_1 + 1 - 2^{k+1}(1 - x)|)(1 - y) \\
&+ \sum_{(1-x,1-y) \in \mathscr{R}_n \cap \mathring{I}_{(k,-1),(m_1,0)}} (1 - |2m_1 + 1 - 2^{k+1}(1 - x)|) y \\
= &\sum_{(x,y) \in \mathscr{R}_n \cap \mathring{I}_{(k,-1),(m_1,0)}} (1 - |2m_1 + 1 - 2^{k+1} x|)
\end{aligned}$$

$$+ \sum_{(1-x,y)\in\mathscr{R}_n\cap\mathring{I}_{(k,-1),(m_1,0)}} (1-|2m_1+1-2^{k+1}(1-x)|) =: S_1+S_2,$$

where we used the obvious equivalences $(x,y)\in\mathscr{R}_n\cap\mathring{I}_{(k,-1),(m_1,0)}$ if and only if $(x,1-y)\in\mathscr{R}_n\cap\mathring{I}_{(k,-1),(m_1,0)}$ as well as $(1-x,y)\in\mathscr{R}_n\cap\mathring{I}_{(k,-1),(m_1,0)}$ if and only if $(1-x,1-y)\in\mathscr{R}_n\cap\mathring{I}_{(k,-1),(m_1,0)}$ in the last step. Since the interval $\mathring{I}_{(k,-1),(m_1,0)}$ is the same as $\mathring{I}_{(k,0),(m_1,0)}$, we obtain $S_1=2^{n-k-1}$ from the first part of Lemma 3. To evaluate S_2 we observe that

$$1-x\in\mathring{I}_{k,m_1}\Leftrightarrow\frac{m_1}{2^k}<1-x<\frac{m_1+1}{2^k}\Leftrightarrow\frac{2^k-1-m_1}{2^k}<x<\frac{2^k-m_1}{2^k}\Leftrightarrow x\in\mathring{I}_{k,\widetilde{m}_1},$$

where we set $\widetilde{m}_1=2^k-1-m_1$. This yields the equivalence of $(1-x,y)\in\mathscr{R}_n\cap\mathring{I}_{(k,-1),(m_1,0)}$ and $(x,y)\in\mathscr{R}_n\cap\mathring{I}_{(k,-1),(\widetilde{m}_1,0)}$. We also find

$$\begin{aligned}|2m_1+1-2^{k+1}(1-x)| &= |2(m_1+1-2^k)-1+2^{k+1}x|\\ &= |-2\widetilde{m}_1-1+2^{k+1}x| = |2\widetilde{m}_1+1-2^{k+1}x|\end{aligned}$$

and hence we obtain

$$S_2=\sum_{(x,y)\in\mathscr{R}_n\cap\mathring{I}_{(k,-1),(\widetilde{m}_1,0)}}(1-|2\widetilde{m}_1+1-2^{k+1}x|)=2^{n-k-1},$$

where we regarded the first part of Lemma 3 again. Altogether, we have

$$\begin{aligned}\mu_{(k,-1),(m_1,0)} &= -\frac{1}{N}2^{-(k+1)}(S_1+S_2)-(-2^{-(2k+3)})\\ &= -2^{-(n+2)}2^{-(k+1)}2^{n-k}+2^{-(2k+3)}=0\end{aligned}$$

with Lemmas 1 and 2, and this part of the proposition is verified. It is clear that the result for $\mu_{(-1,k),(0,m_2)}$ if $k<n$ can be shown analogously.

Finally, we prove (i) and therefore have to analyze the sum

$$T:=\sum_{z\in\widetilde{\mathscr{R}}_n\cap\mathring{I}_{\boldsymbol{j},\boldsymbol{m}}}(1-|2m_1+1-2^{j_1+1}z_1|)(1-|2m_2+1-2^{j_2+1}z_2|),$$

where $\boldsymbol{j}=(j_1,j_2)\in\mathbb{N}_0^2$ with $j_1+j_2<n-1$. We have

$$\begin{aligned}T=&\sum_{(x,y)\in\mathscr{R}_n\cap\mathring{I}_{\boldsymbol{j},\boldsymbol{m}}}(1-|2m_1+1-2^{j_1+1}x|)(1-|2m_2+1-2^{j_2+1}y|)\\ &+\sum_{(x,1-y)\in\mathscr{R}_n\cap\mathring{I}_{\boldsymbol{j},\boldsymbol{m}}}(1-|2m_1+1-2^{j_1+1}x|)(1-|2m_2+1-2^{j_2+1}(1-y)|)\\ &+\sum_{(1-x,y)\in\mathscr{R}_n\cap\mathring{I}_{\boldsymbol{j},\boldsymbol{m}}}(1-|2m_1+1-2^{j_1+1}(1-x)|)(1-|2m_2+1-2^{j_2+1}y|)\end{aligned}$$

$$+ \sum_{(1-x,1-y)\in\mathscr{R}_n\cap\mathring{I}_{j,m}} (1-|2m_1+1-2^{j_1+1}(1-x)|)(1-|2m_2+1-2^{j_2+1}(1-y)|)$$
$$=: T_1 + T_2 + T_3 + T_4.$$

We obtain directly from the second part of Lemma 3 that $T_1 = 2^{n-j_1-j_2-2} + 2^{j_1+j_2-n}$. With the same arguments as in the proof of (iv) we can show

$$T_2 = \sum_{(x,y)\in\mathscr{R}_n\cap\mathring{I}_{j,(m_1,\widetilde{m}_2)}} (1-|2m_1+1-2^{j_1+1}x|)(1-|2\widetilde{m}_2+1-2^{j_2+1}y|),$$
$$T_3 = \sum_{(x,y)\in\mathscr{R}_n\cap\mathring{I}_{j,(\widetilde{m}_1,m_2)}} (1-|2\widetilde{m}_1+1-2^{j_1+1}x|)(1-|2m_2+1-2^{j_2+1}y|),$$
$$T_4 = \sum_{(x,y)\in\mathscr{R}_n\cap\mathring{I}_{j,(\widetilde{m}_1,\widetilde{m}_2)}} (1-|2\widetilde{m}_1+1-2^{j_1+1}x|)(1-|2\widetilde{m}_2+1-2^{j_2+1}y|),$$

where $\widetilde{m}_i = 2^{j_i} - 1 - m_i$ for $i \in \{1, 2\}$. But from this and Lemma 3 we see that $T_2 = T_3 = T_4 = T_1$ and together with Lemmas 1 and 2

$$\mu_{j,m} = \frac{1}{N} 2^{-(j_1+j_2+2)}(T_1+T_2+T_3+T_4) - 2^{-2(j_1+j_2+2)}$$
$$= 2^{-(n+2)} 2^{-(j_1+j_2+2)}(2^{n-j_1-j_2} + 2^{j_1+j_2-n+2}) - 2^{-2(j_1+j_2+2)} = 2^{-2(n+1)}$$

as claimed. The proof of the proposition is complete. □

Now we are able to prove Theorem 1.

Proof We consider any symmetrized Hammersley type point set $\widetilde{\mathscr{R}}_n$ (we do not have to specify the dependence of the s_i on t_i in the definition of $\mathscr{R}_n$). For $\boldsymbol{j} \in \mathbb{N}_{-1}^2$ and $\boldsymbol{m} \in \mathbb{D}_{\boldsymbol{j}}$ let $\mu_{\boldsymbol{j},\boldsymbol{m}}$ be the Haar coefficients of the local discrepancy of $\widetilde{\mathscr{R}}_n$. According to Proposition 1, it suffices to show that for all p, q, r satisfying the conditions in Theorem 1 we have

$$\left(\sum_{\boldsymbol{j}\in\mathbb{N}_{-1}^2} 2^{(j_1+j_2)\left(r-\frac{1}{p}+1\right)q} \left(\sum_{\boldsymbol{m}\in\mathbb{D}_{\boldsymbol{j}}} |\mu_{\boldsymbol{j},\boldsymbol{m}}|^p \right)^{\frac{q}{p}} \right)^{\frac{1}{q}} \lesssim 2^{n(r-1)} n^{1/q}. \tag{3}$$

This yields

$$\|D_{\widetilde{\mathscr{R}}_n} | S_{p,q}^r B([0,1)^2)\| \lesssim 2^{-2(r-1)} 2^{(n+2)(r-1)} (n+2)^{1/q} \lesssim N^{r-1} (\log N)^{1/q}.$$

To verify (3), we split the sum over $\boldsymbol{j}$ in six cases according to Proposition 2 (and thereby applying Minkowski's inequality). We remark that the cases (i), (ii), (iii)

and (v) have already been treated in [5, Sect. 4], since in these cases the bounds on the Haar coefficients of $D_{\mathscr{R}_n}$ are (basically) the same as those for the Haar coefficients of $D_{\widetilde{\mathscr{R}}_n}$. In all cases Hinrichs obtained an upper bound of the form $c2^{n(r-1)}n^{1/q}$ with c independent of n for the whole parameter range $1/p - 1 < r < 1/p$. The only cases where the condition $r \geq 0$ was necessary were (iv) and (vi). However, the symmetrization of $\mathscr{R}_n$ has the effect that the corresponding Haar coefficients of $D_{\widetilde{\mathscr{R}}_n}$ vanish in these two cases, and the result follows. □

Remark 1 Let f be the local discrepancy of the point set $\mathscr{R}_n \cup \mathscr{R}_{n,1}$ and $\mu_{\boldsymbol{j},\boldsymbol{m}}$ for $\boldsymbol{j} \in \mathbb{N}_{-1}^2$ and $\boldsymbol{m} \in \mathbb{D}_{\boldsymbol{j}}$ be the corresponding Haar coefficients. Then one can show that $\mu_{(-1,-1),(0,0)} = 2^{-(n+2)}$ and $\mu_{(-1,k),(0,m_2)} = -2^{-(n+2k+3)} + 2^{-(2n+2)}T_k$ for $k \in \mathbb{N}_0$, $k < n$. Here, $T_k = 1$ if $s_{k+1} = t_{k+1}$ and $T_k = -1$ if $s_{k+1} = 1 - t_{k+1}$ in the definition of $\mathscr{R}_n$. Hence, the proof of Theorem 1 does not work for this class of point sets.

4 Discrepancy in Further Function Spaces and Numerical Integration

As pointed out in [9, 10, 14] one can easily deduce results on the discrepancy of point sets in Triebel–Lizorkin spaces from the discrepancy estimates in Besov spaces. Let $0 < p < \infty$, $0 < q \leq \infty$ and $r \in \mathbb{R}$. The Triebel–Lizorkin space $S_{p,q}^r F(\mathbb{R}^2)$ with dominating mixed smoothness consists of all $f \in \mathscr{S}'(\mathbb{R}^2)$ with finite quasi-norm

$$\|f|S_{p,q}^r F(\mathbb{R}^2)\| = \left\| \left(\sum_{\boldsymbol{k}\in\mathbb{N}_0^2} 2^{r(k_1+k_2)q} |\mathscr{F}^{-1}(\phi_{\boldsymbol{k}}\mathscr{F} f)(\cdot)|^q \right)^{1/q} |L_p(\mathbb{R}^2) \right\|$$

with the usual modification if $q = \infty$. The space $S_{p,q}^r F([0,1)^2)$ can be introduced analogously to $S_{p,q}^r B([0,1)^2)$. For $0 < p, q < \infty$ and $r \in \mathbb{R}$ we have the embeddings

$$S_{\max\{p,q\},q}^r B([0,1)^2) \hookrightarrow S_{p,q}^r F([0,1)^2) \hookrightarrow S_{\min\{p,q\},q}^r B([0,1)^2), \tag{4}$$

which were proven in [10, Corollary 1.13], based on other embedding theorems from [14, Remark 6.28] and [4, Proposition 2.3.7]. From the first embedding together with Theorem 1 we obtain

Corollary 1 *Let* $1 \leq p, q < \infty$ *and* $\frac{1}{\max\{p,q\}} - 1 < r < \frac{1}{\max\{p,q\}}$. *Then the point sets* $\widetilde{\mathscr{R}}_n$ *in* $[0,1)^2$ *with* $N = 2^{n+2}$ *elements satisfy*

$$\|D_{\mathscr{P}}|S_{p,q}^r F([0,1)^2)\| \lesssim N^{r-1}(\log N)^{1/q}.$$

This corollary improves on [9, Theorem 6.1], where Hammersley type point sets in arbitrary base $b \geq 2$ have been considered, by extending again the range of r to negative values. There exist corresponding lower bounds for the norm of the local discrepancy in Triebel–Lizorkin spaces for $\frac{1}{\min\{p,q\}} - 1 < r < \frac{1}{p}$ as shown in [10, Corollary 4.2].

For $1 < p < \infty$ the spaces $S_p^r H([0,1)^2) := S_{p,2}^r F([0,1)^2)$ are called Sobolev spaces with dominating mixed smoothness. By choosing $q = 2$ in Corollary 1 we obtain an analogous result on Sobolev spaces. Further, it is well known that $S_p^0 H([0,1)^2) = L_p([0,1)^2)$. Regarding this fact we derive from Corollary 1 that the symmetrized Hammersley type point sets achieve an L_p discrepancy of order $N^{-1}(\log N)^{1/2}$ for all $p \in [1,\infty)$, which is best possible in the sense of (1). This however is not so surprising, since in [8, Theorem 3] it has been shown that already a Davenport type symmetrization of $\mathscr{R}_n$ achieves the best possible rate of L_p discrepancy for all $p \in [1,\infty)$, i.e. $\|D_{\mathscr{R}_n \cup \mathscr{R}_{n,1}} | L_p([0,1)^2)\| \lesssim N^{-1}(\log N)^{1/2}$. By different means as used in this note, a certain type of symmetrized Hammersley point sets with the optimal order of L_p discrepancy in a prime base b has been studied by Goda [2, Theorem 3], which matches our construction of $\widetilde{\mathscr{R}}_n$ for $b = 2$. We observe that the construction of point sets with the optimal rate of discrepancy in Besov, Triebel–Lizorkin or Sobolev spaces with negative smoothness is even more subtle than to find point sets with the optimal order of L_p discrepancy.

Finally, we would like to add a few words concerning errors of quasi-Monte Carlo (QMC) methods for numerical integration in spaces with dominating mixed smoothness. For a function f in a normed space F of functions on $[0,1)^2$ we would like to approximate the integral $I(f) := \int_{[0,1)^2} f(\boldsymbol{x})\mathrm{d}\boldsymbol{x}$ by a QMC algorithm $Q_N(\mathscr{P}, f) = \frac{1}{N}\sum_{i=1}^N f(\boldsymbol{x}_i)$, where $\mathscr{P} = \{\boldsymbol{x}_1, \ldots, \boldsymbol{x}_N\}$ is a set of N points in the unit square. The minimal worst-case error of QMC algorithms with respect to a class of functions F is defined as

$$\mathrm{err}_N(F) := \inf_{\#\mathscr{P}=N} \sup_{\|f|F\|\le 1} |I(f) - Q_N(\mathscr{P}, f)|.$$

The infimum is extended over all point sets in $[0,1)^2$ with N elements and the supremum is extended over all functions in the unit ball of F. We state a remarkable connection between discrepancy and integration errors in Besov spaces. Let therefore

$$\mathrm{disc}_N(S_{p,q}^r B([0,1)^2)) := \inf_{\#\mathscr{P}=N} \|D_{\mathscr{P}}(\cdot) | S_{p,q}^r B([0,1)^2)\|.$$

It is known that $S_{p',q'}^{1-r} B([0,1)^2)^{\urcorner}$ with $1/p + 1/p' = 1/q + 1/q' = 1$ is the dual space of $S_{p,q}^r B([0,1)^2)$, where $S_{p',q'}^{1-r} B([0,1)^2)^{\urcorner}$ is the class of all functions in $S_{p',q'}^{1-r} B([0,1)^2)$ with zero boundary on the upper and right boundary line. Let $1 \le p, q \le \infty$ ($q < \infty$ if $p = 1$ and $q > 1$ if $p = \infty$) and $1/p < r < 1/p + 1$. Then we have for every integer $N \ge 2$

$$\mathrm{err}_N(S_{p,q}^r B([0,1)^2)^{\urcorner}) \asymp \mathrm{disc}_N(S_{p',q'}^{1-r} B([0,1)^2)), \tag{5}$$

which follows from [14, Theorem 6.11]. This relation leads to the following result:

Theorem 2 *Let* $1 \le p, q \le \infty$ *(*$q < \infty$ *if* $p = 1$ *and* $q > 1$ *if* $p = \infty$*) and* $1/p < r < 1 + 1/p$. *Then for* $N = 2^{n+2}$ *with* $n \in \mathbb{N}$ *we have*

$$\mathrm{err}_N(S^r_{p,q}B([0,1)^2)^\urcorner) \lesssim N^{-r}(\log N)^{1-1/q}.$$

Proof From (5) we have

$$\mathrm{err}_N(S^r_{p,q}B([0,1)^2)^\urcorner) \lesssim \mathrm{disc}_N(S^{1-r}_{p',q'}B([0,1)^2))$$

for $1/p < r < 1 + 1/p$. Theorem 1 yields further

$$\mathrm{disc}_N(S^{1-r}_{p',q'}B([0,1)^2)) \lesssim N^{1-r-1}(\log N)^{1/q'} = N^{-r}(\log N)^{1-1/q}$$

for $1/p' - 1 < 1 - r < 1/p'$. The last condition on r is equivalent to $1/p < r < 1 + 1/p$ and the result follows. □

We remark that there exists a corresponding lower bound on $\mathrm{err}_N(S^r_{p,q}B([0,1)^2)$ which shows that the rate of convergence in this theorem is optimal. The novelty of Theorem 2 is the fact that in the two-dimensional case for $1 < r < 1 + 1/p$ the optimal rate of convergence can be achieved with QMC rules (based on symmetrized Hammersley type point sets). Previously, this has only been shown for the smaller parameter range $1/p < r < 1$ in [10, Theorem 5.6] (but for arbitrary dimensions). The smoothness range, for which the optimal order for the worst-case integration error is achieved, can be further extended if one either considers one-periodic functions only (see [16] for the case $s = 2$ and [7] for a generalization to higher dimensions) or if one allows more general cubature rules that are not necessarily of QMC type. Results in this directions can be found for instance in [16], where Hammersley type point sets were used as integration nodes of non-QMC rules, and [17], where Frolov lattices were proven to yield optimal convergence rates also for higher dimensions and for all $r > 1/p$.

With similar arguments as above we obtain an analogous result on integration errors in Triebel–Lizorkin spaces (and hence in Sobolev spaces).

Corollary 2 *Let* $1 \le p, q \le \infty$ *and* $1/\min\{p,q\} < r < 1 + 1/\min\{p,q\}$. *Then for* $N = 2^{n+2}$ *with* $n \in \mathbb{N}$ *we have*

$$\mathrm{err}_N(S^r_{p,q}F([0,1)^2)^\urcorner) \lesssim N^{-r}(\log N)^{1-1/q}.$$

Proof This result is a consequence of the second embedding in (4), which implies

$$\mathrm{err}_N(S^r_{p,q}F([0,1)^2)^\urcorner) \le \mathrm{err}_N(S^r_{\min\{p,q\},q}B([0,1)^2)^\urcorner),$$

and Theorem 2. □

5 Concluding Remarks

In this note we proved optimal results for the Besov norm of the local discrepancy also in cases where the smoothness parameter r is negative. We only considered the two-dimensional case. However, the situation in higher dimensions s in the case $r < 0$ remains unsolved. According to Triebel, the Besov norm of the local discrepancy of an arbitrary N element point set $\mathscr{P}$ in $[0, 1)^s$ satisfies

$$\left\| D_{\mathscr{P}} | S^r_{p,q} B([0,1)^s) \right\| \gtrsim N^{r-1} (\log N)^{\frac{s-1}{q}},$$

where $1 \leq p, q \leq \infty$ and $\frac{1}{p} - 1 < r < \frac{1}{p}$. For nonnegative smoothness parameters, optimal discrepancy results for point set in $[0, 1)^s$ have been obtained by Markhasin in [10, 11] based on certain digital nets. Considering the ideas in this note, one might wonder whether an 2^s-fold symmetrization of suitable digital nets would yield optimal discrepancy results for $r < 0$ in dimension s. We leave this as an open problem, but remark that a proof might be difficult and technical. Due to the simple structure of the Hammersley point set it is possible to provide exact formulas for the sums in Lemma 3, which appear in the computation of the Haar coefficients. The precise values of these sums were important to show that the essential Haar coefficients of $D_{\widetilde{\mathscr{R}}_n}$ vanish. It is probably much harder to find formulas for similar sums in higher dimensions.

Acknowledgments The author is supported by the Austrian Science Fund (FWF): Project F5509-N26, which is a part of the Special Research Program "Quasi-Monte Carlo Methods: Theory and Applications".

References

1. Davenport, H.: Note on irregularities of distribution. Mathematika **3**, 131–135 (1956)
2. Goda, T.: The b-adic symmetrization of digital nets for quasi-Monte Carlo integration. Unif. Distrib. Theory **12**(1), 1–25 (2017)
3. Halász, G.: On Roth's method in the theory of irregularities of point distributions. Recent Progress in Analytic Number Theory, vol. 2, pp. 79–94. Academic Press, London (1981)
4. Hansen, M.: Nonlinear Approximation and Function Spaces of Dominating Mixed Smoothness. Dissertation, Jena (2010)
5. Hinrichs, A.: Discrepancy of Hammersley points in Besov spaces of dominating mixed smoothness. Math. Nachr. **283**, 478–488 (2010)
6. Hinrichs, A.: Discrepancy, integration and tractability. In: Dick, J., Kuo, F.Y., Peters, G.W., Sloan, I.H. (eds.) Monte Carlo and Quasi-Monte Carlo Methods 2012, pp. 129–172. Springer, Berlin (2013)
7. Hinrichs, A., Markhasin, L., Oettershagen, J., Ullrich, T.: Optimal quasi-Monte Carlo rules on order 2 digital nets for the numerical integration of multivariate periodic functions. Numerische Mathematik (to appear)
8. Hinrichs, A., Kritzinger, R., Pillichshammer, F.: Optimal order of L_p-discrepancy of digit shifted Hammersley point sets in dimension 2. Unif. Distrib. Theory **10**(1), 115–133 (2015)

9. Markhasin, L.: Discrepancy of generalized Hammersley type point sets in Besov spaces with dominating mixed smoothness. Unif. Distrib. Theory **8**(1), 135–164 (2013)
10. Markhasin, L.: Discrepancy and integration in function spaces with dominating mixed smoothness. Diss. Math. **494**, 1–81 (2013)
11. Markhasin, L.: L_p- and $S^r_{p,q}B$-discrepancy of (order 2) digital nets. Acta Arith. **168**, 139–159 (2015)
12. Roth, K.F.: On irregularities of distribution. Mathematika **1**, 73–79 (1954)
13. Schmidt, W.M.: Irregularities of distribution. X. Number Theory and Algebra, pp. 311–329. Academic Press, New York (1977)
14. Triebel, H.: Bases in Function Spaces, Sampling, Discrepancy, Numerical Integration. European Mathematical Society Publishing House, Zürich (2010)
15. Triebel, H.: Numerical integration and discrepancy, a new approach. Math. Nachr. **283**, 139–159 (2010)
16. Ullrich, T.: Optimal cubature in Besov spaces with dominating mixed smoothness on the unit square. J. Complex. **30**, 72–94 (2014)
17. Ullrich, M., Ullrich, T.: The role of Frolov's cubature formula for functions with bounded mixed derivative. SIAM J. Numer. Anal. **54**(2), 969–993 (2016)

A Reduced Fast Construction of Polynomial Lattice Point Sets with Low Weighted Star Discrepancy

Ralph Kritzinger, Helene Laimer and Mario Neumüller

Abstract The weighted star discrepancy is a quantitative measure for the performance of point sets in quasi-Monte Carlo algorithms for numerical integration. We consider polynomial lattice point sets, whose generating vectors can be obtained by a component-by-component construction to ensure a small weighted star discrepancy. Our aim is to significantly reduce the construction cost of such generating vectors by restricting the size of the set of polynomials from which we select the components of the vectors. To gain this reduction we exploit the fact that the weights of the spaces we consider decay very fast.

Keywords weighted star discrepancy · polynomial lattice point sets · quasi-Monte Carlo integration · component-by-component algorithm

1 Introduction

A convenient way to approximate the value of an integral $I_s(F) := \int_{[0,1)^s} F(\boldsymbol{x})\,\mathrm{d}\boldsymbol{x}$ over the s-dimensional unit cube is to use a quasi-Monte Carlo rule of the form

$$Q_{N,s}(F) := \frac{1}{N}\sum_{n=0}^{N-1} F(\boldsymbol{x}_n). \tag{1}$$

R. Kritzinger · M. Neumüller (✉)
Institut für Finanzmathematik und Angewandte Zahlentheorie, Johannes Kepler Universität Linz, Altenbergerstr. 69, 4040 Linz, Austria
e-mail: mario.neumueller@jku.at

R. Kritzinger
e-mail: ralph.kritzinger@jku.at

H. Laimer
Johann Radon Institute for Computational and Applied Mathematics (RICAM), Austrian Academy of Sciences, Altenbergerstr. 69, 4040 Linz, Austria
e-mail: helene.laimer@ricam.oeaw.ac.at

A. B. Owen and P. W. Glynn (eds.), *Monte Carlo and Quasi-Monte Carlo Methods*, Springer Proceedings in Mathematics & Statistics 241,
https://doi.org/10.1007/978-3-319-91436-7_21

The integrand F usually stems from some suitable (weighted) function space and the multiset $\mathscr{P}$ of integration nodes $\boldsymbol{x}_0, \boldsymbol{x}_1, \ldots, \boldsymbol{x}_{N-1}$ in the algorithm $Q_{N,s}(F)$ is chosen deterministically from $[0, 1)^s$. For comprehensive information on quasi-Monte Carlo algorithms consult, e.g., [4, 9, 12]. The quality of a quasi-Monte Carlo rule is for instance measured by some notion of discrepancy. In this paper we consider the weighted star discrepancy, which has been introduced by Sloan and Woźniakowski in [20], exploiting the insight that the weights reflect the influence of different coordinates on the integration error. Let $[s] := \{1, 2, \ldots, s\}$ and consider a weight sequence $\boldsymbol{\gamma} = (\gamma_{\mathfrak{u}})_{\mathfrak{u}\subseteq[s]}$ of nonnegative real numbers, i.e., every group of variables $(x_i)_{i\in\mathfrak{u}}$ is equipped with a weight $\gamma_{\mathfrak{u}}$. Roughly speaking, a small weight indicates that the corresponding variables contribute little to the integration problem. For simplicity, throughout this paper we only consider product weights, defined as follows. Given a non-increasing sequence of positive real numbers $(\gamma_j)_{j\geq 1}$ with $\gamma_j \leq 1$ we set $\gamma_{\mathfrak{u}} := \prod_{j\in\mathfrak{u}} \gamma_j$ and $\gamma_\emptyset := 1$.

Definition 1 Let $\boldsymbol{\gamma} = (\gamma_{\mathfrak{u}})_{\mathfrak{u}\subseteq[s]}$ be a weight sequence and $\mathscr{P} = \{\boldsymbol{x}_0, \ldots, \boldsymbol{x}_{N-1}\} \subseteq [0, 1)^s$ be an N-element point set. The local discrepancy of the point set $\mathscr{P}$ at $\boldsymbol{t} = (t_1, \ldots, t_s) \in (0, 1]^s$ is defined as

$$\Delta(\boldsymbol{t}, \mathscr{P}) := \frac{1}{N} \sum_{n=0}^{N-1} \mathbb{1}_{[\boldsymbol{0},\boldsymbol{t})}(\boldsymbol{x}_n) - \prod_{j=1}^{s} t_j,$$

where $\mathbb{1}_{[\boldsymbol{0},\boldsymbol{t})}$ denotes the characteristic function of $[\boldsymbol{0}, \boldsymbol{t}) := [0, t_1) \times \cdots \times [0, t_s)$. The weighted star discrepancy of $\mathscr{P}$ is then defined as

$$D^*_{N,\boldsymbol{\gamma}}(\mathscr{P}) := \sup_{\boldsymbol{t}\in(0,1]^s} \max_{\emptyset\neq\mathfrak{u}\subseteq[s]} \gamma_{\mathfrak{u}} |\Delta((\boldsymbol{t}_{\mathfrak{u}}, \boldsymbol{1}), \mathscr{P})|,$$

where $(\boldsymbol{t}_{\mathfrak{u}}, \boldsymbol{1})$ denotes the vector $(\tilde{t}_1, \ldots, \tilde{t}_s)$ with $\tilde{t}_j = t_j$ if $j \in \mathfrak{u}$ and $\tilde{t}_j = 1$ if $j \notin \mathfrak{u}$.

A relation between the integration error of quasi-Monte Carlo rules and the weighted star discrepancy is given by the Koksma-Hlawka type inequality (see [20])

$$|Q_{N,s}(F) - I_s(F)| \leq D^*_{N,\boldsymbol{\gamma}}(\mathscr{P}) \|F\|_{\boldsymbol{\gamma}},$$

where $\|\cdot\|_{\boldsymbol{\gamma}}$ is some norm which depends only on the weight sequence $\boldsymbol{\gamma}$ but not on the point set $\mathscr{P}$.

It turns out that lattice point sets (see, e.g., [12, Chap. 5], [9]) and polynomial lattice point sets (see, e.g., [12, Chap. 4], [11], [4, Chap. 10]) are often a good choice as sample points in (1). These two kinds of point sets are strongly connected and have a lot of parallel tracks in their analysis. However, there are some situations were one type of point set is superior to the other in terms of error bounds or the size of the function classes where they yield good results for numerical integration. Thus it is beneficial to have constructions at hand for lattice point sets as well as for polynomial lattice point sets. For a detailed comparison of lattice point sets and polynomial lattice point sets see, e.g., [18]. In this paper we study polynomial lattice point sets, a special class of point sets with low weighted star discrepancy, introduced by Niederreiter in [12, Chap. 4], [11]. For a prime number p, let $\mathbb{F}_p$ be the finite field of order p. We identify $\mathbb{F}_p$ with the set $\{0, 1, \dots, p-1\}$ equipped with the modulo p arithmetic. We denote by $\mathbb{F}_p[x]$ the set of polynomials over $\mathbb{F}_p$ and by $\mathbb{F}_p((x^{-1}))$ the field of formal Laurent series over $\mathbb{F}_p$ with elements of the form $L = \sum_{l=\omega}^{\infty} t_l x^{-l}$, where $\omega \in \mathbb{Z}$ and $t_l \in \mathbb{F}_p$ for all $l \geq \omega$. For a given dimension $s \geq 2$ and some integer $m \geq 1$ we choose a so-called modulus $f \in \mathbb{F}_p[x]$ with $\deg(f) = m$ as well as polynomials $g_1, \dots, g_s \in \mathbb{F}_p[x]$. The vector $\boldsymbol{g} = (g_1, \dots, g_s)$ is called the generating vector of the polynomial lattice point set. Further, we introduce the map $\phi_m : \mathbb{F}_p((x^{-1})) \to [0, 1)$ such that

$$\phi_m\left(\sum_{l=\omega}^{\infty} t_l x^{-l}\right) = \sum_{l=\max\{1,\omega\}}^{m} t_l p^{-l}.$$

With $n \in \{0, 1, \dots, p^m - 1\}$ we associate the polynomial $n(x) = \sum_{r=0}^{m-1} n_r x^r \in \mathbb{F}_p[x]$, as each such n can uniquely be written as $n = n_0 + n_1 p + \cdots + n_{m-1} p^{m-1}$ with digits $n_r \in \{0, 1, \dots, p-1\}$ for all $r \in \{0, 1, \dots, m-1\}$. With this notation, the polynomial lattice point set $\mathscr{P}(\boldsymbol{g}, f)$ is defined as the set of $N := p^m$ points

$$\boldsymbol{x}_n = \left(\phi_m\left(\frac{n(x)g_1(x)}{f(x)}\right), \dots, \phi_m\left(\frac{n(x)g_s(x)}{f(x)}\right)\right) \in [0, 1)^s$$

for $0 \leq n \leq p^m - 1$. See also [4, Chap. 10].

In the following, by $G_{p,m}$ we denote the set of all polynomials g over $\mathbb{F}_p$ with $\deg(g) < m$. Further we define

$$G_{p,m}(f) := \{g \in G_{p,m} \mid \gcd(g, f) = 1\}. \tag{2}$$

For the weighted star discrepancy of a polynomial lattice point set we simply write $D^*_{N,\boldsymbol{\gamma}}(\boldsymbol{g}, f)$.

Niederreiter [12] proved the existence of polynomial lattice point sets with low unweighted star discrepancy by averaging arguments. Generating vectors of good polynomial lattice point sets can be constructed by a component-by-component (CBC) construction. The standard structure of CBC constructions is as follows. We start by setting the first coordinate of the generating vector to 1. After this first step we proceed by increasing the dimension of the generating vector by one in each step until we have a generating vector $(g_1, \dots, g_s)$ of full size s. That is, all previously chosen components stay the same and one new component is added. This new coordinate is chosen from a predefined search set, most commonly from $G_{p,m}(f)$ given by (2). Usually it is determined such that the weighted star discrepancy of the lattice point set, corresponding to the generating vector, consisting of all previously chosen components plus one additional component, is minimized as a function of this last component.

Such constructions were provided in [3] for an irreducible modulus f and in [1] for a reducible f. In these papers, the authors considered the unweighted star discrepancy as well as its weighted version, which we study here. It is the aim of the present paper to speed up these constructions by reducing the search sets for the components of the generating vector $\boldsymbol{g}$ according to each component's importance. It is the nature of product weighted spaces that the components g_j of the generating vector have less and less influence on the quality of the corresponding polynomial lattice point as j increases. Roughly speaking this is due to the weights (γ_j) that are becoming ever smaller with increasing index j. We want to exploit this property in the following way. As the components' influence is decreasing with their indices we want to use less and less time and computational cost to choose these components. To achieve this we choose them from even smaller search sets, which are defined as follows. Let $w_1 \leq w_2 \leq \cdots$ be a non-decreasing sequence of nonnegative integers. This sequence of w_j's is determined in accordance with the weight sequence $\boldsymbol{\gamma}$. Loosely speaking, the smaller γ_j, the bigger w_j is chosen. For $w \in \mathbb{N}_0$ with $w < m$ we define $G_{p,m-w}$ and $G_{p,m-w}(f)$ analogously to $G_{p,m}$ and $G_{p,m}(f)$, respectively. Further we set

$$\mathscr{G}_{p,m-w}(f) := \begin{cases} G_{p,m-w}(f) & \text{if } w < m, \\ \{1 \in \mathbb{F}_p[x]\} & \text{if } w \geq m \end{cases}$$

for any $w \in \mathbb{N}_0$. For $w < m$ these sets have cardinality $p^{m-w} - 1$ in the case of an irreducible modulus f and $p^{m-w-1}(p-1)$ for the special case $f : \mathbb{F}_p \to \mathbb{F}_p, x \mapsto x^m$. We will consider these two cases in what follows. Finally, for $d \in [s]$, we define $\mathscr{G}^d_{p,m-\boldsymbol{w}}(f) := \mathscr{G}_{p,m-w_1}(f) \times \cdots \times \mathscr{G}_{p,m-w_d}(f)$. The idea is to choose the ith component of $\boldsymbol{g}$ of the form $x^{w_i} g_i$, where $g_i \in \mathscr{G}_{p,m-w_i}(f)$, i.e., the search set for the ith component is reduced by a factor $p^{-\min\{w_i,m\}}$ in comparison to the standard CBC construction. We will show that under certain conditions on the weights $\boldsymbol{\gamma}$ and the parameters w_i a polynomial lattice point set constructed according to our reduced CBC construction has a low weighted star discrepancy of order $N^{-1+\delta}$ for all $\delta > 0$. The standard CBC construction (cf. [19]) can be done in $\mathscr{O}(sN^2)$ operations. To speed up the construction, in a first step, making use of ideas from Nuyens and Cools [16, 17] on fast Fourier transformation, the construction cost can be reduced to $\mathscr{O}(sN \log N)$, as for example done in [3]. Combining this with our reduced search sets we obtain a computational cost that is independent of the dimension eventually. Reduced CBC constructions have been introduced first by Dick et al. in [2] for lattice and polynomial lattice point sets with a small worst case integration error in Korobov and Walsh spaces, respectively, and have also been investigated in [7] for lattice point sets with small weighted star discrepancy.

An interesting aspect of the discrepancy of high dimensional point sets is the so-called tractability of discrepancy (see e.g. [13–15]). For $N, s \in \mathbb{N}$ let

$$\mathrm{disc}_\infty(N, s) := \inf_{\substack{\mathscr{P} \subseteq [0,1)^s \\ \#\mathscr{P} = N}} D^*_{N,\boldsymbol{\gamma}}(\mathscr{P})$$

denote the Nth minimal star discrepancy. To introduce the concept of tractability of discrepancy we define the information complexity (also called the inverse of the weighted star discrepancy) as $N^*(s,\varepsilon) := \min\{N \in \mathbb{N} \mid \mathrm{disc}_\infty(N,s) \le \varepsilon\}$. Thus $N^*(s,\varepsilon)$ is the minimal number of points required to achieve a weighted star discrepancy of at most ε. To keep the construction cost of our generating vector low, it is, of course, beneficial to have a small information complexity and thus to stand a chance to have a polynomial lattice point set of small size. We say that we achieve strong polynomial tractability if there exist constants $C, \tau > 0$ such that $N^*(s,\varepsilon) \le C\varepsilon^{-\tau}$ for all $s \in \mathbb{N}$ and all $\varepsilon \in (0,1)$. Roughly speaking, a problem is considered tractable if its information complexity's dependence on s and ε^{-1} is not exponential. Taking weights into account in the definition of discrepancy can sometimes overcome the so-called curse of dimensionality, i.e., an exponential dependence of $N^*(s,\varepsilon)$ on s. We will show that our reduced fast CBC algorithm finds a generating vector $\boldsymbol{g}$ of a polynomial lattice point set that achieves strong polynomial tractability provided that $\sum_{j=1}^{\infty} \gamma_j p^{w_j} < \infty$ with a construction cost of

$$\mathcal{O}\left(N \log N + \min\{s,t\}N + N \sum_{d=1}^{\min\{s,t\}} (m - w_d)p^{-w_d}\right)$$

operations, where $t = \max\{j \in \mathbb{N} \mid w_j < m\}$.

Before stating our main results we would like to discuss a motivating example. Consider first the standard CBC construction as treated in [1, 3], where $w_j = 0$ for all $j \ge 0$. In this case, a sufficient condition for strong polynomial tractability is $\sum_{j=1}^{\infty} \gamma_j < \infty$, which for instance is satisfied for the special choices $\gamma_j = j^{-2}$ and $\gamma_j = j^{-1000}$. However, in the second example the weights decay much faster than in the first. We can make use of this fact by introducing the sequence $\boldsymbol{w} = (w_j)_{j \ge 0}$ such that the condition $\sum_{j=1}^{\infty} \gamma_j p^{w_j} < \infty$ holds, while still achieving strong polynomial tractability (see Corollary 2). This way, we can reduce the size of the search sets for the components of the generating vector if the weights γ_j decay very fast. Consider for example the weight sequence $\gamma_j = j^{-k}$ for some $k > 1$. For $w_j = \lfloor (k-\alpha)\log_p j \rfloor$ with arbitrary $1 < \alpha < k$ we find

$$\sum_{j=1}^{\infty} \gamma_j p^{w_j} \le \sum_{j=1}^{\infty} j^{-k} j^{k-\alpha} = \sum_{j=1}^{\infty} j^{-\alpha} = \zeta(\alpha) < \infty,$$

where ζ denotes the Riemann Zeta function. Observe that for large k, i.e., fast decaying weights, we may choose smaller search sets and thereby speed up the CBC algorithm.

2 A Reduced CBC Construction

In this section we present a CBC construction for the vector $(x^{w_1}g_1, \ldots, x^{w_s}g_s)$ and an upper bound for the weighted star discrepancy of the corresponding polynomial lattice point set.

First note that if $\boldsymbol{g} \in G_{p,m}^s$, then it is known (see [3]) that

$$D_{N,\boldsymbol{\gamma}}^*(\boldsymbol{g}, f) \leq \sum_{\substack{\mathfrak{u}\subseteq[s]\\ \mathfrak{u}\neq\emptyset}} \gamma_{\mathfrak{u}} \left(1 - \left(1 - \frac{1}{N}\right)^{|\mathfrak{u}|}\right) + R_{\boldsymbol{\gamma}}^s(\boldsymbol{g}, f), \tag{3}$$

where in the case of product weights we have

$$R_{\boldsymbol{\gamma}}^s(\boldsymbol{g}, f) = \sum_{\substack{\boldsymbol{h}\in G_{p,m}^s\setminus\{\boldsymbol{0}\}\\ \boldsymbol{h}\cdot\boldsymbol{g}\equiv 0 \bmod f}} \prod_{i=1}^{s} r_p(h_i, \gamma_i). \tag{4}$$

Here, for elements $\boldsymbol{h} = (h_1, \ldots, h_s)$ and $\boldsymbol{g} = (g_1, \ldots, g_s)$ in $G_{p,m}^s$ we define the scalar product by $\boldsymbol{h} \cdot \boldsymbol{g} := h_1 g_1 + \cdots + h_s g_s$. The numbers $r_p(h, \gamma)$ for $h \in G_{p,m}$ and $\gamma \in \mathbb{R}$ are defined as

$$r_p(h, \gamma) = \begin{cases} 1 + \gamma & \text{if } h = 0, \\ \gamma r_p(h) & \text{otherwise,} \end{cases}$$

where for $h = h_0 + h_1 x + \cdots + h_a x^a$ with $h_a \neq 0$ we set $r_p(h) = \frac{1}{p^{a+1}\sin^2\left(\frac{\pi}{p}h_a\right)}$. Thus, in order to analyze the weighted star discrepancy of a polynomial lattice point set it suffices to investigate the quantity $R_{\boldsymbol{\gamma}}^s(\boldsymbol{g}, f)$. This is due to the result of Joe [6], who proved that for any summable weight sequence $(\gamma_j)_{j\geq 1}$ we have

$$\sum_{\substack{\mathfrak{u}\subseteq[s]\\ \mathfrak{u}\neq\emptyset}} \gamma_{\mathfrak{u}} \left(1 - \left(1 - \frac{1}{N}\right)^{|\mathfrak{u}|}\right) \leq \frac{\max(1, \Gamma)\mathrm{e}^{\sum_{i=1}^{\infty}\gamma_i}}{N} \quad \text{with } \Gamma := \sum_{i=1}^{\infty} \frac{\gamma_i}{1 + \gamma_i}.$$

Algorithm 1

Let $p \in \mathbb{P}, m \in \mathbb{N}, f \in \mathbb{F}_p[x]$ and let $(w_j)_{j\geq 1}$ be a non-decreasing sequence of nonnegative integers and consider product weights $(\gamma_j)_{j\geq 1}$. Construct $(g_1, \ldots, g_s) \in \mathcal{G}_{p,m-\boldsymbol{w}}^s(f)$ as follows:

1. Set $g_1 = 1$.
2. For $d \in [s-1]$ assume that $(g_1, \ldots, g_d) \in \mathcal{G}_{p,m-\boldsymbol{w}}^d(f)$ is already found. Choose $g_{d+1} \in \mathcal{G}_{p,m-w_{d+1}}(f)$ such that $R_{\boldsymbol{\gamma}}^{d+1}((x^{w_1}g_1, \ldots, x^{w_d}g_d, x^{w_{d+1}}g_{d+1}), f)$ is minimized as a function of g_{d+1}.
3. Increase d by 1 and repeat the second step until $(g_1, \ldots, g_s)$ is found.

Remark 1 Of course we have $\mathscr{G}^s_{p,m-\mathbf{w}}(f) \subseteq G^s_{p,m}$, and thus in Algorithm 1 it indeed suffices to consider $R^{d+1}_{\boldsymbol{\gamma}}$ rather than the weighted star discrepancy.

In the algorithm above, the search set is reduced for each coordinate of $(g_1, \dots, g_s)$ according to its importance, since with increasing w_j the search set becomes smaller, as the weight γ_j and thus the corresponding component's influence on the quality of the generating vector decreases. For this reason we call Algorithm 1 a reduced CBC algorithm. We will now study Algorithm 1 for different choices of f.

2.1 Polynomial Lattice Point Sets for $f(x) = x^m$

We will now study the interesting case where $f\colon \mathbb{F}_p \to \mathbb{F}_p, x \mapsto x^m$. Throughout the rest of this section we write x^m instead of f to emphasize our special choice of f. Note that for $g \in \mathbb{F}_p((x^{-1}))$ the Laurent series g/f can be easily computed in this case by shifting the coefficients of g m times to the left. This is why the choice x^m for the modulus is the most frequently used in practise. Furthermore, the mathematical analysis of the reduced CBC algorithm is slightly less technical in this case, since the proof of the following discepancy bound requires to compute a sum over all divisors of the modulus f. This is much easier for the special case $f(x) = x^m$ than for a general modulus f. It is the aim of this section to prove the following theorem.

Theorem 1 *Let $\boldsymbol{\gamma} = (\gamma_j)_{j\geq 1}$ be positive real numbers and $\mathbf{w}$ be nonnegative real numbers with $0 = w_1 \leq w_2 \leq \cdots$. Let further $(g_1, \dots, g_s) \in \mathscr{G}^s_{p,m-\mathbf{w}}(x^m)$ be constructed using Algorithm 1. Then we have for every $d \in [s]$*

$$R^d_{\boldsymbol{\gamma}}((x^{w_1}g_1, \dots, x^{w_d}g_d), x^m) \leq \frac{1}{p^m}\prod_{i=1}^{d}\left(1 + \gamma_i + \gamma_i 2p^{\min\{w_i,m\}} m \frac{p^2-1}{3p}\right).$$

As a direct consequence we obtain the following discrepancy estimate.

Corollary 1 *Let $N = p^m$ and $\boldsymbol{\gamma}$, $\mathbf{w}$ and $(g_1, \dots, g_s)$ be as in Theorem 1. Then the polynomial lattice point set $\mathscr{P}((x^{w_1}g_1, \dots, x^{w_s}g_s), x^m)$ has a weighted star discrepancy*

$$\begin{aligned} &D^*_{N,\boldsymbol{\gamma}}\left((x^{w_1}g_1, \dots, x^{w_s}g_s), x^m\right) \\ &\leq \sum_{\substack{\mathfrak{u}\subseteq[s]\\ \mathfrak{u}\neq\emptyset}} \gamma_{\mathfrak{u}}\left(1 - \left(1 - \frac{1}{N}\right)^{|\mathfrak{u}|}\right) + \frac{1}{N}\prod_{i=1}^{s}\left(1 + \gamma_i + \gamma_i 2p^{\min\{w_i,m\}} m \frac{p^2-1}{3p}\right). \quad (5) \end{aligned}$$

Knowing the above discrepancy bound, we are now ready to ask about the size of the polynomial lattice point set required to achieve a weighted star discrepancy not exceeding some ε threshold. In particular, we would like to know how this size depends on the dimension s and on ε.

Corollary 2 *Let $N = p^m$, $\boldsymbol{\gamma}$, and $\mathbf{w}$ be as in Theorem 1 and consider the problem of constructing generating vectors for polynomial lattice point sets with small weighted star discrepancy. Then $\sum_{j=1}^{\infty} \gamma_j p^{w_j} < \infty$ is a sufficient condition for strong polynomial tractability. This condition further implies $D^*_{N,\boldsymbol{\gamma}}((x^{w_1} g_1, \ldots, x^{w_s} g_s), x^m) = \mathcal{O}(N^{-1+\delta})$, with the implied constant independent of s, for any $\delta > 0$, where $(g_1, \ldots, g_s) \in \mathcal{G}^s_{p,m-\mathbf{w}}(x^m)$ is constructed using Algorithm 1.*

Proof Construct a generating vector $(g_1, \ldots, g_s) \in \mathcal{G}^s_{p,m-\mathbf{w}}(x^m)$ by applying Algorithm 1 and consider its weighted star discrepancy, which is bounded by (5). Following closely the lines of the argumentation in [7, Sect. 5] and noticing that $2m \frac{p^2-1}{3p} = \mathcal{O}(\log N)$ we obtain the result. More precisely, provided that the $\gamma_j p^{w_j}$'s are summable, we have a means to construct polynomial lattice point sets $\mathcal{P}(\boldsymbol{g}, f)$ with $D^*_{N,\boldsymbol{\gamma}}(\boldsymbol{g}, f) \leq \varepsilon$, whose sizes grow polynomially in ε^{-1} and are independent of the dimension. As a result the problem is strongly polynomially tractable. The discrepancy result $D^*_{N,\boldsymbol{\gamma}}((x^{w_1} g_1, \ldots, x^{w_s} g_s), x^m) = \mathcal{O}(N^{-1+\delta})$ also follows directly from [7]. □

In order to show Theorem 1 we need several auxiliary results.

Lemma 1 *Let $a \in \mathbb{F}_p[x]$ be monic. Then we have*

$$\sum_{\substack{h \in G_{p,m} \setminus \{0\} \\ a | h}} r_p(h) = (m - \deg(a)) \frac{p^2 - 1}{3p} p^{-\deg(a)}.$$

In particular, for $a = 1$ this formula yields $\sum_{h \in G_{p,m} \setminus \{0\}} r_p(h) = m \frac{p^2-1}{3p}$.

Proof This fact follows from [1, p. 1055] (by setting $\gamma_{d+1} = 1$). The special case $a = 1$ also follows from [3, Lemma 2.2] by setting $s = 1$. □

For our purposes, it is convenient to write $R^s_{\boldsymbol{\gamma}}(\boldsymbol{g}, f)$ from (4) in an alternative way. To this end, we introduce some notation. For a Laurent series $L \in \mathbb{F}_p((x^{-1}))$ we denote by $c_{-1}(L)$ its coefficient of x^{-1}, i.e., its residuum. Further, we set $X_p(L) := \chi_p(c_{-1}(L))$, where χ_p is a non-trivial additive character of $\mathbb{F}_p$. One could for instance choose $\chi_p(n) = \mathrm{e}^{(2\pi \mathrm{i}/p)n}$ for $n \in \mathbb{F}_p$ (see, e.g., [10]). It is clear that $X_p(L) = 1$ if L is a polynomial and that $X_p(L_1 + L_2) = X_p(L_1) X_p(L_2)$ for $L_1, L_2 \in \mathbb{F}_p((x^{-1}))$. From [12, p. 78] we know that for some $q \in \mathbb{F}_p[x]$ we have

$$\sum_{v \in G_{p,m}} X_p\left(\frac{v}{f} q\right) = \begin{cases} p^m & \text{if } f \mid q, \\ 0 & \text{otherwise.} \end{cases} \tag{6}$$

With this, it is an easy task to show the following formula.

Lemma 2 *We have*

$$R_{\boldsymbol{\gamma}}^{s}(\boldsymbol{g}, f) = -\prod_{i=1}^{s}(1+\gamma_i) + \frac{1}{p^m}\sum_{v\in G_{p,m}}\prod_{i=1}^{s}\left(1+\gamma_i+\gamma_i\sum_{h\in G_{p,m}\setminus\{0\}} r_p(h)X_p\left(\frac{v}{f}hg_i\right)\right).$$

Now we study a sum which will appear later in the proof of Theorem 1 and show an upper bound for it.

Lemma 3 *Let $w \in \mathbb{N}_0$ and $v \in G_{p,m}$. Let*

$$Y_{p^m,w}(v, x^m) := \sum_{g\in\mathcal{G}_{p,m-w}(x^m)}\sum_{h\in G_{p,m}\setminus\{0\}} r_p(h)X_p\left(\frac{v}{x^m}hx^w g\right),$$

where x^w denotes the polynomial $f(x) = x^w$. Then we have

$$\frac{1}{\#\mathcal{G}_{p,m-w}(x^m)}\sum_{v\in G_{p,m}}|Y_{p^m,w}(v, x^m)| \le 2p^{\min\{w,m\}}m\frac{p^2-1}{3p}.$$

Proof Let us first assume that $w \ge m$. Then we have $\mathcal{G}_{p,m-w}(x^m) = \{1\}$ and therefore

$$Y_{p^m,w}(v, x^m) = \sum_{h\in G_{p,m}\setminus\{0\}} r_p(h)X_p(vhx^{w-m}) = \sum_{h\in G_{p,m}\setminus\{0\}} r_p(h) = m\frac{p^2-1}{3p}$$

with Lemma 1. Hence, in the case $w \ge m$ we obtain

$$\frac{1}{\#\mathcal{G}_{p,m-w}(x^m)}\sum_{v\in G_{p,m}}|Y_{p^m,w}(v, x^m)| = p^m m\frac{p^2-1}{3p} \le 2p^{\min\{w,m\}}m\frac{p^2-1}{3p}.$$

For the rest of the proof let $w < m$. We abbreviate $\#\mathcal{G}_{p,m-w}(x^m)$ by $\#\mathcal{G}$ and write

$$\frac{1}{\#\mathcal{G}}\sum_{v\in G_{p,m}}|Y_{p^m,w}(v, x^m)| = \frac{1}{\#\mathcal{G}}\sum_{\substack{v\in G_{p,m}\\ x^{m-w}|v}}|Y_{p^m,w}(v, x^m)| + \frac{1}{\#\mathcal{G}}\sum_{\substack{v\in G_{p,m}\\ x^{m-w}\nmid v}}|Y_{p^m,w}(v, x^m)|.$$

In what follows, we refer to the latter sums as

$$S_1 := \frac{1}{\#\mathcal{G}}\sum_{\substack{v\in G_{p,m}\\ x^{m-w}|v}}|Y_{p^m,w}(v, x^m)| \quad\text{and}\quad S_2 := \frac{1}{\#\mathcal{G}}\sum_{\substack{v\in G_{p,m}\\ x^{m-w}\nmid v}}|Y_{p^m,w}(v, x^m)|.$$

We may uniquely write any $v \in G_{p,m} \setminus \{0\}$ in the form $v = qx^{m-w} + \ell$, where $q, \ell \in \mathbb{F}_p[x]$ with $\deg(q) < w$ and $\deg(\ell) < m - w$. Using the properties of X_p it is clear that $Y_{p^m,w}(v, x^m) = Y_{p^m,w}(\ell, x^m)$ and hence

$$S_1 = \frac{1}{\#\mathscr{G}} \sum_{\substack{v \in G_{p,m} \\ x^{m-w} | v}} |Y_{p^m,w}(0, x^m)| = \sum_{\substack{v \in G_{p,m} \\ x^{m-w} | v}} \frac{1}{\#\mathscr{G}} \sum_{g \in \mathscr{G}_{p,m-w}(x^m)} \sum_{h \in G_{p,m} \setminus \{0\}} r_p(h)$$
$$= \sum_{\substack{v \in G_{p,m} \\ x^{m-w} | v}} m \frac{p^2 - 1}{3p} = p^{\min\{w,m\}} m \frac{p^2 - 1}{3p}.$$

We move on to S_2. Let $e(\ell) := \max\{k \in \{0, 1, \ldots, m - w - 1\} : x^k \mid \ell\}$. With this definition we may display S_2 as

$$S_2 = \frac{p^w}{\#\mathscr{G}} \sum_{k=0}^{m-w-1} \sum_{\substack{\ell \in G_{p,m-w} \setminus \{0\} \\ e(\ell) = k}} |Y_{p^m,w}(\ell, x^m)|. \tag{7}$$

We compute $Y_{p^m,w}(\ell, x^m)$ for $\ell \in G_{p,m-w} \setminus \{0\}$ with $e(\ell) = k$. Let μ_p be the Möbius function on the set of monic polynomials over $\mathbb{F}_p$, i.e., $\mu_p : \mathbb{F}_p[x] \to \{-1, 0, 1\}$ and

$$\mu_p(h) = \begin{cases} (-1)^\nu & \text{if } h \text{ is squarefree and has } \nu \text{ irreducible factors,} \\ 0 & \text{else.} \end{cases}$$

We call h squarefree if there is no irreducible polynomial $q \in \mathbb{F}_p[x]$ with $\deg(q) \geq 1$ such that $q^2 \mid h$. The fact that $\mu_p(1) = 1$, $\mu_p(x) = -1$ and $\mu_p(x^i) = 0$ for $i \in \mathbb{N}$, $i \geq 2$, yields the equivalence of $\sum_{t | \gcd(x^{m-w}, g)} \mu_p(t) = 1$ and $\gcd(x^{m-w}, g) = 1$. Therefore we can write

$$Y_{p^m,w}(\ell, x^m) = \sum_{h \in G_{p,m} \setminus \{0\}} r_p(h) \sum_{g \in G_{p,m-w}} X_p\left(\frac{\ell}{x^{m-w}} hg\right) \sum_{t | \gcd(x^{m-w}, g)} \mu_p(t)$$
$$= \sum_{h \in G_{p,m} \setminus \{0\}} r_p(h) \sum_{t | x^{m-w}} \mu_p(t) \sum_{\substack{g \in G_{p,m-w} \\ t | g}} X_p\left(\frac{\ell}{x^{m-w}} hg\right)$$
$$= \sum_{h \in G_{p,m} \setminus \{0\}} r_p(h) \sum_{t | x^{m-w}} \mu_p(t) \sum_{a \in G_{p,m-w-\deg(t)}} X_p\left(\frac{\ell}{x^{m-w}} hat\right)$$
$$= \sum_{h \in G_{p,m} \setminus \{0\}} r_p(h) \sum_{t | x^{m-w}} \mu_p\left(\frac{x^{m-w}}{t}\right) \sum_{a \in G_{p,\deg(t)}} X_p\left(\frac{a}{t} h\ell\right)$$
$$= \sum_{h \in G_{p,m} \setminus \{0\}} r_p(h) \sum_{\substack{t | x^{m-w} \\ t | h\ell}} \mu_p\left(\frac{x^{m-w}}{t}\right) p^{\deg(t)}$$
$$= \sum_{t | x^{m-w}} \mu_p\left(\frac{x^{m-w}}{t}\right) p^{\deg(t)} \sum_{\substack{h \in G_{p,m} \setminus \{0\} \\ t | h\ell}} r_p(h).$$

The equivalence of the conditions $t \mid h\ell$ and $\frac{t}{\gcd(t,\ell)} \mid h$ yields

$$Y_{p^m,w}(\ell, x^m) = \sum_{t \mid x^{m-w}} \mu_p\left(\frac{x^{m-w}}{t}\right) p^{\deg(t)} \sum_{\substack{h \in G_{p,m}\setminus\{0\} \\ \frac{t}{\gcd(t,\ell)} \mid h}} r_p(h).$$

We investigate the inner sum and use Lemma 1 with $a = \frac{t}{\gcd(t,\ell)}$ to find

$$\sum_{\substack{h \in G_{p,m}\setminus\{0\} \\ \frac{t}{\gcd(t,\ell)} \mid h}} r_p(h) = \left(m - \deg\left(\frac{t}{\gcd(t,\ell)}\right)\right) \frac{p^2-1}{3p} p^{-\deg\left(\frac{t}{\gcd(t,\ell)}\right)}.$$

Now we have

$$\begin{aligned}
Y_{p^m,w}(\ell, x^m) =& \frac{p^2-1}{3p} \sum_{t \mid x^{m-w}} \mu_p\left(\frac{x^{m-w}}{t}\right)\left(m - \deg\left(\frac{t}{\gcd(t,\ell)}\right)\right) p^{\deg(\gcd(t,\ell))} \\
=& \frac{p^2-1}{3p} m \sum_{t \mid x^{m-w}} \mu_p\left(\frac{x^{m-w}}{t}\right) p^{\deg(\gcd(t,\ell))} \\
&- \frac{p^2-1}{3p} \sum_{t \mid x^{m-w}} \mu_p\left(\frac{x^{m-w}}{t}\right) \deg\left(\frac{t}{\gcd(t,\ell)}\right) p^{\deg(\gcd(t,\ell))}.
\end{aligned}$$

From the fact that $e(\ell) = k \leq m - w - 1$ we obtain $\gcd(x^{m-w}, \ell) = \gcd(x^{m-w-1}, \ell) = x^k$. This observation leads to

$$\sum_{t \mid x^{m-w}} \mu_p\left(\frac{x^{m-w}}{t}\right) p^{\deg(\gcd(t,\ell))} = p^{\deg(\gcd(x^{m-w},\ell))} - p^{\deg(\gcd(x^{m-w-1},\ell))} = 0$$

and

$$\begin{aligned}
&\sum_{t \mid x^{m-w}} \mu_p\left(\frac{x^{m-w}}{t}\right) \deg\left(\frac{t}{\gcd(t,\ell)}\right) p^{\deg(\gcd(t,\ell))} \\
&= \deg\left(\frac{x^{m-w}}{\gcd(x^{m-w},\ell)}\right) p^{\deg(\gcd(x^{m-w},\ell))} - \deg\left(\frac{x^{m-w-1}}{\gcd(x^{m-w-1},\ell)}\right) p^{\deg(\gcd(x^{m-w-1},\ell))} \\
&= (m-w-k)p^k - (m-w-k-1)p^k = p^k.
\end{aligned}$$

Altogether we have $Y_{p^m,w}(\ell, x^m) = -\frac{p^2-1}{3p} p^k$. Inserting this result into (7) yields

$$S_2 = \frac{p^w}{\#\mathscr{G}} \frac{p^2-1}{3p} \sum_{k=0}^{m-w-1} p^k \sum_{\substack{\ell \in G_{p,m-w}\setminus\{0\} \\ e(\ell)=k}} 1.$$

Since

$$\begin{aligned}
&\#\{\ell \in G_{p,m-w} \setminus \{0\} : e(\ell) = k\} \\
=&\#\{\ell \in G_{p,m-w} \setminus \{0\} : x^k \mid \ell\} - \#\{\ell \in G_{p,m-w} \setminus \{0\} : x^{k+1} \mid \ell\} \\
=&p^{m-w-k} - 1 - (p^{m-w-k-1} - 1) = p^{m-w-k-1}(p-1),
\end{aligned}$$

we have

$$\begin{aligned}
S_2 =& \frac{p^w}{p^{m-w-1}(p-1)} \frac{p^2-1}{3p} \sum_{k=0}^{m-w-1} p^k p^{m-w-k-1}(p-1) \\
=& p^w \frac{p^2-1}{3p}(m-w) \leq p^{\min\{w,m\}} m \frac{p^2-1}{3p}.
\end{aligned}$$

Summarizing, we have shown

$$\frac{1}{\#\mathscr{G}} \sum_{v \in G_{p,m}} |Y_{p^m,w}(v, x^m)| = S_1 + S_2 \leq 2p^{\min\{w,m\}} m \frac{p^2-1}{3p},$$

which completes the proof. □

Now we are ready to prove Theorem 1 using induction on d.

Proof We show the result for $d = 1$. From Lemma 2 we have

$$\begin{aligned}
R^1_{\boldsymbol{\gamma}}((x^{w_1}), x^m) =& -(1+\gamma_1) + \frac{1}{p^m} \sum_{v \in G_{p,m}} \left(1 + \gamma_1 + \gamma_1 \sum_{h \in G_{p,m}\setminus\{0\}} r_p(h) X_p\left(\frac{v}{x^m} h x^{w_1}\right)\right) \\
=& \frac{\gamma_1}{p^m} \sum_{v \in G_{p,m}} \sum_{h \in G_{p,m}\setminus\{0\}} r_p(h) X_p\left(\frac{v}{x^m} h x^{w_1}\right).
\end{aligned}$$

If $w_1 \geq m$, then

$$\begin{aligned}
R^1_{\boldsymbol{\gamma}}((x^{w_1}), x^m) =& \frac{\gamma_1}{p^m} \sum_{v \in G_{p,m}} \sum_{h \in G_{p,m}\setminus\{0\}} r_p(h) = \frac{\gamma_1}{p^m} p^{\min\{w_1,m\}} m \frac{p^2-1}{3p} \\
&\leq \frac{1}{p^m}\left(1 + \gamma_1 + \gamma_1 2 p^{\min\{w_1,m\}} m \frac{p^2-1}{3p}\right).
\end{aligned}$$

If $w_1 < m$, then we can write

$$
\begin{aligned}
R_{\boldsymbol{\gamma}}^1((x^{w_1}), x^m) =& \frac{\gamma_1}{p^m} \sum_{v \in G_{p,m}} \sum_{h \in G_{p,m} \setminus \{0\}} r_p(h) X_p\left(\frac{v}{x^m} h x^{w_1}\right) \\
=& \frac{\gamma_1}{p^m} \sum_{\substack{h \in G_{p,m} \setminus \{0\} \\ x^{m-w_1} | h}} r_p(h) \sum_{v \in G_{p,m}} X_p\left(\frac{v}{x^m} h x^{w_1}\right) \\
&+ \frac{\gamma_1}{p^m} \sum_{\substack{h \in G_{p,m} \setminus \{0\} \\ x^{m-w_1} \nmid h}} r_p(h) \sum_{v \in G_{p,m}} X_p\left(\frac{v}{x^m} h x^{w_1}\right) = \gamma_1 \sum_{\substack{h \in G_{p,m} \setminus \{0\} \\ x^{m-w_1} | h}} r_p(h),
\end{aligned}
$$

where we used (6) in the latter step. We use Lemma 1 with $a = x^{m-w_1}$ to compute

$$
\sum_{\substack{h \in G_{p,m} \setminus \{0\} \\ x^{m-w_1} | h}} r_p(h) = \frac{1}{p^m} p^{w_1} w_1 \frac{p^2-1}{3p} \leq \frac{1}{p^m} p^{\min\{w_1, m\}} m \frac{p^2-1}{3p},
$$

which leads to the desired result also in this case.

Now let $d \in [s-1]$. Assume that we have some $(g_1, \ldots, g_d) \in \mathcal{G}_{p,m-\boldsymbol{w}}^d(x^m)$ such that

$$
R_{\boldsymbol{\gamma}}^d((x^{w_1} g_1, \ldots, x^{w_d} g_d), x^m) \leq \frac{1}{p^m} \prod_{i=1}^d \left(1 + \gamma_i + \gamma_i 2 p^{\min\{w_i, m\}} m \frac{p^2-1}{3p}\right).
$$

Let $g^* \in \mathcal{G}_{p,m-w_{d+1}}(x^m)$ be such that $R_{\boldsymbol{\gamma}}^{d+1}((x^{w_1} g_1, \ldots, x^{w_d} g_d, x^{w_{d+1}} g_{d+1}), x^m)$ is minimized as a function of g_{d+1} for $g_{d+1} = g^*$. Then we have

$$
\begin{aligned}
R_{\boldsymbol{\gamma}}^{d+1}((x^{w_1} g_1, \ldots, x^{w_d} g_d, x^{w_{d+1}} g^*), x^m) &= -(1 + \gamma_{d+1}) \prod_{i=1}^d (1 + \gamma_i) \\
+ \frac{1}{p^m} \sum_{v \in G_{p,m}} \prod_{i=1}^d & \left(1 + \gamma_i + \gamma_i \sum_{h \in G_{p,m} \setminus \{0\}} r_p(h) X_p\left(\frac{v}{x^m} h x^{w_i} g_i\right)\right) \\
\times & \left(1 + \gamma_{d+1} + \gamma_{d+1} \sum_{h \in G_{p,m} \setminus \{0\}} r_p(h) X_p\left(\frac{v}{x^m} h x^{w_{d+1}} g^*\right)\right) \\
=& (1 + \gamma_{d+1}) R_{\boldsymbol{\gamma}}^d((x^{w_1} g_1, \ldots, x^{w_d} g_d), x^m) + L(g^*),
\end{aligned} \tag{8}
$$

where

$$
\begin{aligned}
L(g^*) =& \frac{\gamma_{d+1}}{p^m} \sum_{v \in G_{p,m}} \sum_{h \in G_{p,m} \setminus \{0\}} r_p(h) X_p\left(\frac{v}{x^m} h x^{w_{d+1}} g^*\right) \\
& \times \prod_{i=1}^d \left(1 + \gamma_i + \gamma_i \sum_{h \in G_{p,m} \setminus \{0\}} r_p(h) X_p\left(\frac{v}{x^m} h x^{w_i} g_i\right)\right).
\end{aligned}
$$

A minimizer g^* of $R_{\boldsymbol{\gamma}}^{d+1}((x^{w_1}g_1, \ldots, x^{w_d}g_d, x^{w_{d+1}}g_{d+1}), x^m)$ is also a minimizer of $L(g_{d+1})$. Combining (4) and (8) we obtain that $R_{\boldsymbol{\gamma}}^d(\boldsymbol{g}, f) \in \mathbb{R}$ for all $d \in [s]$. With the ideas in the proof of [3, Theorem 2.7], we see that $L(g) \in \mathbb{R}^+$ for all $g \in \mathcal{G}_{p,m-w_{d+1}}(x^m)$. Thus we may bound $L(g^*)$ by the mean over all $g \in \mathcal{G}_{p,m-w_{d+1}}(x^m)$:

$$
\begin{aligned}
L(g^*) \leq & \frac{1}{\#\mathcal{G}_{p,m-w_{d+1}}(x^m)} \sum_{g_{d+1} \in \mathcal{G}_{p,m-w_{d+1}}(x^m)} L(g_{d+1}) \\
\leq & \frac{\gamma_{d+1}}{p^m} \sum_{v \in G_{p,m}} \frac{1}{\#\mathcal{G}_{p,m-w_{d+1}}(x^m)} \\
& \times \left| \sum_{g_{d+1} \in \mathcal{G}_{p,m-w_{d+1}}(x^m)} \sum_{h \in G_{p,m} \setminus \{0\}} r_p(h) X_p\left(\frac{v}{x^m} h x^{w_{d+1}} g_{d+1}\right) \right| \\
& \times \prod_{i=1}^{d} \left(1 + \gamma_i + \gamma_i \sum_{h \in G_{p,m} \setminus \{0\}} r_p(h) \left| X_p\left(\frac{v}{x^m} h x^{w_i} g_i\right)\right| \right) \\
\leq & \frac{\gamma_{d+1}}{p^m} \prod_{i=1}^{d} \left(1 + \gamma_i + \gamma_i m \frac{p^2-1}{3p}\right) \sum_{v \in G_{p,m}} \frac{|Y_{p^m, w_{d+1}}(v, x^m)|}{\#\mathcal{G}_{p,m-w_{d+1}}(x^m)},
\end{aligned}
$$

where we used the estimate $\left|X_p\left(\frac{v}{x^m} h x^{w_i} g_i\right)\right| \leq 1$ in the last step. With the induction hypothesis and Lemma 3 this leads to

$$
\begin{aligned}
& R_{\boldsymbol{\gamma}}^{d+1}((x^{w_1}g_1, \ldots, x^{w_d}g_d, x^{w_{d+1}}g^*), x^m) \\
\leq & (1+\gamma_{d+1}) \frac{1}{p^m} \prod_{i=1}^{d} \left(1 + \gamma_i + \gamma_i 2 p^{\min\{w_i, m\}} m \frac{p^2-1}{3p}\right) \\
& + \frac{\gamma_{d+1}}{p^m} \prod_{i=1}^{d} \left(1 + \gamma_i + \gamma_i m \frac{p^2-1}{3p}\right) 2 p^{\min\{w_{d+1}, m\}} m \frac{p^2-1}{3p} \\
\leq & \frac{1}{p^m} \prod_{i=1}^{d} \left(1 + \gamma_i + \gamma_i 2 p^{\min\{w_i, m\}} m \frac{p^2-1}{3p}\right) \\
& \times \left(1 + \gamma_{d+1} + \gamma_{d+1} 2 p^{\min\{w_{d+1}, m\}} m \frac{p^2-1}{3p}\right) \\
= & \frac{1}{p^m} \prod_{i=1}^{d+1} \left(1 + \gamma_i + \gamma_i 2 p^{\min\{w_i, m\}} m \frac{p^2-1}{3p}\right).
\end{aligned}
$$

□

The Reduced Fast CBC Construction

So far we have seen how to construct a generating vector $\boldsymbol{g}$ of the point set $\mathcal{P}(\boldsymbol{g}, x^m)$. In fact Algorithm 1 can be made much faster using results of [2, 16, 17]. In this section we are investigating and improving Algorithm 1 and additionally analyzing the computational cost of the improved algorithm.

Walsh functions are a suitable tool for analyzing the computational cost of CBC algorithms for constructing polynomial lattice point sets. Let $\omega = \mathrm{e}^{2\pi i/p}$, $x \in [0, 1)$ and h a nonnegative integer with base p representation $x = x_1/p + x_2/p^2 + \ldots$ and $h = h_0 + h_1 p + \ldots + h_r p^r$, respectively. Then we define

$$\mathrm{wal}_h : [0, 1) \to \mathbb{C},\ \mathrm{wal}_h(x) := \omega^{h_0 x_1 + \ldots + h_r x_{r+1}}.$$

The Walsh function system $\{\mathrm{wal}_h \mid h = 0, 1, \ldots\}$ is a complete orthonormal basis in $L_2([0, 1))$ which has been used in the analysis of the discrepancy of digital nets (an important class of low-discrepancy point sets which contains polynomial lattice point sets) several times before, see for example [3, 5, 8]. For further information on Walsh functions see [4, Appendix A].

Let $d \geq 1$, $N = p^m$. For $P(\boldsymbol{g}, f) = \{\boldsymbol{x}_0, \ldots, \boldsymbol{x}_{p^m-1}\}$ with $\boldsymbol{x}_n = (x_n^{(1)}, \ldots, x_n^{(s)})$ we have the formula (see [3, Sect. 4])

$$\frac{1}{p^m} \sum_{n=0}^{p^m-1} \prod_{i=1}^{s} \mathrm{wal}_{h_i}(x_n^{(i)}) = \begin{cases} 1 & \text{if } \boldsymbol{g} \cdot \boldsymbol{h} \equiv 0 \pmod{f}, \\ 0 & \text{otherwise}, \end{cases} \tag{9}$$

where h_i are nonnegative integers with base p representation $h_i = h_0^{(i)} + h_1^{(i)} p + \ldots + h_r^{(i)} p^r$. We identify these nonnegative integers h_i with the polynomials $h_i(x) = h_0^{(i)} + h_1^{(i)} x + \ldots + h_r^{(i)} x^r$, which are elements of $\mathscr{G}_{p,m}$. The vectors $\boldsymbol{h}$ in (9) are then from $\mathscr{G}_{p,m}^s$ such that $\boldsymbol{h} = (h_1(x), \ldots, h_s(x))$. Equation (9) allows us to rewrite $R_{\boldsymbol{\gamma}}^d(\boldsymbol{g}, x^m)$ in the following way:

$$R_{\boldsymbol{\gamma}}^d(\boldsymbol{g}, x^m) = -\prod_{i=1}^{d}(1 + \gamma_i) + \frac{1}{p^m} \sum_{n=0}^{p^m-1} \prod_{i=1}^{d} \sum_{h=0}^{p^m-1} r_p(h, \gamma_i) \mathrm{wal}_h \left(\phi_m \left(\frac{n x^{w_i} g_i}{x^m} \right) \right).$$

Note that $r_p(h, \gamma)$ is defined as in (4) and we identify the integer in base p representation $h = h_0 + h_1 p + \ldots + h_r p^r$ with the polynomial $h(x) = h_0 + h_1 x + \ldots + h_r x^r$. If we set $\psi(\frac{n x^{w_i} g_i}{x^m}) := \sum_{h=1}^{p^m-1} r_p(h) \mathrm{wal}_h(\phi_m(\frac{n x^{w_i} g_i}{x^m}))$ we get that

$$\begin{aligned} R_{\boldsymbol{\gamma}}^d(\boldsymbol{g}, x^m) &= -\prod_{i=1}^{d}(1 + \gamma_i) + \frac{1}{p^m} \sum_{n=0}^{p^m-1} \prod_{i=1}^{d} \left(1 + \gamma_i + \gamma_i \psi \left(\frac{n x^{w_i} g_i}{x^m} \right) \right) \\ &= -\prod_{i=1}^{d}(1 + \gamma_i) + \frac{1}{p^m} \sum_{n=0}^{p^m-1} \eta_d(n), \end{aligned} \tag{10}$$

where $\eta_d(n) = \prod_{i=1}^{d} \left(1 + \gamma_i + \gamma_i \psi(\frac{n x^{w_i} g_i}{x^m})\right)$.

In [3, Sect. 4] it is proved that we can compute the at most N different values of $\psi(\frac{r}{x^m})$ for $r \in G_{p,m}$ in $\mathscr{O}(N \log N)$ operations.

Let us study one step of the reduced CBC algorithm. Assuming we already have found $(g_1, \ldots, g_d) \in \mathscr{G}_{p,m-\boldsymbol{w}}^d(x^m)$ we have to minimize $R_{\boldsymbol{\gamma}}^{d+1}((x^{w_1} g_1, \ldots, x^{w_{d+1}}$

$g_{d+1}), x^m)$ as a function of $g_{d+1} \in \mathscr{G}_{p,m-w_{d+1}}(x^m)$. If $w_{d+1} \geq m$ then $g_{d+1} = 1$ and we are done. Let now $w_{d+1} < m$. From (10) we have that

$$R_{\boldsymbol{\gamma}}^{d+1}((x^{w_1}g_1, \ldots, x^{w_{d+1}}g_{d+1}), x^m)$$
$$= -\prod_{i=1}^{d+1}(1+\gamma_i) + \frac{1}{p^m}\sum_{n=0}^{p^m-1}\left(1+\gamma_{d+1}+\gamma_{d+1}\psi\left(\frac{nx^{w_{d+1}}g_{d+1}}{x^m}\right)\right)\eta_d(n).$$

In order to minimize $R_{\boldsymbol{\gamma}}^{d+1}((x^{w_1}g_1, \ldots, x^{w_{d+1}}g_{d+1}), x^m)$ it is enough to minimize $T_d(g) := \sum_{n=0}^{p^m-1}\psi(\frac{nx^{w_{d+1}}g}{x^m})\eta_d(n)$. As in [2, Sect. 4] we can represent this quantity using some specific $(p^{m-w_{d+1}-1}(p-1) \times N)$-matrix A and exploiting its additional structure. Let therefore

$$A = \left(\psi\left(\frac{nx^{w_{d+1}}g}{x^m}\right)\right)_{\substack{g\in G_{p,m-w_{d+1}}(x^m),\\ n\in\{0,\ldots,N-1\}}} \text{ and } \boldsymbol{\eta}_d = (\eta_d(0), \ldots, \eta_d(N-1))^\top.$$

First of all observe that we get $(T_d(g))_{g\in G_{p,m-w_{d+1}}(x^m)} = A\boldsymbol{\eta}_d$. Secondly the matrix A is a block matrix and can be written in the following form

$$A = \left(\Omega^{(m-w_{d+1})} \ldots \Omega^{(m-w_{d+1})}\right), \text{ where } \Omega^{(l)} = \left(\psi\left(\frac{nx^{w_{d+1}}g}{x^m}\right)\right)_{\substack{g\in G_{p,m-w_{d+1}}(x^m),\\ n\in\{0,\ldots p^l-1\}}}.$$

If $\boldsymbol{x}$ is any vector of size p^m then we compute

$$A\boldsymbol{x} = \Omega^{(m-w_{d+1})}\boldsymbol{x}_1 + \ldots + \Omega^{(m-w_{d+1})}\boldsymbol{x}_{(p^{w_{d+1}})} = \Omega^{(m-w_{d+1})}(\boldsymbol{x}_1 + \ldots + \boldsymbol{x}_{(p^{w_{d+1}})}),$$

where $\boldsymbol{x}_1$ is the vector consisting of the first $p^{m-w_{d+1}}$ components of $\boldsymbol{x}$, $\boldsymbol{x}_2$ is the vector consisting of the next $p^{m-w_{d+1}}$ components of $\boldsymbol{x}$ and so on. Now we apply the machinery of [16, 17] and get that multiplication with $\Omega^{(m-w_{d+1})}$ can be done in $\mathscr{O}((m-w_{d+1})p^{m-w_{d+1}})$ operations. Summarizing we have:

Algorithm 2

1. Compute $\psi(\frac{r}{x^m})$ for $r \in G_{p,m}$.
2. Set $\eta_1(n) = \psi(\frac{nx^{w_1}g_1}{x^m})$ for $n = 0, \ldots, p^m - 1$.
3. Set $g_1 = 1$, $d = 2$ and $t = \max\{j \in [s] \mid w_j < m\}$. While $d \leq \min\{s, t\}$,

   a. Partition η_{d-1} into p^{w_d} vectors $\eta_{d-1}^{(1)}, \ldots, \eta_{d-1}^{(p^{w_d})}$ of length p^{m-w_d} and let $\eta' = \sum_{i=1}^{p^{w_d}}\eta_{d-1}^{(i)}$.
   b. Let $(T_{d-1}(g))_{g\in G_{p,m-w_d}(x^m)} = \Omega^{(m-w_d)}\eta'$.
   c. Let $g_d = \operatorname{argmin}_g T_{d-1}(g)$.
   d. Let $\eta_d(n) = (1+\gamma_{d-1}+\gamma_{d-1}\psi(\frac{nx^{w_d}g_d}{x^m}))\eta_{d-1}(n)$
   e. Increase d by 1.

4. If $s \geq t$ then set $g_t = g_{t+1} = \ldots = g_s = 1$.

Similar to [2] we obtain from the results in this section the following theorem:

Theorem 2 *Let $N = p^m$ then the cost of Algorithm 2 is*

$$\mathcal{O}\left(N \log N + \min\{s, t\}N + N \sum_{d=1}^{\min\{s,t\}} (m - w_d)p^{-w_d}\right).$$

2.2 Polynomial Lattice Point Sets for Irreducible f

For this section let f be an irreducible polynomial over $\mathbb{F}_p$ with $\deg(f) = m$. The proof of the following result is similar to the proof of [3, Theorem 2.7]. If f is irreducible, then the congruence $\boldsymbol{h} \cdot \boldsymbol{g} \equiv -h_{d+1}x^{w_{d+1}}g_{d+1} \pmod{f}$, which comes from the definition of $R_\gamma(\boldsymbol{g}, f)$, has at most one solution $g_{d+1} \in \mathcal{G}_{p,m-w_{d+1}}(f)$ for fixed $\boldsymbol{g} \in \mathcal{G}^d_{p,m-\boldsymbol{w}}(f)$, $\boldsymbol{h} \in G^d_{p,m}$ and $h_{d+1} \in G_{p,m} \setminus \{0\}$. This fact simplifies the proof compared to the x^m case, as there is no need to deal with the Möbius function again.

Theorem 3 *Let $\boldsymbol{\gamma}$ and $\boldsymbol{w}$ as in Theorem 1 and let $f \in \mathbb{F}_p[x]$ be an irreducible polynomial with $\deg(f) = m$. Let further $(g_1, \ldots, g_s) \subset \mathcal{G}^s_{p,m-\boldsymbol{w}}(f)$ be constructed according to Algorithm 1. Then we have for every $d \in [s]$*

$$R^d_\gamma((x^{w_1}g_1, \ldots, x^{w_d}g_d), f) \leq \frac{1}{p^m} \prod_{i=1}^{d} \left(1 + \gamma_i + \gamma_i p^{\min\{w_i,m\}} m \frac{p+1}{3}\right).$$

As a consequence of (3) and Theorem 3 we obtain analogous results to Corollaries 1 and 2 for an irreducible modulus f.

Acknowledgements We would like to thank Peter Kritzer and Friedrich Pillichshammer for their valuable comments and suggestions which helped to improve our paper. All three authors are supported by the Austrian Science Fund (FWF): Project F5509-N26, Project F5506-N26, Project F5505-N26, where all three projects are part of the Special Research Program "Quasi-Monte Carlo Methods: Theory and Applications".

References

1. Dick, J., Kritzer, P., Leobacher, G., Pillichshammer, F.: Constructions of general polynomial lattice rules based on the weighted star discrepancy. Finite Fields Appl. **13**(4), 1045–1070 (2007)
2. Dick, J., Kritzer, P., Leobacher, G., Pillichshammer, F.: A reduced fast component-by-component construction of lattice points for integration in weighted spaces with fast decreasing weights. J. Comput. Appl. Math. **276**, 1–15 (2015)
3. Dick, J., Leobacher, G., Pillichshammer, F.: Construction algorithms for digital nets with low weighted star discrepancy. SIAM J. Numer. Anal. **43**(1), 76–95 (2005)
4. Dick, J., Pillichshammer, F.: Digital Nets and Sequences. Discrepancy Theory and Quasi-Monte Carlo Integration. Cambridge University Press, Cambridge (2010)

5. Hellekalek, P.: General discrepancy estimates: the Walsh function system. Acta Arith. **67**(3), 209–218 (1994)
6. Joe, S.: Construction of good rank-1 lattice rules based on the weighted star discrepancy. Monte Carlo and Quasi-Monte Carlo Methods 2004, pp. 181–196. Springer, Berlin (2006)
7. Kritzinger, R., Laimer, H.: A reduced fast component-by-component construction of lattice point sets with small weighted star discrepancy. Unif. Distrib. Theory **10**(2), 21–47 (2015)
8. Larcher, G., Pillichshammer, F.: Sums of distances to the nearest integer and the discrepancy of digital nets. Acta Arith. **106**(4), 379–408 (2003)
9. Leobacher, G., Pillichshammer, F.: Introduction to Quasi-Monte Carlo Integration and Applications. Compact Textbook in Mathematics. Birkhäuser/Springer, Cham (2014)
10. Lidl, R., Niederreiter, H.: Introduction to Finite Fields and their Applications, 1st edn. Cambridge University Press, Cambridge (1994)
11. Niederreiter, H.: Low-discrepancy point sets obtained by digital constructions over finite fields. Czechoslov. Math. J. **42(117)**(1), 143–166 (1992)
12. Niederreiter, H.: Random Number Generation and Quasi-Monte Carlo Methods. CBMS-NSF Regional Conference Series in Applied Mathematics, vol. 63. Society for Industrial and Applied Mathematics (SIAM), Philadelphia (1992)
13. Novak, E., Woźniakowski, H.: Tractability of Multivariate Problems. Volume I: Linear Information. EMS Tracts in Mathematics, vol. 6. European Mathematical Society (EMS), Zürich (2008)
14. Novak, E., Woźniakowski, H.: Tractability of Multivariate Problems. Volume II: Standard Information for Functionals. EMS Tracts in Mathematics, vol. 12. European Mathematical Society (EMS), Zürich (2010)
15. Novak, E., Woźniakowski, H.: Tractability of Multivariate Problems. Volume III: Standard Information for Operators. EMS Tracts in Mathematics, vol. 18. European Mathematical Society (EMS), Zürich (2012)
16. Nuyens, D., Cools, R.: Fast algorithms for component-by-component construction of rank-1 lattice rules in shift-invariant reproducing kernel Hilbert spaces. Math. Comput. **75**(254), 903–920 (2006)
17. Nuyens, D., Cools, R.: Fast component-by-component construction of rank-1 lattice rules with a non-prime number of points. J. Complex. **22**(1), 4–28 (2006)
18. Pillichshammer, F.: Polynomial lattice point sets. Monte Carlo and Quasi-Monte Carlo Methods 2010. Springer Proceedings in Mathematics and Statistics, vol. 23, pp. 189–210. Springer, Heidelberg (2012)
19. Sloan, I.H., Reztsov, A.V.: Component-by-component construction of good lattice rules. Math. Comput. **71**(237), 263–273 (2002)
20. Sloan, I.H., Woźniakowski, H.: When are quasi-monte Carlo algorithms efficient for high-dimensional integrals? J. Complex. **14**(1), 1–33 (1998)

Randomized Sobol' Sensitivity Indices

David Mandel and Giray Ökten

Abstract Classical Sobol' sensitivity indices assume the distribution of a model's parameters is known completely for a given model, but this is usually difficult to measure in practical problems. What is measurable is the distribution of parameters for a particular data set, and the Sobol' indices can significantly vary as different data sets are used in the estimation of the parameter distributions. To address this issue, we introduce a hierarchical probabilistic framework where Sobol' sensitivity indices are random variables. An ANOVA decomposition in this hierarchical framework is given. Some analytical examples and an application to interest rate modeling illustrate the use of the randomized Sobol' indices framework.

Keywords Global sensitivity analysis · Quantitative finance · Monte Carlo

1 Introduction

In the classical setup [10, 11], Sobol' indices measure the sensitivity of a mathematical model to its parameters assuming the parameters have a fixed probability distribution. These distributions should accurately reflect the likelihood of the model parameters realizing values in their domain and must be provided in order to implement global sensitivity analysis (GSA). According to Saltelli et al. [9], the distribution for each parameter is assumed to have come from either an expert opinion, who has professional knowledge of the distributions of parameters, or from a calibration, from which distributions are approximated using statistical theory. In particular to the classical setup, the tacit assumption is that the parameter distributions are completely known for a given model, and thus could be viewed as a component of the model.

D. Mandel (✉) · G. Ökten
Florida State University, Tallahassee 32306, FL, USA
e-mail: dmandel@math.fsu.edu

G. Ökten
e-mail: okten@math.fsu.edu

A. B. Owen and P. W. Glynn (eds.), *Monte Carlo and Quasi-Monte Carlo Methods*, Springer Proceedings in Mathematics & Statistics 241,
https://doi.org/10.1007/978-3-319-91436-7_22

There is a practical shortcoming with this approach, however, which is now illustrated by an example from mathematical finance. Consider pricing a US Treasury bill with a maturity of one year (a one-year T-bill) assuming a stochastic short rate model (see Bolder [1] for an introduction to interest rate models). There are a number of models in the literature for the task, but two of the most studied and simplest are the Vasicek [12] and CIR [3] models. Both describe the evolution of the short rate by a stochastic differential equation, which are given in Eqs. (1) and (2). They are parametrized by three variables: $a > 0$, the mean reversion speed; $b > 0$, the long-term mean; and $\sigma > 0$, the volatility. The function $W(t)$ is the standard Brownian motion. In practice, the parameters are estimated by an optimization method that finds the parameters for which the model has the best fit to some historical data. This procedure is known as calibration. The closed-form bond price for either model may then be computed using the parameter estimates (see Brigo and Mercurio [2] for a derivation of the closed-form bond prices under either model).

$$\text{Vasicek:}\quad \mathrm{d}r(t) = a(b - r(t))\mathrm{d}t + \sigma \mathrm{d}W(t)\,, \tag{1}$$

$$\text{CIR:}\quad \mathrm{d}r(t) = a(b - r(t))\mathrm{d}t + \sigma\sqrt{r(t)}\mathrm{d}W(t)\,. \tag{2}$$

The calibration procedure should be considered as part of the model—if the length of data or estimation technique is changed, the model changes as well. Suppose a financial engineer decides to use one year of interest rate data (specifically, yields on one-year US T-bills) in a calibration. For concreteness, assume that it is currently Dec 31, 1974, and a one-year T-bill is to be priced with maturity Dec 31, 1975. Using yields of one-year T-bills from 1974, parameter estimates and standard errors are obtained for each model: the results are displayed in Table 1. (Parameter values assume interest rates are expressed in percent. For example, $b = 7.61$ means the long-term short rate parameter is 7.61%.) The parameter estimation is done using the maximum likelihood estimation (MLE) technique following Duan [4], which implies an asymptotically-normal distribution for the parameter estimators. The point estimate and standard error are exactly the mean and standard deviation, respectively, of the corresponding normal distribution. Together, the mean and standard deviation constitute the *hyperparameters*; these are the parameters of the distribution of the model parameters.

Consider repeating this procedure using 1987 data, and again using 2006 data. The parameter estimates and standard errors for this data are also listed in Table 1. Different point estimates are expected as the US economy is sampled during distinct time periods; however, together with the changing standard errors, this implies different sampling distributions for each parameter estimator. In other words, the hyperparameters describing the parameter distributions are not fixed throughout model applications, and hence a particular parameter distribution is not a fixed component of the model.

How can we perform a global sensitivity analysis (GSA) of the Vasicek and CIR interest rate models? Imagine this is year 1975, and we want to use data from 1974 to price bonds. The first step is calibration as discussed earlier: estimate the parameters

Table 1 Parameter estimates and standard errors for three calibrations

Year	Parameter	Vasicek	CIR
1974	a	4.05 (2.62)	3.36 (0.96)
	b	7.61 (0.46)	2.21 (0.54)
	σ	1.06 (4.53)	1.07 (0.47)
1987	a	4.73 (2.54)	2.72 (0.92)
	b	7.02 (0.33)	1.49 (0.60)
	σ	1.05 (3.65)	1.08 (0.37)
2006	a	6.26 (2.14)	4.32 (1.07)
	b	2.72 (0.08)	0.90 (0.12)
	σ	1.02 (0.94)	1.02 (0.21)

Table 2 Sensitivity indices for three calibrations

Year	Sensitivity	Vasicek	CIR
1974	$\overline{S}_a$	0.57	0.25
	$\overline{S}_b$	0.43	0.76
	$\overline{S}_\sigma$	0.004	0.0
1987	$\overline{S}_a$	0.38	0.23
	$\overline{S}_b$	0.63	0.78
	$\overline{S}_\sigma$	0.002	0.0
2006	$\overline{S}_a$	0.005	0.007
	$\overline{S}_b$	0.99	0.99
	$\overline{S}_\sigma$	0.0002	0.0

a, b, σ of the model using an appropriate method and the data. Using the calibration approach of Duan [4], which is based on maximum likelihood estimation, we obtain the mean and standard errors for each parameter (see Table 1). We also know that the parameter estimators are asymptotically normal. Step 2 is to apply the GSA to the model (function) $\mathscr{F}(a, b, \sigma)$, where $\mathscr{F}$ is the closed-form bond formula for the corresponding Vasicek or CIR model. We compute the upper Sobol' sensitivity indices using Monte Carlo simulation with 10^6 samples (we will discuss Sobol' indices in detail in the next section; see Eq. (6)), assuming that the parameters are independent, and they are normally distributed with the mean and variances that were obtained from the calibration step. Then we repeat this procedure using data from 1987 and 2006, and report the estimated upper Sobol' indices of each parameter, for each interest rate model, in Table 2.

A close look at Table 2 reveals that sensitivity indices vary significantly from year to year. For example, if we consider the sensitivity results from 2006, we may decide to freeze the mean reversion speed a in either model due to its negligible Sobol' upper index, $\overline{S}_a$; however, a is the most influential parameter in the Vasicek model using 1974 parameters, and is certainly non-negligible for CIR in 1974. Thus relying

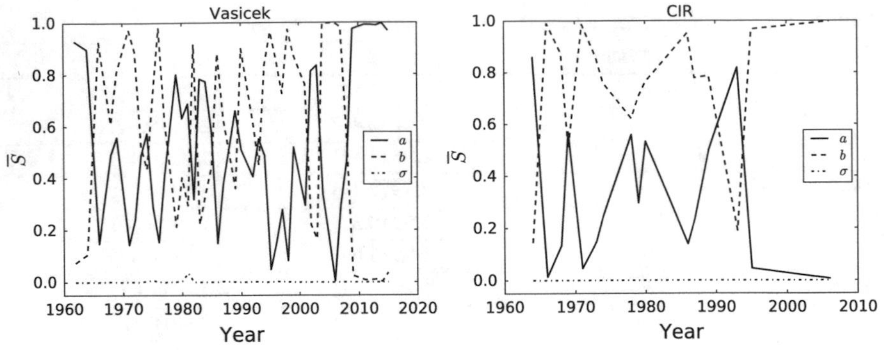

Fig. 1 Upper Sobol' sensitivity indices for Vasicek and CIR models across years

on only one sensitivity analysis would lead to an error. As a side remark, we note that it is not immediately intuitive why the volatility parameter σ appears to be non-influential to either model. Some insight is provided by the fact that doubling the variance of the volatility sampling distribution resulted in only 1% increase in the total model variance on average, indicating the models mute the effect of volatility.

We extend the results presented in Table 2, and compute the upper Sobol' indices by calibrating to one-year US T-bills from 1962 through 2015. Figure 1 plots the upper Sobol' indices as a function of year. (Several calibrations were not included due to potential losses of accuracy in parameter estimation, meaning the Hessian used to approximate the Fisher information matrix was not, or nearly not, positive-definite. See Remillard [8], "Sequential Quasi-Monte Carlo: Introduction for Non-experts, Dimension Reduction, Application to Partly Observed Diffusion Processes", for more details.) A closer look at Fig. 1, and the behavior of the upper Sobol' indices across different years, illustrates the conundrum of classical GSA: conclusions regarding ranking or freezing parameters differ from year-to-year, sometimes significantly. Thus, one cannot make sweeping inferences about the sensitivity of a model in general; rather, one is limited to a particular application (data set) for which one has fixed parameter distributions.

It is this restriction to a particular model application we aim to remove, so that the modeler may feel confident in his/her conclusions regarding the sensitivity patterns of a model in general. Our framework analyzes the Sobol' indices not as constant quantities, but as random variables that take on values for each particular model application. The joint distribution of the upper Sobol' indices will then be used to arrive at more robust conclusions about sensitivity.

Section 2 introduces the hierarchical framework and establishes an ANOVA decomposition used in our analysis. Section 3 provides some analytical examples, which serve to emphasize the need to model the joint distribution of the randomized Sobol' indices (as opposed to marginal distributions). A Monte Carlo algorithm is discussed in Sect. 4. Section 5 applies the framework to the interest rates models presented earlier, and provides practical considerations for randomized Sobol' analysis. In Sect. 6 we summarize our results.

2 Randomized Sobol' Sensitivity Analysis

In the previous section, classical Sobol' indices were shown to be limited to a particular model application (more specifically, to fixed parameter distributions), and are insufficient for generalizing to the model itself. The calibration method, which is used to obtain the parameter distributions, should be viewed as a component of the model. In other words, for the sensitivity indices presented in Table 2, the model is not, "the Vasicek model" (or, the "CIR" model), but rather, "the Vasicek model using MLE on one year of Treasury yields in 1974." If the length, type, or actual data is changed, or a different calibration technique is used, then the sensitivity results will change as well. In our proposed framework—randomized Sobol' sensitivity analysis—the aim is to remove the dependence on the particular data.

2.1 *Hierarchical Framework*

To set up the framework of randomized Sobol' sensitivity analysis, fix a probability space $(\Omega, \mathscr{F}, P)$ and define the random vector of model parameters $\mathbf{X} : \Omega \to \mathbb{R}^d$, where $d \in \mathbb{N}$ is the number of model parameters. In classical GSA, the joint distribution of the model parameters $\Lambda_{\mathbf{X}}$ is defined in terms of the probability measure P; that is, $\Lambda_{\mathbf{X}}(B) = P(\mathbf{X} \in B)$ for all $B \in \mathscr{B}(\mathbb{R}^d)$, where $\mathscr{B}$ denotes the Borel sigma-algebra. Given the evidence that Sobol' indices change across model applications, however, we seek an analysis that both accommodates this fact and allows one to draw general conclusions about a model's sensitivity pattern. With this in mind, introduce the random vector of hyperparameters $\mathbf{Y} : \Omega \to \mathbb{R}^m$, where $m \in \mathbb{N}$ is the number of hyperparameters for the model. For example, in the interest rate models (1) and (2), there are three normally-distributed parameters, each of which have two hyperparameters: the mean and standard deviation of the normal distribution. Thus there are $d = 3$ model parameters and $m = 6$ hyperparameters for these models. Define the joint distribution of hyperparameters as $\Lambda_{\mathbf{Y}}(A) = P(\mathbf{Y} \in A)$ for all $A \in \mathscr{B}(\mathbb{R}^m)$. In the new framework, the distribution of the model parameters is specified only in terms of the conditional probability $P(\mathbf{X} \mid \mathbf{Y})$.

As an example, let $f = f(x_1, \ldots, x_d)$ be a mathematical model and suppose, based on the calibration technique, each model parameter is found to follow an exponential distribution. In the classical sensitivity analysis, we would specify $X_i \sim \exp(\beta_i)$ for fixed $\beta_i > 0$, $i = 1, \ldots, d$. Assuming the random variables $\{X_i\}_{i=1}^d$ are mutually independent, the probability of the event $\{\mathbf{X} \in B\}$ for any $B \in \mathscr{B}(\mathbb{R}^d)$ is computed using the P-measure as

$$P(\mathbf{X} \in B) = \int_B \left(\prod_{i=1}^{d} \frac{1}{\beta_i} e^{-x_i/\beta_i} \right) d\mathbf{x} \,.$$

In the new framework, we instead specify

$$X_i \mid \beta_i = b_i \sim \exp(b_i) \,, \qquad \beta_i \sim \Lambda_{\beta_i} \,, \tag{3}$$

for some distribution Λ_{β_i}, where $i = 1, \ldots, d$. The left-hand side of (3) means X_i is exponentially distributed with mean parameter b_i, given that $\beta_i = b_i$. In this example, $\mathbf{Y} = (\beta_1, \ldots, \beta_d)$ is the random vector of hyperparameters with a realization $\mathbf{y} = (b_1, \ldots, b_d)$, and $m = d$. In this new setup, independence of $\{X_i\}_{i=1}^d$ is in terms of the conditional distribution $\Lambda_{\mathbf{X}|\mathbf{Y}}(\cdot \mid \mathbf{y})$ for each $\mathbf{y} \in \mathbb{R}^m$, i.e., for any $B = B_1 \times \ldots \times B_d \subset \mathbb{R}^d$ and $\mathbf{y} \in \mathbb{R}^m$, we have $\Lambda_{\mathbf{X}|\mathbf{Y}}(B \mid \mathbf{y}) = \prod_{i=1}^d \Lambda_{X_i|\mathbf{Y}}(B_i \mid \mathbf{y})$. Then, the probability of the event $\{\mathbf{X} \in B\}$ is computed using the conditional probability measure $P(\cdot \mid \mathbf{Y})$ as

$$P(\mathbf{X} \in B \mid \mathbf{Y} = \mathbf{y}) = \int_B \left(\prod_{i=1}^d \frac{1}{y_i} \mathrm{e}^{-x_i/y_i} \right) \mathrm{d}\mathbf{x} \,.$$

Thus we have constructed a hierarchical relationship between model calibration and model parameters. Once a calibration has been performed, $\mathbf{Y} = \mathbf{y}$ has been determined, and the joint probability of model parameters changes depending on the realization of the hyperparameters $\mathbf{y}$. If a new calibration is performed on a different data set, a new realization of hyperparameters $\mathbf{Y} = \mathbf{y}'$ for $\mathbf{y}' \neq \mathbf{y}$ will (likely) be obtained, which determines the joint probability of model parameters for this new application of the model.

2.2 *Hierarchical ANOVA*

To measure the variance contributed by each parameter, classical Sobol' indices rely on the ANOVA decomposition of a square integrable function $f : [0, 1]^d \to \mathbb{R}$. To accommodate randomized Sobol' analysis, we will show the ANOVA decomposition may be applied to compositions of the form $(f \circ \mathbf{X}) : \Omega \to \mathbb{R}$, where $\mathbf{X} : \Omega \to \mathbb{R}^d$ is a random vector (of model parameters) and f is square integrable. This is carried out in the state space $(\mathbb{R}^d, \mathscr{B}(\mathbb{R}^d))$ using the change of variables theorem and the conditional distributions $\Lambda_{\mathbf{X}|\mathbf{Y}}$. As described above, the hyperparameters $\mathbf{Y}$ will be random variables, and will be used as an index for a particular model application in which $\mathbf{Y} = \mathbf{y}$. Again, the randomness in the hyperparameters describes the varying distributions of model parameters across applications, whereas the randomness in the model parameters describes the uncertainty inherent in statistical estimation of parameters. To be more specific, the realization of hyperparameters $\mathbf{y}$ will serve as an index to a particular model application, and we will prove an ANOVA decomposition for a particular model application $f^{\mathbf{y}}(\mathbf{X})$.

Let $\mathscr{D} = \{1, \ldots, d\}$ denote the index set for a function $f : \mathbb{R}^d \to \mathbb{R}$ and let $u \subseteq \mathscr{D}$. To be consistent with the GSA literature, we will denote complements of such subsets as $-u = \mathscr{D} \setminus u$. We will write $f(\mathbf{X}_u, \mathbf{x}_{-u})$ to mean the outcome of the function f where $\mathbf{X}_u$ is random and $\mathbf{X}_{-u}$ is fixed at the value $\mathbf{x}_{-u}$.

Theorem 1 *Fix a probability space* $(\Omega, \mathscr{F}, P)$ *and let* $\mathbf{Y} : \Omega \to \mathbb{R}^m$ *be a random vector. For fixed* $\mathbf{y} \in \mathbb{R}^m$*, let* $f^{\mathbf{y}} : \mathbb{R}^d \to \mathbb{R}$ *be square-integrable, and* $X_1, \ldots, X_d$ *be mutually independent, real-valued random variables with finite variance, where independence is conditional on* $\mathbf{Y}$*, i.e., with respect to the joint distribution* $\Lambda(\cdot \mid \mathbf{y})$*. Then for each* $\mathbf{y} \in \mathbb{R}^m$*, there exists a unique representation of* f *as*

$$f^{\mathbf{y}}(\mathbf{X}) = \sum_{u \subseteq \mathscr{D}} f_u^{\mathbf{y}}(\mathbf{X}_u) \tag{4}$$

where each component function $f_u^{\mathbf{y}}$ *satisfies*

$$\int_{\mathbb{R}} f_u^{\mathbf{y}}(x_j, \mathbf{X}_{-\{j\}})\, \Lambda(\mathrm{d}x_j \mid \mathbf{y}) = 0 \tag{5}$$

if $j \in u$.

Proof Once $\mathbf{Y} = \mathbf{y}$, the construction of the components functions is identical to the classical construction [10] if the uniform measure is replaced with the conditional distributions $\Lambda(\cdot \mid \mathbf{y})$. □

In the classical setup, the model variance $D := \mathrm{Var}(f(\mathbf{X}))$ is decomposed into a sum of partial variances $D_u := \mathrm{Var}(f_u(\mathbf{X}_u))$ through the ANOVA decomposition as

$$D = \sum_{u \subseteq \mathscr{D}} D_u \,.$$

With the generalized ANOVA decomposition (4)–(5), the model variance may again be decomposed into a sum of component variances; however, each component variance, as well as the model variance is now a random variable taking on values for each $\mathbf{y}$. Indeed, letting $D^{\mathbf{Y}} = \mathrm{Var}(f(\mathbf{X}) \mid \mathbf{Y})$ and $D_u^{\mathbf{Y}} = \mathrm{Var}(f_u^{\mathbf{Y}}(\mathbf{X}_u) \mid \mathbf{Y})$, we obtain

$$D^{\mathbf{Y}} = \sum_{u \subseteq \mathscr{D}} D_u^{\mathbf{Y}} \,.$$

Clearly, for each fixed model application $\mathbf{Y} = \mathbf{y}$, we get different model variances and component variances, and thus a different decomposition.

In the classical setup, the component functions may be written succinctly in terms of conditional expectations as

$$f_u(\mathbf{X}_u) = E(f(\mathbf{X}) \mid \mathbf{X}_u) - \sum_{v \subsetneq u} E(f(\mathbf{X}) \mid \mathbf{X}_v) \,.$$

In particular, the component functions are only a function of the model parameters $\mathbf{X}$. The lower and upper Sobol' sensitivity indices, $\underline{S}_u$ and $\overline{S}_u$, are then defined by

$$\underline{S}_u = \frac{\mathrm{Var}(E(f(\mathbf{X}) \mid \mathbf{X}_u))}{\mathrm{Var}(f(\mathbf{X}))}, \qquad \overline{S}_u = \frac{E(\mathrm{Var}(f(\mathbf{X}) \mid \mathbf{X}_{-u}))}{\mathrm{Var}(f(\mathbf{X}))}, \tag{6}$$

where $\mathrm{Var}(f(\mathbf{X}))$ is assumed to be nonzero. (See [9] for an intuitive derivation of Sobol' indices as conditional expectations and variances.) The lower indices are also known as closed or first-order effects, and the upper indices are also known as total effects.

In the new framework, component functions may analogously be written as conditional expectations if the hyperparameters are included as conditioning variables:

$$f_u^{\mathbf{Y}}(\mathbf{X}_u) = E(f^{\mathbf{Y}}(\mathbf{X}) \mid \mathbf{X}_u, \mathbf{Y}) - \sum_{v \subsetneq u} E(f^{\mathbf{Y}}(\mathbf{X}) \mid \mathbf{X}_v, \mathbf{Y}) .$$

In the new framework the component functions explicitly depend not only on the model parameters, but also on the hyperparameters $\mathbf{Y}$. Sobol' sensitivity indices in the new framework are the random variables $\underline{S}_u^{\mathbf{Y}}$ and $\overline{S}_u^{\mathbf{Y}}$ defined next.

Definition 1 Let $f : \mathbb{R}^d \to \mathbb{R}$ be a square integrable function and $\mathbf{X} = (X_1, ..., X_d)$ a vector of independent random variables, where the independence is with respect to the conditional joint distribution $\Lambda(\cdot \mid \mathbf{y})$ for each $\mathbf{y} \in \mathbb{R}^m$. Then for any $u \subseteq \mathscr{D}$, the *randomized lower and upper Sobol' indices* are

$$\underline{S}_u^{\mathbf{Y}} := \frac{\mathrm{Var}(E(f(\mathbf{X}) \mid \mathbf{X}_u, \mathbf{Y}))}{\mathrm{Var}(f(\mathbf{X}) \mid \mathbf{Y})}, \tag{7}$$

$$\overline{S}_u^{\mathbf{Y}} := \frac{E(\mathrm{Var}(f(\mathbf{X}) \mid \mathbf{X}_{-u}, \mathbf{Y}))}{\mathrm{Var}(f(\mathbf{X}) \mid \mathbf{Y})}, \tag{8}$$

where it is assumed that $\mathrm{Var}(f(\mathbf{X}) \mid \mathbf{Y} = \mathbf{y}) > 0$ for all $\mathbf{y}$.

The randomized Sobol' indices take on values for each $\mathbf{Y} = \mathbf{y}$; i.e., for each particular model application. The mean of the randomized Sobol' indices with respect to the joint distribution of $\mathbf{Y}$, i.e.,

$$E(\underline{S}_u^{\mathbf{Y}}) = \int_{\mathbb{R}^m} \underline{S}_u^{\mathbf{y}} \, \Lambda(d\mathbf{y}) , \qquad E(\overline{S}_u^{\mathbf{Y}}) = \int_{\mathbb{R}^m} \overline{S}_u^{\mathbf{y}} \, \Lambda(d\mathbf{y}) , \tag{9}$$

can be used as an aggregate measure of model sensitivity.

How do we determine the distribution of $\mathbf{Y}$ in practice? Unlike classical GSA, relying on expert opinion is an unlikely possibility for the distribution of hyperparameters. Instead, such a distribution will likely need to be estimated from multiple calibrations of a model. A practical method is to perform a nonparametric density estimation of the distribution of the Sobol' indices across multiple calibrations of a model. Assuming samples can be drawn from such a density, a Monte Carlo estimate of $E(\underline{S}_u^{\mathbf{Y}})$ and $E(\overline{S}_u^{\mathbf{Y}})$ may be computed. This also has the benefit of circumventing the need for analytical formulas for the randomized Sobol' indices. Details of this procedure are given in Sect. 5.

3 Examples

In this section, we discuss some examples that yield analytical formulas for randomized Sobol' indices. These serve to illustrate the need for modeling the joint distribution of randomized Sobol' indices, as opposed to modeling only the marginal distributions. The formulas for only the first order indices are given (subsets $u \subset \mathscr{D}$ such that $|u| = 1$); it is a widely held belief that in most physical problems, first order terms provide a sufficient characterization of the sensitivity pattern of the model (see, for example, Saltelli et al. [9]). The distribution of $\mathbf{Y}$ is left unspecified for now.

Example 1 Consider the model given by

$$f(\mathbf{X}) = \sum_{i-1}^{d} X_i ,$$

where $\{X_i\}_{i=1}^{d}$ is a collection of mutually independent random variables, each with marginal conditional distribution

$$X_i \mid (M_i, \Sigma_i) = (\mu_i, \sigma_i) \sim \mathscr{N}(\mu_i, \sigma_i^2) , \qquad i = 1, \ldots, d . \tag{10}$$

The random hyperparameters are $\mathbf{Y} = (M_1, \Sigma_1, \ldots, M_d, \Sigma_d)$ for some random variables M_i, and positive random variables Σ_i. It can be shown that the component functions are

$$f_{\emptyset}^{\mathbf{Y}} = \sum_{i=1}^{d} M_i ,$$

$$f_{\{i\}}^{\mathbf{Y}}(X_i) = X_i - M_i ,$$

for $i = 1, \ldots, d$ and $f_u^{\mathbf{Y}} \equiv 0$ for all $|u| \geq 2$. In addition, for $i = 1, \ldots, d$, the first order variances and total variance, respectively, are

$$D_{\{i\}}^{\mathbf{Y}} = \Sigma_i^2 \qquad D^{\mathbf{Y}} = \Sigma_1^2 + \Sigma_2^2 + \ldots + \Sigma_d^2 .$$

Thus for $i = 1, \ldots, d$, the first order randomized Sobol' indices are

$$\underline{S}_{\{i\}}^{\mathbf{Y}} = \overline{S}_{\{i\}}^{\mathbf{Y}} = \frac{D_{\{i\}}^{\mathbf{Y}}}{D^{\mathbf{Y}}} = \frac{\Sigma_i^2}{\Sigma_1^2 + \Sigma_2^2 + \ldots + \Sigma_d^2} . \tag{11}$$

The distribution of the ith randomized Sobol' index is given by the distribution of $\Sigma_i^2/\left(\Sigma_1^2 + \Sigma_2^2 + \ldots + \Sigma_d^2\right)$. Although in this example each randomized Sobol' index is a function only of the variance hyperparameters Σ_i, in general, a randomized Sobol' index may be a function of the full hyperparameter vector $\mathbf{Y}$. Once a distribution is specified for $\mathbf{Y}$, the average randomized Sobol' indices $E(\underline{S}_{\{i\}}^{\mathbf{Y}})$ and $E(\overline{S}_{\{i\}}^{\mathbf{Y}})$ may be computed, either analytically or via Monte Carlo. Note also the dependence of the randomized Sobol' indices on all Σ hyperparameters. This illustrates the need to specify the joint distribution of hyperparameters for randomized Sobol' indices, as opposed to marginal distributions.

Example 2 Consider the function

$$f(X) = \prod_{i=1}^{d} X_i \,, \qquad X_i \mid Y_i = y_i \sim \exp(y_i)\,, \quad i = 1, \ldots, d\,,$$

where the random variables $\{X_i\}_{i=1}^d$ are mutually independent. Here $\mathbf{Y} = (Y_1, \ldots, Y_d)$ for some positive random variables Y_i, $i = 1, \ldots, d$. It can shown that the first order component functions are

$$f_{\emptyset}^{\mathbf{Y}} = \prod_{i=1}^{d} Y_i \,,$$

$$f_{\{j\}}^{\mathbf{Y}}(X_j) = \left(X_j - Y_j\right) \prod_{\substack{i=1 \\ i \neq j}}^{d} Y_i \,,$$

for $j = 1, \ldots, d$, and the first order variances and the total variance are

$$D_{\{j\}}^{\mathbf{Y}} = \prod_{i=1}^{d} Y_i^2 \,, \qquad D^{\mathbf{Y}} = (2^d - 1) \prod_{i=1}^{d} Y_i^2 \,.$$

Therefore the first order randomized Sobol' indices are

$$\underline{S}_{\{j\}}^{\mathbf{Y}} = \frac{1}{2^d - 1} \,, \qquad \overline{S}_{\{j\}}^{\mathbf{Y}} = \frac{2^{d-1}}{2^d - 1} \tag{12}$$

for $j = 1, \ldots, d$. Notice that although the partial and total variances are random variables, the randomized Sobol' indices are constant. This illustrates that in some instances, it may be beneficial to derive formulas of the randomized Sobol' indices in place of numerical approximation; however, as the next example emphasizes, the distribution of randomized Sobol' indices is, of course, dependent on the model parameter distribution.

Example 3 Consider again the function

$$f(X) = \prod_{i=1}^{d} X_i \, ,$$

but now specify the mutually independent random variables X_i as

$$X_i \mid (M_i, \Sigma_i) = (\mu_i, \sigma_i) \sim \mathcal{N}(\mu_i, \sigma_i^2), \quad i = 1, \ldots, d \, .$$

Here $\mathbf{Y} = (M_1, \Sigma_1, \ldots, M_d, \Sigma_d)$, as in the first example. It can be shown that

$$f_{\emptyset}^{\mathbf{Y}} = \prod_{i=1}^{d} M_i \, ,$$

$$f_{\{j\}}^{\mathbf{Y}}(X_j) = (X_j - M_j) \prod_{\substack{i=1 \\ i \neq j}}^{d} M_i \, ,$$

and

$$D_{\{j\}}^{\mathbf{Y}} = \Sigma_j^2 \prod_{\substack{i=1 \\ i \neq j}}^{d} M_i^2 \, , \qquad D^{\mathbf{Y}} = \prod_{i=1}^{d} (\Sigma_i^2 + M_i^2) - \prod_{i=1}^{d} M_i^2 \, ,$$

for $j = 1, \ldots, d$.

The first order randomized Sobol' indices, for $j = 1, \ldots, d$, are

$$\underline{S}_{\{j\}}^{\mathbf{Y}} = \frac{\Sigma_j^2 \prod_{i=1, i \neq j}^{d} M_i^2}{\prod_{i=1}^{d} (\Sigma_i^2 + M_i^2) - \prod_{i=1}^{d} M_i^2} \, , \tag{13}$$

$$\overline{S}_{\{j\}}^{\mathbf{Y}} = \frac{\Sigma_j^2 \prod_{i=1, i \neq j}^{d} (M_i^2 + \Sigma_i^2)}{\prod_{i=1}^{d} (\Sigma_i^2 + M_i^2) - \prod_{i=1}^{d} M_i^2} \, . \tag{14}$$

The distributions of (13) and (14) will likely be difficult to derive analytically for most distributions for **Y**, and hence their expected values will be approximated by Monte Carlo.

4 Monte Carlo Estimates

Even if the randomized Sobol' indices can be computed in closed form as in (11)–(14), it is likely that the expected value of the indices, $E(\underline{S}_u^{\mathbf{Y}})$ and $E(\overline{S}_u^{\mathbf{Y}})$, are analytically intractable, making a Monte Carlo approximation necessary.

If the randomized Sobol' indices are known in closed form and it is possible to sample from $\mathbf{Y}$, then the Monte Carlo estimation of $E(\underline{S}_u^{\mathbf{Y}})$ and $E(\overline{S}_u^{\mathbf{Y}})$ is straightforward, and is not discussed further. Similarly, if the Sobol' indices from multiple calibrations of a model are available, a multivariate kernel density estimator can be used to estimate the joint density of the upper Sobol' indices. We can then draw samples from the kernel density, and estimate expected values using Monte Carlo. An example of this procedure is given in Sect. 5.

An interesting scenario is when the distribution of $\mathbf{Y}$ is provided and the Sobol' indices are analytically intractable or not known in closed form. In this case, both the Sobol' indices and their $\mathbf{Y}$-expected value may be approximated via Monte Carlo using the following algorithm, assuming that we have a method to sample from the necessary distributions. For $i = 1, \ldots, N_1$, let $y^{(i)} \in \mathbb{R}^m$ be a random vector drawn from the joint distribution $\Lambda_{\mathbf{Y}}$. With this fixed $y^{(i)}$, for $j = 1, \ldots, N_2$ draw two independent random vectors $\xi^{(i,j)}, \eta^{(i,j)} \in \mathbb{R}^d$, each from the conditional joint distribution $\Lambda_{\mathbf{X}|\mathbf{Y}}(\cdot \mid y^{(i)})$. Set $\xi^{(i,j)} = (\xi_u^{(i,j)}, \xi_{-u}^{(i,j)})$ and $\eta^{(i,j)} = (\eta_u^{(i,j)}, \eta_{-u}^{(i,j)})$. For each $i = 1, \ldots, N_1$, the standard Monte Carlo algorithms for computation of Sobol' indices may be used to approximate $\underline{S}_u^{y^{(i)}}$ and $\overline{S}_u^{y^{(i)}}$. See Owen [7], Kucherenko et al. [6] or Sobol' [10] for the standard algorithms. We then obtain the Monte Carlo estimates

$$E(\underline{S}_u^{\mathbf{Y}}) \approx \frac{1}{N_1} \sum_{i=1}^{N_1} \underline{S}_u^{y^{(i)}}, \qquad E(\overline{S}_u^{\mathbf{Y}}) \approx \frac{1}{N_1} \sum_{i=1}^{N_1} \overline{S}_u^{y^{(i)}}.$$

The above expected value estimations involve double application of Monte Carlo, and as a result, the computations could be costly. Some possible ideas to speed up the computations are the use of a metamodel (see [9]), and quasi-Monte Carlo simulation.

5 Application to Interest Rate Models

We revisit the interest rate problem from Sect. 1, where the upper Sobol' indices were shown to vary considerably depending on the particular interest rate model application (see Fig. 1). Now we can use the randomized Sobol' framework to analyze the randomized sensitivity indices using statistical tools; in particular, assess the average sensitivity indices of the interest rate models by estimating $E(\underline{S}_{\{i\}}^{\mathbf{Y}})$ and $E(\overline{S}_{\{i\}}^{\mathbf{Y}})$, where $i = a, b, \sigma$. To this end, we first approximate the joint distribution of the upper Sobol' indices using a Gaussian kernel density estimator on the calibrations of the models from 1962–2015. A kernel density estimator was chosen to model the joint density to capture any dependencies among the randomized Sobol' indices; see Givens et al. [5] for details on such estimators. As the dimension increases, the accuracy of the kernel density estimation method will be a point of concern; an analysis of this is left for future work. As an illustration of the density estimation,

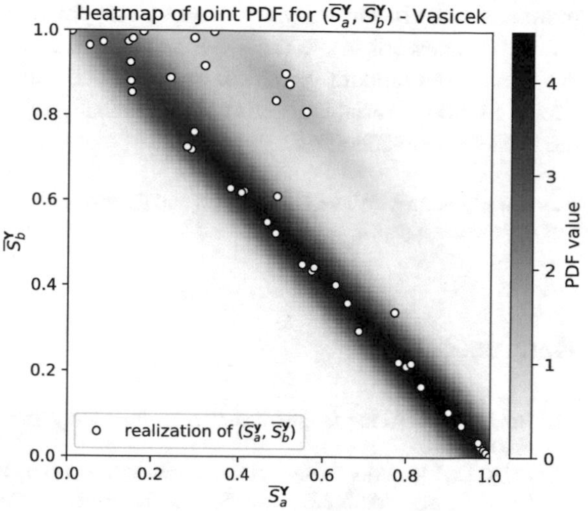

Fig. 2 Heatmap of joint pdf of the pair of randomized upper Sobol' indices $(\overline{S}_a^{\mathbf{Y}}, \overline{S}_b^{\mathbf{Y}})$

Table 3 Approximated mean of randomized Sobol' indices for interest rate models

Sensitivity	Vasicek	CIR
$E(\overline{S}_a^{\mathbf{Y}})$	0.52	0.43
$E(\overline{S}_b^{\mathbf{Y}})$	0.44	0.63
$E(\overline{S}_\sigma^{\mathbf{Y}})$	0.005	0.0

Fig. 2 plots a heatmap of the estimated joint pdf of $(\overline{S}_a^{\mathbf{Y}}, \overline{S}_b^{\mathbf{Y}})$. The circles in this figure depict joint realizations of the given pair of Sobol' indices from the Vasicek model.

Table 3 displays the expected value of the randomized upper Sobol' indices for the Vasicek and CIR models, using the estimated joint density of $(\overline{S}_a^{\mathbf{Y}}, \overline{S}_b^{\mathbf{Y}}, \overline{S}_\sigma^{\mathbf{Y}})$ from the multiple calibrations. The results show that, on average, both the mean reversion speed a and long-term mean b are influential in either model; however, the volatility σ is unimportant. Because these averaged sensitivity indices take into account all applications of the model to one-year T-bills, the modeler should feel confident in freezing the volatility parameters in either model with negligible loss of information.

6 Conclusions

In the classical sensitivity analysis, we usually assume the distribution of a model's parameters is completely known. Using an example from interest rate modeling, we showed how changes in this distribution across different model applications (calibrations) can result in significant changes in the sensitivity pattern of the model parameters. To better analyze changes in model sensitivities, we presented a hierarchical framework where the hyperparameters of a model parameters' distribution are themselves random variables. In this framework, the Sobol' sensitivity indices become random variables, and their statistical properties can offer new insights to the model in consideration.

Acknowledgements We thank Art Owen and the anonymous referee for their valuable comments that improved the paper.

References

1. Bolder, D.J.: Affine term-structure models: Theory and implementation. Technical Report (2001)
2. Brigo, D., Mercurio, F.: Interest Rate Models: Theory and Practice. Springer, New York (2001)
3. Cox, J.C., Ingersoll Jr, J.E., Ross, S.A.: A theory of the term structure of interest rates. Econom. J. Econ. Soc. 385–407 (1985)
4. Duan, J.C.: Maximum likelihood estimation using price data of the derivative contract. Math. Financ. **4**(2), 155–167 (1994)
5. Givens, G.H., Hoeting, J.A.: Computational Statistics, vol. 710. Wiley, New York (2012)
6. Kucherenko, S., Tarantola, S., Annoni, P.: Estimation of global sensitivity indices for models with dependent variables. Comput. Phys. Commun. **183**(4), 937–946 (2012)
7. Owen, A.: Better estimation of small Sobol' sensitivity indices. ACM Trans. Model. Comput. Simul. (TOMACS) **23**(2), 1–17 (2013)
8. Remillard, B.: Statistical Methods for Financial Engineering. CRC Press, Boca Raton (2013)
9. Saltelli, A., Ratto, M., Andres, T., Campolongo, F., Cariboni, J., Gatelli, D., Saisana, M., Tarantola, S.: Global Sensitivity Analysis: The Primer. Wiley, New York (2008)
10. Sobol', I.M.: Sensitivity estimates for nonlinear mathematical models. Math. Model. Comput. Exp. **1**(4), 407–414 (1993)
11. Sobol', I.M.: Global sensitivity indices for nonlinear mathematical models and their Monte Carlo estimates. Math. Comput. Simul. **55**(1), 271–280 (2001)
12. Vasicek, O.: An equilibrium characterization of the term structure. J. Financ. Econ. **5**(2), 177–188 (1977)

Supervised Learning of How to Blend Light Transport Simulations

Hisanari Otsu, Shinichi Kinuwaki and Toshiya Hachisuka

Abstract Light transport simulation is a popular approach for rendering photorealistic images. However, since different algorithms have different efficiencies depending on input scene configurations, a user would try to find the most efficient algorithm based on trials and errors. This selection of an algorithm can be cumbersome because a user needs to know technical details of each algorithm. We propose a framework which blends the results of two different rendering algorithms, such that a blending weight per pixel becomes automatically larger for a more efficient algorithm. Our framework utilizes a popular machine learning technique, regression forests, for analyzing statistics of outputs of rendering algorithms and then generating an appropriate blending weight for each pixel. The key idea is to determine blending weights based on classification of path types. This idea is inspired by the same common practice in movie industries; an artist composites multiple rendered images where each image contains only a part of light transport paths (e.g., caustics) rendered by a specific algorithm. Since our framework treats each algorithm as a black-box, we can easily combine very different rendering algorithms as long as they eventually generate the same results based on light transport simulation. The blended results with our algorithm are almost always more accurate than taking the average, and no worse than the results with an inefficient algorithm alone.

Keywords Light transport simulation · Machine learning · Regression forest

H. Otsu · T. Hachisuka (✉)
The University of Tokyo, 7-3-1 Hongo, Bunkyo-ku, Tokyo, Japan
e-mail: thachisuka@siggraph.org

H. Otsu
e-mail: hotsu@graphics.ci.i.u-tokyo.ac.jp

S. Kinuwaki
Bunkyo-ku, Tokyo, Japan
e-mail: ShinichiKinuwaki@gmail.com

A. B. Owen and P. W. Glynn (eds.), *Monte Carlo and Quasi-Monte Carlo Methods*, Springer Proceedings in Mathematics & Statistics 241,
https://doi.org/10.1007/978-3-319-91436-7_23

1 Introduction

Rendering based on light transport simulation is a popular approach for photorealistic image synthesis. Since such rendering algorithms solve the same governing equations (e.g., the rendering equation [16]), rendering with light transport simulation should give us the same result regardless of the choice of an algorithm. It is however well known that some algorithms are more efficient at rendering certain light transport effects. For example, photon density estimation [5, 14] is often efficient at rendering caustics, and Markov chain Monte Carlo algorithms [12, 31] are considered efficient at resolving complex occlusions.

Because of the varying efficiency of different algorithms on different light transport effects, it is common practice to select an algorithm based on the type of light transport effect that one wants to render. In the movie industry, an artist often decomposes light transport effects into separate images, renders each with a most efficient algorithm, and composites the resulting images into the final one. Selecting appropriate algorithms and composting the results, however, can be difficult and cumbersome tasks. For selection, an artist either needs to know why some algorithms work well for some effects, or briefly tries all the available algorithms to see which one works well. For composition, an artist also needs to pay attention not to double count a certain type of paths such as caustics.

We propose a framework which automates this selection of the algorithms and composition of the resulting images. Our work is inspired by the superhuman performance of recent machine learning algorithms on classification tasks. We apply the same idea to select and composite two different rendering algorithms based on the classification of light transport effects. To be concrete, we use regression forests [1] to learn the relationship between blending weights that minimize the error and the classification of light transport effects. While multiple importance sampling [30] also allows us to blend results of different rendering techniques, the key difference is that our framework treats each rendering algorithms as a black-box. Accordingly, our framework can be easily applied to very different algorithms such as SPPM and MLT without any algorithmic or theoretical modifications for each. To summarize, our contributions are:

- The use of machine learning to automatically blend the results of different rendering algorithms based on path types.
- A blending framework which is independent from how the underlying rendering algorithms work.
- First successful application of regression forests to light transport simulation.

2 Overview

Our goal is to blend the results of two different rendering algorithms such that the error of the blended result is as small as possible. Our algorithm is separated into two phases; the *training* phase and *runtime*. Figure 1 illustrates the algorithm.

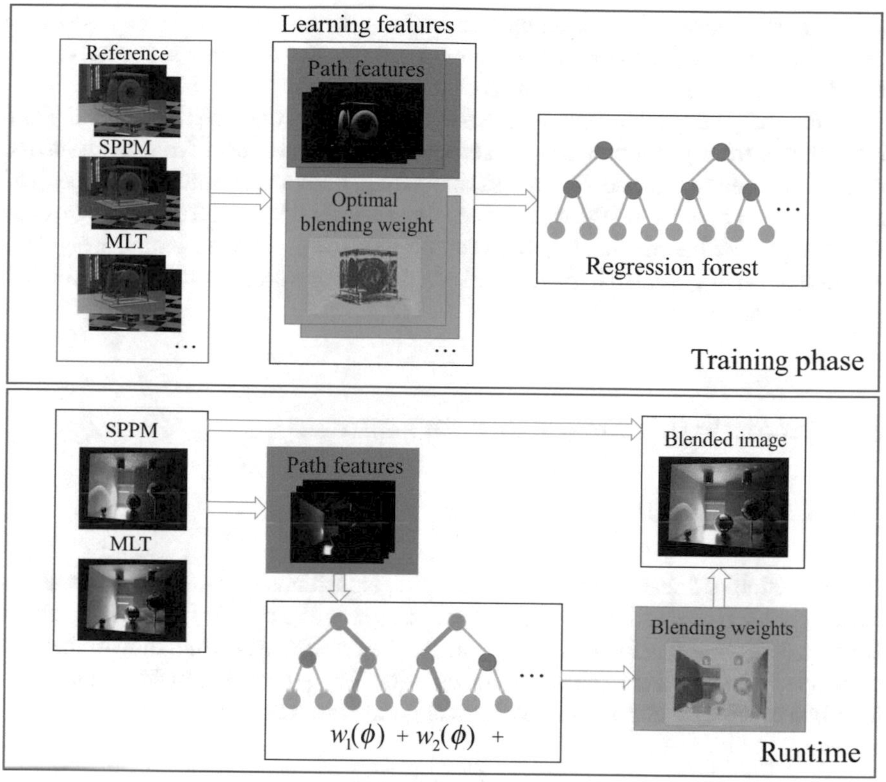

Fig. 1 General idea of blending the results of two different rendering algorithms using regression forests. In the training phase (top), we first calculate the optimal blending weight per pixel, given the reference image and rendered images with different approaches. These weights and the corresponding path features become one training sample for the regression forest for each scene. We iterate this process for various scenes. Our framework thus learns the relationship between input path features and optimal weights during this learning phase. At runtime (bottom), the trained regression forest returns approximated optimal blending weights based on path features of a new scene

In the training phase, we use regression forests [1] (Sect. 4) to learn the relationship between a feature vector of lighting effects extracted from the rendered images and the optimal weights for blending. For each training scene, we render the reference solution, and the two images with both algorithms allocating the same rendering time. Based on the rendered images, we extract *path features* as the relative pixel contributions of different light transport paths according to Heckbert's notation [9]. Modern shader languages often support the same mechanism [3]. We then calculate the *optimal blending weights* based on the reference solution and the results of the two different rendering algorithms. The optimal blending weight is defined such that the error of the blended result is minimized at each pixel. A pair of path features and the optimal blending weight forms one training sample for regression forests.

If a rendering algorithm is based on Monte Carlo methods (which is the case in our experiments), we generate multiple training samples for the same scene in order to avoid the influence of the randomness of rendered images.

At runtime, we use the trained regression forest to approximate the optimal blending weights for a given new scene. The path features extracted from the rendered images are used to traverse the regression forest to obtain the blending weights. The final result is a blended image with the obtained weights. Since the trained regression forest expresses the relationship between path features and the optimal weights, a blended image is expected to have small error, even for a scene that was not included in the training phase.

3 Automatic Blending with Path Features

3.1 Path Features

Our definition of a feature vector for rendering algorithms is inspired by how artists decompose a rendered image into several images with specific lighting effects for each. In order to define the feature vectors, we begin with the formulation of the light transport known as the path integral formulation [28]. According to the formulation, the pixel intensity I observed at each pixel is expressed as

$$I = \int_{\Omega} f(\bar{x}) d\mu(\bar{x}), \tag{1}$$

where $\bar{x}$ is a light transport path, f is the measurement contribution function, and μ is the path measure. Ω is the space of paths of all different path lengths.

The path space Ω can be partitioned into a union of disjoint spaces according to the classification by Heckbert [9]:

$$\Omega = \Omega_{\mathrm{LDE}} \cup \Omega_{\mathrm{LSE}} \cup \Omega_{\mathrm{LDSE}} \cup \Omega_{\mathrm{LSDE}} \cup \cdots, \tag{2}$$

where each Ω_* is a subspace of Ω defined with the paths represented by the Heckbert's notation $*$. For instance, the subspace Ω_{LDSE} with path length 3 is defined as a set of paths $\bar{x} = \mathbf{x}_0\mathbf{x}_1\mathbf{x}_2\mathbf{x}_3$ where $\mathbf{x}_0$ is on a sensor, $\mathbf{x}_1$ is on a diffuse surface, $\mathbf{x}_2$ is on a specular surface, and $\mathbf{x}_3$ is on an emitter. A glossy interaction is classified to either D or S depending on its BRDF.

We thus define a part of the intensity I_* contributed only with the subspace Ω_* as

$$I_* = \int_{\Omega_*} f(\bar{x}) d\mu(\bar{x}). \tag{3}$$

Since the partition in Eq. 2 is disjoint, the pixel intensity I is additive:

$$I = I_{\mathrm{LDE}} + I_{\mathrm{LSE}} + I_{\mathrm{LDSE}} + I_{\mathrm{LSDE}} + \cdots. \tag{4}$$

We thus define the *path features* ϕ as a vector of the intensities I_* relative to I:

$$\phi \equiv \frac{(I_{\mathrm{LDE}}, I_{\mathrm{LSE}}, I_{\mathrm{LDSE}}, I_{\mathrm{LSDE}}, \ldots)}{I}. \tag{5}$$

The definition uses relative intensities such that ϕ is independent from the absolute intensity. We fixed the maximum path length to ten, which makes ϕ a $2^{(10-1)} = 512$ dimensional feature vector. The training phase uses an *estimate* $\hat{\phi}$ instead of the analytical value of ϕ for a given rendering time. We selected the number of dimensions such that all the data fits within the main memory. For instance, the scene rendered with 720p resolution requires a storage of $512 \times 1280 \times 720 \times 4$ bytes ≈ 1.8 GB.

3.2 Optimal Blending Weights

In the training phase, we need to determine the optimal blending weight. This weight is used as an *answer* associated with a path feature vector. A pair of a path feature vector and the optimal blending weight thus becomes a training sample for supervised learning via regression forests.

We define the optimal blending weight w_{opt} that gives the minimum error as

$$w_{\mathrm{opt}} = \underset{w}{\mathrm{argmin}} \left| \left(w\hat{I}_\alpha + (1-w)\hat{I}_\beta \right) - I \right|, \tag{6}$$

where $\hat{I}_\alpha$ and $\hat{I}_\beta$ are the results of two different rendering algorithms α and β respectively, and I is the reference solution. This equation can be easily solved as

$$w_{\mathrm{opt}} = \frac{I - \hat{I}_\beta}{\hat{I}_\alpha - \hat{I}_\beta}. \tag{7}$$

If the solution of Eq. 6 is not in the range of [0, 1], it is clamped to the nearest side such that $w_{\mathrm{opt}} \in [0, 1]$. We apply this clamping such that the blending operation becomes a convex combination of the results. The blended result $w\hat{I}_\alpha + (1-w)\hat{I}_\beta$ is thus guaranteed to be more accurate than one of $\hat{I}_\alpha$ and $\hat{I}_\beta$ since

$$|w\hat{I}_\alpha + (1-w)\hat{I}_\beta - I| \leq \max(|\hat{I}_\alpha - I|, |\hat{I}_\beta - I|) \tag{8}$$

by definition if $w \in [0, 1]$. Intuitively, this clamping process sets $w_{\text{opt}} = 1$ when $\hat{I}_\alpha$ and $\hat{I}_\beta$ both either underestimate or overestimate I and $\hat{I}_\alpha$ is closer to I (vice versa for $\hat{I}_\beta$). If one of the $\hat{I}_\alpha$ and $\hat{I}_\beta$ underestimates and the other overestimates I, we set w_{opt} such that the blended result is exactly equal to I. Note that Eq. 8 only guarantees that an error per pixel does not become worse, not the sum of errors over an image. For example, collecting pixels with worse errors (with $w = 0$ or $w = 1$) still satisfies Eq. 8, but the sum of errors would increase.

The intensities $\hat{I}_\alpha$ and $\hat{I}_\beta$ are the relatively rough estimates of I in practice. If an algorithm is based on Monte Carlo ray tracing, an estimated intensity is an instance of the random variable for each run. Using samples only from a single run of the algorithm causes overfitting to this specific run. For example, it might be that $\hat{I}_\alpha$ happens to be closer to I than $\hat{I}_\beta$ for the single run used in the training phase.

In order to deal with this issue, we use multiple training samples even for the same scene and the same algorithm. In fact, machine learning techniques (including regression forests) are naturally designed for dealing with such variations in the training data.

Problem Statement: Given the definitions above, the goal of our algorithm is to find a function w_{approx} such that

$$w_{\text{opt}} \approx w_{\text{approx}}(\phi), \tag{9}$$

for given path features ϕ and two rendering algorithms α and β. This function w_{approx} basically expresses the preference of the algorithm α over the other algorithm β for paths with a feature vector of ϕ. In order to learn w_{approx}, we use a machine learning algorithm called *regression forests*.

Difference from Multiple Importance Sampling: Conceptually, our proposed blending approach is similar to multiple importance sampling (MIS) [30]. MIS also combines two or more different estimators to improve the efficiency of the combined estimator.

MIS combines multiple sampling strategies by decomposing the measurement contribution function f in Eq. 1 into a weighted sum of M different weights. The estimate of the pixel intensity I by MIS can be written as

$$I = \int_\Omega \sum_{t=1}^{M} w_t(\bar{x}) f(\bar{x}) d\mu(\bar{x}) = \sum_{t=1}^{M} \int_\Omega w_t(\bar{x}) f(\bar{x}) d\mu(\bar{x}) \tag{10}$$

$$\approx \sum_{t=1}^{M} \frac{1}{N_t} \sum_{i=1}^{N_t} w_t(\bar{x}_{t,i}) \frac{f(\bar{x}_{t,i})}{p_t(\bar{x}_{t,i})} \tag{11}$$

where $p_i(\bar{x})$ is the pdf of the ith strategy, $w_i(\bar{x})$ is the weighting function satisfying $\sum_{t=1}^{M} w_i(\bar{x}) = 1$ for all $\bar{x} \in \Omega$ with $f(\bar{x}) \neq 0$, and $w_i(\bar{x}) = 0$ for all $\bar{x} \in \Omega$ with $p_i(\bar{x}) = 0$. In order to use MIS, however, we need to know the probability densities of path sampling techniques for arbitrary sample locations. Such information can be difficult to obtain without modifying an implementation or sometimes impossible due to the formulation of each algorithm.

4 Regression Forests

The basic idea of regression forests is to use a set of binary trees for approximating a multivariate function of the feature vector. This multivariate function expresses the relationship between feature vectors and the corresponding value. Each binary tree is called a *regression tree* where the inner nodes (split nodes) express branching conditions on an input feature vector. Each regression tree takes an input feature vector and outputs a value associated with the corresponding leaf node. Regression forests return the average of the outputs of regression trees as the final output.

4.1 Construction

For the construction of regression forests, we need a large number of training samples which associate feature vectors (a set of path features) and output values (optimal weights). We generate these samples by rendering several training scenes. We then extract the path features and the corresponding optimal weights for each scene. The regression forest is trained to approximate the optimal weights even for a new scene, based only on the path features.

We define a training sample $t \equiv (\phi^t, w^t_{\text{opt}}) \in \mathscr{T}$ as a tuple of path features ϕ^t and the optimal weight w^t_{opt}. $\mathscr{T}$ is a set of all training samples. The construction process begins from the root node of the regression forest. Each step of the construction process recursively splits training samples into left and right nodes. We denote the subset of the training samples in the currently processed node as $T \subseteq \mathscr{T}$ and we start from $T = \mathscr{T}$. The algorithm is similar to a top-down construction of a kd-tree for ray tracing [7, 21].

Node Splitting: The construction process continues splitting the current node until the number of training samples in the current set T is smaller than a threshold, or the depth of the tree has reached the maximum depth. If the recursion terminates, the current node becomes a leaf node. Each leaf node stores the average over the set of the optimal weights in this node as w_{leaf}. This average weight approximates the optimal weight at runtime.

If the recursion continues, we split the current set of samples T into two disjoint subsets T_L and T_R according to a threshold θ and an index k of the path features:

$$T_L(\theta, k) = \{t \in T \mid \phi^t(k) \geq \theta\} \tag{12}$$

$$T_R(\theta, k) = T \setminus T_{\text{L}}(\theta, k) \tag{13}$$

where $\phi^t(k)$ is the kth element of the path features ϕ^t. The threshold θ and the index k at each step are defined as $(\theta, k) = \text{argmax}_{\theta',k'} V(\theta', k', T)$, where $V(\theta', k', T) = \text{Var}(T) - \text{Var}(T_L(\theta', k')) - \text{Var}(T_R(\theta', k'))$. Here $\text{Var}(T)$ is the variance of the optimal weights in T. The function V is used to define the most discriminative pair of the threshold θ and the index of the path feature k according to the variance.

4.2 Runtime

In our framework, we first render a given new scene with two different algorithms $\hat{I}_\alpha$ and $\hat{I}_\beta$ with the same computation time. We also extract the path features ϕ according to the definition by Eq. 5. Using these path features, we can now evaluate each trained regression tree by traversing down the tree according to the branching condition defined in Eq. 12, which eventually reaches a leaf node and the weight w_{leaf} is recorded in the leaf node. By repeating this process for all regression trees in the trained regression forest, we obtain a set of weights w_{leaf} recorded in the leaf nodes for each tree. We define $w_r(\phi)$ as the output of the rth tree in the trained regression forest, given the path features ϕ. The approximated optimal weight $w_{\text{approx}}(\phi)$ with M trees is given as

$$w_{\text{approx}}(\phi) = \frac{1}{M} \sum_{r=1}^{M} w_r(\phi). \tag{14}$$

Blending at each pixel is $w_{\text{approx}}(\phi)\hat{I}_\alpha + (1 - w_{\text{approx}}(\phi))\hat{I}_\beta$. This evaluation process is repeated for all the pixels. The use of forests can alleviate the discontinuity of the resulting weights. Even if one tree suddenly returns a totally different value due to hard classification, it is likely that other trees still return similar weights. As a result, returning weights will be smoothly changing.

4.3 Refinement

A trained regression forest is sometimes too optimized for given training samples. In order to reduce overfitting, we follow the refinement technique for regression forests proposed by Ren et al. [24] and Ladický et al. [19]. The main idea is to split a set of training samples into two subsets and use one for constructing the structure of each tree while using the other for defining the outputs. After the construction step, we first discard the values w_{leaf} assigned to the leaf nodes while keeping the tree structure. The refinement process then updates w_{leaf} using the additional training samples.

For each additional training sample ϕ, we execute the evaluation of the tree until the evaluation process reach to the leaf node. After collecting the set of training samples Φ reached to the leaf node, the updated weight w_r^* can be computed as

$$w_r^* = \frac{1}{|\Phi|} \sum_{(\phi, w_{\text{opt}}) \in \Phi} w_{\text{opt}}(\phi). \tag{15}$$

We iterate this refinement process for each tree in the regression forest using the sample training set for the refinement. Since the training samples are taken from the different portion of the training set independent of the samples assigned for the initial construction, the final weights associated to the leaf node could become more generic, which alleviates overfitting to the initial training set.

5 Results

We selected the two combinations of the rendering algorithms to show the effectiveness of our framework: (1) stochastic progressive photon mapping (SPPM) [4] and Metropolis light transport (MLT) [31] with manifold exploration [12] shown in Fig. 2,

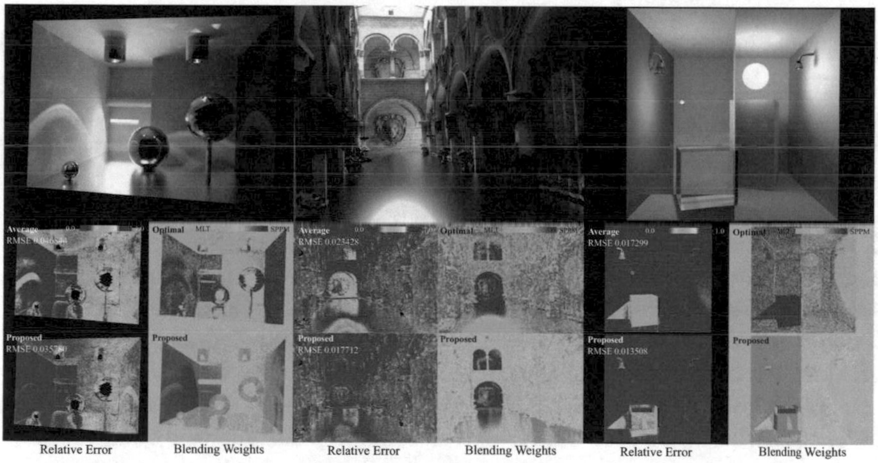

Fig. 2 Equal-time comparison (20 min) of the average and our automatic blending of the images rendered by SPPM [4] and MLT [31] with manifold exploration [12]. We highlighted three scenes with different characteristics from our test cases (box, cryteck-sponza, and water). The top row shows the reference images. The bottom two rows visualize errors, the optimal blending weights, and the output blending weights of our framework. Depending on the types of lighting effects, the optimal blending weights for SPPM and MLT that result in the minimal error vary significantly. Simply taking the average of SPPM and MLT thus produces a suboptimal result in terms of RMS error

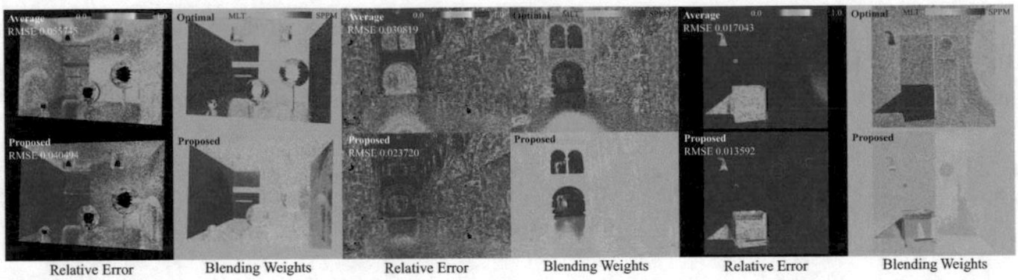

Fig. 3 Comparison of errors, the optimal weights and the approximations by our framework for the combination of BDPT and SPPM. The selection of the scenes and meaning of the images are same as Fig. 2. Similar to the combination of MLT and SPPM, our framework generally captures the preference to the scene according to the characteristics of the scenes, although some difference can be observed, e.g., preference to the scene dominated with specular material is weaker (*box* scene)

(2) SPPM and bidirectional path tracing (BDPT) [20, 29] shown in Fig. 3. We chose these algorithms because both the algorithm and the performance are distinguishably different. One famous characteristic of SPPM is the ability to handle specular-diffuse-specular paths efficiently. A caustic that can be seen through a water surface is an example of such paths. MLT is based on Markov chain Monte Carlo sampling which utilizes a sequence of correlated samples that forms a Markov chain. The sequence of the samples is generated such that the resulting sample distribution follows an arbitrary user-defined target function such as the measurement contribution function. MLT is known to be effective for the scenes with complex occlusion. BDPT can utilize various sampling technique by the combination of paths traced from the sensor and the lights. These sampling techniques are combined with multiple importance sampling [30]. The combination of SPPM and BDPT would exhibit the good trade-off because BDPT is not efficient at handling specular-diffuse-specular paths [18] and while being more efficient at rendering diffuse surfaces [2, 6].

For the implementations of rendering algorithms, we used the Mitsuba renderer [11]. Mutation techniques used for MLT are bidirectional, lens, caustic, multi-chain, and manifold perturbation [12]. All the images except for the reference images are rendered on a machine with Intel Core i7-4720HQ at 2.6 GHz. The training phase is computed with a machine with Intel Core i7-3970X at 3.5 GHz and 16 GB of main memory. We utilized only a single core for rendering in order to alleviate the difference of performance between SPPM and MLT according to the parallelization. In order to facilitate the future work, we publish our implementation on our website.

Training Samples: Our training set consists of 10 scenes with various characteristics in order to cover as many types of paths as possible. We render all the scenes with each rendering algorithm for 5, 10, 15, and 20 min. Each scene is rendered five times, in order to alleviate overfitting as discussed in Sect. 3.2. Given this whole training data, we generate a regression forest for each scene by excluding the scene from the training data. We thus have 10 different regression forests as a result. Each forest is tested against the corresponding scene that was excluded from its training. It is essentially leave-one-out cross-validation in machine learning.

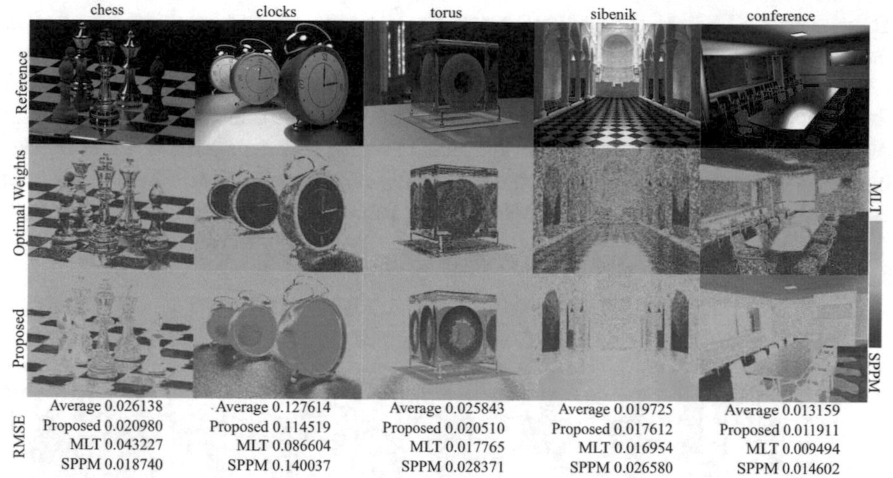

Fig. 4 Comparison of the optimal weights and the approximations by our framework for the selected five scenes combining MLT and SPPM. The first row shows the reference images. The bottom two rows visualize the optimal weights and the approximated weights via trained regression forests. For many scenes, our framework largely reproduces the optimal weights, without any information other than rough estimates of path features per pixel. The RMS errors between the blended images and the references are improved compare to taking the average (Average). We also show RMS errors for MLT and SPPM with the same total rendering time

While it is possible to have a single forest for all the training scenes and test this forest against the same set of scenes, we found that this kind of experiment is prone to overfit to the training scenes. Our regression forest consists of five trees and the maximum depth of each tree is 15. The construction time of the regression forest is 30 min.

Approximated Optimal Weights. Figure 4 shows blending weights and RMS errors for selected five scenes with the combination of SPPM and MLT. Figure 2 shows such results with visualization of the error per pixel for three other scenes. We compare approximated optimal weights via a trained regression forest with the average of five different runs for each scene. The blending weight is fixed to 0.5 when a pixel has no information on path features (e.g., background images). We blended two images rendered by SPPM and MLT by taking the average (Average) or by using the approximated optimal weight per pixel (Proposed). The running time of our framework is less than 50 ms for all the scenes. The storage cost of our regression forest is 100 KB. Both the running time and the storage cost are independent of the geometric complexity of the scenes. We can see that optimal weights and weights suggested by our framework are very similar to each other in almost all the cases. Our framework thus successfully learned the preference of an algorithm only based on path features.

RMS Errors: Figure 5 shows RMS errors for 10 scenes for the combination of MLT and SPPM. We plot RMS errors of MLT, SPPM, their average, and our blended result for each scene with the total rendering time of 20 min for all the methods. The plots are scaled such that the values for the average is one. We can observe that our blending is superior to the average in all scenes. The reduction of error by our blending is larger when the difference of RMS errors between SPPM and MLT is large. Moreover, the blended solution by our framework sometimes outperforms a

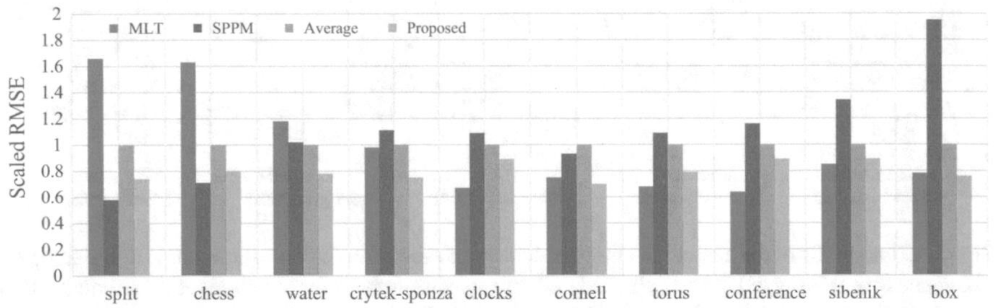

Fig. 5 Scaled RMS errors of MLT, SPPM, Average, and blending with our framework over 10 scenes. All the methods use the total rendering time of 20 min. The average and our blending spends 10 min for both MLT and SPPM, keeping the total rendering time equal to 20 min. We scaled RMS errors such that the average is always one. The scenes are sorted roughly according to the difference of RMS errors between MLT and SPPM

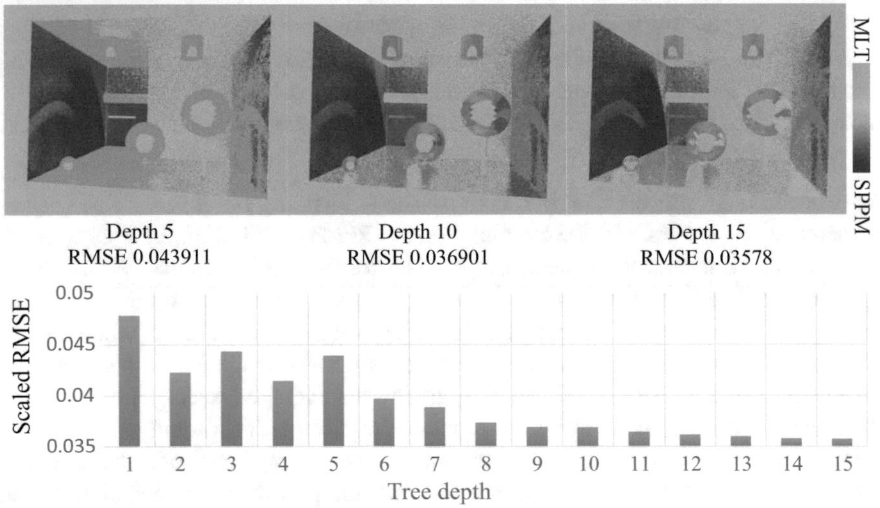

Fig. 6 Visualization of approximated optimal weights for the *box* scene with different tree depth (top) and the corresponding plot of the approximation errors (bottom). The RMS errors of the blended images are shown under each image. As the depth of the tree increases, the color indicating the preference to MLT becomes a bit more explicit, but not significantly after a certain depth. The plot of the variance shows how approximation errors of the optimal weights change according to the tree depth, which also stops converging around the depth of 15

better algorithm with the same total rendering time. Such a result is not trivial since our framework spends only half of the total rendering time for each algorithm. We should also note that just taking the average can in fact increase the error for the same reason (e.g., *Cornell* scene). In contrast, we did not find any such cases using our framework. This result supports that our framework can improve the robustness of light transport simulation in practice.

Effect of Tree Depth: The images in Fig. 6 show the approximated optimal weights for the *box* scene with different depths of the regression trees in the runtime. As the depth increases, we can observe that the preference to each technique becomes more explicit. Yet another observation is that the approximated weights are converged around the tree depth of 15. The graph in Fig. 6 shows the RMS error between the optimal weight and the approximated weight with our framework for this scene. We can observe that the RMS error converges around the depth of 15, and we found that it is similar for the other scenes as well. Along with the saturation of the weights, we thus conservatively set the tree depth to 15 in our experiments.

6 Discussion

6.1 Alternative to Blending

While we found that blending is a practical approach to combine different rendering algorithms, it is tempting to try *selecting* one of the different algorithms instead of *blending* such that we can spend all the allocated rendering time to one algorithm. This alternative solution, however, is not feasible for two major reasons. Firstly, as shown in Fig. 2, a better algorithm can change even within a single image. Even though MLT looks converged in many regions, it can entirely miss certain lighting effects such as specular reflections of caustics. As such, resolving all the effects by a single algorithm can take a significant amount of rendering time as compared to combining the results of two algorithms. Recent work on robust rendering algorithms are based on the same observation [2, 6].

Secondly, defining useful features for this selection is not trivial and algorithm-dependent. In order to select an efficient algorithm for a specific input scene, we would need a feature vector of a whole configuration of the rendering process. This information includes parameters of each rendering algorithm that affects the performance, which in turn makes the whole framework algorithm-dependent. It is also not obvious how to encode input scenes as feature vectors. Unlike images, which contain a set of pixels in a structured manner, scene data contains a set of very different information such as material data, textures, and triangle meshes. There is no single data structure common to all of data necessary to define input scenes. This lack of a common structured input form is a striking differences to applications of machine learning for images.

One might also consider finding a distribution of total rendering time, such that we do not spend too much computation for an algorithm with small weights. This deceivingly obvious improvement, however, is not possible since our regression forest is trained under the assumption that each algorithm spends the same rendering time. Even if we can find such a distribution of rendering time somehow, optimal blending weights are now different from those at the training phase since rendering time for each algorithm is also different. To implement this idea, we would need to have multiple regression forests for all the possible distributions of total rendering time, which is likely infeasible.

6.2 Comparison to Neural Networks

We used regression forests as a machine learning technique to learn the relationship between path features and the optimal blending weights. One possible option is to replace it by neural networks. Given its success in the computer vision community, a deep neural network [10] is a possible candidate. We tested replacing regression forests by a fully-connected four layer's neural network using Caffe [15] on GPU as additional experiments. As shown in Fig. 7, we found that a neural network can achieve similar performance to regression forests. We discarded this approach in the end since even its running time is multiple orders of magnitudes slower (3 min) than regression forests (60 ms) without much improvement in terms of RMS errors.

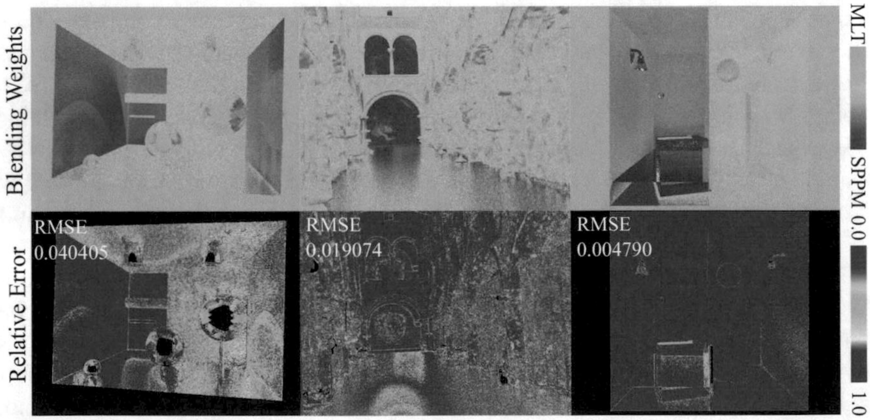

Fig. 7 Approximated blending weights (top) and the relative errors (bottom) for the selected three scenes (*box*, *crytek-sponza*, and *water*) by replacing regression forests via a neural network

Fig. 8 RMS errors for a test scene that is only slightly different from a training scene. This test scene is made by changing the environment light and the camera configuration while retaining the geometry and materials of the *torus* scene in Fig. 2. For this experiment, we used the original *torus* scene for training, and the modified *torus* scene at runtime

6.3 Limitations

Preparing Training Scenes: In general, a machine learning technique needs a large number of training samples to avoid overfitting. While we carefully designed a set of training scenes, it is not guaranteed that the prepared training scenes are indeed sufficient for learning. This situation is in contrast to the computer vision community; there are several standardized large datasets such as ImageNet [25]. Although we used some standard models and scenes often seen in other rendering research, it would be interesting as future work to generate training scenes based on procedural modeling. This procedural modeling should include not only shapes, but also materials, lighting, and camera parameters.

Dependency on Training Scenes: We found that our method works especially well if there are only slight differences between training scenes and test scenes. Figure 8 shows the *torus2* scene which uses the same geometry and materials as the *torus* scene in Fig. 2, but with a slightly different camera configuration and an environment map. For this experiment, we used only the *torus* scene for the training phase, and rendered the *torus2* scene. We can observe that reduction of RMS error is significant in this case. This experiment indicates an interesting use case of our framework in practice: when an artist is modeling a new scene based on existing ones, we can train a regression forest with existing scenes beforehand.

7 Related Work

Light Transport Simulation in Rendering: Since the development of path tracing [16], the number of light transport simulation algorithms have been developed. Among many rendering algorithms, we used the two representative approaches in our tests: SPPM [4] and MLT [31] with manifold exploration [12]. We chose these two approaches because their algorithms are completely different and have different characteristics as rendering algorithms. SPPM works by tracing a number of light paths and estimates density of light path vertices at a visible point through each pixel. MLT on the other hand traces a whole path by a Markov chain from the previously generated path and estimates the histogram of this Markov chain at all the pixels. SPPM is generally considered good at rendering caustics, while MLT is considered efficient at resolving complex visibilities from light sources. Our framework however is not restricted to use very different algorithms, since it is independent of how each algorithm works internally.

Machine Learning in Rendering: Several researchers have already applied machine learning to rendering. One popular application of machine learning in rendering is regression models. Among others, Jacob et al. [13] utilized unsupervised online-learning of a Gaussian mixture model (GMM) to represent a radiance distribution in participating media. Vorba et al. [32] also used online learning of GMM to represent probability density functions for importance sampling. Ren et al. [23] introduced a realtime rendering algorithm using non-linear regression to represent precomputed radiance data. The precomputed radiance data is modeled as a multi-layered neural network [8]. The idea is to learn the relationship between scene configurations and the resulting radiance distribution based on off-line rendering with random attributes. While we also use machine learning for regression, we propose to use machine learning to combine existing rendering algorithms without any modification to them. Our framework thus can be applied on top of any of the previous work mentioned above. More recently, Nalbach et al. [22] showed how to use CNN to approximate screen-space shaders. While the goal of their work is completely different from ours, their work demonstrate the powerful potential of applying machine learning to rendering.

Kalantari et al. [17] recently proposed a image filtering technique to reduce Monte Carlo rendering noise based on the multilayer perception [8]. The idea is to learn the relationship between the scene features such as a shading location or texture values and a set of filtering parameters. Our work is inspired by their successful application and we also use machine learning to find the relationship between path features and the optimal blending weights. The difference is that their work focuses to improve the result of a single image by filtering, while we consider a situation where there are multiple rendering algorithms available for a user.

The aim of our work is to use machine learning to blend the results of different rendering algorithms. Such blending is often done by multiple importance sampling [30], and there have been many recent works on this approach [2, 6]. Our work differs from multiple importance sampling in that we treat each rendering algorithm as a black-box and does not require any detailed algorithmic information such as path probability densities.

Regression Forests: Regression forests [1] are actively used in many applications. One famous example is Kinect body segmentation [26]. By simply fetching neighboring depth values and parse the regression forest, this algorithm can label each pixel by 31 different body parts quite accurately in realtime. For face recognition,

Ren et al. [24] showed that regression forests can be used to detect major features such as eyes, a mouth, and a nose. Tang et al. [27] used regression forests to extract a skeletal hand model from an RGB-depth image.

For applications in computer graphics, Ladický et al. [19] used regression forest for fluid simulation and achieved $\times 200$ speed up. They trained a regression forest via position-based fluid simulation by defining several features around each particle. The trained regression forest is used to update the state of particles at the next time step, without relying on costly simulation. Inspired by the success of regression forests in many applications, we also utilize regression forests instead of a more popular convolution neural network [10]. As far as we know, our work is the first application of regression forests in rendering.

8 Conclusion

We presented a framework to automatically blend results of different light transport simulation algorithms. The key idea is to learn the relationship between a class of light transport paths and the performance of each algorithm on each class. For classification of paths, we introduced a feature vector based on relative contributions from different types of paths according to Heckbert's notation. We then calculate optimal blending weights such that a resulting image has minimal errors on average after blending. Using regression forests, we approximate a function that takes a feature vector of light transport paths and outputs the optimal blending weight per pixel. The resulting framework is independent from how each algorithm works, which makes it easily applicable to different rendering algorithms.

References

1. Breiman, L.: Random forests. Mach. Learn. **45**(1), 5–32 (2001)
2. Georgiev, I., Krivanek, J., Davidovic, T., Slusallek, P.: Light transport simulation with vertex connection and merging. ACM Trans. Graph. (Proceedings of SIGGRAPH Asia) **31**(6) (2012). Article 192
3. Gritz, L., Stein, C., Kulla, C., Conty, A.: Open shading language. ACM SIGGRAPH 2010 Talks (2010). Article 33
4. Hachisuka, T., Jensen, H.W.: Stochastic progressive photon mapping. ACM Trans. Graph. (Proceedings of SIGGRAPH Asia) **28**(5) (2009). Article 141
5. Hachisuka, T., Ogaki, S., Jensen, H.W.: Progressive photon mapping. ACM Trans. Graph. (Proceedings of SIGGRAPH Asia) **27**(5) (2008). Article 130
6. Hachisuka, T., Pantaleoni, J., Jensen, H.W.: A path space extension for robust light transport simulation. ACM Trans. Graph. (Proceedings of SIGGRAPH Asia) **31**(6) (2012). Article 191
7. Havran, V.: Heuristic ray shooting algorithms. Ph.D. thesis, Department of Computer Science and Engineering, Faculty of Electrical Engineering, Czech Technical University in Prague, November (2000)
8. Haykin, S.: Neural Networks: A Comprehensive Foundation, 2nd edn. Prentice Hall PTR, Upper Saddle River (1998)

9. Heckbert, P.S.: Adaptive radiosity textures for bidirectional ray tracing. Comput. Graph. (Proceedings of SIGGRAPH) **24**(4), 145–154 (1990)
10. Hinton, G.E., Osindero, S., Teh, Y.-W.: A fast learning algorithm for deep belief nets. Neural comput. **18**(7), 1527–1554 (2006)
11. Jakob, W.: Mitsuba renderer. http://www.mitsuba-renderer.org (2010)
12. Jakob, W., Marschner, S.: Manifold exploration: a Markov chain Monte Carlo technique for rendering scenes with difficult specular transport. ACM Trans. Graph. (Proceedings of SIGGRAPH) **31**(4) (2012)
13. Jakob, W., Regg, C., Jarosz, W.: Progressive expectation–maximization for hierarchical volumetric photon mapping. Comput. Graph. Forum (Proceedings of the Eurographics Symposium on Rendering) **30**(4) (2011)
14. Jensen, H.W.: Global illumination using photon maps. In: Proceedings of the Eurographics Symposium on Rendering, pp. 21–30 (1996)
15. Jia, Y., Shelhamer, E., Donahue, J., Karayev, S., Long, J., Girshick, R.B., Guadarrama, S., Darrell, T.: Caffe: convolutional architecture for fast feature embedding. CoRR arXiv:1408.5093 (2014)
16. Kajiya, J.T.: The rendering equation. Comput. Graph. (Proceedings of SIGGRAPH) 143–150 (1986)
17. Kalantari, N.K., Bako, S., Sen, P.: A machine learning approach for filtering Monte Carlo noise. ACM Trans. Graph. (TOG) (Proceedings of SIGGRAPH 2015) **34**(4) (2015)
18. Kaplanyan, A.S., Dachsbacher, C.: Path space regularization for holistic and robust light transport. Comput. Graph. Forum (Proceedings of the Eurographics Symposium on Rendering) **32**(2), 63–72 (2013)
19. Ladický, L., Jeong, S., Solenthaler, B., Pollefeys, M., Gross, M.: Data-driven fluid simulations using regression forests. ACM Trans. Graph. (Proceedings of SIGGRAPH Asia) **34**(6), 199 (2015)
20. Lafortune, E., Willems, Y.D.: Bi-directional path-tracing. In: Proceedings of Compugraphics, pp. 145–153 (1993)
21. MacDonald, J.D., Booth, K.S.: Heuristics for ray tracing using space subdivision. Visual Comput. **6**(3), 153–166 (1990)
22. Nalbach, O., Arabadzhiyska, E., Mehta, D., Seidel, H.-P., Ritschel, T.: Deep shading: convolutional neural networks for screen-space shading. Comput. Graph. Forum (Proceedings of the EGSR 2017) **36**(4) (2017)
23. Ren, P., Wang, J., Gong, M., Lin, S., Tong, X., Guo, B.: Global illumination with radiance regression functions. ACM Trans. Graph. (Proceedings of SIGGRAPH) **32**(4), 130:1–130:12 (2013)
24. Ren, S., Cao, X., Wei, Y., Sun, J.: Face alignment at 3000 fps by regressing local binary features. In: IEEE Conference on Computer Vision and Pattern Recognition (CVPR) (2014)
25. Russakovsky, O., Deng, J., Su, H., Krause, J., Satheesh, S., Ma, S., Huang, Z., Karpathy, A., Khosla, A., Bernstein, M., Berg, A.C., Fei-Fei, L.: ImageNet large scale visual recognition challenge. Int. J. Comput. Vision (IJCV) **115**(3), 211–252 (2015). https://doi.org/10.1007/s11263-015-0816-y
26. Shotton, J., Fitzgibbon, A., Cook, M., Sharp, T., Finocchio, M., Moore, R., Kipman, A., Blake, A.: Real-time human pose recognition in parts from a single depth images. In: IEEE Conference on Computer Vision and Pattern Recognition (CVPR) (2011)
27. Tang, D., Yu, T.-H., Kim, T.-K.: Real-time articulated hand pose estimation using semi-supervised transductive regression forests. In: The IEEE International Conference on Computer Vision (ICCV) (2013)
28. Veach, E.: Robust Monte Carlo methods for light transport simulation. Ph.D. thesis, Stanford University, USA (1998). AAI9837162
29. Veach, E., Guibas, L.J.: Bidirectional estimator for light transport. In: Proceedings of the Eurographics Symposium on Rendering, pp. 147–162 (1994)

30. Veach, E., Guibas, L.J.: Optimally combining sampling techniques for Monte Carlo rendering. In: Proceedings of SIGGRAPH '95, pp. 419–428 (1995)
31. Veach, E., Guibas, L.J.: Metropolis light transport. Proceedings of SIGGRAPH **97**, 65–76 (1997)
32. Vorba, J., Karlík, O., Šik, M., Ritschel, T., Křivánek, J.: On-line learning of parametric mixture models for light transport simulation. ACM Trans. Graph. (Proceedings of SIGGRAPH) **33**(4) (2014)

A Dimension-Adaptive Multi-Index Monte Carlo Method Applied to a Model of a Heat Exchanger

Pieterjan Robbe, Dirk Nuyens and Stefan Vandewalle

Abstract We present an adaptive version of the Multi-Index Monte Carlo method, introduced by Haji-Ali, Nobile and Tempone (2016), for simulating PDEs with coefficients that are random fields. A classical technique for sampling from these random fields is the Karhunen–Loève expansion. Our adaptive algorithm is based on the adaptive algorithm used in sparse grid cubature as introduced by Gerstner and Griebel (2003), and automatically chooses the number of terms needed in this expansion, as well as the required spatial discretizations of the PDE model. We apply the method to a simplified model of a heat exchanger with random insulator material, where the stochastic characteristics are modeled as a lognormal random field, and we show consistent computational savings.

Keywords Multi-Index Monte Carlo · Dimension-adaptivity · PDEs with random coefficients

1 Introduction

A key problem in *uncertainty quantification* is the numerical computation of statistical quantities of interest from solutions to models that involve many random parameters and inputs. Areas of application include, for example, robust optimization, risk analysis and sensitivity analysis. A particular challenge is solving problems with a high number of uncertainties, leading to the evaluation of high-dimensional integrals. In that case, classical methods such as polynomial chaos [19, 20] and sparse grids [2] fail, and one must resort to Monte Carlo-like methods. Recently, an efficient class of such Monte Carlo algorithms was introduced by Giles, see [1, 4,

P. Robbe (✉) · D. Nuyens · S. Vandewalle
KU Leuven, Department of Computer Science, NUMA Section, Celestijnenlaan 200A box 2402, 3001 Leuven, Belgium
e-mail: pieterjan.robbe@kuleuven.be

D. Nuyens
e-mail: dirk.nuyens@kuleuven.be

S. Vandewalle
e-mail: stefan.vandewalle@kuleuven.be

A. B. Owen and P. W. Glynn (eds.), *Monte Carlo and Quasi-Monte Carlo Methods*, Springer Proceedings in Mathematics & Statistics 241,
https://doi.org/10.1007/978-3-319-91436-7_24

9, 10]. Central to these *multilevel* algorithms is the use of a hierarchy of numerical approximations or *levels*. By redistributing the available computational budget over these levels, taking into account the bias and variance of the different estimators, the error in the final result is minimized.

A significant extension of the multilevel methodology is the Multi-Index Monte Carlo (MIMC) method, see [14, 15]. MIMC generalizes the scalar hierarchy of levels to a larger, multidimensional hierarchy of *indices*. This is motivated by the observation that in some applications, changing the level of approximation can be done in several ways, for example in time dependent problems where both time step size and spatial resolution can be varied. Each refinement then corresponds to an index in a multidimensional space. The optimal shape of the hierarchy of indices, based on a priori assumptions on the problem, is analyzed in [15]. However, in most practical problems, such knowledge is not available. Hence the need for efficient algorithms that automatically detect important dimensions in a problem. Such adaptivity has also been used for deterministic sparse grid cubature in [7]. We will develop a similar approach for MIMC.

The paper is organized as follows. In Sect. 2, we introduce a particular example of a PDE with random coefficients: the heat equation with random conductivity. The Multi-Index Monte Carlo method and our adaptive variant are presented in Sect. 3. Next, in Sect. 4, we introduce a model for a heat exchanger, in which the heat flow is described by the heat equation with random conductivity. We use our adaptive method to compute expected values of the temperature distribution inside the heat exchanger. We show huge computational savings compared to nonadaptive MIMC. We conclude our work in Sect. 5.

2 The Heat Equation with Random Conductivity

In this section, we study the linear anisotropic steady state heat equation defined on a domain $D \subset \mathbb{R}^m$, with boundary ∂D. The temperature field $T : D \to \mathbb{R} : \boldsymbol{x} \mapsto T(\boldsymbol{x})$ satisfies the partial differential equation (PDE)

$$-\nabla \cdot (k(\boldsymbol{x})\nabla T(\boldsymbol{x})) = F(\boldsymbol{x}) \qquad \text{for } \boldsymbol{x} \in D, \tag{1}$$

with $k(\boldsymbol{x}) > 0$ the thermal conductivity, $F \in L_2(D)$ a source term, and boundary conditions

$$\begin{aligned} T(\boldsymbol{x}) &= T_1(\boldsymbol{x}) & \text{for } \boldsymbol{x} \in \partial D_1, \\ n(\boldsymbol{x}) \cdot (k(\boldsymbol{x})\nabla T(\boldsymbol{x})) &= T_2(\boldsymbol{x}) & \text{for } \boldsymbol{x} \in \partial D_2, \end{aligned}$$

where ∂D_1 and ∂D_2 are two disjoint parts of ∂D such that $\partial D = \partial D_1 \cup \partial D_2$. Here, $n(\boldsymbol{x})$ denotes the exterior unit normal vector to D at $\boldsymbol{x} \in \partial D_2$.

Consider now the case where Eq. (1) has a conductivity modeled as a random field, i.e., $k : D \times \Omega \to \mathbb{R} : (\boldsymbol{x}, \omega) \mapsto k(\boldsymbol{x}, \omega)$ also depends on an event ω of a probability space $(\Omega, \mathscr{F}, P)$. Then, the solution $T(\boldsymbol{x}, \omega)$ is also a random field and solves almost surely (a.s.)

$$\begin{aligned} -\nabla \cdot (k(\boldsymbol{x}, \omega)\nabla T(\boldsymbol{x}, \omega)) &= F(\boldsymbol{x}) && \text{for } \boldsymbol{x} \in D \text{ and } \omega \in \Omega, \\ T(\boldsymbol{x}, \omega) &= T_1(\boldsymbol{x}) && \text{for } \boldsymbol{x} \in \partial D_1, \\ n(\boldsymbol{x}) \cdot (k(\boldsymbol{x}, \omega)\nabla T(\boldsymbol{x}, \omega)) &= T_2(\boldsymbol{x}) && \text{for } \boldsymbol{x} \in \partial D_2. \end{aligned} \tag{2}$$

For simplicity, we only study the PDE subject to deterministic boundary conditions.

In what follows, we will develop efficient methods to approximate the expected value

$$I(g(\omega)) := \mathbb{E}[g(\omega)] = \int_\Omega g(\omega)\, \mathrm{d}P(\omega),$$

where $g(\omega) = f(T(\cdot, \omega))$ is called the quantity of interest. Typical examples of $g(\omega)$ include the value of the temperature at a certain point, the mean value in (a subdomain of) D, or a flux through (a part of) the boundary ∂D.

A commonly used model for the conductivity $k(\boldsymbol{x}, \omega)$ in (2) is a lognormal random field, i.e.,

$$k(\boldsymbol{x}, \omega) = \exp(Z(\boldsymbol{x}, \omega)),$$

where Z is an underlying Gaussian random field with given mean and covariance. The exponential ensures that the condition $k(\boldsymbol{x}, \omega) > 0$ is satisfied for all $\boldsymbol{x} \in D$ and $\omega \in \Omega$, a.s.

In the following, we recall some details about Gaussian random fields that can be found in literature, such as [16, 17]. A Gaussian random field $Z(\boldsymbol{x}, \omega)$ is a random field where every vector $z = (Z(\boldsymbol{x}_i, \omega))_{i=1}^M$ follows a multivariate Gaussian distribution with given covariance function for every $\boldsymbol{x}_i \in D$ and $M \in \mathbb{N}$. Specifically, we write $z \sim \mathscr{N}(\boldsymbol{\mu}, \Sigma)$, with $\mu_i = \mu(\boldsymbol{x}_i)$ the mean, and with $\Sigma_{i,j} = C(\boldsymbol{x}_i, \boldsymbol{x}_j) := \operatorname{cov}(Z(\boldsymbol{x}_i, \omega), Z(\boldsymbol{x}_j, \omega))$ for every $\boldsymbol{x}_i, \boldsymbol{x}_j \in D$, and C the covariance function.

An example of such a covariance function is the *Matérn* covariance

$$C(\boldsymbol{x}_i, \boldsymbol{x}_j) = \sigma^2 \frac{1}{2^{\nu-1}\Gamma(\nu)} \left(\sqrt{2\nu}\frac{\|\boldsymbol{x}_i - \boldsymbol{x}_j\|_p}{\lambda}\right)^\nu K_\nu\left(\sqrt{2\nu}\frac{\|\boldsymbol{x}_i - \boldsymbol{x}_j\|_p}{\lambda}\right), \quad \boldsymbol{x}_i, \boldsymbol{x}_j \in D, \tag{3}$$

where Γ is the Gamma function and K_ν is the modified Bessel function of the second kind. The parameter λ is the correlation length, σ^2 is the (marginal) variance, and ν is the smoothness of the random field.

Samples of the Gaussian random field can be computed via the *Karhunen–Loève* (KL) *expansion*

$$Z(\boldsymbol{x}, \omega) = \mu(\boldsymbol{x}) + \sum_{r=1}^{\infty} \sqrt{\theta_r} f_r(\boldsymbol{x}) \xi_r(\omega). \tag{4}$$

In this expansion, the $\xi_r(\omega)$, $r \geq 1$, are independent standard normally distributed random numbers and f_r and θ_r are the solutions to the eigenvalue problem

$$\int_D C(\boldsymbol{x}_i, \boldsymbol{x}_j) f_r(\boldsymbol{x}_j) \mathrm{d}\boldsymbol{x}_j = \theta_r f_r(\boldsymbol{x}_i), \quad \boldsymbol{x}_i, \boldsymbol{x}_j \in D, \tag{5}$$

where the eigenfunctions f_r need to be normalized for (4) to hold. With every event $\omega \in \Omega$ we can associate the (infinite-dimensional) vector $\boldsymbol{\xi}(\omega) = (\xi_r(\omega))_{r \geq 1}$ and, hence, a realization of the random field $k(\boldsymbol{x}, \omega)$. There exist other methods to generate samples of a random field with given covariance function, such as circulant embedding [11, 16]. Here we choose the KL expansion because of the *best approximation property* described below.

In practice, the infinite sum in (4) must be truncated after a finite number of terms s, that is, $\boldsymbol{\xi}(\omega)$ must be truncated to a vector of finite length. The KL expansion gives the best (in MSE sense) s-term approximation of the random field if the eigenvalues are ordered in decreasing magnitude [8, 16]. The value of s to reach a certain accuracy depends on the decay rate of the eigenvalues θ_r. The more terms are retained in the expansion, the better the approximation of the random field, but also, the more costly the expansion. This cost involves both the composition of the sum in (4), and the (numerical) solution of the eigenvalue problem (5). When a lot of terms are required to model the random field, i.e., when the decay of θ_r is slow, this cost can no longer be ignored compared to the cost of solving the deterministic PDE in every sample of (2). Hence, it is necessary to construct algorithms that take advantage of the *best approximation property*, and only increase the number of KL terms when required. In Sect. 3.3 below, we present an algorithm for such a dimension-adaptive construction of the KL expansion.

3 The Multi-Index Monte Carlo Method

In Sects. 3.1 and 3.2 we introduce the Multi-Index Monte Carlo (MIMC) method which was presented and analyzed in [15]. Following that, in Sect. 3.3 we discuss an adaptive version of the method based on techniques used in generalized sparse grids, see [2, 7]. See also [12] for the *combination technique* on which MIMC is based.

3.1 Properties of Monotone Sets

The formulation of the MIMC method uses the notion of *indices* $\boldsymbol{\ell} \in S$ and *index sets* $\mathscr{I} \subseteq S$, where $S := \mathbb{N}_0^d = \{\boldsymbol{\ell} = (\ell_i)_{i=1}^d : \ell_i \in \mathbb{N}_0\}$, with $\mathbb{N}_0 = \{0, 1, 2, \ldots\}$ and $d \geq 1$. A *monotone set* is a nonempty set $\mathscr{I} \subseteq S$ such that for all

$$\boldsymbol{\tau} \le \boldsymbol{\ell} \in \mathcal{I} \Rightarrow \boldsymbol{\tau} \in \mathcal{I}, \tag{6}$$

where $\boldsymbol{\tau} \le \boldsymbol{\ell}$ means $\tau_j \le \ell_j$ for all j, see [3]. Property (6) is also known as *downward closedness*. An index set that is monotone is also called a downward closed or *admissible* index set. In the remainder of the text, the index set $\mathcal{I}$ will always be constructed in such a way that it is an admissible index set.

Using the definition of the Kronecker sequence $\boldsymbol{e}_i := (\delta_{ij})_{j=1}^d$, a monotone set $\mathcal{I}$ can also be defined using the property

$$\big(\boldsymbol{\ell} \in \mathcal{I} \quad \text{and} \quad \ell_i \neq 0\big) \quad \Rightarrow \quad \boldsymbol{\ell} - \boldsymbol{e}_i \in \mathcal{I} \quad \text{for all } i = 1, 2, \ldots, d.$$

In other words, for every index $\boldsymbol{\ell} \neq (0, 0, \ldots)$ in a monotone set, all indices with a smaller (but positive) entry in a certain direction are also included in the set. In the following, we also use the concept of *forward neighbors* of an index $\boldsymbol{\ell}$, i.e., all indices $\{\boldsymbol{\ell} + \boldsymbol{e}_i : i = 1, 2, \ldots, d\}$, and *backward neighbors* of an index $\boldsymbol{\ell}$, i.e., all indices $\{\boldsymbol{\ell} - \boldsymbol{e}_i : i = 1, 2, \ldots, d\}$.

Examples of monotone sets are rectangles

$$R(\boldsymbol{\ell}) := \{\boldsymbol{\tau} \in S : \boldsymbol{\tau} \le \boldsymbol{\ell}\}$$

and simplices

$$T_{\boldsymbol{\rho}}(L) := \{\boldsymbol{\tau} \in S : \boldsymbol{\rho} \cdot \boldsymbol{\tau} \le L\},$$

with $\boldsymbol{\rho} \in \mathbb{R}_+^d$ and where $\cdot$ denotes the usual Euclidean scalar product in $\mathbb{R}^d$.

3.2 Formulation

We briefly review the basics of the MIMC method and indicate some of its properties.

Consider the approximation of the expected value of a quantity of interest g,

$$I(g) := \mathbb{E}[g] = \int_{\Omega} g \, \mathrm{d}P,$$

by an N-point Monte Carlo estimator

$$Q(g) := \frac{1}{N} \sum_{n=0}^{N-1} g(\omega_n).$$

Here, the ω_n, $n = 0, 1, \ldots$ refer to N random samples from the probability space Ω. Hence, the estimator itself is also a random quantity. In our application, the quantity of interest g cannot be evaluated exactly, and we need to resort to discretizations $g_{\boldsymbol{\ell}}$, where the different components of $\boldsymbol{\ell} = (\ell_1, \ldots, \ell_d)$ are different discretization levels of those quantities that need discretization. Note that the dimensionality of the integral s and the number of discretization dimensions d are not to be confused.

For a given index $\boldsymbol{\ell}$, define the difference operator in a certain direction i, denoted by Δ_i, as

$$\Delta_i g_{\boldsymbol{\ell}} := \begin{cases} g_{\boldsymbol{\ell}} - g_{\boldsymbol{\ell}-\boldsymbol{e}_i} & \text{if } \ell_i > 0, \\ g_{\boldsymbol{\ell}} & \text{otherwise,} \end{cases} \qquad i = 1, \ldots, d.$$

The MIMC estimator involves a tensor product $\Delta := \Delta_1 \otimes \cdots \otimes \Delta_d$ of difference operators, where the difference is taken with respect to all backward neighbors of the index $\boldsymbol{\ell}$.

Using this definition, the MIMC estimator for $I(g)$ can be formulated as

$$Q_L(g) := \sum_{\boldsymbol{\ell} \in \mathscr{I}(L)} Q(\Delta g_{\boldsymbol{\ell}}) = \sum_{\boldsymbol{\ell} \in \mathscr{I}(L)} \frac{1}{N_{\boldsymbol{\ell}}} \sum_{n=0}^{N_{\boldsymbol{\ell}}-1} (\Delta_1 \otimes \cdots \otimes \Delta_d)\, g_{\boldsymbol{\ell}}(\omega_{\boldsymbol{\ell},n}), \tag{7}$$

where $\mathscr{I}(L)$ is an admissible index set. The parameter L governs the size of the index set.

Note that the Multilevel Monte Carlo (MLMC) estimator from [4, 9, 10] is a special case of the MIMC estimator, where $d = 1$. That is, the summation involves a loop over a range of scalar levels ℓ, and there is no tensor product involved:

$$Q_L^{(\mathrm{ML})}(g) := \sum_{\ell=0}^{L} Q(\Delta g_\ell) = \sum_{\ell=0}^{L} \frac{1}{N_\ell} \sum_{n=0}^{N_\ell-1} \Delta g_\ell(\omega_{\ell,n}).$$

For convenience, we use the following shorthand notation: $E_{\boldsymbol{\ell}} := |\mathbb{E}[\Delta g_{\boldsymbol{\ell}}]|$ for the absolute value of the mean and $V_{\boldsymbol{\ell}} := \mathbb{V}[\Delta g_{\boldsymbol{\ell}}]$ for the variance. By $W_{\boldsymbol{\ell}}$ we denote the amount of computational work to compute a single realization of the difference $\Delta g_{\boldsymbol{\ell}}$. The total work of estimator (7) is

$$\text{Total Work} = \sum_{\boldsymbol{\ell} \in \mathscr{I}(L)} W_{\boldsymbol{\ell}} N_{\boldsymbol{\ell}}. \tag{8}$$

In (7), one still has the freedom to choose the index set $\mathscr{I}(L)$ and the number of samples $N_{\boldsymbol{\ell}}$ at each index $\boldsymbol{\ell}$. In the following, we will show how these two parameters can be quantified.

The objective is to find an index set $\mathscr{I}(L)$ and sample sizes $N_{\boldsymbol{\ell}}$ such that (7) achieves a *mean square error* (MSE) smaller than a prescribed tolerance ε^2, with the lowest possible cost. From standard statistical analysis, it is known that the MSE can be expressed as a sum of a stochastic error and a discretization error, i.e.,

$$\mathbb{E}\left[(Q_L(g) - I(g))^2\right] = \mathbb{E}\left[(Q_L(g) - \mathbb{E}[Q_L(g)])^2\right] + (\mathbb{E}[Q_L(g)] - I(g))^2. \tag{9}$$

The first term in (9) is the variance of the estimator, which, by independence of the events $\omega_{\ell,n}$, is given by

$$\mathbb{V}[Q_L(g)] = \sum_{\ell \in \mathscr{I}(L)} \frac{V_\ell}{N_\ell}. \tag{10}$$

It can be reduced by increasing the number of samples N_ℓ. The second term in (9) is the square of the bias. It can be reduced by augmenting the index set $\mathscr{I}(L)$. A sufficient condition to ensure an MSE smaller than ε^2, is that both terms in (9) are smaller than $\varepsilon^2/2$:

$$\mathbb{V}[Q_L(g)] = \mathbb{E}\left[(Q_L(g) - \mathbb{E}[Q_L(g)])^2\right] \leq \varepsilon^2/2, \quad \text{and} \tag{C1}$$

$$|\mathbb{E}[Q_L(g)] - I(g)| \leq \varepsilon/\sqrt{2}. \tag{C2}$$

As in [5, 15], we will also use an alternative error splitting, based on a splitting parameter. The value of this parameter is then computed using a Bayesian approach. This alternative splitting will also be used in our numerical experiments later.

The error splitting in (9) will prove to be essential in the algorithm presented below. Since the total error is the sum of two independent contributions, we can solve for both unknowns N_ℓ and $\mathscr{I}(L)$ independently. Minimizing the total cost subject to the statistical constraint (C1) will give the optimal number of samples. Minimizing the total cost subject to the bias constraint (C2) will yield the optimal shape of the index set. When using these optimal values for N_ℓ, $\mathscr{I}(L)$, and the error splitting parameter, the cost of the MIMC estimator is minimal, for a given value of ε^2.

3.2.1 Minimizing the Stochastic Error: Optimal Number of Samples

Consider an MIMC estimator with a sufficiently large index set $\mathscr{I}(L)$, such that the bias constraint (C2) is satisfied. Then, one still has to decide the number of samples for each $\ell \in \mathscr{I}(L)$. This freedom can be used to minimize the cost of the MIMC estimator (8) while assuring that the statistical constraint (C1) is satisfied, i.e.,

$$\min_{N_\ell \in \mathbb{R}_+} \sum_{\tau \in \mathscr{I}(L)} N_\tau W_\tau \tag{11}$$

$$\text{s.t.} \sum_{\tau \in \mathscr{I}(L)} \frac{V_\tau}{N_\tau} \leq \frac{\varepsilon^2}{2}.$$

This minimization problem can be solved using Lagrange multipliers. The optimal number of samples at each index such that the total cost is minimized, is

$$N_{\boldsymbol{\ell}} = \frac{2}{\varepsilon^2}\sqrt{\frac{V_{\boldsymbol{\ell}}}{W_{\boldsymbol{\ell}}}} \sum_{\boldsymbol{\tau}\in\mathscr{I}(L)} \sqrt{V_{\boldsymbol{\tau}} W_{\boldsymbol{\tau}}} \quad \text{for all } \boldsymbol{\ell} \in \mathscr{I}(L). \tag{12}$$

In practice, this number is rounded up to the nearest integer number of samples. Also, sample variances and estimates for the cost can be used to replace the true variance $V_{\boldsymbol{\ell}}$ and true cost $W_{\boldsymbol{\ell}}$ at each index. Using (12), we can rewrite the total cost of the MIMC estimator as

$$\text{Total Work} = \frac{2}{\varepsilon^2}\left(\sum_{\boldsymbol{\ell}\in\mathscr{I}(L)} \sqrt{V_{\boldsymbol{\ell}} W_{\boldsymbol{\ell}}}\right)^2. \tag{13}$$

3.2.2 Minimizing the Discretization Error: Optimal Index Sets

The most simple multi-index method considers indices that are contained in cubes $\mathscr{I}(L) = R((L, L\ldots))$ or simplices $\mathscr{I}(L) = T_{(1,1\ldots)}(L)$. It is possible to extend the latter to the class of general simplices $T_{\boldsymbol{\rho}}(\boldsymbol{\ell})$. An a priori analysis could then identify important directions in the problem and choose a suitable vector $\boldsymbol{\rho}$. However, this approach suffers from two drawbacks. First, such an analysis may be difficult or prohibitively expensive. Furthermore, it is possible that the class of general simplices is inadequate to represent the problem under consideration, especially when mixed directions are involved. In our estimator, we will allow general monotone index sets in the summation (7). The algorithm we designed adaptively detects important directions in the problem. By a careful construction of the corresponding admissible index set, we hope to achieve an estimator for which the MSE, for a given amount of work, is at least as small as for these classical constructions. Note that as with all adaptive algorithms, the algorithm could be fooled by a quantity of interest for which it seems there is no benefit of extending the index set at some point, and for which essential contributions are hidden at an arbitrary further depth in the index set.

Since the index set is finite, the discretization error is equal to the sum of all neglected contributions, i.e.,

$$|\mathbb{E}[Q_L(g)] - I(g)| = \left|\sum_{\boldsymbol{\ell}\notin\mathscr{I}(L)} \mathbb{E}[\Delta g_{\boldsymbol{\ell}}]\right| \leq \sum_{\boldsymbol{\ell}\notin\mathscr{I}(L)} E_{\boldsymbol{\ell}}.$$

Similar to (11), we search for the index set that minimizes the (square root of the) total amount of work (13). Here, we impose that the bias constraint (C2) is satisfied, i.e.,

$$\min_{\mathscr{I}(L)\subseteq S} \sum_{\boldsymbol{\ell}\in\mathscr{I}(L)} \sqrt{V_{\boldsymbol{\ell}} W_{\boldsymbol{\ell}}}$$
$$\text{s.t.} \sum_{\boldsymbol{\ell}\notin\mathscr{I}(L)} E_{\boldsymbol{\ell}} \leq \varepsilon/\sqrt{2}.$$

This problem can be formulated as a binary knapsack problem by assigning a *profit indicator* to each index. Define this profit as the ratio of the error contribution and the work contribution, i.e.,

$$P_{\boldsymbol{\ell}} = \frac{E_{\boldsymbol{\ell}}}{\sqrt{V_{\boldsymbol{\ell}} W_{\boldsymbol{\ell}}}}, \tag{14}$$

see [15]. A binary knapsack problem is a knapsack problem where the number of copies of each kind of item is either zero or one, i.e., we either include or exclude an index $\boldsymbol{\ell}$ from the set $\mathscr{I}(L)$. In the next section, we introduce an adaptive greedy algorithm that solves this knapsack problem, where the profits $P_{\boldsymbol{\ell}}$ are used as item weights.

3.3 An Adaptive Method

The goal is to find an admissible index set such that the corresponding MSE is as small as possible subject to an upper bound on the amount of work. Starting from index $(0, 0, \ldots)$, we will successively add indices to the index set such that (a) the resulting index set remains monotone and (b) the error is reduced as much as possible. That is, we require $\mathscr{I}(0) = \{(0, 0, \ldots)\}$ and $\mathscr{I}(L) \subseteq \mathscr{I}(L+1)$ for all $L \geq 0$. Using the definition of profit above, we can achieve this by always adding the index with the highest profit to the index set. An algorithm that uses this strategy in the context of dimension-adaptive quadrature using sparse grids is presented in [7]. We recall the main ideas below.

The complete algorithm is sketched in Algorithm 1. We assume the current index set $\mathscr{I}$ is partitioned into two disjoint sets, containing the *active* indices $\mathscr{A}$ and *old* indices $\mathscr{O}$, respectively. The active set $\mathscr{A}$ contains all indices for which none of their forward neighbors are included in the index set $\mathscr{I} = \mathscr{A} \cup \mathscr{O}$. These indices form the boundary of the index set $\mathscr{I}$ and will actively be adapted in the algorithm. The old index set $\mathscr{O}$ contains all other indices of the index set, they have at least one forward neighbor in $\mathscr{I} = \mathscr{A} \cup \mathscr{O}$. Equivalently, this means that all backward neighbors of an index in $\mathscr{I} = \mathscr{A} \cup \mathscr{O}$ are always in $\mathscr{O}$, which means $\mathscr{I}$ and $\mathscr{O}$ are admissible index sets. Initially, we set $\mathscr{O} = \varnothing$ and $\mathscr{A} = \{(0, 0, \ldots)\}$. In every iteration of the adaptive algorithm, the index $\bar{\boldsymbol{\ell}}$ with the largest profit $P_{\bar{\boldsymbol{\ell}}}$ is selected from the active set $\mathscr{A}$. This index is moved from the active set to the old set. Next, all forward neighbors $\boldsymbol{\tau}$ of $\bar{\boldsymbol{\ell}}$ are considered. If the neighbor is admissible in the old index set $\mathscr{O}$, the index is added to the active set $\mathscr{A}$. A number of warm-up samples are taken at index $\boldsymbol{\tau}$ to be

Algorithm 1 Dimension-Adaptive Multi-Index Monte Carlo

```
ℓ := (0, ..., 0)
𝒪 := ∅
𝒜 := {ℓ}
P_ℓ := 0
repeat
  Select index ℓ̄ from 𝒜 with largest profit P_ℓ̄
  𝒜 := 𝒜 \ {ℓ̄}
  𝒪 := 𝒪 ∪ {ℓ̄}
  for k in 1, 2, ..., d do
    τ := ℓ̄ + e_k
    if τ − e_j ∈ 𝒪 for all j = 1, 2, ..., d for which τ_j > 0 then
      𝒜 := 𝒜 ∪ {τ}
      Take N* warm-up samples at index τ
      Set Q_τ := Q(Δg_τ)
      Estimate V_τ by (10) and E_τ by |Q_τ|
    end if
  end for
  for ℓ ∈ 𝒪 ∪ 𝒜 do
    Compute optimal number of samples N_ℓ using (12)
    Ensure that at least min(2, ⌈N_ℓ⌉) samples are taken at each index ℓ and re-evaluate Q_ℓ
    Update the estimate of V_ℓ and E_ℓ
    (Re)compute profit indicator P_ℓ using (14)
  end for
until Σ_{ℓ∈𝒜} |Q_ℓ| < ε/√2
return Σ_{ℓ∈𝒪∪𝒜} Q_ℓ
```

used in the evaluation of (12). After that, we ensure that at least $N_{\boldsymbol{\ell}}$ samples are taken at all indices in the index set $\mathscr{I} = \mathscr{A} \cup \mathscr{O}$. Using the updated samples, the profit indicators, as well as the estimates for $V_{\boldsymbol{\ell}}$ and $E_{\boldsymbol{\ell}}$, are recomputed for all indices in $\mathscr{I}$. The algorithm continues in the next iteration by selecting the index with the now largest profit, until the condition on the discretization error (C2) is satisfied. Similar to the approach in [15], we use the heuristic bias estimate

$$\left| \sum_{\boldsymbol{\ell} \notin \mathscr{I}(L)} \mathbb{E}[\Delta g_{\boldsymbol{\ell}}] \right| \approx \sum_{\boldsymbol{\ell} \in \mathscr{A}} |Q(\Delta g_{\boldsymbol{\ell}})| \,. \tag{15}$$

Thus, the absolute value of the Monte Carlo estimators for the differences associated with the indices in the active set $\mathscr{A}$ act as an estimate for the bias. Finally, note that as soon as an index is added to the active set $\mathscr{A}$, it is also used in the evaluation of (7). Indeed, it does not make sense to take samples at these indices only to evaluate the profit indicator, and then exclude these samples in the evaluation of the telescoping sum.

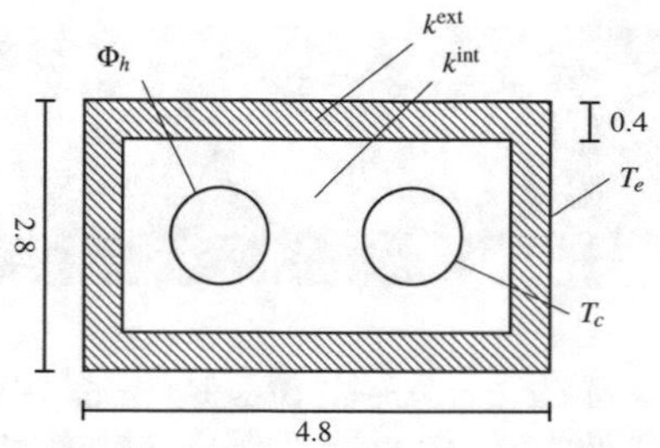

Fig. 1 Setup for the heat exchanger problem. Hot fluid flows through the left-hand pipe, where a constant heat flux $\boldsymbol{\Phi}_h$ is applied. The cooling fluid in the right-hand pipe has a constant temperature T_c. The exterior temperature is T_e. The conductivity of the interior conducting material is k^{int}, while the conductivity of the exterior insulating material is k^{ext}

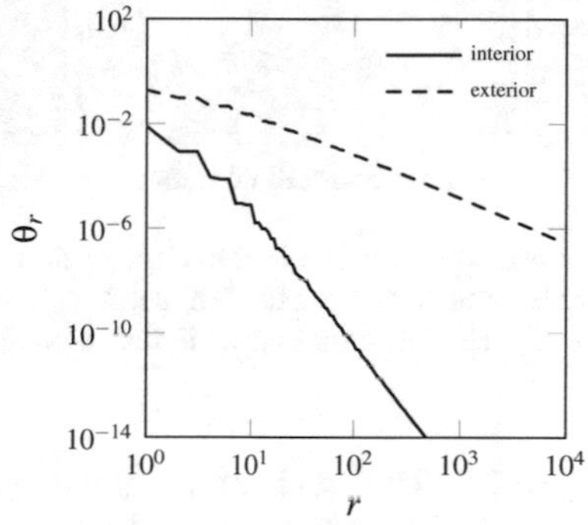

Fig. 2 Decay of the eigenvalues θ^{int} and θ^{ext}

4 A Simple Model for a Heat Exchanger

We study the behavior of the adaptive algorithm by applying it to the heat equation with random conductivity from (2). A numerical example, using the strongly simplified model for a heat exchanger from [17] is presented below. Note that this example, including the choice of its stochastic characteristics, is used for numerical illustration purposes only.

4.1 The Model

We refer to Fig. 1 for a visualization of the description in this section. A two-dimensional heat exchanger consists of a rectangular piece of material perforated by two circular holes. The first hole contains a hot fluid that injects heat at a constant and known rate $\Phi_h = 125/\pi$, and the second hole contains a cooling fluid at

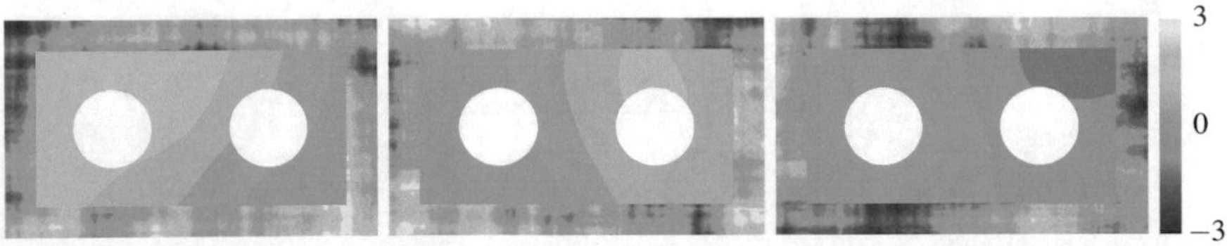

Fig. 3 Example realizations of the (zero-mean) Gaussian fields Z^{int} and Z^{ext} used in the heat exchanger problem. The insulator material has a lower correlation length and smoothness, and a higher variance. The associated conductivity is $k = \exp(Z^{\text{int}} \cup Z^{\text{ext}})$. The number of terms used in the KL expansions for Z^{int} and Z^{ext} is 512 and 8 192, respectively

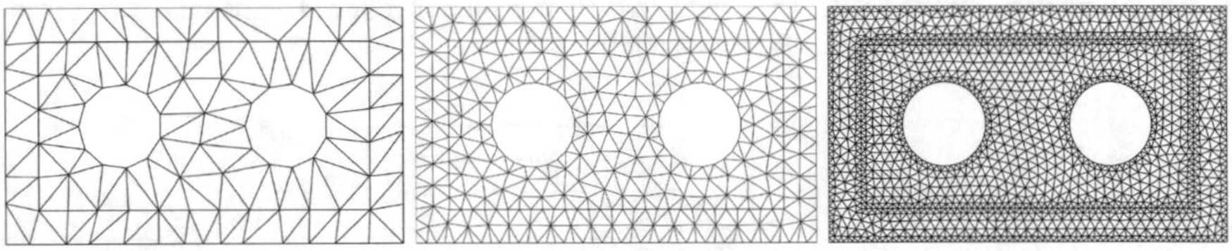

Fig. 4 Some finite-element meshes used in the heat exchanger problem. The coarsest mesh has 102 points (144 elements), the intermediate mesh has 309 points (640 elements), and the fine mesh has 1 619 points (2 887 elements). The finest mesh used in the simulations is not shown

a constant temperature $T_c = 7.5$. The conductivity of the heat exchanger material is modeled as a lognormal random field $k^{\text{int}} = \exp(Z^{\text{int}})$, where Z^{int} is a Gaussian random field with mean $\mu^{\text{int}} = 0$ and Matérn covariance with correlation length $\lambda^{\text{int}} = 1$, standard deviation $\sigma^{\text{int}} = \sqrt{0.1}$, norm $p = 1$ and smoothness $\nu^{\text{int}} = 1$, see (3).

A layer of insulator material is added to the heat exchanger, to thermally insulate it from its surroundings, which has a constant temperature $T_e = 20$. The conductivity of the insulator material is modeled as a lognormal random field $k^{\text{ext}} = \exp(Z^{\text{ext}})$, where Z^{ext} is a Gaussian random field with mean $\mu^{\text{ext}} = \log(0.01)$ and Matérn covariance with correlation length $\lambda^{\text{ext}} = 0.3$, standard deviation $\sigma^{\text{ext}} = 1$, norm $p = 1$ and smoothness $\nu^{\text{ext}} = 0.5$.

Samples of both random fields are generated using a truncated KL expansion, see (4). Figure 2 shows the decay of the two-dimensional eigenvalues for both the conductor (interior) and insulator (exterior) material. These eigenvalues and corresponding eigenfunctions are computed once for the maximal number of terms allowed in the expansion. Every realization of the conductivity $k = \exp(Z)$ is formed using a sample of the Gaussian random fields Z^{int} and Z^{ext}. Three samples of the (Gaussian) random field Z are shown in Fig. 3. Note that Z^{int} only varies mildly in comparison to Z^{ext}.

For the spatial discretization, we use eleven different nonnested finite-element (FE) meshes with an increasing number of elements. For every mesh, the number of points is roughly twice the number of points of its predecessor. That way, the size of the finite-element system matrix doubles between successive approximations. The coarsest mesh has 102 points (144 elements), and the finest mesh has 94 614 points (186 268 elements). Three examples are shown in Fig. 4.

The heat flow through the exchanger is described by (2), with source term $F := 0$. As a quantity of interest, we consider the value of the temperature at the leftmost point on the boundary of the hot fluid pipe. As shown in Fig. 6, this corresponds to the highest expected temperature in the heat exchanger. Note that we have made sure that this point is included on every FE mesh, to avoid an interpolation error.

4.2 Numerical Results

We set up an adaptive MIMC algorithm with three refinement dimensions, i.e. $d = 3$. The first dimension corresponds to the spatial discretization, the second dimension is the number of terms in the KL expansion of the conductor material, and the last dimension is used for the number of terms in the KL expansion of the insulator material. The number of terms in either KL expansion doubles between subsequent approximations, similar to the connection between the different spatial discretizations. If the effect of adding more KL terms to the approximation of the quantity of interest was known in advance, one could derive the optimal relation between the different approximations, similar to [13]. This relation will, amongst others, depend on the decay rate of the eigenvalues of the KL expansion, hence, it will be different for the insulator and conductor material. However, in the absence of this knowledge, doubling the number of terms (a geometric relation, following [13]) seems an obvious thing to do. Note that the slow eigenvalue decay rate for the insulator material in Fig. 2 is reflected in the number of terms used in the coarsest approximation: index $(\cdot, 0, 0)$ corresponds to an approximation using $s_0^{\text{int}} = 4$ terms in the KL expansion of the heat exchanger material and $s_0^{\text{ext}} = 64$ terms in the expansion of the insulator material.

In practice, we do not start the algorithm from index $(0, \ldots, 0)$ as is indicated in Algorithm 1, but start with an index set $T_{(1,1,1)}(2)$, to ensure the availability of robust estimates for the profit indicator on the coarsest approximations.

The total cost of the computation of $G_{\boldsymbol{\ell}}$ is equal to the sum of the cost of composing the random field using the KL expansion and the cost of the finite-element computation. For a given index $\boldsymbol{\ell} = (\ell_1, \ell_2, \ell_3)$, we assume that there are $\texttt{elements}(\ell_1)$ elements and $\texttt{nodes}(\ell_1)$ nodes in the discretization. The KL expansions at that index use $s_0^{\text{int}} 2^{\ell_2}$ terms for the conductor and $s_0^{\text{ext}} 2^{\ell_3}$ terms for the insulator. We propose the cost model

$$C_1(\texttt{elements}(\ell_1))(s_0^{\text{int}} 2^{\ell_2} + s_0^{\text{ext}} 2^{\ell_3}) + C_2(\texttt{nodes}(\ell_1))^{\gamma}, \qquad (16)$$

for some suitable constants C_1, C_2 and γ. We numerically found the values $C_1 =$ 1.596e-8, $C_2 =$ 1.426e-6 and $\gamma = 1.664$. There is no cost involved in computing the quantity of interest $G_{\boldsymbol{\ell}}$ from the solution $T(\boldsymbol{x}, \cdot)$, since no interpolation is required. The cost $W_{\boldsymbol{\ell}}$ of computing a single sample of $\Delta G_{\boldsymbol{\ell}}$ can be computed by expansion

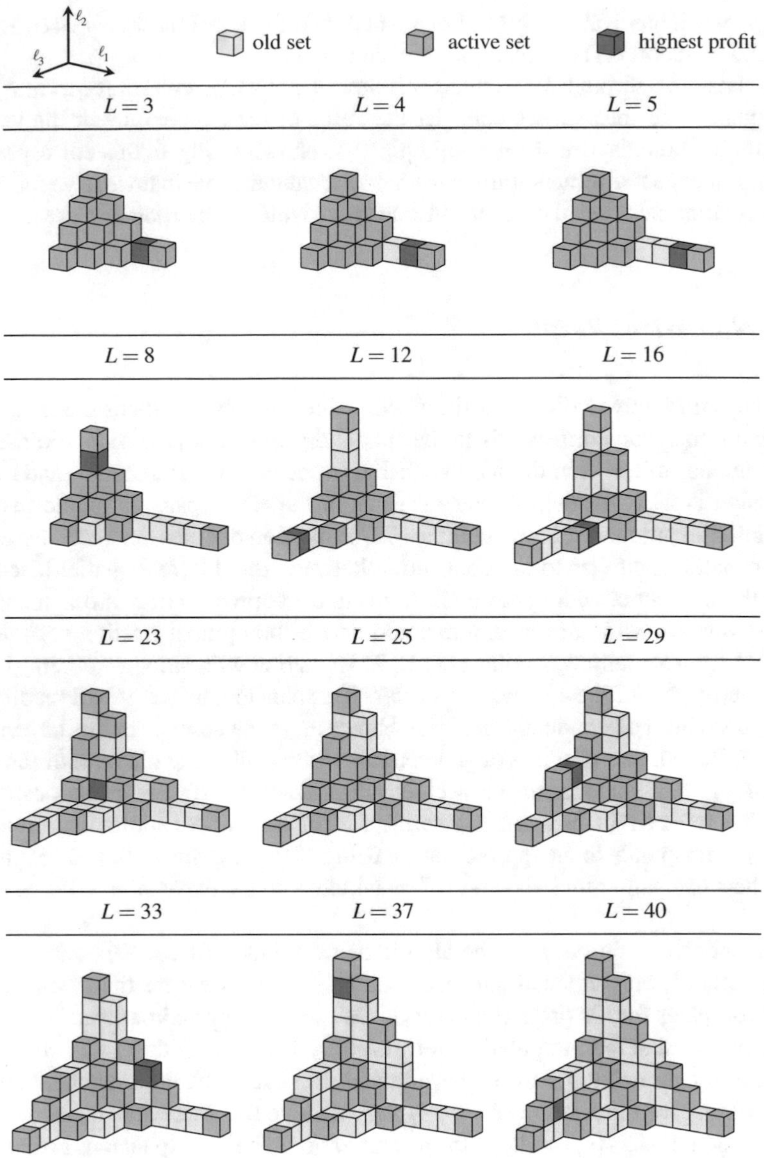

Fig. 5 Examples of nontrivial index sets in the heat exchanger problem for selected iterations in the adaptive algorithm

of the tensor product $\Delta = \Delta_1 \otimes \Delta_2 \otimes \Delta_3$. Note that it is also possible to use actual simulation times as measures for the cost. However, since the cost estimate appears in the profit indicator, and thus determines the shape of the index set, one should ensure that the estimates are stable and reliable. Finally, we use the continuation approach from [5] and run the MIMC algorithm for a sequence of larger tolerances than required, to obtain more accurate estimates of the sample variances, and to avoid having to take warm-up samples at every index.

We run our adaptive algorithm for a relative tolerance of $\varepsilon_{\text{rel}} = 1 \cdot 10^{-3}$. Note that Algorithm 1 is formulated in terms of an *absolute* tolerance ε. We adapt for (estimated) relative tolerances by using the current estimate for the expected value of the quantity of interest as a scaling factor. The mean value of the quantity of interest was computed by our algorithm as $Q_L(g) = 133.71$ with a standard error of 10.45 in $L = 40$ iterations. The standard deviation of the estimator is 0.0975, and the estimated bias is 0.0782, giving a total (root mean square) error (RMSE) estimate of $0.125 < \varepsilon_{\text{rel}} \cdot Q_L(g)$. Figure 5 shows the shape of the index set for some selected iterations. We see that the adaptive algorithm mainly exploits the spatial resolutions, until the addition of more spatial levels is estimated to be too expensive ($L = 8$). After that, the approximations for the conductor and insulator material are improved up to 256 and 8 192 terms respectively. From $L = 25$ and beyond, the *mixed directions* that improve the approximation for both conductor and insulator material, and the approximation for the conductor material and the mesh refinement, are activated. Observe that the final shape of the index set ($L = 40$) is far from trivial, and is also not immediately representable by an anisotropic simplex.

Finally, we investigate the performance of our adaptive method compared to standard MIMC with the common choice of simplices $T_{(1,1,1)}(L)$ as index sets. The mean value, RMSE and runtime for all tolerances are shown in Table 1. All simulations are performed on a 2.6 GHz Intel Xeon CPU with 64GB of RAM. Observe that both methods converge to the same value. The adaptive algorithm outperforms the non-adaptive MIMC method for all values of ε_{rel} considered. Note that we are not able to solve for smaller tolerances using the nonadaptive MIMC method, because of the increasing memory requirement of the available spatial resolutions.

The adaptive algorithm is not limited to scalar quantities of interest. It is also possible to include multiple quantities of interest in a single simulation. We then take the worst value of the profit over all quantities considered to compute the next iterate, see [10]. As an example, Fig. 6 shows the mean value of the temperature in the heat exchanger on a mesh with 1 524 elements, for a relative tolerance of $1 \cdot 10^{-3}$. The highest expected temperature is located at a point on the boundary of the hot fluid pipe, opposite to the pipe containing the coolant fluid. This is what might have been anticipated from physical considerations, assuming that the heat flux Φ_h is large enough to heat the material around the left-hand pipe to a temperature higher than T_e. The effect of the insulator material is obvious from the large temperature gradient present at the left side of the insulator.

Table 1 Mean, RMSE and running time for MIMC on the left, and for adaptive MIMC on the right. The nonadaptive version uses simplices $T_{(1,1,1)}(L)$ as index set

	Nonadaptive MIMC			Adaptive MIMC		
ε_{rel}	Mean	RMSE	Time (s)	Mean	RMSE	Time (s)
2.890e-2	135.20	1.948e-0	622	133.77	2.107e-0	591
1.927e-2	134.46	1.293e-0	1 175	132.56	1.545e-0	1 228
1.285e-2	134.10	1.100e-0	2 667	132.47	1.412e-0	1 228
8.564e-3	133.04	9.767e-1	11 951	132.38	1.115e-0	7 034
5.710e-3	133.76	4.430e-1	30 552	133.20	4.744e-1	20 725
3.806e-3	133.74	4.321e-1	38 997	133.61	1.027e-1	28 335
2.538e-3	133.72	3.521e-1	92 223	133.63	1.723e-1	81 711
1.692e-3	133.73	2.789e-1	257 698	133.70	1.396e-1	233 458

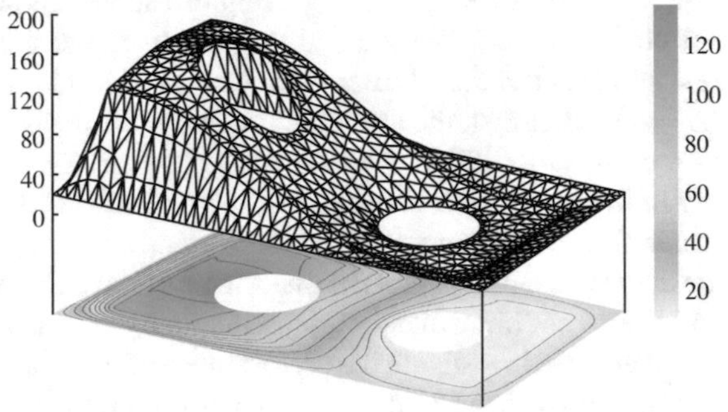

Fig. 6 Mean temperature field of the heat exchanger on a 890-point mesh as an example of a nonscalar quantity of interest

5 Discussion and Future Work

We have presented a dimension-adaptive Multi-Index Monte Carlo (MIMC) method for the approximation of the expected value of a quantity of interest that is a function of the solution of a PDE with random coefficients. The method, which can be seen as a generalization of the classical MIMC method, automatically finds important directions in the problem. These directions are not limited to spatial dimensions only, as is demonstrated by a numerical experiment. We have demonstrated an efficient implementation of the method, based on a similar construction used in dimension-adaptive integration with sparse grids.

The adaptive algorithm is particularly interesting when the optimal shape of the MIMC index set is unknown or nontrivial, since it does not require a priori knowledge of the structure of the problem. In these situations, the method may include or exclude certain indices to achieve an estimator that minimizes computational effort needed to obtain a certain tolerance.

Finally, adaptivity can be used in combination with other techniques, such as *Quasi-Monte Carlo*, see [6], or [18] for the multi-index setting. We expect similar gains as outlined in this paper.

References

1. Barth, A., Schwab, C., Zollinger, N.: Multi-level Monte Carlo finite element method for elliptic PDEs with stochastic coefficients. Numer. Math. **119**(1), 123–161 (2011)
2. Bungartz, H.J., Griebel, M.: Sparse grids. Acta Numer. **13**, 147–269 (2004)
3. Chkifa, A., Cohen, A., Schwab, C.: High-dimensional adaptive sparse polynomial interpolation and applications to parametric PDEs. Found. Comput. Math. **14**(4), 601–633 (2014)
4. Cliffe, K.A., Giles, M.B., Scheichl, R., Teckentrup, A.L.: Multilevel Monte Carlo methods and applications to elliptic PDEs with random coefficients. Comput. Vis. Sci. **14**(1), 3–15 (2011)
5. Collier, N., Haji-Ali, A.L., Nobile, F., Schwerin, E., Tempone, R.: A continuation multilevel Monte Carlo algorithm. BIT Numer. Math. **55**(2), 399–432 (2014)
6. Dick, J., Kuo, F.Y., Sloan, I.H.: High-dimensional integration: the quasi-Monte Carlo way. Acta Numer. **22**, 133–288 (2013)
7. Gerstner, T., Griebel, M.: Dimension-adaptive tensor-product quadrature. Computing **71**(1), 65–87 (2003)
8. Ghanem, R.G., Spanos, P.D.: Stochastic Finite Elements: A Spectral Approach. Springer, New York (1991)
9. Giles, M.B.: Multilevel Monte Carlo path simulation. Oper. Res. **56**(3), 607–617 (2008)
10. Giles, M.B.: Multilevel Monte Carlo methods. Acta Numer. **24**, 259–328 (2015)
11. Graham, I.G., Kuo, F.Y., Nuyens, D., Scheichl, R., Sloan, I.H.: Quasi-Monte Carlo methods for elliptic PDEs with random coefficients and applications. J. Comput. Phys. **230**(10), 3668–3694 (2011)
12. Griebel, M., Schneider, M., Zenger, C.: A combination technique for the solution of sparse grid problems. In: de Groen, D., Beauwens, R. (eds.) Iterative Methods in Linear Algebra 1991, pp. 263–281. Elsevier, Amsterdam (1992)
13. Haji-Ali, A.L., Nobile, F., von Schwerin, E., Tempone, R.: Optimization of mesh hierarchies in multilevel Monte Carlo samplers. Stoch. Part. Differ. Equ: Anal. Comput. **4**(1), 76–112 (2016)
14. Haji-Ali, A.L., Nobile, F., Tamellini, L., Tempone, R.: Multi-index stochastic collocation for random PDEs. Comput. Methods Appl. Mech. Eng. **306**, 95–122 (2016)
15. Haji-Ali, A.L., Nobile, F., Tempone, R.: Multi-index Monte Carlo: when sparsity meets sampling. Numer. Math. **132**(4), 767–806 (2016)
16. Lord, G.J., Powell, C.E., Shardlow, T.: An Introduction to Computational Stochastic PDEs. Cambridge University Press, Cambridge (2014)
17. Le Maître, O.L., Knio, O.M.: Spectral Methods for Uncertainty Quantification with Applications to Computational Fluid Dynamics. Springer Science and Business Media, Berlin (2010)
18. Robbe, P., Nuyens, D., Vandewalle, S.: A Multi-index Quasi-Monte Carlo algorithm for lognormal diffusion problems. SIAM J. Sci. Comput. **39**(5), S851–S872 (2017)
19. Xiu, D.: Fast numerical methods for stochastic computations: a review. Commun. Comput. Phys. **5**(2–4), 242–272 (2009)
20. Xiu, D., Karniadakis, G.E.: Modeling uncertainty in steady state diffusion problems via generalized polynomial chaos. Comput. Methods Appl. Mech. Eng. **191**(43), 4927–4948 (2002)

Towards Real-Time Monte Carlo for Biomedicine

Shuang Zhao, Rong Kong and Jerome Spanier

Abstract Monte Carlo methods provide the "gold standard" computational technique for solving biomedical problems but their use is hindered by the slow convergence of the sample means. An exponential increase in the convergence rate can be obtained by adaptively modifying the sampling and weighting strategy employed. However, if the radiance is represented globally by a truncated expansion of basis functions, or locally by a region-wise constant or low degree polynomial, a bias is introduced by the truncation and/or the number of subregions. The sheer number of expansion coefficients or geometric subdivisions created by the biased representation then partly or entirely offsets the geometric acceleration of the convergence rate. As well, the (unknown amount of) bias is unacceptable for a gold standard numerical method. We introduce a new unbiased estimator of the solution of radiative transfer equation (RTE) that constrains the radiance to obey the transport equation. We provide numerical evidence of the superiority of this Transport-Constrained Unbiased Radiance Estimator (T-CURE) in various transport problems and indicate its promise for general heterogeneous problems.

Keywords Monte Carlo simulations · Transport-constrained radiance estimators

S. Zhao
Donald Bren School of Information and Computer Sciences,
University of California @ Irvine, 3019 Donald Bren Hall, Irvine, CA 92697, USA
e-mail: shz@ics.uci.edu

R. Kong
Hyundai Capital America, 3161 Michelson Drive, Irvine, CA 92612, USA
e-mail: kongr413@yahoo.com

J. Spanier (✉)
Beckman Laser Institute and Medical Clinic, University of California @ Irvine, 1002 Health Sciences Road, Irvine, CA 92612, USA
e-mail: jspanier@uci.edu

A. B. Owen and P. W. Glynn (eds.), *Monte Carlo and Quasi-Monte Carlo Methods*, Springer Proceedings in Mathematics & Statistics 241,
https://doi.org/10.1007/978-3-319-91436-7_25

1 Introduction

Monte Carlo simulation has provided the "gold standard" numerical method for solving biomedical problems for the past thirty years [22]. Nevertheless, its slow convergence (at the rate $N^{-1/2}$ where N equals sample size) inhibits use of Monte Carlo on a routine basis. Instead, diffusion-based numerical methods are often used because of their superior speed of execution, even though they may provide very poor descriptions of the radiant light field in many situations. Consequently, there has been a lot of interest in accelerating the convergence of Monte Carlo simulations, especially within the biomedical community, where accuracy is a primary focus.

Conventional density function estimation methods [6, 16, 17] are widely used with success where photorealism—not image perfection—is the goal. Such methods have revolutionized the rendering of scenes for electronic games and movies [7]. Density estimation methods avoid the need to represent the radiance in a functional expansion, but they introduce the need for "smoothing parameters" which also causes a bias in the density estimator. This precludes convergence to the exact solution and is unacceptable as a gold standard method for biomedicine or biology. The question we then asked was: Can any of these ideas be used in such a way that the radiance it produces actually satisfies the governing radiative transport equation? If so, might that produce a candidate to serve as a gold standard for biomedical simulations? That investigation has led to the publication [12] and to this paper.

2 Radiative Transport Fundamentals

Before proceeding with this line of thinking we want to establish our notation and clarify our goals.

The rigorous transport of light in tissue usually begins with the integro-differential form of the equation which is then transformed to the integral form [18] of the RTE:

$$L(\boldsymbol{P}) = \int_{\Gamma} K(\boldsymbol{P}' \to \boldsymbol{P})\, L(\boldsymbol{P}')\, \mathrm{d}\boldsymbol{\omega}' \mathrm{d}\rho + S(\boldsymbol{P}), \tag{1}$$

where $\boldsymbol{P} := (\boldsymbol{r}, \boldsymbol{\omega})$, $\boldsymbol{P}' := (\boldsymbol{r}', \boldsymbol{\omega}')$, $\boldsymbol{r}' := \boldsymbol{r} - \rho\boldsymbol{\omega}$ and

$$K(\boldsymbol{P}' \to \boldsymbol{P}) := \frac{\mu_s(\boldsymbol{r}')}{\mu_t(\boldsymbol{r}')}\, f(\boldsymbol{r}';\ \boldsymbol{\omega}' \to \boldsymbol{\omega})\, T(\boldsymbol{r}' \to \boldsymbol{r};\ \boldsymbol{\omega}), \tag{2}$$

$$T(\boldsymbol{r}' \to \boldsymbol{r};\ \boldsymbol{\omega}) := \mu_t(\boldsymbol{r}')\, \mathrm{e}^{-\int_0^{\|\boldsymbol{r}-\boldsymbol{r}'\|} \mu_t(\boldsymbol{r}-\tau\boldsymbol{\omega})\mathrm{d}\tau}, \tag{3}$$

with source function

$$S(P) := \mathrm{e}^{-\int_0^R \mu_t(\boldsymbol{r}-\tau\boldsymbol{\omega})\mathrm{d}\tau} Q_0(\boldsymbol{r} - R\boldsymbol{\omega}, \boldsymbol{\omega}) + \int_0^R \mathrm{e}^{-\int_0^\rho \mu_t(\boldsymbol{r}-\tau\boldsymbol{\omega})\mathrm{d}\tau} Q(\boldsymbol{r} - \rho\boldsymbol{\omega}, \boldsymbol{\omega})\, \mathrm{d}\rho. \tag{4}$$

Appendix 7 provides details on how Eq. (1) arises from the integro-differential equation (24) in the time-independent case.

To complete the mathematical description,

- $\Gamma := V \times \mathbb{S}^2$ ($V \subseteq \mathbb{R}^3$) denotes the *phase space* of vectors $(\boldsymbol{r}, \boldsymbol{\omega})$;
- μ_s and μ_a are respectively the *scattering* and *absorption* coefficients;
- $\mu_t := \mu_s + \mu_a$ is the *total attenuation* coefficient;
- f is the *single-scattering phase function* (that scatters photons from direction $\boldsymbol{\omega}'$ to $\boldsymbol{\omega}$ at location $\boldsymbol{r}'$);
- L denotes *photon radiance*.

2.1 *Role of Eqs.* (1)–(4) *in Generating Samples*

We now indicate how the Eqs. (1)–(4) play a role in generating photon biographies; i.e., samples drawn from our sample space $\mathbb{B}$.

Figure 1 depicts a hypothetical photon biography that is launched from the light source at the left, makes collisions at the locations $\boldsymbol{r}_1$, $\boldsymbol{r}_2$, $\boldsymbol{r}_3$ and $\boldsymbol{r}_4$, then exits the tissue at the detector on the right. If we assume that there is no internal volumetric source, ($Q \equiv 0$ in Eq. (4)), but there is a nonzero source of light Q_0 on the boundary, then the launch position and direction $\boldsymbol{P}_0 = (\boldsymbol{r}_0, \boldsymbol{\omega}_0)$ are drawn by sampling Q_0, while the first collision location $\boldsymbol{r}_1$ is drawn by sampling the exponential probability density function with exponent $\int_0^{\|\boldsymbol{r}_0-\boldsymbol{r}_1\|} \mu_t(\boldsymbol{r}_0 - \tau\boldsymbol{\omega}_0)\,\mathrm{d}\tau$ (see Eq. (3)). Provided that the photon is scattered at $\boldsymbol{r}_1$ (with probability μ_s/μ_t), the direction $\boldsymbol{\omega}_0$ is scattered into the direction $\boldsymbol{\omega}_1$ by sampling from the single-scattering phase function $f(\boldsymbol{r}_1; \boldsymbol{\omega}_0 \to \boldsymbol{\omega}_1)$. This process of locating successive collision points and unit directions continues until the photon biography $\boldsymbol{P}_0 = (\boldsymbol{r}_0, \boldsymbol{\omega}_0)$, $\boldsymbol{P}_1 = (\boldsymbol{r}_1, \boldsymbol{\omega}_1), \ldots$ either terminates by absorption (with probability $1 - \mu_s/\mu_t$) at some collision point or escapes from the tissue, either at the detector or elsewhere on ∂V.

2.2 *Equivalence Between Physical/Analytic and Stochastic Models*

The **physical/analytic RTE model** consists of the equations of radiative transport in tissue, together with a linear functional L_i of the solution L of the RTE for each detector:

$$L_i = \int_\Gamma d_i(\boldsymbol{r}, \boldsymbol{\omega}) L(\boldsymbol{r}, \boldsymbol{\omega})\, \mathrm{d}\boldsymbol{r}\, \mathrm{d}\boldsymbol{\omega}, \quad (i = 1, \ldots, d).$$

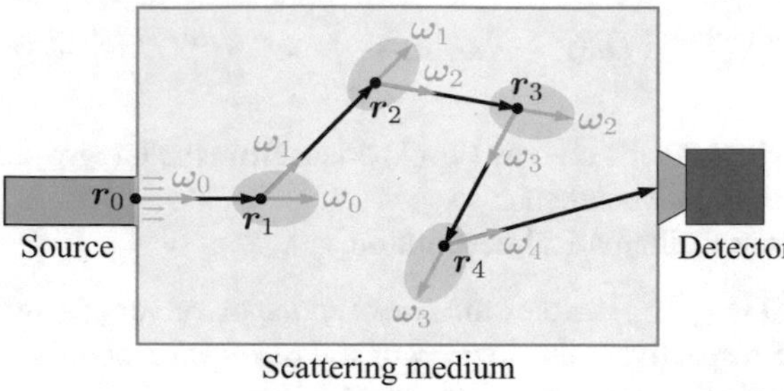

Fig. 1 Illustration of a photon biography comprising four collision points at $\boldsymbol{r}_1, \boldsymbol{r}_2, \boldsymbol{r}_3, \boldsymbol{r}_4$. When entering a collision $\boldsymbol{r}_i$, the direction $\boldsymbol{\omega}_{i-1}$ of the photon changes to $\boldsymbol{\omega}_i$ according to the single-scattering phase function at $\boldsymbol{r}_i$ (indicated as green ellipses)

The **stochastic/probabilistic RTE model** used to characterize the Monte Carlo solution of this system consists of a probability measure space $\mathbb{B}$, a set of measurable subsets $\mathcal{F}$ of $\mathbb{B}$, and a probability measure $\mathbb{M}$ on $\mathbb{B}$ together with d random variables $\Xi_i : \mathbb{B} \to \mathbb{R}$ for $i = 1, \ldots, d$, each of which describes the contribution (tally) of any photon biography $\bar{b}$ to the detector L_i. Here,

- $\mathbb{B}$ is the *sample space* of all possible photon biographies $\bar{b}$ (that are termed as *light transport paths* in computer graphics [20], as illustrated in Fig. 1);
- $\Xi_i(\bar{b})$ is the *tally/score* associated with biography $\bar{b}$ for detector i;
- $\mathbb{E}[\Xi_i]$ is the expected value of Ξ_i with respect to $\mathbb{M}$;
- V is the physical domain of the phase space Γ.

If the measure $\mathbb{M}$ on $\mathbb{B}$ is induced by the analog simulation (by launching photons according to the physical source Q, transporting them from $\boldsymbol{r}'$ to $\boldsymbol{r}$ along $\boldsymbol{\omega}$ by sampling T, absorbing them at $\boldsymbol{r}$ with probability $1 - \mu_s/\mu_t$, scattering them at $\boldsymbol{r}$ with probability μ_s/μ_t and changing their direction from $\boldsymbol{\omega}'$ to $\boldsymbol{\omega}$ by sampling f), it is the case that for $i = 1, 2, \ldots, d$:

$$\mathbb{E}[\Xi_i] = \int_{\mathbb{B}} \Xi_i(\bar{b})\,\mathrm{d}\mathbb{M}(\bar{b}) = \int_{\Gamma} d_i(\boldsymbol{r}, \boldsymbol{\omega}) L(\boldsymbol{r}, \boldsymbol{\omega})\,\mathrm{d}\boldsymbol{r}\,\mathrm{d}\boldsymbol{\omega} = L_i. \tag{5}$$

The equality (5) establishes that the probabilistic model (the left-hand side) and the analytic model (the right-hand side) both represent the quantities L_i being estimated.

The same equality shows that each Ξ_i is a **theoretically unbiased** estimator of L_i for every N:

$$\frac{1}{N}\sum_{j=1}^{N} \Xi_i(\bar{b}_j) \xrightarrow{N\to\infty} \int_{\mathbb{B}} \Xi_i(\bar{b})\,\mathrm{d}\mathbb{M}(\bar{b}) = L_i. \tag{6}$$

The symbol Ξ is reserved here for a random variable on the space $\mathbb{B}$ of all possible photon biographies $\bar{b}$ (i.e., the sample space), N is the total number of photons released from the source, and $\mathbb{M}$ denotes the probability measure induced on $\mathbb{B}$ by the process associated with generating $\bar{b} \in \mathbb{B}$. That is, the measure $\mathbb{M}$ is constructed from the choice of probability density functions used to launch, transport, scatter and absorb each biography $\bar{b}$, whether these are analog or not.

3 Our Goal and Current Status

Our previous research (see [8–11, 14]) on adaptive Monte Carlo algorithms for radiative transport problems resulted in the development of several geometrically convergent algorithms for global transport solutions L. By geometric convergence, we mean

$$E_s < \lambda E_{s-1} < \lambda^s E_0, \qquad (0 < \lambda < 1),$$

where s is the stage number and E_s is the sth stage error; e.g.,

$$E_s = \left\| L(\boldsymbol{P}) - \tilde{L}^s(\boldsymbol{P}) \right\|_\infty,$$

and $\tilde{L}^s(\boldsymbol{P})$ is an approximation obtained in the sth stage to $L(\boldsymbol{P})$, the solution of the radiative transport equation (RTE). The geometric convergence means that the *rate of convergence* of the approximate solution $\tilde{L}^s(\boldsymbol{P})$ to the solution $L(\boldsymbol{P})$ is exponentially greater than the central limit theorem-constrained rate of non-adaptive methods. However, taking into account *both* variance *and* time, our true goal for adaptive methods is to exponentially increase the computational efficiency

$$\text{Eff} := \frac{1}{\text{Var} \times T},$$

when compared with non-adaptive Monte Carlo, where Var is the estimator variance, and T denotes total computation time.

We have demonstrated geometric convergence using both correlated sampling and importance sampling as the stage-to-stage variance reduction mechanisms. Our algorithms, as well as others developed at Los Alamos [1–3], also achieve geometric convergence but each faces implementation challenges and limitations. For example, for Sequential Correlated Sampling (SCS), the evaluation of the residual (i.e., the RTE equation error) and its use in generating a distributed source for each new adaptive stage creates unavoidable new sources of approximation errors. However, SCS is fast and very robust because each adaptive stage produces a correction to the estimate of the solution obtained from all of the previous stages. For Adaptive Importance Sampling (AIS), there is both a cost and loss of precision involved in sampling from the complex importance-modified expressions that result from altering the kernel K at each adaptive stage. On the plus side, AIS is very powerful and seems to produce the most rapid error reduction per adaptive stage of those adaptive methods we know.

Table 1 Comparison of the convergence characteristics of GWAS and AIS when each is used to estimate the solution of a 1D bidirectional RTE for which an exact solution is known

Method	S	Est.	$\|R\|$	σ^2	t	Rel. Eff
Exact	–	0.8964537768861041	–	–	–	∞
GWAS	60	0.8964537768857454	5.95×10^{-13}	4.29×10^{-21}	203,940	1.142×10^{15}
AIS	20	0.8964537768868207	5.36×10^{-12}	6.08×10^{-19}	743,100	2.212×10^{12}

In [19] we introduced a new adaptive Monte Carlo method—Generalized Weighted Analog Sampling (GWAS)—for the solution of RTEs. The idea behind GWAS is to combine the power of importance sampling with strategies that loosen the restrictions associated with sampling from importance-modified transport kernels. In this way, we hoped to combine rapid error reduction with fast algorithm execution in order to exponentially increase the computational efficiency. The price we pay for the flexibility of GWAS is that it biased. The fact that GWAS is biased (though asymptotically unbiased) greatly complicates the proof that it produces geometrically convergent estimates of RTE solutions. However, we have recently proved a new theorem that establishes that geometric convergence does obtain for GWAS [13]. As well, GWAS is able to provide increased computational efficiency compared with AIS, as we showed in [13]. The complete numerical results are provided in this recent publication, but here we repeat the table that summarizes this behavior.

We note that GWAS has a much higher computational efficiency than AIS because of its speed of execution. Note, too, that even though the variance of GWAS is more than 100 times as large as that of AIS, the efficiency of GWAS is more than 5,000 times that of AIS.

Each of these three adaptive methods:

1. generates biographies in stages, each of which consists of the same number of biographies;
2. applies variance reduction (correlated sampling, importance sampling, and GWAS) in each stage, linking stage s output to stage $(s + 1)$ input in an intrinsic way;
3. makes use of an analytic representation of the radiance.

Detailed examination of the behavior of these three adaptive algorithms reveals that *the need to represent the RTE solution by means of a* ***formula*** *introduces* ***bias*** *in its adaptive estimates*. This, in turn, prevents each algorithm from achieving unlimited precision as the number of adaptive stages tends to infinity. Thus, even though we don't *need* unlimited precision in order to make our adaptive algorithms useful, introducing an unknown amount of bias in our estimates falls short of our goal to create a new gold standard simulation tool for adoption by the biomedical community. In fact, the central obstacle to creating a real-time transport-rigorous Monte Carlo simulator is estimator bias (Table 1).

4 The Role of Bias in Estimating Radiance

The **theoretical bias** of an estimator $\Xi(\bar{b})$ of a linear functional

$$I = \int_{V\times\mathbb{S}^2} d(\boldsymbol{r}, \boldsymbol{\omega})\, L(\boldsymbol{r}, \boldsymbol{\omega})\, \mathrm{d}\boldsymbol{r}\, \mathrm{d}\boldsymbol{\omega},$$

of the radiance $L(\boldsymbol{r}, \boldsymbol{\omega})$ defined on the sample space $\mathbb{B}$ is

$$\mathbb{E}[\Xi(\bar{b})] - I = \int_{\mathbb{B}} \Xi(\bar{b})\, \mathrm{d}\mathbb{M}(\bar{b}) - I.$$

Theoretical bias introduces a component of *systematic* error in the mean integrated square error:

$$\begin{aligned}
\text{MISE} &= \mathbb{E}\left[\int_{\mathbb{B}} (\Xi(\bar{b}) - I)^2\, \mathrm{d}\mathbb{M}(\bar{b})\right] \\
&= \int_{\mathbb{B}} \left(\mathbb{E}[\Xi(\bar{b})] - I + \Xi(\bar{b}) - \mathbb{E}[\Xi(\bar{b})]\right)^2 \mathrm{d}\mathbb{M}(\bar{b}) \qquad (7) \\
&= \int_{\mathbb{B}} \text{BIAS}^2[\Xi(\bar{b})]\, \mathrm{d}\mathbb{M}(\bar{b}) + \int_{\mathbb{B}} \text{Var}[\Xi(\bar{b})]\, \mathrm{d}\mathbb{M}(\bar{b}).
\end{aligned}$$

Here $\mathbb{E}[]$ is the expected value operator and $\bar{b} \in \mathbb{B}$ is a photon biography (i.e., a sample). In contrast with theoretical bias, we will say that **computational bias** results from the accumulation of small errors due to the computer's limited precision. Computational bias is unavoidable in most cases. However, with sufficient care, computational sources of error can be controlled and estimated, whereas the source of error from theoretical bias is largely unknown and therefore much more difficult to estimate and to control.

The first term of the right hand side of (7) is the integral of the squared bias, while the second term is the integrated variance. Thus, for biased estimators it is necessary to control *both* the bias *and* the variance to exhibit geometric convergence.

Our approach to the avoidance of biased RTE estimators was to see whether the biased estimators often used in the graphics community could be improved or modified sufficiently to serve as the engine of a "gold standard" RTE solver. One of the conventional tools used for achieving realistic-looking scenes rapidly is **kernel density estimation** [17].

Kernel density estimation is a non-parametric method (i.e., no assumptions are made about the unknown underlying pdf) for recovering an unknown probability density function $f(x)$ by drawing samples $x_1, x_2, \ldots, x_n$ that are distributed according to $f(x)$. The kernel estimator with kernel k (satisfying $\int_{-\infty}^{\infty} k(x)\, \mathrm{d}x = 1$) produces the estimate $\tilde{f}(x)$ of the pdf $f(x)$ according to

$$\tilde{f}(x) = \frac{1}{nh} \sum_{i=1}^{n} k\left(\frac{x - X_i}{h}\right). \tag{8}$$

where $X_1, \ldots, X_n$ are samples drawn independently from $f(x)$ and h is the window width (or smoothing parameter) that controls the influence of the kernel k near each sample point X_i. The kernel k can be chosen in a variety of ways: for example, as a standard Gaussian density

$$k_G(x) = \frac{1}{\sqrt{2\pi}} e^{-x^2/2},$$

if x ranges over the entire real line, or as the Epanechnikov density

$$k_E(x) = \begin{cases} 6\left(\frac{1}{4} - x^2\right) & x^2 < \frac{1}{4}; \\ 0 & \text{otherwise}, \end{cases}$$

or in various other ways. Kernel Density Estimation is consistent: i.e., $\tilde{f}_h(x)$ converges to $f(x)$ as the number of samples n increases without limit, provided that the smoothing parameter h tends to 0 in such a way that the product hn tends to ∞. This last requirement means roughly that there are sufficiently many samples within the support sets of the kernel as the smoothing parameter is reduced.

Instead of applying conventional (unconstrained) density estimation to the RTE, we took the approach of constraining our method; i.e., to relate the RTE solution expansion directly to the random walks actually generated, treating these as the "samples" of the Monte Carlo simulation. Indeed, the sample space $\mathbb{B}$ is defined in this way [18], and the new Transport-Constrained Unbiased Radiance Estimator (T-CURE) does exactly this: it describes the expected contribution to the RTE solution at *any* point of phase space from *each* collision point of *every* photon biography. This creates an *unbiased representation* of the global RTE solution for all sample sizes that requires no smoothing parameters nor any special treatment of boundaries.

In the following section we will show how the T-CURE estimator can be derived as an extension of the conventional collision estimator [18].

4.1 T-CURE

We return to the integral equation characterized by Eq. (1), together with Eqs. (2), (3) and (4). The scattering integrals appearing on the right hand side of Eq. (1) are functions defined on the problem phase space Γ that are closely related to the RTE solution itself. We establish an unbiased estimator for those functions (hence, the RTE solutions) that extends the "conventional" collision estimators [18] to produce estimates of the entire RTE solution.

First, however, we consider the problem of computing the inner product of L, the solution of Eq. (1), and some $S^* : \Gamma \to \mathbb{R}$:

$$I = \langle S^*, L \rangle := \int_\Gamma S^*(\boldsymbol{r}, \boldsymbol{\omega}) L(\boldsymbol{r}, \boldsymbol{\omega}) \, \mathrm{d}\boldsymbol{\omega} \, \mathrm{d}\boldsymbol{r}. \tag{9}$$

Instead of estimating Eq. (9) directly, one can equally well solve its dual problem which leads to the same answer I:

$$I = \langle S, L^* \rangle := \int_\Gamma S(\boldsymbol{r}, \boldsymbol{\omega}) L^*(\boldsymbol{r}, \boldsymbol{\omega}) \, \mathrm{d}\boldsymbol{\omega} \, \mathrm{d}\boldsymbol{r}, \tag{10}$$

where L^* is the solution to the adjoint integral equation:

$$L^*(\boldsymbol{P}) = \int_\Gamma K^*(\boldsymbol{P} \to \boldsymbol{P}') \, L^*(\boldsymbol{P}') \, \mathrm{d}\rho \, \mathrm{d}\boldsymbol{\omega}' + S^*(\boldsymbol{P}), \tag{11}$$

with K^* being the *adjoint* of K satisfying $K^*(\boldsymbol{P} \to \boldsymbol{P}') = K(\boldsymbol{P}' \to \boldsymbol{P})$.

The conventional collision estimators for Eqs. (9) and (10) are

$$\eta(\bar{b}) := \frac{S(\boldsymbol{P}_1)}{p_1(\boldsymbol{P}_1)} \sum_{i=1}^{k} S^*(\boldsymbol{P}_i) \quad \text{and} \quad \eta^*(\bar{b}) := \frac{S^*(\boldsymbol{P}_1)}{p_1^*(\boldsymbol{P}_1)} \sum_{i=1}^{k} S(\boldsymbol{P}_i), \tag{12}$$

where $\bar{b} := (\boldsymbol{P}_1, \boldsymbol{P}_2, \ldots, \boldsymbol{P}_k)$ is a biography consisting of collision points $\boldsymbol{P}_i = (\boldsymbol{r}_i, \boldsymbol{\omega}_i)$ and k is the number of collisions made by $\bar{b}$ in the interior of the physical domain V of Γ. For the estimator η in Eq. (12), $\bar{b}$ is created using the following random walk process. The first collision $\boldsymbol{P}_1$ is drawn from a pre-defined density p_1 (which is normally selected to be proportional to S). At each collision $\boldsymbol{P}_i$, the next state, which can either be the next collision $\boldsymbol{P}_{i+1}$ or the termination of the random walk (leading to $k = i$), is determined using K^*. Similarly, when η^* is used, the photon biography $\bar{b}$ should be generated in the adjoint manner by sampling $\boldsymbol{P}_1$ based on S^* and the next state at collision $\boldsymbol{P}_i$ using K.

These conventional collision estimators (12) can be highly inefficient when S or S^* are concentrated in small subsets of Γ and vanishes everywhere else. This, unfortunately, is usually the case in biomedicine since many applications involve optical sources and/or detectors with small physical sizes (e.g., lasers and optical fibers). In particular, when the support of S^* is small, $S^*(\boldsymbol{P}_i)$ will be zero with high probability, making the estimator η inefficient. On the other hand, when S vanishes almost everywhere in the phase space, η will offer limited efficiency. Further, S and S^* can contain delta functions, making direct evaluations of $S^*(\boldsymbol{P}_i)$ in η and $S(\boldsymbol{P}_i)$ in η^* problematic.

To ease this problem, we introduce T-CURE estimators as an extension of the conventional collision estimators (12). For notational convenience, define operators $\mathscr{K}$ and $\mathscr{K}^*$ as functionals on $h : \Gamma \to \mathbb{R}$ as

$$(\mathscr{K}h)(\boldsymbol{P}) := \int_{\Gamma} K(\boldsymbol{P}' \to \boldsymbol{P})\, h(\boldsymbol{P}')\, \mathrm{d}\boldsymbol{\omega}'\, \mathrm{d}\rho, \tag{13}$$

$$(\mathscr{K}^*h)(\boldsymbol{P}) := \int_{\Gamma} K^*(\boldsymbol{P} \to \boldsymbol{P}')\, h(\boldsymbol{P}')\, \mathrm{d}\rho\, \mathrm{d}\boldsymbol{\omega}', \tag{14}$$

where the use of ρ is the same as in Eqs. (1)–(4). Then, Eqs. (1) and (11) respectively simplify to

$$L = \mathscr{K}L + S, \tag{15}$$

$$L^* = \mathscr{K}^*L^* + S^*. \tag{16}$$

It follows that

$$\langle S^*, L\rangle = \langle S^*,\ \mathscr{K}L + S\rangle = \langle S^*, H\rangle + \langle S^*, S\rangle, \tag{17}$$

$$\langle S, L^*\rangle = \langle S,\ \mathscr{K}^*L^* + S^*\rangle = \langle S, H^*\rangle + \langle S, S^*\rangle, \tag{18}$$

with $H := \mathscr{K}L$ and $H^* := \mathscr{K}^*L^*$. In Eqs. (17) and (18), $\langle S^*, S\rangle$ contains only known quantities and can be easily evaluated. Thus, we now focus on estimating the remaining terms $\langle S^*, H\rangle$ and $\langle S, H^*\rangle$.

By respectively applying $\mathscr{K}$ and $\mathscr{K}^*$ to both sides of Eqs. (15) and (16), we have

$$H = \mathscr{K}H + \mathscr{K}S, \tag{19}$$

$$H^* = \mathscr{K}^*H^* + \mathscr{K}^*S^*. \tag{20}$$

Notice that Eqs. (19) and (20) differ from Eqs. (15) and (16) only by the source terms. Therefore, $\langle S^*, H\rangle$ and $\langle S, H^*\rangle$ can be estimated using Eq. (12) with updated source terms, yielding

$$\begin{aligned} \eta_{\mathrm{NE}}(\bar{b}) &:= \frac{S(\boldsymbol{P}_1)}{p_1(\boldsymbol{P}_1)} \sum_{i=1}^{k} (\mathscr{K}^*S^*)(\boldsymbol{P}_i), \\ \eta^*_{\mathrm{NE}}(\bar{b}) &:= \frac{S^*(\boldsymbol{P}_1)}{p^*_1(\boldsymbol{P}_1)} \sum_{i=1}^{k} (\mathscr{K}S)(\boldsymbol{P}_i). \end{aligned} \tag{21}$$

Thanks to the integrals involved in $\mathscr{K}$ and $\mathscr{K}^*$, as defined in Eqs. (13) and (14), $\mathscr{K}S$ and $\mathscr{K}^*S^*$ generally have much greater supports than S and S^*, making our T-CURE estimators (21) significantly more effective than the conventional ones (12) when S and S^* vanish almost everywhere in the phase space.

Since both K and S (as well as K^* and S^*) are known, we can, in principle, evaluate these new extended next-event estimators (21) exactly. In particular, given Eqs. (13) and (14), it holds that

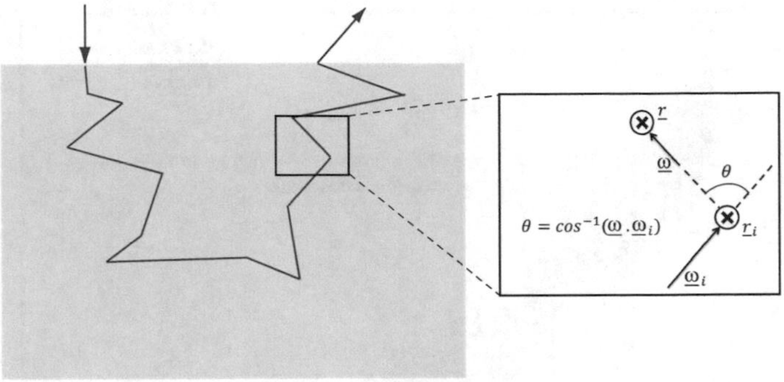

Fig. 2 Illustration of T-CURE Mechanism: Photon biography $\bar{b}$ moving in direction ω_i collides at $\boldsymbol{r}_i$ in the blow-up at the right. Our estimator η_{NE} from Eq. (21) involves evaluating $\mathscr{K}^* S^*$ at each collision $(\boldsymbol{r}_i, \boldsymbol{\omega}_i)$. We estimate $(\mathscr{K}^* S^*)(\boldsymbol{r}_i, \boldsymbol{\omega}_i)$ by selecting some $\boldsymbol{\omega} \in \mathbb{S}^2$ and $\boldsymbol{r} \in V$ based on $K^*(\boldsymbol{r}_i, \boldsymbol{\omega}_i \to \boldsymbol{r}, \boldsymbol{\omega})\, S^*(\boldsymbol{r}_i, \boldsymbol{\omega}_i)$ but independent of $(\boldsymbol{r}_{i+1}, \boldsymbol{\omega}_{i+1})$

$$(\mathscr{K}^* S^*)(\boldsymbol{r}_i, \boldsymbol{\omega}_i) := \int_{\Gamma} K^*(\boldsymbol{r}_i, \boldsymbol{\omega}_i \to \boldsymbol{r}_i + \rho\boldsymbol{\omega}, \boldsymbol{\omega})\, S^*(\boldsymbol{r}, \boldsymbol{\omega})\, \mathrm{d}\rho\, \mathrm{d}\boldsymbol{\omega}, \tag{22}$$

$$(\mathscr{K} S)(\boldsymbol{r}_i, \boldsymbol{\omega}_i) := \int_{\Gamma} K(\boldsymbol{r}_i - \rho\boldsymbol{\omega}_i, \boldsymbol{\omega} \to \boldsymbol{r}_i, \boldsymbol{\omega}_i)\, S(\boldsymbol{r}, \boldsymbol{\omega})\, \mathrm{d}\boldsymbol{\omega}\, \mathrm{d}\rho. \tag{23}$$

In practice, the integrals in Eqs. (22) and (23) can be evaluated analytically or numerically depending on the exact forms of S and S^*. Lastly, the evaluation of the T-CURE tallies from every collision at $(\boldsymbol{r}_i, \boldsymbol{\omega}_i)$ can be accomplished either on-the-fly or by post-processing all of the biography collision points saved from the generation of the photon biographies. This is because the T-CURE tally at $(\boldsymbol{r}_i, \boldsymbol{\omega}_i)$ is independent of the process that generates the next collision at $(\boldsymbol{r}_{i+1}, \boldsymbol{\omega}_{i+1})$. See Fig. 2 for an illustration of this process.

Besides offering superior computational efficiency, our new T-CURE estimators also enjoy the following properties:

- They require the imposition of no mesh on the phase space, so they provide the basis for plotting or otherwise displaying features of the RTE solution over any desired mesh, or of several such, based on a single set of biographies.
- They can be implemented either with "on-the-fly" computation or, after generating a "smallish" number N_0 of biographies, by post-processing key data stored from the N_0 "baseline" set of biographies, or by a combination of these two methods.

The power of T-CURE estimation derives from replacing occasional contributions to reflectance from biographies that are *actually* detected (i.e., the terminal estimate of reflectance) by sums of analytic formulas over *all* the collisions of biographies, whether or not they ultimately reach the detector. Not surprisingly, this replacement often (though not always!) achieves variance reduction in estimates of reflectance.

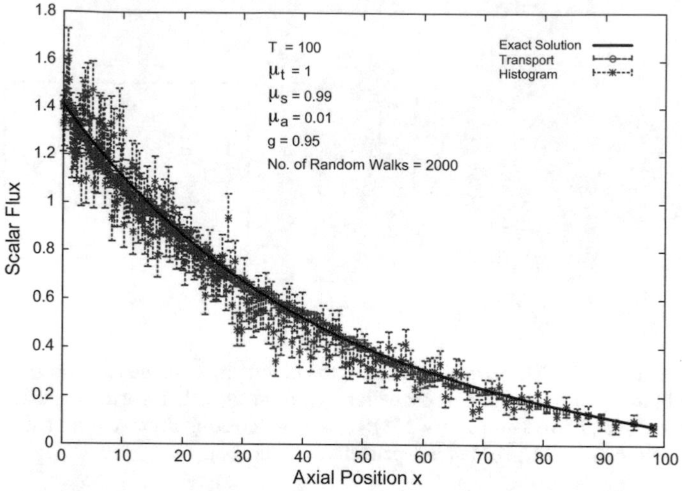

Fig. 3 Comparison of transport-constrained density estimation with both histogram and Epanechnikov density estimators. The exact solution of this 1D problem is known and constitutes the black curve, while the two more conventional density estimators are indicated in blue and green. Estimated standard deviations are shown as error bars

Previously, we have shown that T-CURE is unbiased for all sample sizes, in sharp contrast with the conventional (unconstrained) density estimators [12]. As well, T-CURE is roughly an order of magnitude more accurate than the conventional ones found in the statistical literature [6, 15, 16]. Figure 3 (from our prior work [12]) illustrates these gains in a 1-D model RTE problem that plots the (scalar) intensity of the light field against distance, x, from the light source.

5 Recent Numerical Studies: T-CURE in Multi-dimensions

The comparison of T-CURE performance with that of the histogram and the Epanechnikov density estimators was motivated by the desire to compare its computational efficiency as a constrained "density estimator" with that of the more conventional unconstrained density estimators that are widely used by the graphics community. The bidirectional problem is especially simple since it involves only one spatial dimension and two discrete scattering directions. In order to investigate how the T-CURE estimator behaves when applied to more challenging problems, we turned to multilayer tissue problems. For all of these numerical experiments we used input data typical of normal cervical tissue. The tissue was represented as two layers: a top epithelial layer and the stromal layer below. Optical data for this 5-D problem is shown in the table:

Layer	Optical data			Layer thickness
	μ_s	μ_a	g	
Epithelium	~80/cm [5]	0.12/cm	0.95	360 μm [21]
Stroma	150/cm [4]	0.15–1.2/cm[a]	0.88	∞

[a] Particular value for μ_a depends on whether there is Hb absorption at that wavelength

Intuitively, we do not expect the performance of T-CURE to degrade significantly as the dimension of the underlying phase space increases. This is because the overhead associated with T-CURE depends mainly on the *number of collisions, which is determined by the optical properties, not the dimension of the phase space.*

Next we illustrate the use of T-CURE to estimate spatially resolved reflectances in two tissue problems. Both experiments used the two-layer data for normal cervical tissue shown in the table above but ignore possible refractive index mismatches at the interfaces (Fig. 4).

Note especially the graininess of the terminal estimator and also how the information degrades as the source-detector distances increase. This graininess contrasts with the much greater smoothness of the T-CURE images, even at 10.0 cm from the source where the signal is quite low. Note also the scales along the y-axis of the images. Even though the magnitude of the signal falls by about two orders of magnitude as the source and detector radii shrink, the T-CURE 95% relative confidence interval sizes remain below 3.5% over the entire range of source-detector distances, while the terminal 95% confidence interval sizes grow from 15 to 100% as the s-d distances grow from 1.0 to 10.0 cm (Fig. 5).

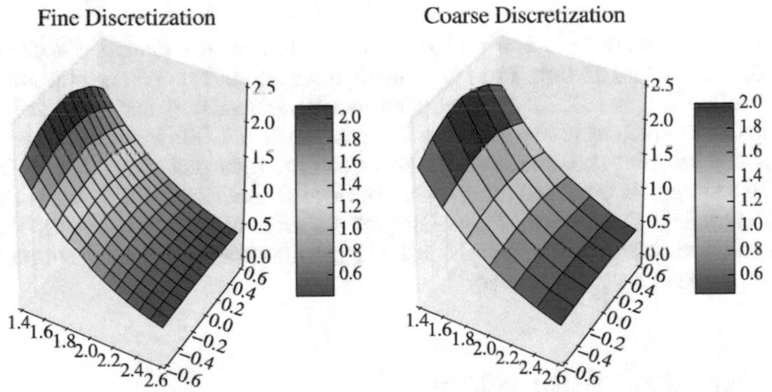

Fig. 4 5-D Cervical Tissue Problem. This example displays another advantage of T-CURE when compared to more traditional estimators: one can generate a "smallish" initial set of biographies, show the output from that, and then refine the output mesh by post-processing the initial set without generating any new biographies. For example, the plot on the right was produced by processing biographies "on the fly" (6 min), while the refined plot on the left was obtained by post-processing stored data from these biographies (few seconds)

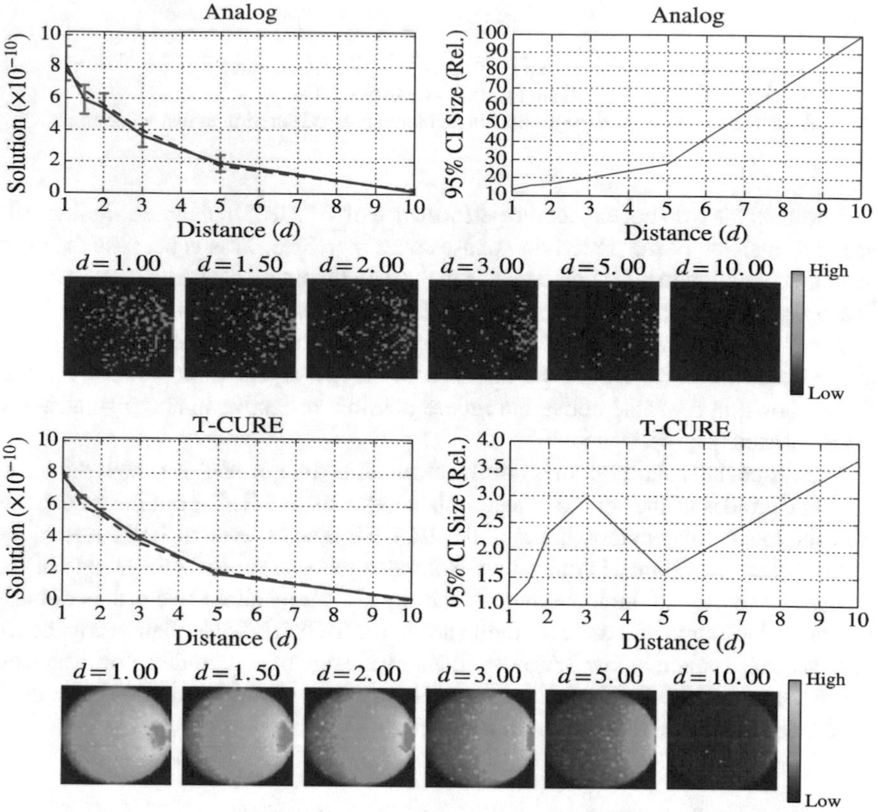

Fig. 5 In this problem, there is a single source at (0, 0, 0) and six detectors spaced at distances 1.0, 1.5, 2.0, 3.0, 5.0 and 10.0 cm from the source. Both the source and detector are discs with identically small radii $R = 0.0025$ cm. We applied both the terminal (top) and our T-CURE (bottom) estimators to this problem by letting them run for approximately 1700 s for each data point. In both experiments, the left plot shows estimation results and the right plot shows the relative size of 95% confidence intervals (obtained using twice the standard deviation of the sample data). These confidence intervals are also drawn in the left plots (as error bars). Lastly, the red dashed lines indicate the solution obtained using the other estimator, demonstrating that our analog and T-CURE estimators converge to the same answer

6 Summary and Future Work

We have advanced our earlier work [12] by examining the computational efficiency of T-CURE on multidimensional, heterogeneous (multilayer) tissue problems involving 5 independent variables: 3 for position and 2 for unit direction. We observe that the advantages of T-CURE are maintained in the higher dimensional cases. We believe that this new estimator has the potential to be the computational engine of an adaptive (geometrically convergent) algorithm.

We will turn our attention next to the fully general multilayer tissue problem, including the possibility of refractive index mismatches at the layer interfaces. This introduces new challenges that call for a strategy that controls the run time consumption caused by photon biographies that can cross refractive-index-mismatched layer interfaces in both directions a very large number of times. Provided that this degradation of the computational power of the adaptive algorithm can be controlled, the resulting T-CURE-based adaptive algorithm should serve as a new gold standard Monte Carlo solver for biomedical problems.

Acknowledgements The third author gratefully acknowledges partial support from award numbers: P41RR001192 from the National Center for Research Resources and P41EB015890 from the National Institute of Biomedical Imaging and Bioengineering.

The content of this paper is solely the responsibility of the authors and does not necessarily represent the official views of the National Center For Research Resources, National Institute of Biomedical Imaging and Bioengineering, or the National Institutes of Health.

7 Appendix: Transport Equations

The time independent transport equation can be written as

$$(\boldsymbol{\omega}\cdot\nabla)L+\mu_t(\boldsymbol{r})L(\boldsymbol{r},\boldsymbol{\omega})=\int_{\mathbb{S}^2}\mu_s(\boldsymbol{r})f(\boldsymbol{r}';\ \boldsymbol{\omega}'\to\boldsymbol{\omega})\,L(\boldsymbol{r},\boldsymbol{\omega}')\,\mathrm{d}\boldsymbol{\omega}'+Q(\boldsymbol{r},\boldsymbol{\omega}),\tag{24}$$

for all $r\in V\subseteq\mathbb{R}^3$ and

$$L(\boldsymbol{r},\boldsymbol{\omega})=Q_0(\boldsymbol{r},\boldsymbol{\omega}),\tag{25}$$

for all $\boldsymbol{r}\in\partial V$ and $\boldsymbol{\omega}\in\mathbb{S}^2$ satisfying $\boldsymbol{\omega}\cdot\boldsymbol{n}(\boldsymbol{r})<0$ as the boundary condition. In Eqs. (24) and (25), $\boldsymbol{r}:=(x,y,z)$ and $\boldsymbol{n}(\boldsymbol{r})$ denotes the outward unit normal on the boundary ∂V at $\boldsymbol{r}$. Note that

$$\boldsymbol{\omega}=\omega_x\boldsymbol{i}+\omega_y\boldsymbol{j}+\omega_z\boldsymbol{k}=\sin\theta\cos\phi\,\boldsymbol{i}+\sin\theta\sin\phi\,\boldsymbol{j}+\cos\theta\,\boldsymbol{k},\tag{26}$$

$$\boldsymbol{\omega}\cdot\nabla=\sin\theta\cos\phi\frac{\partial}{\partial x}+\sin\theta\sin\phi\frac{\partial}{\partial y}+\cos\theta\frac{\partial}{\partial z},\tag{27}$$

we now convert Eq. (24) to an integral equation using the method of characteristics. Consider the following characteristic system for Eq. (24):

$$\frac{\mathrm{d}x}{\mathrm{d}\rho}=-\omega_x,\qquad\frac{\mathrm{d}y}{\mathrm{d}\rho}=-\omega_y,\qquad\frac{\mathrm{d}z}{\mathrm{d}\rho}=-\omega_z.\tag{28}$$

Solving Eq. (28) produces

$$x = x_0 - \omega_x \rho, \qquad y = y_0 - \omega_y \rho, \qquad z = z_0 - \omega_z \rho. \tag{29}$$

We can then write Eq. (24) in the following form:

$$-\frac{\mathrm{d}L(\boldsymbol{r}', \boldsymbol{\omega})}{\mathrm{d}\rho} + \mu_t(\boldsymbol{r}')L(\boldsymbol{r}', \boldsymbol{\omega}) = \int_{\mathbb{S}^2} \mu_s(\boldsymbol{r}')\, f(\boldsymbol{r}';\ \boldsymbol{\omega}' \to \boldsymbol{\omega})\, L(\boldsymbol{r}',\ \boldsymbol{\omega}')\, \mathrm{d}\omega' + Q(\boldsymbol{r}', \boldsymbol{\omega}), \tag{30}$$

where $\boldsymbol{r}' := \boldsymbol{r}_0 - \rho\boldsymbol{\omega}$, which can in turn be rewritten as:

$$\begin{aligned} &-\frac{\mathrm{d}}{\mathrm{d}\rho}\left[\mathrm{e}^{-\int_0^\rho \mu_t(\boldsymbol{r}_0-\tau\boldsymbol{\omega})\,\mathrm{d}\tau} L(\boldsymbol{r}',\ \boldsymbol{\omega})\right] \\ &= \mathrm{e}^{-\int_0^\rho \mu_t(\boldsymbol{r}_0-\tau\boldsymbol{\omega})\,\mathrm{d}\tau}\left[\int_{\mathbb{S}^2} \mu_s(\boldsymbol{r}')f(\boldsymbol{r}';\ \boldsymbol{\omega}' \to \boldsymbol{\omega})L(\boldsymbol{r}',\ \boldsymbol{\omega}')\,\mathrm{d}\omega' + Q(\boldsymbol{r}', \boldsymbol{\omega})\right], \end{aligned} \tag{31}$$

We now integrate the two sides of Eq. (30) with respect to ρ from $\rho = 0$ to $\rho = R$ and replace $\boldsymbol{r}_0$ with $\boldsymbol{r}$ to produce:

$$\begin{aligned} &L(\boldsymbol{r}, \boldsymbol{\omega}) \\ &= \int_0^R \mathrm{e}^{-\int_0^\rho \mu_t(\boldsymbol{r}-\tau\boldsymbol{\omega})\,\mathrm{d}\tau}\left[\mu_s(\boldsymbol{r}')\int_{\mathbb{S}^2} f(\boldsymbol{r}';\ \boldsymbol{\omega}' \to \boldsymbol{\omega})\, L(\boldsymbol{r}', \boldsymbol{\omega}')\mathrm{d}\omega' + Q(\boldsymbol{r}', \boldsymbol{\omega})\right]\mathrm{d}\rho \\ &+ \mathrm{e}^{-\int_0^R \mu_t(\boldsymbol{r}-\tau\boldsymbol{\omega})\,\mathrm{d}\tau}\, Q_0(\boldsymbol{r} - R\boldsymbol{\omega},\ \boldsymbol{\omega}). \end{aligned} \tag{32}$$

or

$$L(\boldsymbol{P}) = \int_\Gamma K(\boldsymbol{P}' \to \boldsymbol{P})\, L(\boldsymbol{P}')\,\mathrm{d}\omega'\,\mathrm{d}\rho + S(\boldsymbol{P}), \tag{33}$$

where $\boldsymbol{P} := (\boldsymbol{r}, \boldsymbol{\omega})$, $\boldsymbol{P}' := (\boldsymbol{r}', \boldsymbol{\omega}')$,

$$\begin{aligned} K(\boldsymbol{P}' \to \boldsymbol{P}) &:= \frac{\mu_s(\boldsymbol{r}')}{\mu_t(\boldsymbol{r}')}\, f(\boldsymbol{r}'; \boldsymbol{\omega}' \to \boldsymbol{\omega})\, T(\boldsymbol{r}' \to \boldsymbol{r};\ \boldsymbol{\omega}), \\ T(\boldsymbol{r}' \to \boldsymbol{r};\ \boldsymbol{\omega}) &:= \mu_t(\boldsymbol{r}')\, \exp\left(-\int_0^{\|\boldsymbol{r}-\boldsymbol{r}'\|} \mu_t(\boldsymbol{r} - \tau\boldsymbol{\omega})\,\mathrm{d}\tau\right), \\ S(\boldsymbol{P}) &:= \mathrm{e}^{-\int_0^R \mu_t(\boldsymbol{r}-\tau\boldsymbol{\omega})\,\mathrm{d}\tau}\, Q_0(\boldsymbol{r} - R\boldsymbol{\omega},\ \boldsymbol{\omega}) \\ &\quad + \int_0^R \mathrm{e}^{-\int_0^\rho \mu_t(\boldsymbol{r}-\tau\boldsymbol{\omega})\,\mathrm{d}\tau}\, Q(\boldsymbol{r} - \rho\boldsymbol{\omega},\ \boldsymbol{\omega})\mathrm{d}\rho. \end{aligned}$$

Eqs. (32) and (33) are the integral forms of the RTE that we seek.

References

1. Baggerly, K., Cox, D., Picard, R.: Exponential convergence of adaptive importance sampling for Markov chains. J. Appl. Probab. **37**(2), 342–358 (2000)
2. Booth, T.: Exponential convergence on a continuous Monte Carlo transport problem. Nucl. Sci. Eng. **127**(3), 338–345 (1997)
3. Booth, T.: Adaptive importance sampling with a rapidly varying importance function. Nucl. Sci. Eng. **136**(3), 399–408 (2000)
4. Chang, V.T.C., Cartwright, P.S., Bean, S.M., Palmer, G.M., Bentley, R.C., Ramanujam, N.: Quantitative physiology of the precancerous cervix in vivo through optical spectroscopy. Neoplasia **11**(4), 325–332 (2009)
5. Collier, T., Arifler, D., Malpica, A., Follen, M., Richards-Kortum, R.: Determination of epithelial tissue scattering coefficient using confocal microscopy. IEEE J. Sel. Top. Quantum Electron. **9**(2), 307–313 (2003)
6. Devroye, L.: A Course in Density Estimation. Progress in Probability. Birkhauser, Boston (1987)
7. Jensen, H.W.: Realistic Image Synthesis Using Photon Mapping. AK Peters Ltd, Wellesley (2001)
8. Kong, R.: Transport problems and Monte Carlo methods. Ph.D. thesis, Claremont Graduate University (1999)
9. Kong, R., Ambrose, M., Spanier, J.: Efficient, automated Monte Carlo methods for radiation transport. J. Comput. Phys. **227**(22), 9463–9476 (2008)
10. Kong, R., Spanier, J.: A new proof of geometric convergence for general transport problems based on sequential correlated sampling methods. J. Comput. Phys. **227**(23), 9762–9777 (2008)
11. Kong, R., Spanier, J.: Geometric convergence adaptive Monte Carlo algorithms for radiative transport problems based on importance sampling methods. Nucl. Sci. Eng. **168**(3), 197–225 (2011)
12. Kong, R., Spanier, J.: Transport-constrained extensions of collision and track length estimators for solutions of radiative transport problems. J. Comput. Phys. **242**(0), 682–695 (2013). https://doi.org/10.1016/j.jcp.2013.02.023, http://www.sciencedirect.com/science/article/pii/S0021999113001423
13. Kong, R., Spanier, J.: A new proof of geometric convergence for the adaptive generalized weighted analog sampling (GWAS) method. Monte Carlo Methods Appl. **22**(3), 161–196 (2016)
14. Lai, Y., Spanier, J.: Adaptive importance sampling algorithms for transport problems. Monte Carlo and Quasi-MonteCarlo Methods 1998, pp. 273–283. Springer, Berlin (1999)
15. Rao, C.R.: Linear Statistical Inference and Its Applications. Wiley, New York (1973)
16. Scott, D.: Multivariate Density Estimation: Theory. Practice and Visualization. Wiley Series in Probability and Mathematical Statistics. Wiley, New York (1990)
17. Silverman, B.: Density Estimation for Statistics and Data Analysis. Chapman and Hall, London (1986)
18. Spanier, J., Gelbard, E.: Monte Carlo Principles and Neutron Transport Problems. Wesley, New York (1969). (reprinted by Dover Publications, Inc. 2008)
19. Spanier, J., Kong, R.: A new adaptive method for geometric convergence. Monte Carlo and Quasi-MonteCarlo Methods 2002, pp. 439–449. Springer, Berlin (2004)
20. Veach, E.: Robust Monte Carlo methods for light transport simulation. Ph.D. thesis, Stanford, CA, USA (1998). AAI9837162
21. Walker, D., Brown, B., Blackett, A., Tidy, J., Smallwood, R.: A study of the morphological parameters of cervical squamous epithelium. Physiol. Meas. **24**(1), 121 (2003)
22. Wilson, B.C., Adam, G.: A Monte Carlo model for the absorption and flux distributions of light in tissue. Med. Phys. **10**(6), 824–830 (1983)

Rates of Convergence and CLTs for Subcanonical Debiased MLMC

Zeyu Zheng, Jose Blanchet and Peter W. Glynn

Abstract In constructing debiased multi-level Monte Carlo (MLMC) estimators, one must choose a randomization distribution. In some algorithmic contexts, an optimal choice for the randomization distribution leads to a setting in which the mean time to generate an unbiased observation is infinite. This paper extends the well known efficiency theory for Monte Carlo algorithms in the setting of a finite mean for this generation time to the infinite mean case. The theory draws upon stable law weak convergence results, and leads directly to exact convergence rates and central limit theorems (CLTs) for various debiased MLMC algorithms, most particularly as they arise in the context of stochastic differential equations. Our CLT theory also allows simulators to construct asymptotically valid confidence intervals for such infinite mean MLMC algorithms.

Keywords Monte Carlo estimator efficiency · Central limit theorems
Subcanonical convergence rates · Infinite mean generation time · Infinite variance

1 Introduction

In comparing Monte Carlo algorithms, a key result in the literature concerns the efficiency trade-off between the variance of an estimator, and the computer time required to compute that estimator. In particular, suppose that a quantity $z = E(X)$ is to be computed. The associated Monte Carlo estimator is constructed by generating independent and identically distributed (iid) copies $X_1, X_2, \ldots$ of X; the computer time required to generate X_i is given by τ_i, a positive random variable (rv). The

Z. Zheng (✉) · J. Blanchet · P. W. Glynn
Stanford University, Stanford, CA 94305, USA
e-mail: zyzheng@stanford.edu

J. Blanchet
e-mail: jblanchet@stanford.edu

P. W. Glynn
e-mail: glynn@stanford.edu

A. B. Owen and P. W. Glynn (eds.), *Monte Carlo and Quasi-Monte Carlo Methods*, Springer Proceedings in Mathematics & Statistics 241,
https://doi.org/10.1007/978-3-319-91436-7_26

(X_i, τ_i) pairs are then iid in i, where X_i and τ_i are generally correlated. Given a computer time budget c, let $N(c)$ be the number of X_i's generated in c units of computer time, so that $N(c) = \max\{n \geq 0 : \tau_1 + \cdots + \tau_n \leq c\}$. The estimator for z that is available with computational budget c is then $\bar{X}_{N(c)}$, where $\bar{X}_n = n^{-1}(X_1 + \cdots + X_n)$. It is well known that when $E(\tau_1) < \infty$ and $\mathrm{Var}(X_1) < \infty$, the central limit theorem (CLT)

$$c^{1/2}\left(\bar{X}_{N(c)} - z\right) \Rightarrow \sqrt{E(\tau_1) \cdot \mathrm{Var}(X_1)}\, N(0, 1) \tag{1}$$

holds as $c \to \infty$, where $\Rightarrow$ denotes weak convergence, and $N(0, 1)$ denotes a normal rv with mean 0 and variance 1. With (1) in hand, one can now compare the efficiency of different algorithms (as associated with two rv's X and Y for which $E(X) = z = E(Y)$) for a given (large) computational budget c. The result (1) is discussed in [7], but is worked out in much greater detail in [5]. In the latter reference, the theory focuses on settings in which $E(\tau_1) < \infty$; an extension to $\mathrm{Var}(X_1) = \infty$ can also be found there.

In this paper, we extend this efficiency framework to the setting in which $E(\tau_1) = \infty$. As we shall argue in Sect. 3, this extension is useful in some applications of debiased MLMC; see [10]. In particular, there are various debiased MLMC algorithms which lead naturally to $E(\tau_1) = \infty$; such algorithms are believed to converge at a rate slower than the "canonical" $c^{-1/2}$ rate associated with (1), so that they exhibit "subcanonical rates." However, theoretical analysis of such algorithms has been hampered by the fact that no analog to (1) exists when $E(\tau_1) = \infty$. For example, much of the theory on subcanonical MLMC establishes upper bounds on the rate of convergence, but not lower bounds. Such lower bounds would follow automatically, in the presence of an analog to (1). Other references which study estimators based on Multilevel Monte Carlo (MLMC) via weak convergence methods include [1, 6, 9] (but they do not analyze debiased estimators, nor do they focus on the subcanonical case studied here).

It is worth noting that the act of terminating a debiased computation at a fixed computational budget inevitably introduces bias. This bias is theoretically inevitable, since any part of the sample space for the underlying random variables that takes more computation than provided by the budget can not be sampled within the given budget. Fortunately, the bias of the estimators discussed here typically goes to zero rapidly; see [4]. Furthermore, in the limit theorems described in this paper, the bias is always of smaller order than the sampling variability, as suggested by the fact that the limit random variables in all our theorems have mean zero.

This paper establishes limit theory for such subcanonical rate algorithms in Sect. 2 for both the case in which X has finite variance (Theorems 1 and 2) and when X is in the domain of attraction of a finite mean stable law (Theorem 3). Section 3 shows how the theory applies specifically to the debiased MLMC setting, and provides theory that slightly improves upon the known convergence rates for such algorithms in the stochastic differential equation context, and shows how the theory can be used to obtain asymptotically valid confidence intervals for such infinite mean procedures.

2 The Key Limit Theorems When $E(\tau_1) = \infty$

In the setting in which $E(\tau_1) = \infty$, limit theory for sums and averages typically fail to hold unless one makes strong assumptions about the tail behavior of τ_1. Consequently, we now require that τ_1 satisfy the tail condition:

A1. There exists $\alpha \in (0, 1]$ and a slowly varying function $L(\cdot)$ such that

$$P(\tau_1 > x) = x^{-\alpha} L(x)$$

for $x \geq 0$.

Remark 1 We note that a function $L(\cdot)$ is said to be *slowly varying* if for each $q > 0$, $L(qx)/L(x) \to 1$ as $x \to \infty$.

The assumption A1 is a strong requirement on the tail of τ_1 that comes close to asserting that τ_1 has a parametric-type Pareto tail. For typical Monte Carlo algorithms, there is no reason to believe that A1 will hold. However, in the debiased MLMC setting, the simulator must specify a randomization that strongly controls the distribution of τ_1. In this specific context, the randomization can be designed so that $\text{Var}(X_1) < \infty$, with A1 describing the tail behavior of τ_1; see Sect. 3 for further discussion. (Requiring that $\text{Var}(X_1) < \infty$ simplifies the construction of confidence intervals and the development of sequential procedures; see [10]).

In view of the above, we will focus first on the case where A1 holds with $\text{Var}(X_1) < \infty$. The case in which $\alpha = 1$ is qualitatively different from the case in which $\alpha \in (0, 1)$. As it turns out, the most important applications of our theory in Sect. 3 concern the $\alpha = 1$ setting. Consequently, we start with this case. We assume here that $L(\cdot)$ takes the specific form

$$L(x) = a(\log x)^{\gamma} (\log \log x)^{\delta} \tag{2}$$

for $x \geq x_0$ and $a > 0$. If $\gamma < -1$ or if $\gamma = -1$ with $\delta < -1$, $E(\tau_1) < \infty$ and so this is covered by the theory presented in [5]. We therefore restrict our analysis to the case where $\gamma > -1$ or $\gamma = -1$ with $\delta \geq -1$.

Let $S_\alpha(\sigma, \beta, \mu)$ be a stable rv with index α, scale parameter σ, skewness parameter β, and shift parameter μ, with corresponding characteristic function

$$E(\exp(i\theta S_\alpha(\sigma, \beta, \mu))) = \begin{cases} \exp\left(-\sigma^\alpha |\theta|^\alpha (1 - i\beta(\text{sign}\,\theta)\tan(\pi\alpha/2)) + i\mu\theta\right), & \alpha \neq 1; \\ \exp\left(-\sigma |\theta| (1 + i\beta\frac{2}{\pi}(\text{sign}\,\theta)\log(|\theta|)) + i\mu\theta\right), & \alpha = 1. \end{cases}$$

Theorem 1 *Suppose* $\sigma^2 = \text{Var}(X_1) < \infty$. *If* $\alpha = 1$ *and* $L(\cdot)$ *is as in (2), then*

$$\sqrt{\frac{c}{r(c)}} \left(\bar{X}_{N(c)} - z\right) \Rightarrow \sigma N(0, 1)$$

as $c \to \infty$, *where*

$$r(c) = \begin{cases} \frac{a}{1+\gamma}(\log c)^{1+\gamma}(\log\log c)^{\delta}, & \gamma > -1; \\ \frac{a}{1+\gamma}(\log\log c)^{1+\delta}, & \gamma = -1 < \delta; \\ a \log\log\log c, & \gamma = -1 = \delta. \end{cases}$$

Proof We start by noting that Theorem 4.5.1 of [11] implies that

$$\frac{\sum_{i=1}^{n} \tau_i - m_n}{c_n} \Rightarrow S_1(1, 1, 0) \tag{3}$$

as $n \to \infty$, where $(c_n : n \geq 1)$ is any sequence for which

$$\frac{nL(c_n)}{c_n} \to \frac{2}{\pi} \tag{4}$$

as $n \to \infty$, and $(m_n : n \geq 1)$ is chosen as

$$m_n = nc_n E(\sin(\tau_1/c_n)). \tag{5}$$

Given (2), (4) is satisfied by setting

$$c_n = \frac{\pi a}{2} n(\log n)^{\gamma}(\log\log n)^{\delta}.$$

As for m_n, fix $w > 0$ and write

$$m_n = nc_n\left(E(\sin(\tau_1/c_n)I(\tau_1 \leq wc_n)) + E(\sin(\tau_1/c_n)I(\tau_1 > wc_n))\right),$$

where $I(A)$ denotes the indicator function which is 1 when A occurs and 0 otherwise. Note that

$$nc_n\left|E(\sin(\tau_1/c_n)I(\tau_1 > wc_n))\right| \leq nc_n P(\tau_1 > wc_n) = O(c_n) \tag{6}$$

as $n \to \infty$, where $O(a_n)$ denotes any sequence for which $(|O(a_n)|/a_n : n \geq 1)$ is bounded.

On the other hand,

$$\begin{aligned} &nc_n E(\sin(\tau_1/c_n)I(\tau_1 \leq wc_n)) \\ &= nc_n E\left(\int_0^{\tau_1/c_n} \cos(y)dy I(\tau_1 \leq wc_n)\right) \\ &= nc_n \int_0^w \cos(y)P(yc_n < \tau_1 \leq wc_n)dy \\ &= nc_n \int_0^w \cos(y)P(\tau_1 > yc_n)dy - nc_n \sin(w)P(\tau_1 > wc_n) \end{aligned}$$

$$= nc_n \int_0^w \cos(y) P(\tau_1 > yc_n) dy + O(c_n) \tag{7}$$

as $n \to \infty$. But

$$n \cos(w) \int_0^{wc_n} P(\tau_1 > y) dy \le nc_n \int_0^w \cos(y) P(\tau_1 > yc_n) dy \le n \int_0^{wc_n} P(\tau_1 > y) dy \tag{8}$$

for $w \in [0, \pi/2]$. The upper and lower bounds in (8) follows from a change-of-variable arguement and the fact that $\cos(w) \le \cos(y)$ for any $0 \le y \le w \le \pi/2$. We shall argue below that

$$\int_0^{wc_n} P(\tau_1 > y) dy \sim r(c_n) \tag{9}$$

as $n \to \infty$, where we write $a_n \sim b_n$ as $n \to \infty$ whenever $a_n/b_n \to 1$ as $n \to \infty$. It is easily verified that $nr(c_n)/c_n \to \infty$, so it follows from (6), (7), and (8) that

$$\cos(w) \le \underline{\lim}_{n\to\infty} \frac{m_n}{nr(c_n)} \le \overline{\lim}_{n\to\infty} \frac{m_n}{nr(c_n)} \le 1. \tag{10}$$

By sending $w \to 0$ in (10), we conclude that

$$m_n \sim nr(c_n)$$

as $n \to \infty$.

Because $m_n/c_n \to \infty$, (3) implies that

$$\frac{1}{m_n} \sum_{i=1}^{n} \tau_i \Rightarrow 1$$

as $n \to \infty$, from which we find that

$$\frac{1}{c\eta} \sum_{i=1}^{\lfloor c\eta/r(c\eta) \rfloor} \tau_i \Rightarrow 1 \tag{11}$$

as $c \to \infty$ for any $\eta > 0$, where $\lfloor x \rfloor$ is the floor of x. If we choose $\eta = 1 + \varepsilon$ and $\eta = 1 - \varepsilon$ in (11), and use the fact that $r(\cdot)$ is slowly varying, we are led to the conclusion that

$$\frac{N(c)r(c)}{c} \Rightarrow 1 \tag{12}$$

as $c \to \infty$.

Donsker's theorem implies that

$$\sqrt{\frac{c}{r(c)}}\left(\bar{X}_{\lfloor tc/r(c)\rfloor} - z\right) \Rightarrow \sigma\,\frac{B(t)}{t}$$

as $c \to \infty$ in $D(0,\infty)$, where $B(\cdot)$ is standard Brownian motion; see [2]. A standard random time change argument (see, for example, Sect. 14 of [2]) then proves that

$$\sqrt{\frac{c}{r(c)}}\left(\bar{X}_{N(c)} - z\right) \Rightarrow \sigma\, B(1)$$

as $c \to \infty$ proving our theorem.

It remains only to prove (9). For $\gamma > -1$, write

$$\int_0^{wc_n} P(\tau_1 > y)dy = \int_{wc_n^\varepsilon}^{wc_n} P(\tau_1 > y)dy + \int_0^{wc_n^\varepsilon} P(\tau_1 > y)dy$$

for $1 > \varepsilon > 0$. On $[wc_n^\varepsilon, wc_n]$, $\log\log y/\log\log wc_n \to 1$ uniformly in y, so

$$\int_{wc_n^\varepsilon}^{wc_n} P(\tau_1 > y)dy \sim a(\log\log c_n)^\delta \int_{wc_n^\varepsilon}^{wc_n} \frac{(\log y)^\gamma}{y}dy \sim a(\log\log c_n)^\delta\,\frac{(\log c_n)^{\gamma+1}}{\gamma+1} \tag{13}$$

as $n \to \infty$. On the other hand,

$$\begin{aligned}\int_{x_0\vee 1}^{wc_n^\varepsilon} P(\tau_1 > y)dy &\le a(\log\log wc_n^\varepsilon)\int_{x_0\vee 1}^{wc_n^\varepsilon} \frac{(\log y)^\gamma}{y}dy \\ &= a(\log\log wc_n^\varepsilon)\frac{(\varepsilon\log c_n + \log w)^{\gamma+1}}{\gamma+1}.\end{aligned} \tag{14}$$

Since the right-hand side of (14) can be made arbitrarily small relative to the right-hand side of (13), by choosing ε small enough, we obtain (9) for $\gamma > -1$.

As for the cases where $\gamma = -1$, $\bar{F}(\cdot)$ can then be exactly integrated, and the exact integration yields the rest of (9). □

We turn next to the case where $\alpha \in (0, 1)$. To simplify our discussion, we assume here that the algorithm has been designed so that $L(x) \equiv a$ for $x \ge x_0$. For $0 < \alpha < 1$, define the constant C_α as

$$C_\alpha = \frac{1-\alpha}{\Gamma(2-\alpha)\cos(\pi\alpha/2)}; \tag{15}$$

here $\Gamma(\cdot)$ is the gamma function.

Theorem 2 *Suppose* $\sigma^2 = \mathrm{Var}(X_1) < \infty$ *and assume* $\alpha \in (0, 1)$. *Set* $\kappa = (a/C_\alpha)^{1/\alpha}$. *Then,*

$$(c/\kappa)^{\alpha/2}\left(\bar{X}_{N(c)} - z\right) \Rightarrow \frac{\sigma\, B(\nu_\alpha)}{\nu_\alpha}$$

as $c \to \infty$, where ν_α is independent of the standard Brownian motion B and has the distribution of $1/S_\alpha(1, 1, 0)^\alpha$.

Proof We start by noting that Theorem 4.5.3 of [11] implies that

$$Y_n(\cdot) \triangleq \frac{\sum_{i=1}^{\lfloor n\cdot \rfloor} \tau_i}{c_n} \Rightarrow Y_\alpha(\cdot)$$

as $n \to \infty$ in $D[0, \infty)$, where $Y_\alpha = (Y_\alpha(t) : t \geq 0)$ is a Lévy process with $Y_\alpha(1) \stackrel{\mathscr{D}}{=} S_\alpha(1, 1, 0)$ and $\stackrel{\mathscr{D}}{=}$ means "equality in distribution." The constants c_n are given by

$$c_n = (a/C_\alpha)^{1/\alpha} n^{1/\alpha} = \kappa n^{1/\alpha}.$$

Let

$$Z_n(\cdot) = \frac{\sum_{i=1}^{\lfloor n\cdot \rfloor} X_i - ze(n\cdot)}{\sqrt{n}},$$

where $e(t) = t$. We will now prove that Z_n is asymptotically independent of Y_n as $n \to \infty$. To establish the independence, we will "Poissonify." Specifically, let $R = (R(t) : t \geq 0)$ be a unit rate Poisson process with associated event times $(T_n : n \geq 1)$. Put

$$\tilde{Z}_n(t) = Z_n(R(nt)/n), \quad \tilde{Y}_n(t) = Y_n(R(nt)/n)$$

and set

$$\tilde{Z}_n^{(1)}(t) = \sum_{i=1}^{R(nt)} (X_i - z) I(\tau_i \leq a_n)/n^{1/2},$$

$$\tilde{Y}_n^{(1)}(t) = \sum_{i=1}^{R(nt)} \tau_i I(\tau_i \leq a_n)/c_n,$$

$$\tilde{Z}_n^{(2)}(t) = \sum_{i=1}^{R(nt)} (X_i - z) I(\tau_i > a_n)/n^{1/2},$$

$$\tilde{Y}_n^{(2)}(t) = \sum_{i=1}^{R(nt)} \tau_i I(\tau_i > a_n)/c_n.$$

Because of the Poissonification, $\tilde{Z}_n^{(1)}$ is independent of $\tilde{Y}_n^{(2)}$. Note that

$$E\left(\sup_{0 \leq s \leq t} \tilde{Y}_n^{(1)}(s)\right) = E\left(\tilde{Y}_n^{(1)}(t)\right) \leq \frac{ntE(\tau_1 I(\tau_1 \leq a_n))}{c_n}.$$

If we choose $a_n = n^{1/(2\alpha)-1/2}$, we find that $nE(\tau_1 I(\tau_1 \leq a_n))/c_n \to 0$, so that

$$\sup_{0\le s\le t} \tilde{Y}_n^{(1)}(s) \Rightarrow 0$$

as $n \to \infty$. Similarly, Kolmogorov's inequality implies that

$$\sup_{0\le s\le t} |\tilde{Z}_n^{(2)}(s)| \Rightarrow 0$$

as $n \to \infty$. Since

$$\tilde{Z}_n = \tilde{Z}_n^{(1)} + \tilde{Z}_n^{(2)} \Rightarrow \sigma B$$

and

$$\tilde{Y}_n = \tilde{Y}_n^{(1)} + \tilde{Y}_n^{(2)} \Rightarrow Y_\alpha$$

as $n \to \infty$ in $D[0, \infty)$,

$$(\tilde{Z}_n, \tilde{Y}_n) \Rightarrow (\sigma B, Y_\alpha)$$

as $n \to \infty$, where B is independent of Y_α. We now recover Z_n and Y_n via the representation

$$Z_n(t) = \tilde{Z}_n(T_{\lfloor nt\rfloor}/n),$$
$$Y_n(t) = \tilde{Y}_n(T_{\lfloor nt\rfloor}/n).$$

Since $T_{\lfloor n\cdot\rfloor}/n \Rightarrow e(\cdot)$ in $D[0, \infty)$, it follows that

$$(Z_n, Y_n) \Rightarrow (\sigma B, Y_\alpha) \tag{16}$$

as $n \to \infty$; see Theorem 13.2.2 of [11].

If f is a bounded continuous function on $D[0, \infty)$, (16) implies that

$$E\left(f(Z_{\lfloor (c/\kappa)^\alpha\rfloor})I(Y_{\lfloor (c/\kappa)^\alpha\rfloor}(y) > 1)\right) \to E\left(f(\sigma B)I(Y_\alpha(y) > 1)\right) \tag{17}$$

as $c \to \infty$, since $Y_\alpha(y)$ is a continuous rv (so its distribution is continuous); see [3, 11]. But

$$\{Y_{\lfloor (c/\kappa)^\alpha\rfloor}(y) > 1\} = \left\{ \sum_{i=1}^{\lfloor (c/\kappa)^\alpha y\rfloor} \tau_i > \kappa(\lfloor (c/\kappa)^\alpha\rfloor)^{1/\alpha} \right\}$$

and hence (17) implies that

$$E\left(f(Z_{\lfloor (c/\kappa)^\alpha\rfloor})I\left(\frac{N(c)}{(c/\kappa)^\alpha} < y\right)\right) \to E(f(\sigma B))P(Y_\alpha(y) > 1) \tag{18}$$

as $c \to \infty$. Also,

$$Y_{\lfloor (c/\kappa)^\alpha\rfloor}(y) = Y_{\lfloor (c/\kappa)^\alpha y\rfloor}(1)(y^{1/\alpha} + o(1))$$

(where $o(a_n)$ is a function for which $o(a_n)/a_n \to 0$ as $n \to \infty$), so that

$$\begin{aligned} E(f(Z_{\lfloor (c/\kappa)^\alpha y \rfloor})I(Y_{\lfloor (c/\kappa)^\alpha \rfloor}(y) > 1)) &\to E(f(\sigma B))P(Y_\alpha(1) > y^{-1/\alpha}) \\ &= E(f(\sigma B))P(S_\alpha(1,1,0) > y^{-1/\alpha}) \\ &= E(f(\sigma B))P(\nu_\alpha < y) \end{aligned} \tag{19}$$

as $c \to \infty$. Combining (18) and (19), we have that

$$E\left(f(Z_{\lfloor (c/\kappa)^\alpha y \rfloor})I\left(\frac{N(c)}{(c/\kappa)^\alpha} < y\right)\right) \to E\left(f(\sigma B)\right) P(\nu_\alpha < y)$$

as $c \to \infty$, so that

$$\left(Z_{\lfloor (c/\kappa)^\alpha \rfloor}, \frac{N(c)}{(c/\kappa)^\alpha}\right) \Rightarrow (\sigma B, \nu_\alpha)$$

as $c \to \infty$, where ν_α is independent of σB. The continuous mapping principle, based on a time substitution, then yields the theorem. □

We finish this section with a brief discussion of the rate of convergence of Monte Carlo algorithms in the setting in which $\mathrm{Var}(X_1) = \infty = E(\tau_1)$, when both X_1 and τ_1 lie in the domain of attraction of a stable law. Of course, we need $E(|X_1|) < \infty$ in order that $z = E(X_1)$ be well-defined, so we are considering here a stable index ρ for X_1 lying in the interval $(1, 2)$. To simplify our exposition, we postulate that X_1 is in the normal domain of attraction of $S_\rho(1, \beta, 0)$, so that

$$P(|X_1| > x) \sim bx^{-\rho} \tag{20}$$

as $x \to \infty$, where $b > 0$.

Let $Y_\rho = (Y_\rho(t) : t \geq 0)$ be the Lévy process in which $Y_\rho(1) \stackrel{\mathcal{D}}{=} S_\rho(1, \beta, 0)$.

Theorem 3 *Suppose that X_1 lies in the domain of attraction of $S_\rho(1, \beta, 0)$ and satisfies (20).*

(a) If τ_1 satisfies the hypotheses of Theorem 1, then

$$\left(\frac{c}{r(c)}\right)^{1-1/\rho} (\bar{X}_{N(c)} - z) \Rightarrow d\, Y_\rho(1)$$

as $n \to \infty$, where $d = (b/C_\rho)^{1/\rho}$.

(b) If τ_1 satisfies the hypotheses of Theorem 2 (so that $\alpha \in (0, 1)$), then

$$\left(\frac{c}{\kappa}\right)^{\alpha(1-1/\rho)} (\bar{X}_{N(c)} - z) \Rightarrow \frac{d\, Y_\rho(\nu_\alpha)}{\nu_\alpha}$$

as $c \to \infty$, where Y_ρ is independent of ν_α.

Proof We note that under our hypotheses,

$$\left(\frac{c}{r(c)}\right)^{1-1/\rho}\left(\bar{X}_{\lfloor \frac{c}{r(c)}\cdot\rfloor}-z\right)\Rightarrow\frac{d\,Y_\alpha(\cdot)}{e(\cdot)}$$

in $D[0,\infty)$. We now utilize (12) and the stochastic continuity of Y_ρ to apply the continuous mapping principle, thereby obtaining (a).

For part (b), we argue as in Theorem 2 that

$$\left(n^{1-1/\rho}\left(\bar{X}_{\lfloor n\cdot\rfloor}-z\right),\frac{\sum_{i=1}^{\lfloor n\cdot\rfloor}\tau_i}{n^{1/\alpha}}\right)\Rightarrow\left(\frac{d\,Y_\rho(\cdot)}{e(\cdot)},\kappa Y_\alpha(\cdot)\right)$$

as $n\to\infty$, where Y_ρ and Y_α are independent and Y_α is as in Theorem 2. It follows that

$$\left((c/\kappa)^{\alpha(1-1/\rho)}\left(\bar{X}_{\lfloor (c/\kappa)^\alpha\cdot\rfloor}-z\right),\frac{N(c)}{(c/\kappa)^\alpha}\right)\Rightarrow\left(\frac{d\,Y_\rho(\cdot)}{e(\cdot)},\nu_\alpha\right)$$

as $c\to\infty$, where ν_α is independent of Y_ρ. We finish the proof with a continuous mapping argument based on use of the obvious composition mapping. □

We remark that this theorem is more challenging to apply in the Monte Carlo setting, than are Theorems 1 and 2, because it requires verifying that X_1 is in the domain of attraction of a stable law.

3 Applications to Debiased MLMC

Suppose that $z=E(W)$, where W can not be simulated in finite time, but an approximating sequence $(W_n : n\geq 1)$ is available, in which the W_n's can be simulated in finite time. In particular, suppose that W_n converges to W in L^2, so that

$$\|W_n-W\|_2\to 0$$

as $n\to\infty$, where $\|U\|_2=\sqrt{E(U^2)}$ for a generic rv U.

Set $W_0=0$ and put $\Delta_i=W_i-W_{i-1}$ for $i\geq 1$. Then, under appropriate regularity conditions (see below),

$$X=\sum_{i=1}^{M}\frac{\Delta_i}{P(M\geq i)} \tag{21}$$

is an unbiased estimator for z, when M is generated independently of the Δ_i's. Specifically, Theorem 1 of [10] shows that if

$$\sum_{n=1}^{\infty} \frac{\|W_{n-1} - W\|_2^2}{P(M \geq n)} < \infty,$$

then X is unbiased, and

$$E(X^2) = \sum_{n=1}^{\infty} \frac{\|W_{n-1} - W\|_2^2 - \|W_n - W\|_2^2}{P(M \geq n)}.$$

An important application of such "debiased MLMC" estimator is numerical computation for stochastic differential equations (SDE's). In that context, the simplest and most natural approximation to W is to use the sequence $\{W_n : n \geq 1\}$ obtained by Euler discretization of the SDE. Specifically, we let W_n be the Euler discretization to W associated with a time step of order 2^{-n}, and couple the W_n's via the use of a common driving Brownian motion across all the approximations in n. If we do this, it is known in significant generality that for problems involving Lipschitz functions of the final value, $\|W_n - W\|_2^2 = O(2^{-n})$ as $n \to \infty$; see [8].

Hence, a sufficient condition on the distribution of M ensuring that $\text{Var}(X) < \infty$ is to choose M so that

$$\sum_{n=1}^{\infty} \frac{2^{-n}}{P(M \geq n)} < \infty. \tag{22}$$

However, we also need to consider the computer time τ for generating X. If we take the (reasonable) view that generating a discretization with time step 2^{-n} takes computational effort 2^n, then $\tau = 2^M$. So,

$$P(\tau > x) = P(M > \lfloor \log_2 x \rfloor).$$

Suppose we now choose M so that $P(M > n) = 2^{-\alpha n}$ for $n \geq n_0$, with $\alpha \in (0, 1)$; this choice of α guarantees that $\text{Var}(X) < \infty$. Hence, $P(\tau > x) = 2^{-\lfloor \log_2 x \rfloor \alpha}$ for x sufficiently large. However, τ does not have a regular varying tail, so the theory of Sect. 2 does not directly apply. But we can always choose to randomly delay the completion time of X. Specifically suppose that we start by generating τ so that

$$P(\tau > x) = x^{-\alpha} \tag{23}$$

for $x \geq 1$. With τ in hand, we set $M = \lfloor \log_2 \tau \rfloor$. Note that $P(M > i) = P(\tau \geq 2^i)$ so $P(M \geq i) = P(\tau \geq 2^{i-1}) = 2^{(1-i)\alpha}$ for $i \geq 1$. We now delay the completion of X from time $2^M = 2^{\lfloor \log_2 \tau \rfloor}$ to time τ. With this convention in place, our theory applies and Theorem 2 establishes that

$$(c/\kappa)^{\alpha/2} \left(\bar{X}_{N(c)} - z\right) \Rightarrow \frac{\sigma B(\nu_\alpha)}{\nu_\alpha}$$

as $c \to \infty$, where $\kappa = C_\alpha^{-1/\alpha}$. Hence, the rate of convergence of $\bar{X}_{N(c)}$ to z is of order $c^{-\alpha/2}$ with this choice of randomization for M.

The above CLT-type theorem also allows us to construct confidence intervals for z in this setting in which $E(\tau) = \infty$. In particular, if we select $\tilde{a}$ such that $P\left(-\tilde{a} \le B(\nu_\alpha)/\nu_\alpha \le \tilde{a}\right) = 0.9$ (say), then the interval

$$\left[\bar{X}_{N(c)} - \tilde{a}\hat{\sigma}(c)\left(\frac{\kappa}{c}\right)^{\alpha/2},\ \bar{X}_{N(c)} + \tilde{a}\hat{\sigma}(c)\left(\frac{\kappa}{c}\right)^{\alpha/2}\right] \tag{24}$$

is an asymptotic 90% confidence interval for z, where $\hat{\sigma}(c)$ is the sample standard deviation estimator given by

$$\hat{\sigma}(c) = \sqrt{\frac{1}{N(c)-1}\sum_{i=1}^{N(c)}\left(X_i - \bar{X}_{N(c)}\right)^2}. \tag{25}$$

Other choices for the randomization distribution are also possible. In the case α, suppose that we generate τ so that

$$P(\tau > x) = x^{-1}(\log x)^\gamma(\log\log x)^\delta$$

for x sufficiently large. Again, we let $M = \lfloor \log_2 \tau \rfloor$ and again note that $P(M > i) = P(\tau \ge 2^{i-1})$ for $i \ge 1$. In order that $\mathrm{Var}(X) < \infty$, we choose either $\gamma > 1$ or $\gamma = 1$ with $\delta > 1$. Applying Theorem 1, we find that

$$\sqrt{\frac{c}{r(c)}}\left(\bar{X}_{N(c)} - z\right) \Rightarrow \sigma\, N(0,1) \tag{26}$$

as $c \to \infty$, where $r(c) = (1+\gamma)^{-1}(\log c)^{1+\gamma}(\log\log c)^\delta$. The best convergence rate is attained when $\gamma = 1$ with $\delta > 1$ but close to 1. In this case, the exact convergence rate is of order $c^{-1/2}(\log c)(\log\log c)^{\delta/2}$, and the computational budget required to obtain an accuracy ε is of order $\varepsilon^{-2}(\log(1/\varepsilon))^2(\log\log(1/\varepsilon))^\delta$ with $\delta > 1$. This complexity estimate for debiased MLMC is slightly better than that provided by Proposition 4 of [10], in which the estimate takes the form $\varepsilon^{-2}(\log(1/\varepsilon))^q$ with $q > 2$.

As for the case where τ is chosen so that (23) holds, confidence intervals for z can again be generated. The CLT (26) implies that if $\tilde{a}$ is chosen so that $P(-\tilde{a} \le N(0,1) \le \tilde{a}) = 0.9$, then

$$\left[\bar{X}_{N(c)} - \tilde{a}\hat{\sigma}(c)\sqrt{\frac{r(c)}{c}},\ \bar{X}_{N(c)} + \tilde{a}\hat{\sigma}(c)\sqrt{\frac{r(c)}{c}}\right]$$

is an asymptotic 90% confidence interval for z, as $c \to \infty$.

If one prefers an analysis in which no delay in generating X is introduced, one can observe that $\tau/2 \le 2^M \le \tau$ when $M = \lfloor \log_2 \tau \rfloor$. If $\underset{\sim}{N}(c) = \max\{n \ge 0 : \sum_{i=1}^n \tau_i \le c\}$ and we model the time required to generate X_i as 2^{M_i}, then $\underset{\sim}{N}(c) \le N(c) \le \underset{\sim}{N}(2c)$ for $c \ge 0$. Furthermore, when $\alpha = 1$ so that Theorem 1 applies, then

$$\sqrt{\frac{c}{r(c)}}|\bar{X}_{N(c)} - z| \le \sqrt{\frac{c}{r(c)}}|\bar{X}_{\underset{\sim}{N}(c)} - z|\frac{\underset{\sim}{N}(2c)}{\underset{\sim}{N}(c)} + \sqrt{\frac{c}{r(c)}}\frac{|\sum_{k=\underset{\sim}{N}(c)}^{N(c)}(X_i - z)|}{N(c)}. \tag{27}$$

Now, Theorem 1 applies to $(c/r(c))^{1/2}|\bar{X}_{\underset{\sim}{N}(c)} - z|$ and so is stochastically bounded (i.e. tight) in c. In addition, the proof of Theorem 1 shows that both $\underset{\sim}{N}(c)r(c)/c$ and $\underset{\sim}{N}(2c)r(c)/c$ are tight, so that the first term on the right-hand side of (27) is stochastically bounded. Furthermore, Kolmogorov's maximal inequality and $\mathrm{Var}(X) < \infty$ imply that

$$\sqrt{\frac{r(c)}{c}}|\sum_{k=\underset{\sim}{N}(c)}^{N(c)}(X_i - z)| \le \sqrt{\frac{r(c)}{c}} \max_{\underset{\sim}{N}(c) \le k \le \underset{\sim}{N}(2c)} |\sum_{i=\underset{\sim}{N}(c)}^{k}(X_i - z)|$$

is stochastically bounded, so that the tightness of $N(c)r(c)/c$ yields the stochastic boundedness of the left-hand side of (27).

This proves that $\bar{X}_{N(c)}$ (with no delay introduced) does indeed converge to z at a rate that is at most $(r(c)/c)^{1/2}$ as $c \to \infty$. Note, however, that we can not get an asymptotic confidence interval for z directly from this bounding argument.

4 A Numerical Example

In this section, we implement a debiased MLMC estimator with finite variance and infinite expected computer time and use our theory to construct asymptotically valid confidence intervals. We consider an option pricing problem in the SDE context, where the underlying diffusion process obeys the SDE

$$dX(t) = rX(t)dt + \sigma X(t)dB(t),$$

in which the parameters are the interest rate $r = 0.05$, volatility $\sigma = 0.2$ and initial asset price $X(0) = 100$. We focus on pricing a European call option with payoff $\max(X(t) - K, 0)$ at maturity $t = 1$ at three different strike prices $K = 90, 100, 110$. We implement the debiased MLMC estimator described in Sect. 3 Eq. (21), in which the approximating sequence $(W_n : n \ge 1)$ is obtained by Euler discretization with step size $2^{-n}t$ and the integer-valued randomization M is chosen as $P(M > n) = 2^{-2\alpha n}$ for $n \ge 1$ with $\alpha = 1/2$. We delay the completion time such that it has a regular

Table 1 Computational result for a debiased MLMC estimator with $E(\tau) = \infty$ and $\mathrm{Var}(X) < \infty$

Strike price	True value	Computation budget	Debiased MLMC estimator	
			C.I.	Coverage (%)
$K = 90$	16.70	20,000	16.75 ± 4.46	89.0 ± 1.62
		80,000	16.73 ± 3.21	89.4 ± 1.60
		3,20,000	16.71 ± 2.29	90.1 ± 1.55
$K = 100$	10.45	20,000	10.41 ± 3.72	88.9 ± 1.63
		80,000	10.43 ± 2.70	88.8 ± 1.67
		3,20,000	10.43 ± 1.83	89.3 ± 1.60
$K = 110$	6.04	20,000	6.01 ± 2.95	87.9 ± 1.69
		80,000	6.09 ± 2.13	88.9 ± 1.63
		3,20,000	6.05 ± 1.52	91.2 ± 1.47

varying tail $P(\tau > x) = x^{-\alpha}$. Our theory applies and Theorem 2 shows that

$$(c/\kappa)^{1/4}\left(\bar{X}_{N(c)} - z\right) \Rightarrow \frac{\sigma\, B(\nu_\alpha)}{\nu_\alpha}$$

as $c \to \infty$, where $\kappa = C_\alpha^{-1/\alpha}$ and C_α is defined in Eq. (15). This result establishes an exact convergence rate of order $c^{-1/4}$ for the estimator and allows us to construct confidence intervals following the procedure in Sect. 3; see Eqs. (24) and (25). For each strike price, we implement the algorithm with computational budget $c = 10000$, 20000, 80000 and 320000. Finally, in each experiment, we construct an approximate 90% confidence interval for the mean, based on the limit distribution above, and then run 1000 independent replications of each experiment.

In Table 1, we report the computational results. The columns labeled C.I. report the average midpoint of the 1000 intervals, together with the average confidence interval half-width, again averaged over the 1000 replications. The columns labeled Coverage report 90% confidence intervals (based on the normal approximation) for the percentage of the 1000 replications in which the confidence interval contains the true option price. As shown in the table, the confidence intervals are asymptotically valid. We further note that this debiased MLMC estimator (with parameter $\alpha = 1/2$) demonstrates a convergence rate of order $c^{-1/4}$, as the length of the confidence interval roughly halves when the sample size is multiplied by a factor of sixteen. This result agrees with the exact convergence rate established by our CLT.

Acknowledgements The authors gratefully acknowledge the support of the National Science Foundation under Award Number DMS-1320158.

References

1. Alaya, M.B., Kebaier, A.: Central limit theorem for the multilevel Monte Carlo Euler method. Ann. Appl. Probab. **25**(1), 211–234 (2015)
2. Billingsley, P.: Convergence of Probability Measures. Wiley, New York (1968)
3. Bingham, N.H., Goldie, C.M., Teugels, J.L.: Regular Variation, vol. 27. Cambridge University Press, Cambridge (1989)
4. Glynn, P.W., Heidelberger, P.: Bias properties of budget constrained simulations. Oper. Res. **38**(5), 801–814 (1990)
5. Glynn, P.W., Whitt, W.: The asymptotic efficiency of simulation estimators. Oper. Res. **40**(3), 505–520 (1992)
6. Giorgi, D., Lemaire, V., Pagès, G.: Limit theorems for weighted and regular multilevel estimators. Monte Carlo Methods Appl. **23**(1), 43–70 (2017)
7. Hammersley, J., Handscomb, D.: Monte Carlo Methods. Methuen, London (1964)
8. Kloeden, P.E., Planten, E.: Numerical Solution of Stochastic Differential Equations. Springer, Berlin (1992)
9. Lemaire, V., Pagès, G.: Multilevel Richardson-Romberg extrapolation. Bernoulli **23**(4A), 2643–2692 (2017)
10. Rhee, C.-H., Glynn, P.W.: Unbiased estimation with square root convergence for SDE models. Oper. Res. **63**(5), 1026–1043 (2015)
11. Whitt, W.: Stochastic-Process Limits: An Introduction to Stochastic-Process Limits and Their Application to Queues. Springer Science & Business Media, New York (2002)

Printed in the United States
By Bookmasters